DIFFERENTIAL TOPOLOGY

Exploring the full scope of differential topology, this comprehensive account of geometric techniques for studying the topology of smooth manifolds offers a wide perspective on the field. Building up from first principles, concepts of manifolds are introduced, supplemented by thorough appendices giving background on topology and homotopy theory. Deep results are then developed from these foundations through in-depth treatments of the notions of general position and transversality, proper actions of Lie groups, handles (up to the h-cobordism theorem), immersions and embeddings, concluding with the surgery procedure and cobordism theory.

Fully illustrated and rigorous in its approach, little prior knowledge is assumed, and yet growing complexity is instilled throughout. This structure offers advanced students and researchers an accessible route into the wide-ranging field of differential topology.

C. T. C. Wall is Emeritus Professor in the Division of Pure Mathematics at the University of Liverpool. During his career he has held positions at Oxford and Cambridge and been invited as a major speaker to numerous conferences in Europe, the USA, and South America. He was elected Fellow of the Royal Society in 1969.

CAMBRIDGE STUDIES IN ADVANCED MATHEMATICS

All the titles listed below can be obtained from good booksellers or from Cambridge University Press. For a complete series listing, visit: www.cambridge.org/mathematics.

Already published

118 G. W. Anderson, A. Guionnet & O. Zeitouni *An introduction to random matrices*
119 C. Perez-Garcia & W. H. Schikhof *Locally convex spaces over non-Archimedean valued fields*
120 P. K. Friz & N. B. Victoir *Multidimensional stochastic processes as rough paths*
121 T. Ceccherini-Silberstein, F. Scarabotti & F. Tolli *Representation theory of the symmetric groups*
122 S. Kalikow & R. McCutcheon *An outline of ergodic theory*
123 G. F. Lawler & V. Limic *Random walk: A modern introduction*
124 K. Lux & H. Pahlings *Representations of groups*
125 K. S. Kedlaya *p-adic differential equations*
126 R. Beals & R. Wong *Special functions*
127 E. de Faria & W. de Melo *Mathematical aspects of quantum field theory*
128 A. Terras *Zeta functions of graphs*
129 D. Goldfeld & J. Hundley *Automorphic representations and L-functions for the general linear group, I*
130 D. Goldfeld & J. Hundley *Automorphic representations and L-functions for the general linear group, II*
131 D. A. Craven *The theory of fusion systems*
132 J. Väänänen *Models and games*
133 G. Malle & D. Testerman *Linear algebraic groups and finite groups of Lie type*
134 P. Li *Geometric analysis*
135 F. Maggi *Sets of finite perimeter and geometric variational problems*
136 M. Brodmann & R. Y. Sharp *Local cohomology (2nd Edition)*
137 C. Muscalu & W. Schlag *Classical and multilinear harmonic analysis, I*
138 C. Muscalu & W. Schlag *Classical and multilinear harmonic analysis, II*
139 B. Helffer *Spectral theory and its applications*
140 R. Pemantle & M. C. Wilson *Analytic combinatorics in several variables*
141 B. Branner & N. Fagella *Quasiconformal surgery in holomorphic dynamics*
142 R. M. Dudley *Uniform central limit theorems (2nd Edition)*
143 T. Leinster *Basic category theory*
144 I. Arzhantsev, U. Derenthal, J. Hausen & A. Laface *Cox rings*
145 M. Viana *Lectures on Lyapunov exponents*
146 J.-H. Evertse & K. Győry *Unit equations in Diophantine number theory*
147 A. Prasad *Representation theory*
148 S. R. Garcia, J. Mashreghi & W. T. Ross *Introduction to model spaces and their operators*
149 C. Godsil & K. Meagher *Erdős–Ko–Rado theorems: Algebraic approaches*
150 P. Mattila *Fourier analysis and Hausdorff dimension*
151 M. Viana & K. Oliveira *Foundations of ergodic theory*
152 V. I. Paulsen & M. Raghupathi *An introduction to the theory of reproducing kernel Hilbert spaces*
153 R. Beals & R. Wong *Special functions and orthogonal polynomials (2nd Edition)*
154 V. Jurdjevic *Optimal control and geometry: Integrable systems*
155 G. Pisier *Martingales in Banach spaces*
156 C. T. C. Wall *Differential topology*
157 J. C. Robinson, J. L. Rodrigo & W. Sadowski *The three-dimensional Navier–Stokes equations*

Differential Topology

C. T. C. WALL
University of Liverpool

CAMBRIDGE
UNIVERSITY PRESS

University Printing House, Cambridge CB2 8BS, United Kingdom

One Liberty Plaza, 20th Floor, New York, NY 10006, USA

477 Williamstown Road, Port Melbourne, VIC 3207, Australia

314-321, 3rd Floor, Plot 3, Splendor Forum, Jasola District Centre, New Delhi - 110025, India

79 Anson Road, #06-04/06, Singapore 079906

Cambridge University Press is part of the University of Cambridge.

It furthers the University's mission by disseminating knowledge in the pursuit of education, learning and research at the highest international levels of excellence.

www.cambridge.org
Information on this title: www.cambridge.org/9781107153523

First published 2016
Reprinted 2017

A catalogue record for this publication is available from the British Library

ISBN 978-1-107-15352-3 Hardback

Contents

Introduction

Differential topology, like differential geometry, is the study of smooth (or 'differential') manifolds. There are several equivalent versions of the definition: a common one is the existence of local charts mapping open sets in the manifold M^m to open sets in $\mathbb{R}^m$, with the requirement that coordinate changes are smooth, i.e. infinitely differentiable.

If M and N are smooth manifolds, a map $f : M \to N$ is called smooth if its expressions by the local coordinate systems are smooth. This leads to the concept of smooth embedding. If $f : M \to N$ and $g : N \to M$ are smooth and inverse to each other, they are called diffeomorphisms: we can then regard M and N as copies of the same manifold. If f and g are merely continuous and inverse to each other, they are homeomorphisms. Thus homeomorphism is a cruder means of classification than diffeomorphism.

The notion of smooth manifold gains in concreteness from the theorem of Whitney that any smooth manifold M^m may be embedded smoothly in Euclidean space $\mathbb{R}^n$ for any $n \geq 2m + 1$, and so may be regarded as a smooth submanifold of $\mathbb{R}^n$, locally defined by the vanishing of $(n - m)$ smooth functions with linearly independent differentials. An important example is the unit sphere S^{n-1} in $\mathbb{R}^n$. The disc D^n bounded by S^{n-1} is an example of the slightly more general notion of manifold with boundary.

Whitney's result is more precise: it states that (if M is compact) embeddings are dense in the space of all maps $f : M^m \to \mathbb{R}^n$, suitably topologised, provided $n \geq 2m + 1$, and more generally the same holds for maps $M^m \to N^n$ for any manifold N of dimension n. Other 'general position' results include the fact that if $m > p + q$, a map $f : P^p \to M^m$ will in general avoid any union of submanifolds of M of dimension $\leq q$. These results can be deduced from the general transversality theorem, which also applies to permit detailed study of the local forms of singularities of smooth maps.

One of the ultimate aims of differential topology is the classification up to diffeomorphism of (say, compact) smooth manifolds, and while this is algorithmically impossible in dimensions ≥ 4 on account of the corresponding result for finitely presented groups, we can perform it in some cases of interest. The technique is to reduce first to a problem in homotopy theory, and solve that using algebraic techniques. A basic requirement is a reasonably intrinsic way to describe manifolds: this is provided by a handle presentation.

Another central question is the possibility or otherwise of finding an embedding of a given manifold V^v in a given manifold M^m of larger dimension. Whitney himself found a key technique in the first tricky case $m = 2v$, and his idea was extended to general results in a range, roughly $m > \frac{3}{2}v$.

Classification results are accompanied by theorems giving methods of constructing manifolds: here we prescribe the homotopy type (which must satisfy Poincaré duality) and further 'normal' structure, apply transversality to construct something, and then endeavour to perform surgery to obtain the desired result.

Classification up to diffeomorphism is very fine, and only available in a few cases. The equivalence relation given by cobordism is much cruder, but is generally applicable and computable. Extensive calculations are available, and indeed through these, differential topology feeds back as a tool in pure homotopy theory.

Although the foundations have much in common with differential geometry, we approached the subject from a background in algebraic topology, and this book is written from that viewpoint. The study of differential topology stands between algebraic geometry and combinatorial topology. Like algebraic geometry, it allows the use of algebra in making local calculations, but it lacks rigidity: we can make a perturbation near a point without affecting what happens far away. While the classification results are close to those for combinatorial manifolds, the differential structure gives access to a rich source of techniques.

While the notion of differentiable manifold had gradually evolved over a century, differential topology as a subject was to a large extent begun by Whitney, with a major paper [175] in 1936 which, as well as clarifying the notion of 'differentiable manifold', established several foundational results. He obtained further important results in [176] and [177]. Spectacular new ideas were introduced in 1954 by Thom [150] on cobordism and in 1956 by Milnor [92] on differential structures on S^7. From then on, the pace of development was rapid, with contributions by numerous mathematicians. The author personally was inspired by lectures and writings of Milnor.

In a somewhat separate development, there was great progress in studying group actions. The solution of Hilbert's fifth problem [106], while independent of the study of smooth actions, gave impetus to the whole area. Major results were established by Montgomery, Mostow, and others in papers (for example, [104], [111], [112], [105]) in the 1950s. The publication of the seminar notes [20] was a landmark. The paper [119] extended key results to the case of proper actions. By 1960 this topic had been absorbed in the mainstream of geometric topology.

Many of the central problems in the topology of manifolds had been solved (or reduced to problems in homotopy theory) by 1970: in §7.8 we describe how to approach diffeomorphism classification and give some examples, and in Theorem 6.4.8 we give a result dealing with smooth embeddings. As a result, the focus of current research gradually shifted elsewhere.

The original draft of this book was written at a time when differential topology was new and exciting, and there were no books on the topic. While there now exist introductory accounts and books on particular areas of differential topology, there does not seem to be any other that does justice to the breadth of the subject.

This book falls roughly in two halves: introductory chapters with general techniques, then four chapters, each including a major result. There are also two appendices.

We begin in Chapter 1 with the definitions of smooth manifold, manifold with boundary, and tangent bundle. We give equivalent formulations of the definition, and go on to techniques for piecing together local constructions, which are fundamental for much that follows.

It is often convenient to regard a manifold as formed by fitting pieces together, and we deal with several aspects of this process in Chapter 2. We introduce and establish the main results about tubular neighbourhoods, which form the main pieces. We give the necessary details about cutting and glueing, including a discussion of corners and how to straighten them.

Chapter 3 opens with basic definitions of Lie groups and of group actions, and some basic properties. The key to the geometric description of the actions is the notion of slice. The existence of slices was established in [104] for actions of compact groups, and was extended to proper group actions by Palais [119]. Slices lead to local models for actions, which allow us to extend many of the results of the first chapters of this book to the case of group actions: leading notably to existence of invariant metrics and (with some necessary restrictions) equivariant embeddings in Euclidean space. We go on to define and study the stratification of a proper G-manifold by orbit types, which to some extent reduces the classification problem for actions to problems not involving the group, and illustrate by discussing the case when there are at most two orbit types.

In Chapter 4 we treat 'general position' arguments, which are of frequent use in constructions. We can begin with the naive idea that one can push a k-dimensional subset and a v-dimensional one apart in a manifold of dimension $n > k + v$. However – and this is one area where differential topology is much richer than piecewise linear or 'pure' topology – we can apply the same basic idea at the level of jets to study singularities. The key underlying result is the transversality theorem. This whole subject has developed enormously, particularly after the work of John Mather in the sequence of papers [88]. We have tried to steer a middle course, keeping to fairly direct arguments, obtaining details on the results wanted elsewhere in this book, and giving a brief introduction to the study of singularities.

If M is a manifold with boundary ∂M and $f : S^{r-1} \times D^{m-r} \to \partial M$ an embedding, the union $M \cup_f h^r$ of M and $D^r \times D^{m-r}$, with the copies of $S^{r-1} \times D^{m-r}$ identified by f, is said to be obtained from M by attaching an r-handle (some care is necessary at the corner $S^{r-1} \times S^{m-r-1}$). A handle can be studied via the embedding of the sphere $f \mid (S^{r-1} \times \{0\})$, and extending to a tubular neighbourhood. Any compact manifold admits a decomposition into finitely many handles. In Chapter 5 we develop handle theory up to the central result, the h-cobordism theorem. Here we have taken the approach of forming a manifold by glueing pieces together, rather than manipulating a function on a fixed manifold: the latter is in some ways more elegant, but the former seems more perspicuous. The h-cobordism theorem is the key result enabling classification of manifolds up to diffeomorphism, and we illustrate with a few examples of explicit diffeomorphism classifications. The absence of any such result for 4-manifolds means that no such classifications exist here. In the detailed treatment we restrict to the simply connected case, but describe briefly in a final section how to modify the theorem for the general case.

In Chapter 6, on immersions and embeddings, we include an account of Smale–Hirsch theory, which gives a reduction of the classification of immersions to a homotopy problem. We then describe in full Whitney's method of removing self-intersections of an n-manifold in a $(2n)$-manifold, and Haefliger's extension of the method to obtain a full theory of embeddings and immersions in the metastable range, giving (when they apply) necessary and sufficient conditions for a given map to be homotopic to an embedding (or immersion) or for two homotopic embeddings to be diffeotopic.

Next, Chapter 7 gives a full account of the theory of surgery (in the simply connected case), with a number of applications. This restriction allows a much simpler presentation than in my book [167], closer to the original papers, but the approach is the same. Sections are included on the relevant pieces of quadratic algebra, and on Poincaré complexes and maps of degree 1. A section on homotopy theory of Poincaré complexes includes a discussion of Spivak's theorem and its uses, and a brief account of Brown's treatment of the Kervaire invariant.

Finally, in Chapter 8, we tackle the topic of cobordism, describe the main geometrical ideas, and show how to build up cobordism groups, rings, and bordism as a homology theory. We also give accounts of calculations of unoriented, unitary, oriented, and (perhaps rather ambitiously) special unitary bordism. Here we suppress many details which would require an extensive knowledge of homotopy theory; even so, much more is demanded of the reader than in earlier chapters. A final section ties together much of the preceding with an account of homotopy spheres and their embedding in the standard sphere.

Each chapter opens with a summary of its contents and concludes with a 'Notes' section consisting of historical remarks, key references, and notes on additional developments.

There are two appendices. Appendix A opens in §A.1 and §A.2 with a summary of useful results from analytic topology; §A.3 gives the results we need about proper group actions; and §A.4 offers a treatment of the requisite results on the topology of mapping spaces.

I attempt a bird's eye view of homotopy theory in Appendix B: here I aim to include the necessary definitions with (I hope) enough connecting material to make them intelligible, but cannot attempt a full exposition. In §B.1, I give basic terminology and describe the general framework for homotopy notions. The next section §B.2 gives definitions and basic properties of (mostly classical) Lie groups and classifying spaces. In §B.3, I list a number of calculations including those to which reference is made in the main text. Finally, §B.4 gives very brief introductions to skeletons, connected covers, Eilenberg–MacLane spaces and cohomology operations, and spectra.

The focus of this book is on the geometric techniques required for the study of the topology of smooth manifolds. One important tool I do not use is the calculus of differential forms, and its application (de Rham) to calculating real cohomology: for an account in this spirit I refer to the book [23] by Bott and Tu. I also eschew the technical details required for the comparisons of different kinds of structure: differential vs. combinatorial, C^∞ vs. C^r or vs. real analytic: these questions are not considered here, though I do pay some attention to comparing smooth and topological structures. Although I introduce, and use, Riemannian metrics, I am not concerned with properties of the metric, so feel free to choose a convenient metric when required. Symplectic structures are outside the scope of this book: the methods devised in the last 30 years for their study are of a different nature to those studied here.

In keeping with the original setting, I assume elementary analysis, but quote (with references) some results from analysis that are needed. I will take basic topological ideas and results as understood (there is a very brief account in §A.1).

I also assume a certain background in algebraic topology, though this is not needed in Chapters 1–4, and in Chapter 5 only basic homology theory is used. All chapters are to a large extent independent (in particular, Chapter 6 is independent of Chapter 5, so the forward references do not give problems); in general they are ordered so that later chapters use an increasing knowledge of homotopy theory.

The first draft of most of the book (Chapters 1–5, 7) was a series of duplicated notes based on seminars in the early 1960s. Chapters 1, 2, and 4 were originally based on a seminar held in Cambridge 1960–61. For the original notes, it seemed desirable to elaborate the foundations considerably beyond the point from which the lectures started, and the notes expanded accordingly. For these, I am indebted to all the Cambridge topology research students of the time for participating in the seminar, in particular to P. Baxandall, and to Steve Gersten for considerable assistance in writing up. For Chapters 1 and 2 this book remains fairly close to the original notes. However for Chapter 4, the area has developed enormously in the interim, particularly after Mather's work. So I have rewritten most; in doing so I have tried to steer a middle course, keeping to fairly direct arguments, but obtaining details on the results wanted elsewhere.

The original notes for Chapter 3 were issued a few years later, with thanks to Peter Whitham. They were an attempt to pull together results from several sources to get a coherent theory. The main source was the volume [20]. This focussed on topological (rather than smooth) actions; indeed its first section was on homology manifolds. Thus much of the emphasis on my notes was also on questions of analytic topology. The account presented in this chapter is thus completely rewritten: not only does it go well beyond the content of my old notes, but has a very different emphasis.

Chapter 5 on handle decompositions, leading up to the h-cobordism theorem, is based on lectures given and seminars held in Oxford (1962) and Cambridge (1964). Thanks are due to numbers of the then research students for their participation; in particular to the late Charles Thomas and to Denis Barden. I am indebted to Shu Otsuka for rendering the original notes of these chapters into LaTeX, and to Iain Rendall for drawing the diagrams.

The remainder of the book has been newly written. Chapter 6 follows the plan I had formed back in the 1960s. Chapter 7, although much simpler in detail, was informed by the same philosophy as my book [167]. The first part of Chapter 8 is based on my old seminar notes 'Cobordism: geometric theory' issued in Liverpool about 1965, but I felt that to give the chapter substance it was also necessary to include some significant calculations.

Thanks are due to Andrew Ranicki for encouraging me to turn the old notes into a book: a time-consuming, but agreeable task.

1

Foundations

If we start from the notions of curve – of dimension 1, locally like the line $\mathbb{R}$ of real numbers, and surface – of dimension 2, locally like the plane $\mathbb{R}^2$, the general term is 'manifold'. We begin with perhaps the most elegant form of the definition, but will prove it equivalent to other versions.

We say that a function F defined on $\mathbb{R}^n$ (or on an open subset thereof) is *smooth* if it admits continuous partial derivatives of *all* orders. We use the term 'smooth' in this sense throughout.

In the opening section, we begin with the definition of smooth manifold, introduce the bump function, and proceed to the construction of partitions of unity. We then discuss connectedness.

Probably the most important property distinguishing smooth from topological manifolds is the existence of tangent vectors. Again we begin with a formal definition, then give alternative ways to view the concept. We introduce smooth maps, and discuss concepts of submanifold and embedding.

The tangent vectors to a smooth manifold form a vector bundle, so we next introduce the notions of Lie group and of fibre bundle, and establish the existence of a Riemannian structure on any smooth manifold.

An essential tool in the study of smooth manifolds is the integration of smooth vector fields. This becomes effective when combined with the use of partitions of unity to construct vector fields. We show how to reformulate the basic theorem asserting the existence solutions of ordinary differential equations in geometrical terms to yield flows on smooth manifolds.

Finally we extend the concept of smooth manifold to that of manifold with boundary, and establish the existence of a collar neighbourhood of the boundary.

1.1 Smooth manifolds

A *smooth m-manifold* is a Hausdorff topological space M^m with a family $\mathcal{F} = \mathcal{F}_M$ of continuous real-valued functions defined on M and satisfying the following conditions:

(M1) $\mathcal{F}$ is local. If $f : M \to \mathbb{R}$ is such that each point of M has a neighbourhood in which f agrees with a function of $\mathcal{F}$, then $f \in \mathcal{F}$.

(M2) $\mathcal{F}$ is differentiably closed. If $f_1, \ldots, f_k \in \mathcal{F}$, and F is a smooth function on $\mathbb{R}^n$, then $F(f_1, \ldots, f_k) \in \mathcal{F}$.

(M3) $(M, \mathcal{F})$ is locally Euclidean. For each point $P \in M$, there are m functions $f_1, \ldots, f_m \in \mathcal{F}$ such that $Q \mapsto (f_1(Q), \ldots, f_m(Q))$ gives a homeomorphism of a neighbourhood U of P in M onto an open subset V of $\mathbb{R}^m$. Every function $f \in \mathcal{F}$ coincides on U with $F(f_1, \ldots, f_m)$, where F is a smooth function on V.

(M4) M is a countable union of compact subsets.

We call functions $f \in \mathcal{F}$ *smooth functions* of M, and the mapping defined in (M3) (or, by abuse of language, the set U) a *coordinate neighbourhood* of P. It follows from (M2) that sums, products, and constant multiples of smooth functions are also smooth.

The integer m is called the *dimension* of the manifold M.

We now give some simple examples of smooth manifolds.

The empty set is a smooth m-manifold (the definition is vacuously satisfied).

Euclidean space $\mathbb{R}^m$, with smooth functions taken in the ordinary sense, is a smooth m-manifold. Condition (M1) is trivial; (M2) follows from the rule for differentiating a composite (a function of a function); for (M3), since the coordinate functions are smooth, we take the identity map; and $\mathbb{R}^m$ is the union of the compact subsets given by $\|x\| \leq n$.

The disjoint union of a finite or countable set of smooth m-manifolds is another. Define a function to be smooth if the induced function on each part is so; the conditions are then all trivial.

Let O be an open subset of $\mathbb{R}^m$. Write $\mathcal{G}_O$ for the restriction to O of functions of $\mathcal{F}_{\mathbb{R}^m}$; $\mathcal{F}_O$ for the set of functions locally agreeing with a function of $\mathcal{G}_O$. Then since O is open in $\mathbb{R}^m$, $(O, \mathcal{G}_O)$ satisfies conditions (M1), (M3); $(O, \mathcal{F}_O)$ satisfies them and also condition (M2).

For each positive integer i, consider the sets $D^m_x(\sqrt{m}/i)$[1] such that all the coordinates of ix are integers and which are contained in O. There are only countably many of these. For any $y \in O$, some $\mathring{D}^m_y(\delta) \subset O$. Choose $i > 2\sqrt{m}/\delta$. Then some x with $ix \in \mathbb{Z}^m$ is within a distance $\sqrt{m}/i$ of y, and

$$y \in D^m_x(\sqrt{m}/i) \subset D^m_y(2\sqrt{m}/i) \subset \mathring{D}^m_y(\delta) \subset O.$$

[1] For this notation and others, see the Index of Notations on p. 340.

Thus the chosen sets cover M, so (M4) also holds, and O is a smooth manifold.

More generally, let M be any smooth m-manifold and O be an open subset of M. Again write $\mathcal{G}_O$ for the restriction to O of functions of $\mathcal{F}_M$; $\mathcal{F}_O$ for the set of functions locally agreeing with a function of $\mathcal{G}_O$. We see as above that (M1)-(M3) hold. Now M is covered by coordinate charts, so any compact subset is covered by finitely many; hence M is covered by countably many charts U_α. Thus O is the union of countably many sets $O \cap U_\alpha$, each of which can be regarded as an open set in Euclidean space, so by the preceding paragraph is a countable union of compact sets. Thus (M4) also holds, and the structure of smooth m-manifold on M induces such a structure on O. We call O an *open submanifold* of M.

Let $M_1^{m_1}, M_2^{m_2}$ be smooth manifolds. Then the topological product $N^{m_1+m_2} = M_1^{m_1} \times M_2^{m_2}$ has a natural structure of smooth manifold. For let π_1, π_2 denote projections on the factors. Then for $f_1 \in \mathcal{F}_{M_1}$, $f_2 \in \mathcal{F}_{M_2}$, we define $f_1 \circ \pi_1$, $f_2 \circ \pi_2$ to belong to $\mathcal{F}_N$; any smooth functions of a finite set of these; and any function locally agreeing with one of these functions. This definition ensures that conditions (M1) and (M2) are satisfied. But so is (M3), for it now follows that if $\varphi_1 : U_1 \to \mathbb{R}^{m_1}$, $\varphi_2 : U_2 \to \mathbb{R}^{m_2}$ are coordinate neighbourhoods in M_1 and M_2, then $\varphi_1 \times \varphi_2 : U_1 \times U_2 \to \mathbb{R}^{m_1+m_2}$ can be taken as a coordinate neighbourhood in $M_1 \times M_2$. And (M4) follows since (see §A.2) the product of two compact sets is compact.

The first tool for working with our definition is a bump function. Define first a function B_1 on $\mathbb{R}$ by:

$$B_1(x) = \begin{cases} \exp\left(\frac{1}{(x(x-1))}\right) & \text{if} \quad 0 \leq x \leq 1, \\ 0 & \text{otherwise.} \end{cases}$$

Then B_1 is smooth, non-negative, and differs from zero when $0 < x < 1$. The *bump function* $Bp(x)$ is now given by

$$Bp(x) = \int_0^x B_1(t)dt \bigg/ \int_0^1 B_1(t)dt.$$

Since $B_1(x)$ is smooth, so is $Bp(x)$. Also

$$\begin{aligned} Bp(x) &= 0 \quad \text{if} \quad x \leq 0, \\ 0 < Bp(x) &< 1 \quad \text{if} \quad 0 < x < 1, \quad \text{and} \\ Bp(x) &= 1 \quad \text{if} \quad x \geq 1. \end{aligned}$$

The bump function is illustrated in Figure 1.1. Although we have given an explicit construction, the above are the essential properties of the bump

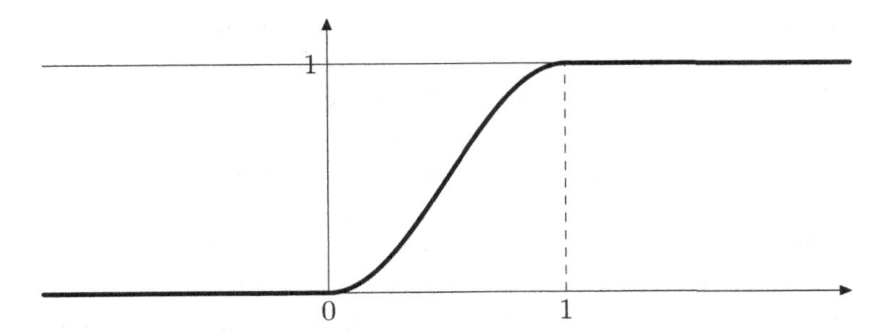

Figure 1.1 The bump function

function. We also note that since $B_1(1-x) = B_1(x)$ we have $Bp(1-x) = 1 - Bp(x)$; we also have $Bp'(x) > 0$ if $0 < x < 1$. We now have

Proposition 1.1.1 *Let $\varphi : U \to V$ be a coordinate neighbourhood of $P \in M$, and let F be a smooth function on V. Then there is a function $f \in \mathcal{F}$, agreeing with $F \circ \varphi$ in a neighbourhood of P, and zero outside U.*

Proof Without loss of generality, let $\varphi(P) = 0$. Since V is a neighbourhood of 0, we can find $r > 0$ with $\mathring{D}^m(3r) \subset V$. Define $\Phi(x) = Bp(2 - r^{-1}\|x\|)$. Then $\Phi(x) = 1$ for $\|x\| \leq r$, $\Phi(x) = 0$ for $\|x\| \geq 2r$, and Φ is a smooth function on $\mathbb{R}^m$, hence also on V, since Bp is smooth, and $\|x\|$ is smooth except at 0. Then $F\Phi$ is also smooth on V, and $F(x)\Phi(x) = 0$ if $\|x\| \geq 2r$. We define a function f on M by:

$$f(P) = \begin{cases} F(\varphi(P))\Phi(\varphi(P)) & \text{if} \quad P \in M \\ 0 & \text{otherwise.} \end{cases}$$

Then, by (M2), $f \in \mathcal{F}$, and f agrees with $f \circ \varphi$ in $\varphi^{-1}(D^m(r))$. □

Another commonly given version of the definition of manifold is as follows. For a Hausdorff topological space M, a *chart* is a homeomorphism of an open subset U of M onto an open subset V of $\mathbb{R}^m$. A collection of charts $\{\varphi_\alpha : U_\alpha \to V_\alpha\}$ is an *atlas* if the open sets U_α cover M. For any pair of charts, set $V_{\alpha,\beta} := \varphi_\alpha(U_\alpha \cap U_\beta)$; then there is a homeomorphism $\psi_{\alpha,\beta} : V_{\alpha,\beta} \to V_{\beta,\alpha}$ between open sets of Euclidean space induced by $\varphi_\beta \circ \varphi_\alpha^{-1}$. Then we say that the atlas is smooth if each $\psi_{\alpha,\beta}$ is smooth.

Lemma 1.1.2 *M is a (smooth) manifold if and only if it has a (smooth) atlas with countably many charts.*

Proof If $(M, \mathcal{F})$ satisfies (M1)–(M3), the coordinate neighbourhoods form a smooth atlas. Each compact subset of M is covered by open charts, hence by a finite subset; it follows from (M4) that a countable number of charts cover M.

Conversely, given a smooth atlas, we define $\mathcal{F}$ by letting $f \in \mathcal{F}$ if, for each α, $f \circ \varphi_\alpha^{-1}$ is a smooth function on V_α. It is immediate that this satisfies (M1) and (M2). As to (M3), for each $P \in M$, choose α with $P \in U_\alpha$; then the functions $x_i \circ \varphi_\alpha^{-1}$ are defined near P and, by the proposition, there are functions f_i, smooth on U_α, vanishing outside a neighbourhood of P, and agreeing with these on a smaller neighbourhood. Extend f_i to a function on M vanishing outside U_α. Then $f_i \in \mathcal{F}$, and $(f_1, \ldots, f_m)$ have the desired property. Now (M4) follows since each coordinate neighbourhood is a countable union of compact sets. □

Using the notion of atlas, we now give further important examples of smooth manifolds. If V is a vector space over $\mathbb{R}$, with O the origin, the projective space $P(V)$ is the quotient of $V \setminus \{O\}$ by the equivalence relation $\mathbf{v} \sim k\mathbf{v}$ for $0 \neq k \in \mathbb{R}$. For $(x_0, \ldots, x_n) \neq 0 \in \mathbb{R}^{n+1}$ we write $(x_0 : \ldots : x_n)$ for its image in $P^n(\mathbb{R}) := P(\mathbb{R}^{n+1})$ (since it is given by the ratios of the x_i). We define an atlas for $P^n(\mathbb{R})$ by taking open sets U_i given by $x_i \neq 0$ and defining $\varphi_i : U_i \to \mathbb{R}^n$ by $\varphi_i(x_0 : \ldots : x_n) := (x_0x_i^{-1}, \ldots, x_nx_i^{-1})$ (with the ith term $x_ix_i^{-1}$ omitted). The coordinate transformations are multiplications by $x_ix_j^{-1}$ on each coordinate, so are smooth. We use the same notations with $\mathbb{C}$ in place of $\mathbb{R}$, giving the complex projective space $P^n(\mathbb{C})$.

For the next tools we will need condition (M4), and begin with a general result. First observe that any manifold is locally Euclidean, and hence locally compact.

Proposition 1.1.3 *Suppose that X is locally compact and a countable union of compact subsets. Then we can find compact subsets C_n and open subsets $B_{n+\frac{1}{2}}$ such that $X = \bigcup_n C_n$ and for all $n \geq 1$, $C_n \subset B_{n+\frac{1}{2}} \subset C_{n+1}$.*

Proof Suppose X the union of compact sets A_n. Define $C_1 := A_1$. Now suppose inductively C_n defined. Since X is locally compact, each $x \in (C_n \cup A_{n+1})$ has an open neighbourhood U_x with compact closure. These open sets cover the compact set $C_n \cup A_{n+1}$, so we can choose finitely many of them which together cover this set. Define $B_{n+\frac{1}{2}}$ to be the union of these open sets, and C_{n+1} to be its closure: this is a finite union of compact sets, so is compact. Finally since $X = \bigcup_n A_n$ and $A_n \subseteq C_n$, $X = \bigcup_n C_n$. □

Theorem 1.1.4 *For any manifold M^m, we can find a set of coordinate neighbourhoods $\varphi_\alpha : U_\alpha \to \mathring{D}^m(3)$ for M^m such that*

(i) The sets $\varphi_\alpha^{-1}(\mathring{D}^m)$ cover M.

(ii) Each $P \in M$ has a neighbourhood in M which meets only a finite number of sets U_α, i.e., the U_α are locally finite.

Moreover, the covering by the U_α may be chosen to refine any given covering of M.

Proof Choose subsets C_n and $B_{n+\frac{1}{2}}$ of X as in Proposition 1.1.3. Any $x \in X$ belongs to some $C_n \setminus C_{n-1}$, so the open set $B_{n+\frac{1}{2}} \setminus C_{n-1}$ is a neighbourhood of x: we may choose a coordinate neighbourhood $U_\alpha = \varphi_\alpha^{-1}(\mathring{D}^m(3))$ of x inside this, and also contained in one of the open sets of the given covering. The neighbourhoods $\varphi_\alpha^{-1}(\mathring{D}^m)$ cover the compact set $C_{n+1} \setminus B_{n-\frac{1}{2}}$, so we may choose a finite subcovering. The collection of these for all n covers M and refines the given open covering.

Now any $y \in X$ has $B_{m+\frac{1}{2}} \setminus C_{m-1}$ as a neighbourhood for some m, and this meets $B_{n+\frac{1}{2}} \setminus C_{n-1}$ only if $|m-n| \leq 1$, hence meets only finitely many of the U_α. □

The *support* of a continuous function ψ on M is the set $\{x \in M \mid \psi(x) \neq 0\}$. A set of non-negative continuous functions $\{\psi_\alpha\}$ on M is called a *partition of unity* if their supports U_α form a locally finite covering of M and $\sum_\alpha \psi_\alpha(P) = 1$. (Local finiteness is not strictly necessary, but ensures convergence and continuity of the infinite sum.) If we are given an open covering $\{U_\beta\}$ of M, the partition $\{\psi_\alpha\}$ of unity is said to be *subordinate* to the covering if the support of each ψ_α is contained in some set U_β of the covering.

If the ψ_α are smooth, we have a smooth partition of unity. These will be key to numerous constructions. It will be useful if we can say that if f_α is a smooth function defined on U_α, then the function equal to $f_\alpha \psi_\alpha$ on U_α and zero elsewhere is smooth on M. This holds if moreover the support of ψ_α is contained in a closed set in the interior of U_α. If this holds for each α, we will say that the partition $\{\psi_\alpha\}$ of unity is *strictly subordinate* to the covering.

Theorem 1.1.5 *For any open covering $\mathcal{V}$ of a smooth manifold M, there is a smooth partition of unity strictly subordinate to it.*

Proof By Theorem 1.1.4 there is a locally finite refinement of $\mathcal{V}$ by a set of coordinate neighbourhoods $\varphi_\alpha : U_\alpha \to \mathring{D}^m(3)$ such that the $\varphi_\alpha^{-1}(\mathring{D}^m)$ cover M. For each α, set

$$\Psi_\alpha(P) = \begin{cases} Bp(2 - \|x\|) & \text{if} \quad P \in U_\alpha, \phi_\alpha(P) = x, \\ 0 & \text{otherwise.} \end{cases}$$

As in the proof of Proposition 1.1.1, $\Psi_\alpha(P)$ is smooth. The above properties imply that for each $P \in M$, there is an α with $\Psi_\alpha(P) = 1$, and that each $P \in M$

has a neighbourhood on which all but a finite number of functions Ψ_α vanish. Hence the function $\Sigma(P) = \sum_\alpha \Psi_\alpha(P)$ can be defined, and is everywhere smooth. Thus the functions $\psi_\alpha(P) = \Psi_\alpha(P)/\Sigma(P)$ give a partition of unity.

The support of Ψ_α, hence also that of ψ_α, is $\varphi_\alpha^{-1}(D^m(2))$, which is in the interior of U_α, so the partition of unity is strictly subordinate to the given covering. □

The next result is of the same type, but needs more work. It will be needed for Theorem 4.6.1 and Lemma 4.6.2. A slight modification of the argument gives a corresponding result for subsets of Cartesian powers M^r with $r > 2$.

Lemma 1.1.6 *(i) There is a countable collection of pairs of disjoint compact sets (K_α, K'_α) in M such that for any $P, P' \in M$ with $P \neq P'$ there exists α with $P \in K_\alpha$ and $P' \in K'_\alpha$.*

(ii) Let U be an open neighbourhood of the diagonal $\Delta(M)$ in $M \times M$. Then we can find pairs of disjoint compact sets (L_β, L'_β) in M such that for any $(P, P') \in (M \times M \setminus U)$ there exists β with $P \in L_\beta$ and $P' \in L'_\beta$ and moreover such that $\{L_\beta, L'_\beta\}$ is locally finite.

Proof (i) Let δ_α be a partition of unity constructed as in Theorem 1.1.4. Then the closure K_α of the support of δ_α is compact. Given $P, P' \in M$, either (a) there exists α with $P, P' \in K_\alpha$ or (b) we can choose $P \in K_\alpha \setminus K_{\alpha'}$, $P' \in K_{\alpha'} \setminus K_\alpha$.

In case (a), the points P, P' lie in the same coordinate patch. Here we have a problem in Euclidean space, and the disjoint pairs of $\mathring{D}^n_x(\sqrt{m}/i)$ (where ix has integer coordinates) give what we want.

To deal with (b), define compact sets by $K_{\alpha,n} := \{P \in M \mid \delta_\alpha(P) \geq \frac{1}{n}\}$ for each α and $n \geq 2$. Then if $n > \delta_\alpha(P)^{-1}$ and similarly for n', P and P' lie in the disjoint sets $K_{\alpha,n}$, $K_{\alpha',n'}$.

(ii) First choose a locally finite cover $\{C_\gamma\}$ of M by compact sets – for example, the sets $C_{n+1} \setminus B_{n-\frac{1}{2}}$ of Proposition 1.1.3. For $(P, P') \in (M \times M \setminus U)$, choose sets of this cover with $P \in C_\gamma$, $P' \in C_{\gamma'}$. If $C_\gamma, C_{\gamma'}$ are disjoint, we can choose these as our pair (L_β, L'_β). If not, $K = C_\gamma \cup C_{\gamma'}$ is compact, hence so is $K \times K \setminus U$; thus it is covered by finitely many of the pairs $K_\alpha \times K'_\alpha$, and we choose these as our (L_β, L'_β).

Since each C_γ meets only finitely many other such sets, it meets only finitely many of the chosen L_β and L'_β. □

We return to partitions of unity, which are an essential tool in numerous proofs. As first applications, we can approximate continuous functions by smooth functions.

Proposition 1.1.7 *(i) Let f be a continuous positive function on M. Then we can find a smooth function g, with $0 < g(P) < f(P)$ for all $P \in M$.*

(ii) For any continuous function f on M and any $\varepsilon > 0$ there exists a smooth function h on M with $|h(P) - f(P)| < \varepsilon$ for every $P \in M$.

(iii) If $f : M \to \mathbb{R}$ is continuous, $\varepsilon > 0$, and F is a closed subset of M such that f is smooth on some open set $U \supset F$, we can find h such that also $h = f$ on a neighbourhood of F.

Proof (i) Let $\{\psi_\alpha\}$ be a smooth partition of unity, and choose $\delta_\alpha > 0$ less than the infimum of f on the support of ψ_α (since the support has compact closure, this infimum cannot be zero). Then $g := \sum_\alpha (\delta_\alpha \psi_\alpha)$ has $g(P) > 0$ since $\psi_\alpha(P) > 0$ for some α; on the other hand, for each α with $\psi_\alpha(P) > 0$ we have $\delta_\alpha < f(P)$, so $g(P) < \sum \psi_\alpha(P) f(P) = f(P)$.

(ii) The sets $\{P \in M \mid n < \frac{2}{\varepsilon} f(P) < n + 2\}$ form an open cover of M. By Theorem 1.1.5, we may choose a smooth partition $\{(U_\alpha, \psi_\alpha)\}$ of unity strictly subordinate to this cover. For each α choose $P_\alpha \in U_\alpha$. Now the function $h := \sum_\alpha f(P_\alpha)\psi_\alpha$ is well defined and smooth. Any $Q \in M$ belongs to U_α for a finite, non-empty set of α, and as each such U_α is contained in one of the sets of the original cover, $f(Q)$ and $f(P_\alpha)$ lie in the same interval of length ε. Thus we have $f(Q) - f(P_\alpha) < \varepsilon$, so

$$
\begin{aligned}
|h(Q) - f(Q)| &= \left| \sum_\alpha (f(P_\alpha) - f(Q))\psi_\alpha \right| \\
&\leq \sum_\alpha |(f(P_\alpha) - f(Q)|\psi_\alpha \\
&< \sum_\alpha \varepsilon \psi_\alpha = \varepsilon.
\end{aligned}
$$

(iii) As well as the open sets U_α, which we may take disjoint from F, we now choose open sets $U_\beta \subset U$ which cover U, and set $f_\beta := f$. Piecing together using a partition of unity now yields the result. □

This approximation technique is very useful. It is also flexible: we will show in Proposition 2.3.4 that the target can be any smooth manifold.

Our next topic is connectedness of smooth manifolds. A smooth map $\alpha : \mathbb{R} \to M$ is called a *path* in M. Two points P, Q in M are called *connected in M* if there is a path in M whose image contains P and Q.

Lemma 1.1.8 *Connectedness in M is an equivalence relation.*

Proof By definition, the relation is symmetric. It is reflexive, since a constant map is a path. To prove transitivity, first observe that if $h : I \to M$ is a smooth path, the normalised path $N(h) : \mathbb{R} \to M$ given by $N(h)(t) = h(Bp(t))$ is smooth. If now $h : (I, 0, 1) \to (M, P, Q)$ and $k : (I, 0, 1) \to (M, Q, R)$ are smooth paths, a smooth path joining P to R is given by setting $H(t) = N(h)(2t)$ for $0 \le t \le \frac{1}{2}$ and $H(t) = N(k)(2t - 1)$ for $\frac{1}{2} \le t \le 1$. □

The equivalence classes are called the *components* of M.

Lemma 1.1.9 *(i) Each equivalence class is open and closed in M.*

(ii) A subset of M is open and closed if and only if it is a union of equivalence classes.

Proof (i) If $\varphi : U \to V$ is a coordinate neighbourhood of P such that V is convex, every point of U can be joined to P using the path corresponding to the straight line in V (suitably parametrised). Hence an equivalence class contains a neighbourhood of each of its points, so is open.

Since each equivalence class is the complement of the union of the other equivalence classes, it is closed in M.

(ii) Sufficiency follows by (i). For necessity, observe that since $\mathbb{R}$ is connected, any path which meets an open and closed subset is contained in it, so such a subset is saturated for the equivalence relation. □

It follows that M is *connected* in the usual sense if it only has one component. We also see that for smooth manifolds, connection and connection by smooth paths are equivalent. A component of M, being open, is an open submanifold; and M is the disjoint union of all its components. Thus to study M, it suffices to take the components separately; we shall frequently do this.

1.2 Smooth maps, tangent vectors, submanifolds

Let M^m, V^v be smooth manifolds. A mapping $\varphi : M \to V$ is called *smooth* if for each $f \in \mathcal{F}_V$, $f \circ \varphi \in \mathcal{F}_M$.

In view of (M3) this is equivalent to the requirement that each transformation of coordinates induced by φ between coordinate neighbourhoods in M and in V be smooth in the usual sense. The above definition is more convenient: for example, the following are immediate.

If $\varphi_1 : M_1 \to M_2$ and $\varphi_2 : M_2 \to M_3$ are smooth, then so is $\varphi_2 \circ \varphi_1 : M_1 \to M_3$.

If O is an open submanifold of M, $i : O \subset M$ is smooth.

A bijective correspondence $\varphi : M^m \to V^m$ between two smooth manifolds is a *diffeomorphism* if both φ and φ^{-1} are smooth. M^m and V^m are called *diffeomorphic*.

Thus a diffeomorphism is a bijective correspondence between the two manifolds under which smooth functions correspond: this is the equivalence relation which classifies manifolds.

A *tangent vector* at a point P of a smooth manifold M is a derivation on $\mathcal{F}$ to $\mathbb{R}$. In detail, a tangent vector at $P \in M$ is a mapping $\xi : \mathcal{F} \to \mathbb{R}$ which satisfies:

$$(i) \text{ if } a_1, a_2 \in \mathbb{R}, f_1, f_2 \in \mathcal{F}, \text{ then } \xi(a_1 f_1 + a_2 f_2) = a_1 \xi(f_1) + a_2 \xi(f_2); \tag{1.2.1}$$

$$(ii) \text{ if } f_1, f_2 \in \mathcal{F}, \text{ then } \xi(f_1 f_2) = \xi(f_1) f_2(P) + f_1(P) \xi(f_2). \tag{1.2.2}$$

We next study the set of all tangent vectors to M. Since sums and real multiples of tangent vectors at P are also tangent vectors at P, the tangent vectors to M at P form a vector space: we call it the *tangent space* $T_P M$ to M at P.

If $p : U \to M$ is a smooth path (U open in $\mathbb{R}$), with $p(0) = P$, the expression $\xi(f) = \frac{d}{dt} p(f(t))|_{t=0}$ is defined, and the map $\xi : \mathcal{F} \to \mathbb{R}$ satisfies (i) and (ii), so $\xi \in T_P M$. We call ξ the tangent to the path. Thus tangent vectors correspond to displacement along the manifold.

Let $\varphi : U \to V \subset \mathbb{R}^m$ be a coordinate neighbourhood of P with $\varphi(P) = 0$. Let $x_1, \ldots, x_m$ be coordinates in $\mathbb{R}^m$. Then for each $f \in \mathcal{F}$, $F := f \circ \varphi^{-1}$ is a smooth function on V, so $\frac{\partial}{\partial x_i}(f) := \left.\frac{\partial F}{\partial x_i}\right|_0$ is well defined. Then $\frac{\partial}{\partial x_i}$ is a tangent vector at P: condition (i) is clear, and (ii) follows by the rule for differentiating a product. We will prove that the $\frac{\partial}{\partial x_i}$ form a basis for $T_P M$; first, however, we need a lemma, which will be used again.

Lemma 1.2.3 *Let f be a smooth function on an open convex subset V of $\mathbb{R}^m$ containing 0, and let $f(0) = 0$. Then there exist further smooth functions f_i ($1 \le i \le m$) on V such that $f(x) = \sum_1^m x_i f_i(x)$. Moreover, if f is a smooth function of additional parameters a_j, we may suppose that f_i also are.*

Proof We may write

$$f(x) = f(x) - f(0) = \int_0^1 \frac{\partial f(tx)}{\partial t} dt.$$

But $\frac{\partial f(tx)}{\partial t} = \sum_1^m x_i \frac{\partial f}{\partial x_i}(tx)$. Substituting this gives $f(x) = \sum_1^m x_i f_i(x)$, where $f_i(x) := \int_0^1 \frac{\partial f}{\partial x_i}(tx) dt$. The last part also follows. □

Theorem 1.2.4 *The tangent vectors $\frac{\partial}{\partial x_1}, \ldots, \frac{\partial}{\partial x_m}$ form a basis for $T_P M$.*

Proof We first remark that a tangent vector is essentially local in nature: if $f = g$ in a neighbourhood U of P, and ξ is a tangent vector at P, then $\xi(f) = \xi(g)$. For by Proposition 1.1.1, we can find a function Φ on M, equal to 1 in a neighbourhood of P, and zero outside U. Then $\Phi f = \Phi g$, and so $f - g = (f - g)(1 - \Phi)$. Thus

$$\xi(f) - \xi(g) = \xi(f - g) = \xi(f - g)(1 - \Phi(P)) + (f(P) - g(P))\xi(1 - \Phi) = 0.$$

Hence it is sufficient to consider only functions defined and smooth in U, where $\varphi : U \to V$ is a coordinate neighbourhood of P with V convex; it will be simpler to speak directly of functions on V.

For any smooth function f on V, by Lemma 1.2.3, we can put

$$f(x) = f(0) + \sum x_i f_i(x).$$

For any tangent vector ξ at P, then,

$$\begin{aligned}\xi(f) &= \xi(f(0)) + \sum \xi(x_i f_i)\\ &= f(0)\xi(1) + \sum \xi(x_i) f_i(0) + \sum x_i(0)\xi(f_i).\end{aligned}$$

But $\xi(1) = \xi(1 \cdot 1) = 1 \cdot \xi(1) + \xi(1) \cdot 1 = 2\xi(1)$, and so $\xi(1) = 0$. Thus

$$\xi(f) = \sum \xi(x_i) f_i(0).$$

In particular

$$\frac{\partial}{\partial x_j}(f) = \sum \frac{\partial}{\partial x_j}(x_i) f_i(0) = \sum \delta_{ij} f_i(0) = f_j(0).$$

Thus $\xi(f) = \sum \xi(x_i)\frac{\partial f}{\partial x_i}$, and as this is true for all f, $\xi = \sum \xi(x_i)\frac{\partial}{\partial x_i}$. Hence the $\frac{\partial}{\partial x_i}$ span $T_P M$. Since $\frac{\partial}{\partial x_i}(x_j) = \delta_{ij}$, they are linearly independent. Hence they form a basis. □

For example, we may identify the tangent space to $\mathbb{R}^m$ at any point a with $\mathbb{R}^m$ itself, by identifying $\sum_i k_i \partial/\partial x_i$ with the vector $(k_1, \ldots, k_m)$. In particular, $T_a\mathbb{R}$ is identified with $\mathbb{R}$.

Now let $\varphi : M^m \to V^v$ be a smooth mapping, and let $\varphi(P) = Q$. The *differential* of φ at P, $d\varphi_P : T_P M \to T_Q V$ is defined by:

$$d\varphi_P(\xi)(f) = \xi(f \circ \varphi) \quad \text{for} \quad \xi \in T_P M, f \in \mathcal{F}_V.$$

Since f, φ are smooth, so is $f \circ \varphi$, so the right-hand side is defined. Then $d\varphi_P(\xi)$ is a derivation since ξ is. Clearly, $d\varphi_P$ is a linear mapping of $T_P M$ to $T_Q V$.

If $f \in \mathcal{F}_M$, then $f : M^m \to \mathbb{R}$ is a smooth mapping, so for any $P \in M$, we have $df_P : T_P M \to T_{f(P)}\mathbb{R} = \mathbb{R}$. Since df_P is linear, it is an element of the dual vector space $T_P^\vee M$ to $T_P M$. Now, if $x_1, \dots, x_m$ are local coordinates at P, we have

$$dx_i(\partial/\partial x_j) = \partial x_i/\partial x_j = \delta_{ij}$$

so the dx_j form the basis of $T_P^\vee M$ dual to the basis $\partial/\partial x_i$ of $T_P M$.

Theorem 1.2.5 (Inverse Function Theorem) *Let $f_1, \dots, f_n$ be smooth functions defined in a neighbourhood of $O \in \mathbb{R}^n$, and suppose $\left|\frac{\partial f_i}{\partial x_j}\right| \neq 0$ at O. Then $(f_1, \dots, f_n)$ defines a diffeomorphism of some neighbourhood U of O on an open subset of $\mathbb{R}^n$.*

A proof can be found, for example, in [40, Theorem 10.2.1].

We can now give a simple test for coordinate neighbourhoods of a point.

Corollary 1.2.6 *Let M^n be a smooth manifold; $f_1, \dots, f_n$ be smooth functions on M, $P \in M$. The f_i may be taken as coordinate functions for a coordinate neighbourhood of P if and only if the df_i form a basis for $T_P M^\vee$.*

Proof Let $\varphi : U \to \mathbb{R}^n$ be a coordinate neighbourhood of P. Then the $f_i \circ \varphi^{-1}$ are smooth functions on a neighbourhood of $\varphi(P) \in \mathbb{R}^n$; by the theorem, they define a diffeomorphism of some such neighbourhood if and only if the Jacobian determinant $|\frac{\partial(f_i \circ \varphi^{-1})}{\partial x_j}| \neq 0$ at $\varphi(P)$. But the elements of this matrix are just the coefficients in the df_i of basis elements dx_j of $T_P M^\vee$. □

Theorem 1.2.7 (Implicit Function Theorem) *Let $f_1, \dots, f_r$ be smooth functions defined in a neighbourhood of $O \in \mathbb{R}^{r+s}$ and suppose the determinant formed by their partial derivatives with respect to $x_1, \dots, x_r$ is non-zero at O. Then there are r smooth functions $g_1, \dots, g_r$ defined in a neighbourhood of $O \in \mathbb{R}^s$ such that within some neighbourhood of $O \in \mathbb{R}^{r+s}$, a point satisfies $f_i(P) = 0$ $(1 \leq i \leq r)$ if and only if it satisfies*

$$x_i = g_i(x_{r+1}, \dots, x_{r+s}) \quad (1 \leq i \leq r).$$

Proof It follows from the hypothesis that the map defined by

$$(f_1, \dots, f_r, x_{r+1}, \dots, x_{r+s})$$

on a neighbourhood of O satisfies the hypothesis of the Inverse Function Theorem 1.2.5. Hence by that result, there is a smooth inverse map. We may write this map as $(h_1, \dots, h_r, x_{r+1}, \dots, x_{r+s})$. The result now follows on setting $g_i(x_{r+1}, \dots, x_{r+s}) := h_i(0, \dots, 0, x_{r+1}, \dots, x_{r+s})$. □

A subset M^m of a smooth manifold N^n is a *submanifold* (of dimension m and *codimension* $n - m$) if, for each point $P \in M$, there is a coordinate neighbourhood $\varphi : U \to \mathbb{R}^n$ of P in N such that $U \cap M = \varphi^{-1}(\mathbb{R}^m)$.

By Corollary 1.2.6, an equivalent requirement is that in a neighbourhood of each point of M, M is defined by the vanishing of $(n - m)$ functions with linearly independent differentials. For in the case above, M is defined by the vanishing of the last $(n - m)$ coordinate functions; while by that corollary, any set of functions with linearly independent differentials can be taken as functions of a coordinate neighbourhood. If M is a closed subset of N, we call it a *closed submanifold*.

A submanifold M^m of N^n has a natural induced structure of smooth m-manifold: the existence of coordinate neighbourhoods for M and the fact that overlaps are smooth follow immediately from the definition.

Lemma 1.2.8 *If M is a closed submanifold of N, $\mathcal{F}_M$ consists of the restrictions to M of the functions of $\mathcal{F}_N$.*

Proof We have an open covering of N consisting of charts U_α as in the definition of submanifold, and the subset $U_0 := N \setminus M$. By Theorem 1.1.5 we can pick a smooth partition of unity $(\{\delta_\alpha\}, \delta_0)$ strictly subordinate to this covering. For each $f \in \mathcal{F}_M$, the restriction $f \mid M \cap U_\alpha$ of f to U_α extends to a smooth function f_α on U_α using projection in the chart. Now $\sum \delta_\alpha f_\alpha$ is a smooth extension of f. □

If M is not closed, we can construct smooth functions on M that do not even extend to continuous functions on N: the simplest example is $N = \mathbb{R}$, $M = \{x \mid x > 0\}$ with $f(x) = x^{-1}$.

Many important examples of manifolds occur as submanifolds of Euclidean or projective space, often given (at least locally) by equations with linearly independent differentials: for example, we have the unit sphere $S^{n-1} \subset \mathbb{R}^n$ defined by $\|x\|^2 = 1$; in particular, the unit circle S^1.

There are plenty of examples of smooth manifolds.

Lemma 1.2.9 *Any finite simplicial complex X is homotopy equivalent to a smooth manifold.*

This result is proved by first embedding X in Euclidean space of high enough dimension, then taking a 'regular' neighbourhood N of X, which is a compact manifold with boundary, containing X in its interior, and having X as (strong) deformation retract, and then rounding the corner to make N a smooth manifold (for details see [71]). Characterising homotopy types of compact manifolds without boundary is much more delicate: we will turn to this in §7.8.

A map $f : V \to M$ between two smooth manifolds will be called a *smooth embedding* if $f(V)$ is a submanifold W of M, and f induces a diffeomorphism of V on W, where W has the induced structure. This is more stringent than the notion of (topological) embedding, where only a homeomorphism is required.

A map $f : V \to M$ between two smooth manifolds is called an *immersion* if f is smooth and, for each $P \in V$, $df_P : T_P V \to T_{f(Q)}M$ is injective. The following criterion uses the notion of proper map, which is defined and studied in §A.2.

Proposition 1.2.10 *(i) A map $f : V \to M$ is a smooth embedding if and only if it is both a (topological) embedding and an immersion.*

(ii) A map $f : V \to M$ is an embedding as a closed submanifold if and only if it is injective, proper, and an immersion.

Proof (i) It follows from the definition that if f is a smooth embedding, it is an embedding. To see that it is an immersion at P, choose a coordinate neighbourhood at $Q = f(P)$, with $x_1, \ldots, x_m$ the coordinate functions on M at Q, and such that $f(V)$ is given locally by $x_{\upsilon+1} = \ldots = x_m = 0$. By definition of the induced structure, $x_1 \circ f, \ldots, x_\upsilon \circ f$ define a coordinate neighbourhood of P in V say $y_i = x_i \circ f$. But then $df(\partial/\partial y_i) = \partial/\partial x_i$ and so df has rank υ at Q.

For the converse, let $f : V^\upsilon \to N^n$ be a smooth immersion and an embedding with image W. Let $P \in V$, $f(P) = Q$, and choose a coordinate neighbourhood $\varphi : U \to \mathbb{R}^m$ of Q in M such that $df^*(dx_1), \ldots, df^*(dx_\upsilon)$ form a basis for $T_P^\vee V$ - this is possible since f is an immersion. Write $y_i = x_i \circ f$: then since $dy_1, \ldots, dy_\upsilon$ form a basis for $T_P^\vee V$ by Corollary 1.2.6, $y_1, \ldots, y_\upsilon$ may be taken as coordinates in a neighbourhood of P. Since the other y_i are smooth functions, by the definition of smooth manifold we can write $y_i = g_i(y_1, \ldots, y_\upsilon)$ $(\upsilon < i \leq m)$ in a neighbourhood of P in V. Since f is an embedding, $x_i = g_i(x_1, \ldots, x_\upsilon)$ in a neighbourhood of Q in W. Thus W is locally defined by vanishing of the $n - \upsilon$ smooth functions $x_i - g_i(x_1, \ldots, x_\upsilon)$, which clearly have linearly independent differentials. So W is a submanifold, and it now follows that f defines a diffeomorphism of V on W.

(ii) follows since by Lemma A.2.3, a map is proper and injective if and only if it is an embedding as a closed subset. □

We have a hierarchy of conditions on a smooth map $f : V \to M$: proper embedding $\Rightarrow$ smooth embedding $\Rightarrow$ injective immersion $\Rightarrow$ immersion. None of the implications can be reversed: we now offer examples, which are illustrated in Figure 1.2.

The inclusion in $\mathbb{R}$ of $\{x \in \mathbb{R} \mid x > 0\}$ is a smooth embedding which is not proper; another example is a curve $(e^{-t}\cos t, e^{-t}\sin t)$ spiralling in to the origin in the plane.

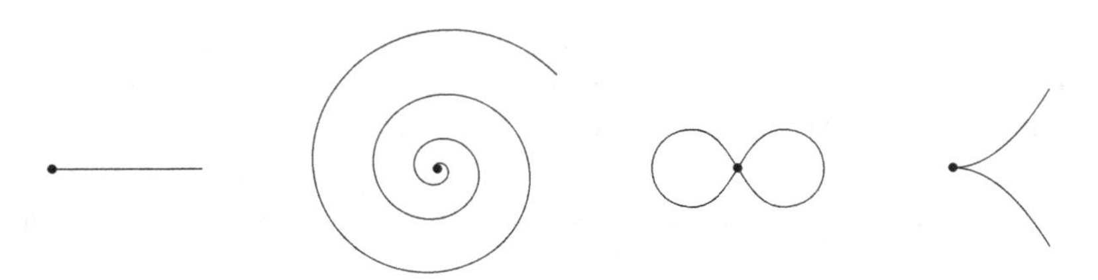

Figure 1.2 Examples which fail to give embeddings

The parametrisation $f(\theta) = (\sin(\frac{1}{2}\theta), \sin(\theta))$ defines a figure eight curve in the plane, with equation $y^2 = 4x^2(1 - x^2)$. As the differential df is nowhere zero, f is an immersion, but it is not injective. As θ runs from -2π to 2π, the point $f(\theta)$ starts at $(0, 0)$, describes a loop in $x < 0$ returning to the origin at $\theta = 0$, then describes a loop in $x > 0$.

However, if we take $t = \tan(\frac{1}{4}\theta)$ as parameter for the same curve, we have a map given by $g(t) = \left(\frac{2t}{1+t^2}, \frac{4t(1-t^2)}{(1+t^2)^2}\right)$. As t goes from $-\infty$ to $+\infty$, θ increases from -2π to 2π, so g is an injective immersion, but not an embedding.

The map $h : \mathbb{R} \to \mathbb{R}^2$ defined by $h(t) = (t^2, t^3)$ (a cusp) is a (topological) embedding which is not an immersion.

Theorem 1.2.11 *Any compact manifold M^m can be imbedded in a Euclidean space.*

Proof Let $\{\varphi_i : U_i \to \mathring{D}^m(3)\}$ be the coordinate neighbourhoods constructed in Theorem 1.1.4: since they are locally finite, and M compact, there are only a finite number. Also as in Theorem 1.1.5, let $\Phi_i(P) = Bp(2 - \|\varphi_i(P)\|)$ for P in the range of φ_i, 0 otherwise. Now define functions f_{ij} by

$$\begin{aligned} f_{i0}(P) &= \Phi_i(P) \\ f_{ij}(P) &= \Phi_i(P)x_j(\varphi_i(P)) \quad P \text{ in range of } \varphi_i \\ &= 0 \qquad \text{otherwise.} \end{aligned}$$

Then the f_{ij} are all smooth functions of P; if the range of i is $1 \leq i \leq N$, there are $(m+1)N$ of them, so they define a smooth map $F : M^m \to \mathbb{R}^{(m+1)N}$. We assert that F is an embedding: since M is compact, it suffices by Proposition 1.2.10 to prove that F is injective and an immersion.

Since the $\varphi_i^{-1}(\mathring{D}^m(1))$ cover M, each $P \in M$ belongs to at least one of them. But in this set, $\Phi_i = 1$, $f_{ij}(P) = x_j(\varphi_i(P))$, and so the df_{ij} with $j > 0$ form a basis for $T_P^\vee M$. Thus $dF_P : T_P M \to T_{f(P)}\mathbb{R}^{(m+1)N}$ is injective, and so F is an immersion.

If $F(P) = F(Q)$, and $P \in \varphi_i^{-1}(\mathring{D}^m(1))$, then $1 = \Phi_i(P) = f_{i0}(P)$, and so $1 = f_{i0}(Q) = \Phi_i(Q)$, and $Q \in \varphi_i^{-1}(\mathring{D}^m(1))$ also. But in this set, we can take the $f_{ij}(= x_j)$ as coordinates. Since these have the same values for P and Q, we have $P = Q$. Thus F is also injective. $\square$

Here we have presented a shortcut to the result: we will give a sharper statement in Theorem 4.2.2 and will see in Corollary 4.7.8 that embeddings are dense in the space of smooth maps between any manifolds M^m and V^v with $v > 2m$; if M is not compact we need V non-compact and must restrict to proper maps.

1.3 Fibre bundles

A map $\pi : T \to M$ is the projection of an *n-vector bundle* if M can be covered by open sets U_α such that

(i) There are homeomorphisms $\varphi_\alpha : U_\alpha \times \mathbb{R}^n \to \pi^{-1}(U_\alpha)$ such that, for all $x \in U_\alpha, y \in \mathbb{R}^n, \pi\varphi_\alpha(x, y) = x$.

(ii) For each pair (α, β) there is a continuous map $g_{\alpha\beta} : U_\alpha \cap U_\beta \to GL_n(\mathbb{R})$ such that, for all $x \in U_\alpha \cap U_\beta,\ y \in \mathbb{R}^n,\ \varphi_\beta(x, y) = \varphi_\alpha(x, g_{\alpha\beta}(x).y)$.

The space M is called the *base space* of the bundle, and T is its total space; $\mathbb{R}^n$ is the *fibre*; more precisely, the fibre over $m \in M$ is the preimage $\pi^{-1}(m)$.

If $\pi : T \to M$ is a vector bundle, and $V \subset T$ is such that $\pi \mid V$ is a vector bundle with $\pi^{-1}(x) \cap V$ a vector subspace of $\pi^{-1}(x)$ for each $x \in M$, then V is called a subbundle of T.

More generally, we can define fibre bundles. A *Lie group* is a smooth manifold G, which is also a group, such that the group operations $g \mapsto g^{-1}$, $(g, h) \mapsto gh$ are smooth maps $G \to G, G \times G \to G$. A *smooth action* of a Lie group G on a smooth manifold M is a smooth map $\phi : G \times M \to M$ which is a group action, i.e. which satisfies the identity $\phi(g_1, f(g_2, x)) = \phi(g_1 g_2, x)$. If the action is understood, it is frequently denoted by a dot: thus $\phi(g, x)$ becomes $g.x$. We will discuss Lie groups and smooth actions more fully in §3.

Given a smooth action of G on F, we define $\pi : T \to B$ to be the projection of a *smooth fibre bundle* with structure group G and fibre F if B can be covered by open sets U_α such that

(i) There are homeomorphisms $\varphi_\alpha : U_\alpha \times F \to \pi^{-1}(U_\alpha)$ such that, for all $x \in U_\alpha, y \in F, \pi\varphi_\alpha(x, y) = x$.

(ii) For each pair (α, β) there is a continuous map $g_{\alpha\beta} : U_\alpha \cap U_\beta \to G$ such that for $x \in U_\alpha \cap U_\beta,\ y \in F,\ \varphi_\beta(x, y) = \varphi_\alpha(x, g_{\alpha\beta}(x).y)$.

The simplest example is a product $T = M \times F$ given by a single chart: this is called a *trivial bundle*.

The structure of a bundle is determined by the maps $g_{\alpha\beta}$; two bundles with the same $g_{\alpha\beta}$ but different fibres are called *associated*. If the $g_{\alpha\beta}$ all have images in a subgroup G' of G, we say that the group of the bundle *reduces* to G', and we say that a bundle with group G', together with an isomorphism to the given bundle, defines a *reduction* of the structure group from G to G'. A map $\chi : M \to T$ is called a *cross-section* if $\pi \circ \chi = 1$.

A trivial vector bundle $\pi : T \to M$ is isomorphic to a product $M \times \mathbb{R}^n$. In this case each fibre of π is isomorphic to $\mathbb{R}^n$, each unit vector e_i of $\mathbb{R}^n$ defines a section $E_i : M \to M \times \mathbb{R}^n \cong T$, and for each $x \in M$ the vectors $E_i(x)$ give a basis of the vector space $\pi^{-1}(x)$ or, as one sometimes says, a *framing* of this vector space. Conversely, a set of sections of a vector bundle π defining a framing of each fibre gives an isomorphism $T \to M \times \mathbb{R}^n$, which may be called a framing or a trivialisation of the bundle; it is also a reduction of the structure group of π to the trivial group.

Given two vector bundles $\xi_1 = (\pi_1 : T_1 \to M)$ and $\xi_2 = (\pi_2 : T_2 \to M)$ over the same base space M we can construct a new vector bundle $\xi = \xi_1 \oplus \xi_2$ over M, called the direct sum or *Whitney sum* of the bundles ξ_1 and ξ_2: its fibre over any $m \in M$ is the direct sum of the fibres of ξ_1 and ξ_2 over m. In particular, the direct sum of ξ with a trivial line bundle is called the *suspension* of ξ. Two vector bundles ξ_1 and ξ_2 are said to be *stably isomorphic* if there exist a trivial bundle η and an isomorphism $\xi_1 \oplus \eta \cong \xi_2 \oplus \eta$.

If we have two fibre bundles $\pi_1 : T_1 \to M_1$, $\pi_2 : T_2 \to M_2$ with the same group G and fibre F, a *G-bundle map* is given by maps $f : T_1 \to T_2$, $b : M_1 \to M_2$ with $\pi_2 \circ f = b \circ \pi_1$ such that if $U_\alpha \subset M_1$ and $V_\beta \subset M_2$ are open sets as above, there exists a continuous map $g_{\alpha,\beta} : U \cap b^{-1}(V) \to G$ such that for $x \in (U \cap b^{-1}(V))$, $y \in F$ we have $\varphi_\beta(b(x), y) = f(\varphi_\alpha(x, g_{\alpha,\beta}(x).y))$.

The total space T of a smooth fibre bundle admits a natural structure as smooth manifold such that the maps φ_α are diffeomorphisms on open submanifolds. For if we use these to define coordinate neighbourhoods, then we have smooth transformations of coordinates on the intersections.

The reason for introducing these concepts at this point is that the set of all tangent vectors to a smooth manifold M has a natural structure of a vector bundle.

Write $\mathbb{T}(M) = \cup\{T_PM : P \in M\}$ for the set of all tangent vectors to M. Define $\pi : \mathbb{T}(M) \to M$ by $\pi(T_PM) = P$. Let $H_\alpha : U_\alpha \to V_\alpha$ be a set of local coordinate systems, with the U_α covering M, and for $P \in U_\alpha$, $v \in \mathbb{R}^m$, define $\varphi_\alpha(P, v)$ as the tangent vector at P determined by $\sum v_i\partial/\partial x_i$. Then for each α, the mapping $\varphi_\alpha : U_\alpha \times \mathbb{R}^m \to \pi^{-1}(U_\alpha)$ is bijective. On $U_\alpha \cap U_\beta$, denoting the two systems of coordinates by x^α, x^β; we have, by the usual transformation rule,

$$\partial/\partial x_i^\beta = \sum_\alpha (\partial x_j^\alpha/\partial x_i^\beta)(\partial/\partial x_j^\alpha),$$

so we define $g_{\alpha\beta} : U_\alpha \cap U_\beta \to GL_m(\mathbb{R})$ by

$$g_{\alpha\beta}(Q) = \left(\frac{\partial x_j^\alpha}{\partial x_i^\beta}\right)_Q.$$

Then $g_{\alpha\beta}$ is a smooth mapping, and satisfies the condition above. Now take the φ_α (or rather their inverses) as coordinate neighbourhoods, and thus define on $\mathbb{T}(M)$ the structure of smooth manifold, which in particular gives it a topology, with the φ_α homeomorphisms. Thus we have a smooth vector bundle.

We say that $\pi : \mathbb{T}(M) \to M$ is the *tangent bundle* to M. Write $\mathbb{T}^0(M)$ for the zero cross-section, i.e. the set of zero tangent vectors. In general, a smooth cross-section of $\mathbb{T}(M)$ is called a *vector field* on M. We can identify the set of smooth vector fields on M with the set of derivations from $\mathcal{F}_M$ to itself, for if ξ is such a derivation then for each $P \in M$, $f \mapsto \xi(f)(P)$ is a tangent vector at P.

Any bundle associated to $\mathbb{T}(M)$ via a linear representation of $GL_m(\mathbb{R})$ is called a *tensor bundle* (and a points of it are tensors, whose type is determined by the representation). The bundle $\mathbb{T}^\vee(M)$ given by the adjoint representation is the *bundle of differential 1-forms* on M^m; its fibre over P is the dual space $T_P^\vee M$ to $T_P M$.

The bundle whose fibre over P is the set of all positive definite quadratic forms on $T_P M$ is called the *Riemann bundle*, and any smooth cross-section of it a *Riemannian structure* on M. In local coordinates this takes the form $\sum_1^m g_{i,j}(x)dx_i dx_j$.

We now prove the fundamental.

Theorem 1.3.1 *Every smooth manifold M^m has a Riemannian structure.*

Proof Let $\{U_\alpha\}$ be an open covering such that we have charts $\varphi_\alpha : U_\alpha \to \mathbb{R}^m$ (see, for example, Theorem 1.1.4). Let Ψ_α be a partition of unity strictly subordinate to this cover. Now $\mathbb{R}^m$ has the standard Euclidean Riemannian structure: $\sum_{i=1}^m dx_i^2$. We write $ds^2 = \sum_\alpha \Psi_\alpha(\sum_{i=1}^m d(x_i \circ \varphi_\alpha)^2)$. Since the U_α are locally finite, the sum is defined; since the partition was strictly subordinate to the cover, the sum is smooth. Since a linear combination of positive definite quadratic forms is again positive definite, ds^2 is everywhere positive definite. Thus it defines a Riemannian structure on M^m. □

Given a Riemannian structure on M^m, we can choose orthonormal bases in the fibres of $\mathbb{T}(M)$ by applying the Gram–Schmidt orthogonalisation process. This will modify the maps $\varphi_\alpha : U_\alpha \times \mathbb{R}^m \to \pi^{-1}(U_\alpha)$ so as to preserve the inner product on the fibres. Indeed, consider φ_α as a map $\varphi : \mathbb{R}^m \to \mathbb{R}^m$ depending on certain parameters, and set $\varphi'(e_i) = \sum_{j \leq i} \lambda_{ij} \varphi(e_j)$, where the λ_{ij} with $j < i$ are chosen inductively to make the $\varphi'(e_i)$ orthogonal and the $\lambda_{ii} > 0$ so as to make the $\varphi'(e_i)$ unit vectors. Then the λ_{ij} are also smooth functions of the parameters.

A Riemannian structure on M^m determines a reduction of the group of the tangent bundle to the orthogonal group O_m; conversely, a reduction to O_m corresponds to a Riemannian structure. We also observe that the choice of an inner product on $T_P M$ allows us to identify $T_P M$ with $T_P^\vee M$. For a Riemannian manifold, we shall usually do this.

M^m is called *orientable* if the group of the tangent bundle is reducible to $GL_m^+(\mathbb{R})$, *oriented* if the group is so reduced. Since the coordinate transformations were given by the matrices $(\partial x_j^\alpha / \partial x_i^\beta)$, the condition is that all the Jacobian determinants are positive. The total space of the bundle associated to the tangent bundle with fibre $GL_m(\mathbb{R})/GL_m^+(\mathbb{R}) = \mathbb{Z}_2$ is a double covering $\tilde{M}$ of M, called the *orientation covering*. Its projection on M, together with coordinate neighbourhoods of M, can be taken as coordinate neighbourhoods, so $\tilde{M}$ is a smooth manifold. By the definition, all the Jacobians occurring are positive, so this manifold is orientable.

If M itself is orientable, $\tilde{M}$ consists of two copies of M; if M is connected and non-orientable, $\tilde{M}$ is connected. If M is non-orientable, we can find a closed chain of coordinate neighbourhoods, each overlapping the next, such that the number of negative Jacobians is odd.

We can specify an orientation of M at a point P by giving an isomorphism of $\mathbb{R}^m$ on $T_P M$, or equivalently, an ordered basis $(e_1, \ldots, e_m)$ of $T_P M$; another basis defines the same orientation if the determinant of the basis change is positive.

If M has a Riemannian structure, an orientation gives a reduction of the group of the tangent bundle from O_m to SO_m.

1.4 Integration of smooth vector fields

We have already seen that a smooth path in a manifold has a tangent vector at each of its points. We now show that, conversely, a tangent vector field can be integrated to give a deformation (family of paths) in the manifold. This is an essential technique for constructing deformations.

The key is Picard's existence theorem for differential equations.

Theorem 1.4.1 (Existence Theorem for Ordinary Differential Equations) *Let U be an open subset of $\mathbb{R}^n$, K a compact subset of U. Given a system of equations $\frac{d\mathbf{x}}{dt} = \mathbf{X}(\mathbf{x})$, where $\mathbf{X}$ is a smooth function on U to $\mathbb{R}^n$, then for some $\varepsilon > 0$ there exists a unique smooth function $\mathbf{x} = g(\mathbf{x}_0, t) = g_t(\mathbf{x}_0)$ on $K \times E$ to U, where E is the set $|t| < \varepsilon$, satisfying the equation, and such that $\mathbf{x}_0 = g_0(\mathbf{x}_0)$.*

A proof is given in [40, Theorem 10.4.5].

We next translate this from the language of analysis to that of geometry, and then see how to reformulate it. First write $\mathbf{X} = (X_1, \ldots, X_n)$ and define a vector field ξ on U by $\xi = \sum_i X_i \frac{\partial}{\partial x_i}$. Then the given equation becomes $\xi(x_i) = X_i$.

For any smooth function f on U, and $\mathbf{x} = \varphi_t(\mathbf{x}_0)$, we have

$$\frac{df(\mathbf{x})}{dt} = \sum_i \frac{\partial f}{\partial x_i}\frac{dx_i}{dt} = \sum_i X_i \frac{\partial f}{\partial x_i} = \xi(f),$$

so this relation is not restricted to f being a coordinate function x_i.

We define a *flow* on a smooth manifold M^m as a map $\varphi : V \to M \times \mathbb{R}$ with V some neighbourhood of $M \times \{0\}$ in $M \times \mathbb{R}$, where we write $\varphi_t(P)$ for $\varphi(P, t)$, such that

(i) $\varphi_0(P) = P$ for all $P \in M$,

(ii) $\varphi_s(\varphi_t(P)) = \varphi_{s+t}(P)$ whenever both are defined.

A flow gives rise to a vector field ξ on M as follows. For $f \in \mathcal{F}_M$, $P \in M$, we set

$$\xi_P(f) = \lim_{t\to 0} \frac{f(\varphi_t(P)) - f(P)}{t} = \left.\frac{d}{dt} f(\varphi_t(P))\right|_{t=0}.$$

It is clear that ξ_P is a tangent vector to M at P, and that ξ_P varies smoothly with P, so that ξ is a vector field. Substituting $P = \varphi_s(Q)$, and using (ii), it follows that

$$\xi_{\varphi_s(Q)}(f) = \frac{d}{dt} f(\varphi_{t+s}(Q))|_{t=0} = \frac{d}{dt} f(\varphi_t(Q))|_{t=s}.$$

We now show that any vector field defines a flow.

Theorem 1.4.2 *Let M^m be a smooth manifold, ξ a vector field on M. Then there is a flow $\varphi : U \to M \times \mathbb{R}$ giving rise to ξ, and any two such flows agree on some neighbourhood of $M \times \{0\}$.*

Proof Any $P \in M$ lies in a compact set K contained in the interior of some V, where $H : V \to U$ is a coordinate neighbourhood. In U, write ξ in local coordinates as $\sum_1^n X_i(x)\partial/\partial x_i$, and consider the system $\frac{dx_i}{dt} = X_i(x)$. Apply Theorem 1.4.1: we find $\varepsilon > 0$, and a smooth function $x = g(x_0, t)$ for $x_0 \in K$, $|t| < \varepsilon$, uniquely determined by the equation. We define φ_t in V by this relation in U.

The fact that the functions defined by different coordinate neighbourhoods agree on the intersection follows by the uniqueness, and the fact that the equations solved are simply derived from each other by change of variables.

The functions $\varphi_{s+t}(P) \to g(\mathbf{x}_0, s+t)$ satisfy the same equation, with initial value $g(\mathbf{x}_0, s)$. By the uniqueness, $g(\mathbf{x}_0, s+t) = g(g(\mathbf{x}_0, s), t)$, i.e., $\varphi_{s+t}(P_0) = \varphi_t \varphi_s(P_0)$, at least on some neighbourhood of $s = t = 0$ in $M \times \mathbb{R}^2$. □

We have seen that, for each point $P \in M$, $\varphi_t(P)$ is defined for $t = 0$, and that if it is defined for $t = t_0$, then it can be uniquely defined in some neighbourhood $|t - t_0| < \varepsilon$. There is thus a bound $B_P \leq \infty$ such that $\varphi_t(P)$ is defined for all $0 \leq t < B_P$, but no further.

Lemma 1.4.3 *Either $B_P = \infty$ or the map $[0, \infty) \to M$ given by $t \mapsto \varphi_t(P)$ is proper.*

Proof We need to show that if $B_P < \infty$ and K is a compact subset of M, then the set of t with $\varphi_t(P) \in K$ is compact, i.e. that it has an upper bound strictly less than B_P.

It follows from Theorem 1.4.2 and Corollary A.2.4 that there is a number $\varepsilon > 0$ such that $\varphi_t(Q)$ is defined for all $Q \in K$ and all t with $|t| < \varepsilon$. Suppose there exists $t > B_P - \varepsilon$ with $Q = \varphi_t(P) \in K$. Then it follows that the definition of $\varphi_t(P)$ extends beyond $t = B_P$, contradicting our hypothesis. □

One sometimes wishes to solve an equation of the form $\frac{dx}{dt} = X(x, t)$. This is not essentially different in nature: merely take t as an additional coordinate, with $\frac{dt}{dt} = 1$. In geometrical terms, we have a 'time-dependent vector field' $\xi(t)$ defined on M, and treat this as a vector field $\xi + \partial_t$ on $M \times \mathbb{R}$, i.e. a vector field on $M \times \mathbb{R}$ whose projection on $\mathbb{R}$ is equal to ∂_t. The corresponding flow $\varphi : V \to (M \times \mathbb{R}) \times \mathbb{R}$ then has the property that whenever $\varphi_s(P, t)$ is defined, its second component is equal to $s + t$.

If we have a flow on M defined on the whole of $M \times \mathbb{R}$ and satisfying $\varphi_s(\varphi_t(P)) = \varphi_{s+t}(P)$ everywhere, then each φ_t is a smooth map $M \to M$ and has an inverse map φ_{-t}, hence is a diffeomorphism. The map $\varphi : M \times \mathbb{R} \to M$ thus defines a differentiable group action of the additive group $\mathbb{R}$ on M, often called a 1-parameter group of diffeomorphisms of M. In general, a vector field on M is called *complete* if it generates a 1-parameter group of diffeomorphisms of M. We collect some simple sufficient conditions for completeness.

Proposition 1.4.4 *(i) If M^m is compact, each vector field on M is complete.*

(ii) The constant vector field $\partial/\partial t$ on $\mathbb{R}$ is complete.

(iii) If ξ is a complete vector field on V, and M is any manifold, $\xi \oplus 0$ is complete on $V \times M$.

(iv) If ξ is complete, and ξ' agrees with ξ outside a compact subset of M, then ξ' is also complete.

(v) If M has a complete metric, any bounded vector field is complete.

Proof (i) follows from Lemma 1.4.3, since there are no proper maps $[0, \infty) \to M$ if M is compact.

(ii) and (iii) are trivial.

(iv) $\varphi_t(P)$ is defined for all P and t, by hypothesis; since ξ and ξ' differ only on a compact set, there is an $\varepsilon > 0$ such that $\varphi'_t(P)$ is defined for all P and all $|t| < \varepsilon$. It follows that it is defined for all t.

(v) Since ξ is bounded, there is a uniform bound $\rho(\varphi_t(P), \varphi_s(P)) < A|s - t|$. Thus as t converges to any limit B, the points $\varphi_t(P)$ form a Cauchy sequence, so converge since the metric is complete. Thus a limit value B_P as in Lemma 1.4.3 cannot exist. □

1.5 Manifolds with boundary

We now extend the notion of manifold by considering manifolds with boundary. In the sequel these will play as much part as the manifolds already defined; we have merely deferred the definition till this point to help concentrate ideas.

N^n is a smooth *manifold with boundary*, or bounded manifold, if it satisfies all the defining conditions of a smooth manifold, with the exception that we allow coordinate neighbourhoods to map onto open sets in either $\mathbb{R}^n$ or $\mathbb{R}^n_+$, where $\mathbb{R}^n_+ := \{(x_1, \dots, x_n) \in \mathbb{R}^n \mid x_1 \geq 0\}$.

Since we will not always include the phrase 'with boundary', we also use the term *closed manifold* for a compact manifold without boundary (the phrase 'open manifold' is sometimes used for a non-compact manifold without boundary).

A point is a *boundary point* of N if its image by the chart lies on the boundary $\{x_1 = 0\}$ of $\mathbb{R}^n_+$: it is clear that this property is preserved on change of coordinate neighbourhood. The set of such points is the *boundary* of N, which we always denote by ∂N. The restrictions of coordinate charts give ∂N the structure of a smooth manifold of dimension $n - 1$. We write $\mathring{N} := N \setminus \partial N$, the 'interior' of N. This is a manifold, an open submanifold of N. A simple example of manifold with boundary is the unit disc D^n, with boundary $\partial D^n = S^{n-1}$.

The concept of smooth function on a manifold with boundary is clarified by the following.

Theorem 1.5.1 (Whitney's Extension Theorem) *Let f be a smooth function defined on the open set $x_1 > 0$ of $\mathbb{R}^n$, and suppose that f and all its partial* *derivatives extend to continuous functions on* $\mathbb{R}^n_+$. *Then there is a smooth function* g *on* $\mathbb{R}^n$ *which agrees with* f *in its range of definition.*

Whitney's proof, which establishes results of much greater generality, can be found in his paper [173].

A function on $\mathbb{R}^n_+$ is called smooth if it satisfies the equivalent conditions of the theorem. With this as the definition on a chart, we extend to a definition of smooth functions and maps on manifolds with boundary in general.

A diffeomorphism between manifolds with boundary is a smooth bijection whose inverse is smooth. Necessarily, the two boundaries correspond.

We also say N^n is a *manifold with corner* if it satisfies the defining conditions for a smooth manifold, except that coordinate neighbourhoods may map into open sets in any of $\mathbb{R}^n$, $\mathbb{R}^n_+$ and $\mathbb{R}^n_{++}$, where $\mathbb{R}^n_{++}$ denotes the set of points $(x_1, \ldots, x_n) \in \mathbb{R}^n$ with $x_1 \geq 0$, $x_2 \geq 0$. Topologically, as opposed to differentiably, N is a manifold with boundary; its boundary ∂N consists of points corresponding to $x_1 = 0$ (in $\mathbb{R}^n_+$) or to $x_1 x_2 = 0$ (in $\mathbb{R}^n_{++}$). Points corresponding to $x_1 = x_2 = 0$ in $\mathbb{R}^n_{++}$ form the *corner* $\angle N$, which is a smooth manifold of dimension $n - 2$.

If M_1, M_2 are manifolds with boundary, products of coordinate neighbourhoods of M_1 and M_2 give coordinate neighbourhoods in $M_1 \times M_2$ which (up to a permutation of coordinates) are appropriate for a manifold N with corner. We have $\partial(M_1 \times M_2) = (\partial M_1 \times M_2) \cup (M_1 \times \partial M_2)$ and $\angle(M_1 \times M_2) = \partial M_1 \times \partial M_2$. In this, as in most other important cases, $\angle N$ separates ∂N into two parts; of course this is always true locally.

The discussion of orientability and orientations for manifolds with boundaries (and perhaps corners) is essentially the same as before. However, at boundary points $P \in \partial N$, we must distinguish between *inward-* and *outward-pointing* tangent vectors: in terms of a coordinate neighbourhood of P, these are vectors $\Sigma \lambda_i \partial / \partial x_i$ with $\lambda_1 > 0$ resp. $\lambda_1 < 0$. If $\lambda_1 = 0$, we call the vector tangent to the boundary; indeed, if $i : \partial N \to N$ is the inclusion map, such vectors form the image of di, so do come from tangent vectors of ∂N. If $p : \mathbb{R}^+ \to N$ is a path with $p(0) = P$, we see by considering local coordinates that the tangent to p at P has $\lambda_1 \geq 0$. Thus the terminology is independent of the choice of local coordinates. Boundaries of manifolds and submanifolds are pictured in Figure 1.3.

In the presence of boundaries or corners, there are various corresponding extensions of the notion of submanifold. A subset M of a manifold N with boundary is a *submanifold* if it satisfies the same conditions as when N is not bounded, except that the coordinate neighbourhood φ may map U to $\mathbb{R}^n$ or $\mathbb{R}^n_+$. Thus in a neighbourhood of a point of M, the pair (N, M) is locally like

Figure 1.3 Boundaries of manifolds and submanifolds

$(\mathbb{R}^n, \mathbb{R}^m)$ or $(\mathbb{R}^n_+, \mathbb{R}^m_+)$. Geometrically, we can say that M meets ∂N transversely (for the general notion of transversality, see §4). M has an induced structure of manifold with boundary, just as above, and we have $\partial M = M \cap \partial N$. The definition includes the case when ∂M is empty, and M is disjoint from ∂N; then M is a submanifold of $\mathring{N}$. A result corresponding to Proposition 1.2.10 continues to hold.

As before, submanifolds which are not closed may have bad behaviour. We usually require the condition $\bar{M} \cap \partial N = M \cap \partial N$, which excludes such examples as $N = \{(x, y) \in \mathbb{R}^2 \mid y \geq 0\}$, $M = \{(0, y) \in \mathbb{R}^2 \mid y > 0\}$.

We could go on to consider further cases where the pair (N, M) is modelled on *any* product of pairs $(\mathbb{R}, \mathbb{R})$, $(\mathbb{R}, \mathbb{R}^+)$, $(\mathbb{R}, 0)$, $(\mathbb{R}^+, \mathbb{R}^+)$, $(\mathbb{R}^+, 0)$, but restrict to the following.

If N^n is a manifold, perhaps with boundary, we define a closed subset M^m to be a *closed submanifold with boundary* of N^n if each point of M^m has a neighbourhood U in N and a smooth chart $\varphi : U \to \mathbb{R}^n$ with $\varphi(U)$ an open set in $\mathbb{R}^n$ or $\mathbb{R}^n_+$ and $\varphi(U \cap M)$ its intersection with $\mathbb{R}^m$ or with $\mathbb{R}^m \cap \{x \mid x_2 \geq 0\}$. Thus in the case when N has a boundary we allow M to have a corner, and $\angle M$ divides ∂M into $M \cap \partial N$ and the closure of $M \cap \mathring{N}$.

The results on vector fields and flows extend as follows to manifolds with boundary. First, the local existence theorem adapts as follows.

Lemma 1.5.2 *If U is open in $\mathbb{R}^n_+$, $K \subset U$ compact, and ξ a smooth vector field on U, inward pointing along $U \cap \partial\mathbb{R}^n_+$, then for some $\varepsilon > 0$ there is a map $\varphi : K \times [0, \varepsilon] \to U$ with $\partial\varphi(x, t)/\partial t = \xi$ and $\varphi(x, 0) = x$ for all $x \in K$.*

For the global case, if ξ is a vector field on M, inward pointing at all points of ∂M, it follows as for Theorem 1.4.2 that there is a flow $\varphi : V \to M \times \mathbb{R}_+$ for some neighbourhood V of $M \times \{0\}$ in $M \times \mathbb{R}^+$.

Now suppose more generally that along some components of M, whose union we denote by $\partial_- M$, ξ is inward pointing, and along the rest (forming $\partial_+ M$), it is outward pointing. Then for each $P \in M$ we have $\varphi(x, t)$ defined for t in some interval in $\mathbb{R}$ containing 0 and with end points A_P, B_P say, and we have

Lemma 1.5.3 *Either $B_P = \infty$ or $\varphi(P, B_P) \in \partial_+ M$ or the map $[0, \infty) \to M$ given by $t \mapsto \varphi_t(P)$ is proper.*

The following result is the first step towards the construction of diffeomorphisms.

Theorem 1.5.4 *Suppose M a compact manifold with boundary $\partial_- M \cup \partial_+ M$, ξ a vector field on M, pointing inward on $\partial_- M$ and outward on $\partial_+ M$, and $f : M \to \mathbb{R}$ such that $\xi(f) > 0$ on M. Then $M \cong \partial_- M \times I$.*

Proof Integrating ξ gives a flow φ which is defined on a neighbourhood of $\partial_- M \times \{0\}$ in $\partial_- M \times \mathbb{R}_+$. Now apply Lemma 1.5.3. Since M is compact, there is no proper map $[0, \infty) \to M$. Since $\xi(f) > 0$ on the compact manifold M, it has a positive lower bound c, so $f(\varphi_{t+T}(P)) \geq f(\varphi_t(P)) + cT$, and as f also must be bounded, the case $B_P = \infty$ is ruled out. Thus each orbit of the flow can terminate only on $\partial_+ M$ and likewise (as t decreases) on $\partial_- M$. There is thus a smooth positive function g on $\partial_- M$ such that for $P \in \partial_- M$, the flow is defined at (P, t) if and only if $0 \leq t \leq g(P)$. Now the map $(P, t) \mapsto \varphi(P, \frac{t}{g(P)})$ gives a diffeomorphism of $\partial_- M \times I$ on M. □

If N is a manifold with boundary, a *collar neighbourhood* of ∂N in N is an embedding $\psi : \partial N \times I \to N$ as submanifold with boundary, extending the projection of $\partial N \times 0$ on ∂N. The use of collars will often enable us, when discussing manifolds with boundary, to avoid special difficulties arising at the boundary. We now establish their existence.

Theorem 1.5.5 *For every manifold with boundary, the boundary has a collar neighbourhood.*

Proof Each point $P \in \partial N$ lies in the domain of a coordinate neighbourhood U_α with a map $\varphi_\alpha : U_\alpha \to \mathring{D}^n_+$. We may suppose these chosen so that the ∂U_α cover ∂N. Hence the U_α together with $U_0 := N \setminus \partial N$ form an open cover of N. By Theorem 1.1.5 we can pick a strictly subordinate locally finite smooth partition of unity δ_α, δ_0.

We next construct a vector field on N which is inward pointing along ∂N. The vector field $\partial/\partial x_1$ on U^+ corresponds under φ_α to a smooth vector field ξ_α on U_α, which is inward pointing. Then $\delta_\alpha \xi_\alpha$ gives a smooth vector field on N, vanishing outside U_α. Now consider the smooth vector field $\xi := \sum \delta_\alpha \xi_\alpha$ on N. Each point P of ∂N lies in the support of some δ_α, so in the chart φ_α, the coefficient of $\partial/\partial x_1$ in $\delta_\beta \xi_\beta$ at P is non-negative for every β and positive for α, hence ξ is inward pointing at P.

We can now integrate ξ on some neighbourhood of $N \times \{0\}$ in $N \times \mathbb{R}^+$ to give a map to N. We are only interested in the restriction ψ to a neighbourhood

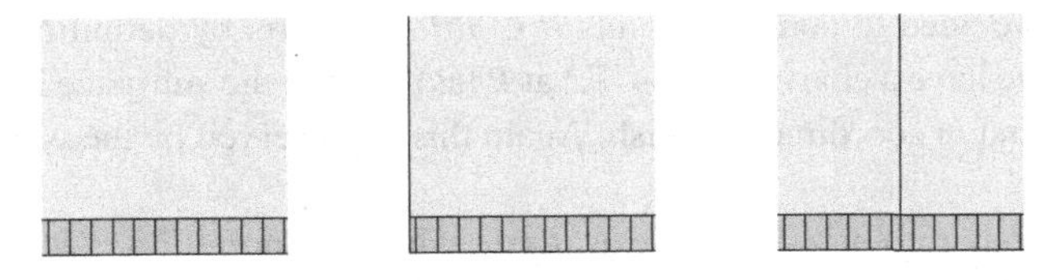

Figure 1.4 Collars of manifolds and submanifolds

W_0 of $\partial N \times \{0\}$ in $\partial N \times \mathbb{R}^+$. Along $\partial N \times \{0\}$ the map is the inclusion of ∂N in N, and by (iii) of the theorem, the derivative with respect to t is the vector field ξ. Since ξ is inward pointing, it follows from Theorem 1.2.5 that the map ψ is a local diffeomorphism. By Corollary A.2.6 there is a (smaller) neighbourhood W_1 of $\partial N \times \{0\}$ on which ψ is an embedding. By Proposition 1.1.7 (i) we can choose a smooth positive function g on ∂N such that W_1 contains $W_2 := \{(x, t) \in \partial N \times \mathbb{R} \mid x \in \partial N,\ 0 \leq t \leq g(x)\}$.

The map $(x, t) \to \psi(x, tg(x))$ now gives the desired collar neighbourhood. □

Extensions of the argument enable us to establish the existence of collars compatible with corners and submanifolds. It will be convenient to introduce the following terminology. For M a manifold with corner, a subset Q of ∂M is a *smooth part* if $\partial Q = Q \cap \angle M$. Thus the interior of Q is a union of connected components of $\partial M \setminus \angle M$.

Proposition 1.5.6 *(i) For N a manifold with corner and Q a smooth part of ∂N, there is a smooth embedding of $Q \times I$ in N giving a neighbourhood of Q in N.*

(ii) For N a manifold with boundary, M a submanifold, there is a collar neighbourhood of ∂N whose restriction to $\partial M \times I$ gives a collar neighbourhood for ∂M.

(iii) For N a manifold with boundary, M a submanifold with boundary, so $\angle M$ separates ∂M into $\partial_0 M := M \cap \partial N$ and $\partial_1 M$, there is a collar neighbourhood of ∂N whose restriction to $\partial_0 M \times I$ gives a collar neighbourhood as in (i).

Collars of manifolds and submanifolds are illustrated in Figure 1.4.

Proof Once we have constructed suitable vector fields in coordinate neighbourhoods, the piecing together using partitions of unity and integration of the vector field to give a local diffeomorphism proceeds in just the same way as above. But the local vector field can also be taken as $\partial/\partial x_1$ in each case.

For (i) it is sufficient to consider a chart $\varphi : U \to \mathbb{R}^n$ at $P \in \partial Q$ taking $U \cap (\partial N \setminus Q)$ to $x_2 = 0$ and $U \cap Q$ to $x_1 = 0$. Integrating $\partial/\partial x_1$ gives translation in the x_1 direction, which preserves $\partial N \setminus Q$.

For (ii) we need to consider points $P \in \partial M$. But here, by definition of submanifold, we have a chart $\varphi : U \to \mathbb{R}^n$ at P taking M to the subspace $\mathbb{R}^m$ where all but the first m coordinates vanish. Again this is preserved by the vector field $\partial/\partial x_1$.

For (iii), other points of $\partial_0 M$ are as in (ii), while at a point $P \in \angle M$, we have a chart $\varphi : U \to \mathbb{R}^n$ taking M to the subset of $\mathbb{R}^m$ with $x_1 \geq 0$ and $x_2 \geq 0$, and the same vector field remains suitable. □

1.6 Notes on Chapter 1

§1.1 The concept of manifold gradually evolved during the nineteenth century, beginning with the cases of curves and surfaces in Euclidean space, with successive steps taken by Riemann (who considered the n dimensional case) and Poincaré (who introduced charts). Manifolds not considered as subsets of Euclidean space first appeared in 1931 in the book [156] by Veblen and Whitehead; see also Weyl [172]. A decisive step was taken by Whitney [175] in 1936, who was the first to prove that any abstract manifold could be regarded as a manifold embedded in Euclidean space.

The use of atlases allows several variations of the definition giving related concepts: for example, instead of requiring the coordinate transformations $\psi_{\alpha,\beta}$ to be smooth, we could have required them merely to be continuous, giving topological manifolds; or to have all partial derivatives of order $\leq r$ defined and continuous, giving C^r-manifolds; or had charts as open subsets of $\mathbb{C}^r$ with holomorphic coordinate transformations, giving complex manifolds.

Any smooth atlas defines a smooth structure; conversely, the set of all smooth charts is a unique maximal atlas, and we could take 'maximal atlas' as the basic concept.

Alternatively, for each $P \in M$, we can write $\mathcal{F}_P$ for the ring of germs at P of elements of $\mathcal{F}$. The rings $\mathcal{F}_P$ fit to give a sheaf, and we can recover $\mathcal{F}$ from the sheaf of rings $\mathcal{F}_P$ as the ring of global sections. Axiom (M1) is part of the definition of sheaf; (M2)-(M3) easily translate into axioms on the sheaf.

Since our main interest is in compact manifolds (where the proofs are easier), the reader new to the subject can afford to ignore most of the references to topology, though of course the model example $\mathbb{R}^n$ is not compact.

It can be shown that in the presence of axioms (M1-M3), the following further conditions are equivalent for smooth manifolds which are connected (more generally if the set of components is (at most) countable):

M is a countable union of compact sets (the above condition (M4)),

Every open covering of M has a locally finite refinement (M is paracompact),

M has a countable base of open sets,
There is an embedding of M in Euclidean space,
The topology of M is metrisable.

We have seen in Theorem 1.1.4 that the first condition implies the second. The third follows since as M is covered by coordinate neighbourhoods, so any compact subset of M is contained in the union of finitely many, M is covered by countably many coordinate neighbourhoods. Since $\mathbb{R}^n$, and hence any open subset, has a countable base of open sets, the same follows for M. The other conditions follow from Theorems 1.2.11 and 2.1.1.

More general results of this kind are also known for topological spaces satisfying appropriate local conditions.

Examples satisfying (M1-M3) but not (M4) can be constructed, but such examples do not occur naturally. It is hard to obtain results of interest about such objects, and we do not consider them further.

§1.2 Lemma 1.2.3 is due to Marston Morse.

Proofs of Theorem 1.2.5 can be found in any good book on analysis, for example in [40].

§1.3 We give here merely the definitions necessary for the first two chapters of this book. Smooth group actions will be more fully treated in Chapter 3.

We refer the reader to Steenrod's book [144] for a systematic account of fibre bundles: this is the classic exposition. Many others have appeared since; another good reference is [77]. See also Appendix B.

§1.4 Proofs of Theorem 1.4.1 can also be found in any good book on analysis: in [40] both Theorems 1.2.5 and 1.4.1 are obtained as simple applications of the Contraction Mapping Theorem. Another reference is Hurewicz [75, 2.5]. The little book [83] gives slick treatments of all the topics up to this point, in a somewhat abstract framework.

§1.5 Manifolds with boundary were, I believe, first introduced by Poincaré.

2

Geometrical tools

We can regard a compact smooth manifold as built up by glueing together smaller pieces, which are easier to analyse. In this chapter we begin the description of this process. After obtaining some basic results on Riemannian metrics, we study geodesics for such metrics. The key result is that any two nearby points are joined by a unique shortest geodesic. This leads us to study the way in which a closed submanifold lies in a manifold: we describe the structure of a neighbourhood of the submanifold as having the form of a tube.

A diffeotopy, or differentiable isotopy, can be considered either as deforming the embedding of one manifold in another or as an embedding of a product with I. If the deformation can be extended to the whole manifold, the two embeddings are equivalent. The diffeotopy extension theorem asserts that under certain conditions, this extension is possible; it may thus be looked on as a uniqueness theorem. We apply this result to obtain a uniqueness theorem for tubular neighbourhoods, which enables us to pass from knowledge of the structure of a compact submanifold M of a manifold N to knowledge of a neighbourhood of M: the only extra piece of information needed is the structure of the normal bundle $\mathbb{N}(N/M)$. This contributes to the general aim of building up global results from merely local ones.

We define inverse procedures for straightening a corner, to yield a manifold with boundary, and for introducing corners: it will be useful in Chapter 5 to be able to effectively ignore corners.

Finally we discuss glueing and the inverse process of cutting: these are simple geometrical constructions which, given some smooth manifolds (perhaps with boundaries and corners) and additional data where necessary, give rise to new manifolds. On account of their perspicuity, these methods are traditional in describing the topology of surfaces, and they remain a very powerful tool in higher dimensions.

2.1 Riemannian metrics

We recall that if M^m is a smooth manifold, the bundle over M associated to the tangent bundle and whose fibre over P is the set of all positive definite quadratic forms on T_PM is called the Riemann bundle, and any cross-section of it a Riemannian structure on M; in local coordinates this takes the form $\sum_{i,j=1}^{m} g_{i,j}(x)dx_i dx_j$.

We saw in Theorem 1.3.1 that every smooth manifold M^m has a Riemannian structure. Such a structure induces an inner product on each T_PM, which we use to introduce the notion of length of tangent vectors. A (smooth) *path* in M is a smooth map p to M with source $\mathbb{R}$ or an interval contained in $\mathbb{R}$. For a path p, we define the *length* of p between two of its points by

$$l(p) = \int_a^b \frac{ds}{dt} dt,$$

where $(ds/dt)^2 = \sum_{i,j} g_{i,j}(dx_i/dt)(dx_j/dt)^2$, the derivatives being taken along the path. We set

$$\rho(P, Q) = \inf\{l(p) : p \text{ a path joining } P \text{ to } Q\};$$

this is defined if and only if P, Q are in the same component of M. We could also, for example, define $\rho(P, Q) = 1$ whenever P and Q are in different components, but the case of interest is when M is connected.

We call ρ the *Riemannian metric*: we now show that it *is* a metric.

Theorem 2.1.1 *The function ρ defines a metric on M which induces the given topology on M.*

Proof The triangle inequality follows since, as in Lemma 1.1.8, we can (up to re-parametrising, which does not alter length) combine smooth paths from P to Q and from Q to R to give a smooth path from P to R. That $\rho(P, Q) = 0$ implies $P = Q$ follows from the argument below.

To show that the metric induces the given topology, we need to establish that, for any point $P \in M$,
(i) any neighbourhood of P in M contains $\{Q \in M \mid \rho(P, Q) < A\}$ for some A,
(ii) any such set is a neighbourhood of P.

Choose a coordinate neighbourhood $\varphi : U \to \mathbb{R}^m$ with $\varphi(P) = O$. By a linear change of coordinates in $\mathbb{R}^m$, we can reduce the matrix $\big(g_{i,j}(P)\big)$ to the identity, so at P the metric ds^2 agrees with the Euclidean metric $\sum_1^n dx_i^2$. Hence there is a neighbourhood of P on which the ratio is bounded:

$$\frac{1}{2}\sum_1^n dx_i^2 \leq \sum_{i,j} g_{i,j}(x)dx_i dx_j \leq 2\sum_1^n dx_i^2$$

for $\|x\| < A$, say.

Thus if p is a path in M with $\varphi(p) \subset \mathring{D}^n(A)$, and $l(\varphi(p))$ denotes the length of $\varphi(p)$ in the Euclidean metric, $\frac{1}{2}l(\varphi(p)) \leq l(p) \leq 2l(\varphi(p))$.

Now (ii) follows since, if $B \leq A$, then for any $Q = \varphi^{-1}(x)$ with $\|x\| < \frac{B}{2}$, taking the path p_3 such that $\varphi(p_3)$ is the straight segment from O to x gives

$$\rho(P, Q) \leq l(p_3) \leq 2l(\varphi(p_3)) < B,$$

so the set $\{Q \mid \rho(P, Q) < B\}$ contains the neighbourhood $\varphi^{-1}\{\mathring{D}^m(\frac{1}{2}B)\}$.

As to (i), first note that if $\varphi(Q) = x$ with $\|x\| < \frac{A}{2}$, and p_1 is a path from Q with $\varphi(p_1)$ leaving $\mathring{D}^m(A)$, then $l(\varphi(p_1)) \geq \frac{A}{2}$, hence $l(p_1) \geq \frac{A}{4}$. Thus for any path p_2 from P to Q with $\varphi(p_2)$ leaving $\mathring{D}^m(A)$, we have $l(p_2) \geq \frac{A}{2}$.

Now for any $B < \frac{A}{4}$, since any path p from P with $l(p) < B$ is contained in $\varphi^{-1}\{\mathring{D}^m(A)\}$, it follows that $D := \{Q \in M \mid \rho(P, Q) < B\}$ is also contained in this region; and now since we need only consider paths p in this region, and $l(\varphi(p)) < 2l(p)$, D is contained in $\varphi^{-1}\{\mathring{D}^m(2B)\}$. □

The basic results about Riemannian metrics: existence of a Riemannian structure, and the definition and properties of a metric: apply without essential change also to manifolds with boundary.

Next let V^v be a submanifold of a smooth manifold M^m. If $P \in V$, the inclusion $i : V \to M$ induces $di : T_PV \to T_PM$ of rank v, hence the dual map $di^* : T_P^\vee M \to T_P^\vee V$ also has rank v, and its kernel has rank $(m - v)$.

The kernel of $di^* : T_P^\vee M \to T_P^\vee V$ is called the *normal space* to V in M at P; we will denote it by $N_P(M/V)$. The union of these normal spaces is the *normal bundle* $\mathbb{N}(M/V)$ of V in M. We must check that the normal bundle is indeed a vector bundle over V. Let $\varphi : U \to \mathbb{R}^m$ be a coordinate neighbourhood of P in M with $U \cap V = \varphi^{-1}(\mathbb{R}^v)$; then in $U \cap V$ we may take $dx_{v+1}, \ldots, dx_m$ as a basis for the normal space. These give the local product maps φ_α required of a fibre bundle; as with the tangent bundle, the maps $g_{\alpha\beta}$ come from Jacobians on change of coordinates.

A Riemannian structure on M induces one on V. The distinction between $T_P^\vee M$ and T_PM disappears, and in this case we can regard $\mathbb{N}(M/V)$ as a sub-bundle of the restriction $\mathbb{T}(M)|V$ of $\mathbb{T}(M)$ to V.

Proposition 2.1.2 *$\mathbb{T}(M)|V$ is the Whitney sum of $\mathbb{N}(M/V)$ and $\mathbb{T}(V)$,*

Proof Since all the above bundles are defined, and the latter two are sub-bundles of the first, it is sufficient to verify that at each point the fibre of the first is the direct sum of the latter two. Since we have a positive definite inner product, it will be sufficient to verify that the fibre $N_p(M/V)$ of $\mathbb{N}(M/V)$ over P is the orthogonal complement of the fibre $T_P V$ of $\mathbb{T}(V)$ in the fibre $T_P M$ of $\mathbb{T}(M)$, or that it is the annihilator of $T_P V$ in $T_P^\vee M$. But since di^* is dual to di, the kernel of di^* is certainly the annihilator of the image of di. □

We say that a submanifold V of M meets ∂M *orthogonally* if the normal vectors to V and ∂M at each point of ∂V are perpendicular.

Lemma 2.1.3 *Let M be a manifold with boundary, V a submanifold. Then M has a Riemannian metric in which V meets ∂M orthogonally.*

Proof We construct a metric just as in Theorem 1.3.1; the only point to watch is that V meets ∂M orthogonally in each of the partial metrics to be fitted together. But since V is a submanifold, at a point of ∂V, there is a coordinate map of an open set of (M, V) to $(\mathbb{R}^n_+, \mathbb{R}^m_+)$, and the Euclidean metric will do. Now when we fit these together, V continues to meet ∂M orthogonally. □

2.2 Geodesics

For a connected manifold M^m with a Riemannian structure, we have already defined the length of a path and the distance function as the infimum of lengths of paths, and shown in Theorem 2.1.1 that the infimum $\rho(P, Q)$ of lengths of paths joining P to Q is a metric defining the topology on M.

We now focus attention on the paths minimising this distance. Recall that the length of a path $p : U \to M$ (U open in $\mathbb{R}$) between two of its points is defined by $l(p) := \int_a^b \frac{ds}{dt} dt$, where $(ds/dt)^2 = \sum_{i,j} g_{i,j}(dx_i/dt)(dx_j/dt)^2$, the derivatives being taken along the path. We now define the *energy* of p by

$$E(p) := (b - a) \int_a^b \left(\frac{ds}{dt}\right)^2 dt.$$

Then a *geodesic* is defined to be a smooth path $p : U \to M$ giving an extremal value to the energy between any two of its points.

By Schwarz' inequality,

$$l(p)^2 = \left(\int_a^b \frac{ds}{dt} dt\right)^2 \leq \int_a^b dt \int_a^b \left(\frac{ds}{dt}\right)^2 dt = (b - a) \int_a^b \left(\frac{ds}{dt}\right)^2 dt = E(p),$$

with equality if and only if ds/dt is constant, so that the curve is parametrised proportionately to arc length. Since any curve can be parametrised by arc length, the geodesic gives an extremal value also to the length of the path.

Proposition 2.2.1 *In local coordinates, geodesics are defined by equations*

$$\frac{d^2x_i}{dt^2} + \sum_{j,k} \Gamma^i_{jk} \frac{dx_j}{dt}\frac{dx_k}{dt} = 0.$$

Proof Euler's equation for the variational problem of minimising the integral of $G := \sum_{j,k} g_{jk} \frac{dx_j}{dt}\frac{dx_k}{dt}$ is $\frac{\partial G}{\partial x_r} = \frac{d}{dt}\left(\frac{\partial G}{\partial y_r}\right)$, where $y_r = \frac{dx_r}{dt}$. This gives

$$\begin{aligned}
\sum_{j,k} \frac{\partial g_{jk}}{\partial x_r}\frac{dx_j}{dt}\frac{dx_k}{dt} &= \frac{d}{dt}\left(2\sum_j g_{rj}\frac{dx_j}{dt}\right) \\
&= 2g_{rj}\frac{d^2x_j}{dt^2} + 2\frac{\partial g_{rj}}{\partial x_k}\frac{dx_j}{dt}\frac{dx_k}{dt} \\
&= 2g_{rj}\frac{d^2x_j}{dt^2} + \frac{dx_j}{dt}\frac{dx_k}{dt}\left(\frac{\partial g_{rj}}{\partial x_k} + \frac{\partial g_{rk}}{\partial x_j}\right),
\end{aligned}$$

where in the last step we use symmetry under the interchange of j and k. If g^{ij} is the inverse matrix to $g_{i,j}$, multiply by g^{ir}, sum over r and simplify:

$$\frac{d^2x_i}{dt^2} + \frac{1}{2}\sum_r g^{ir}\left(\frac{\partial g_{rj}}{\partial x_k} + \frac{\partial g_{rk}}{\partial x_j} - \frac{\partial g_{jk}}{\partial x_r}\right)\frac{dx_j}{dt}\frac{dx_k}{dt} = 0.$$

The coefficient of the last term is usually abbreviated to Γ^i_{jk}. □

Theorem 2.2.2 *For any point $Q \in M$, we can find a neighbourhood V of Q in M and an $\varepsilon > 0$ such that for any $P \in V$, and $\upsilon \in T_P(M)$ with $\|\upsilon\| < \varepsilon$, there is a unique geodesic $p(t)$ with*

$$p(0) = P, \qquad \left.\frac{d}{dt}p(t)\right|_{t=0} = \upsilon.$$

This is defined for $|t| < 2$, stays in V, and depends smoothly on p, υ, t.

Proof We take a coordinate neighbourhood φ on M at Q mapping onto $\mathring{D}^m(3)$ and apply the Existence Theorem for Ordinary Differential Equations (Theorem 1.4.1). Consider the system

$$\left.\begin{aligned} dx_i/dt &= y_i \\ dy_i/dt &= \Gamma^i_{jk}(x)y_jy_k \end{aligned}\right\},$$

where $x \in \mathring{D}^m(3)$, $\|y\| < 3$ corresponds to the U of that theorem, and $x \in D^m(2)$, $\|y\| \leq 2$ to its K. Then for some $\varepsilon > 0$, we find a unique solution $x = f(x_0, y_0, t)$ for all $\|x_0\| \leq 2$, $\|y_0\| < 2$, $|t| < \varepsilon$ depending smoothly on all its arguments, and lying in $\|x\| < 3$. Lifting to V by φ^{-1}, this gives a geodesic in M.

To deduce the theorem, we need only change parameter by $t' = \frac{2}{\varepsilon}t$; this has the effect of multiplying the initial $\frac{d}{dt}p(t)$ by the inverse factor, and so altering the condition $\|v\| \leq 2$ to $\|v\| \leq \varepsilon$. □

It is worth emphasising that though the argument involved defining a flow in the tangent bundle $\mathbb{T}(M)$, the geodesic itself is a path in M.

As for flows in general, the local existence and uniqueness of geodesics given by Theorem 2.2.2 does not imply global existence, but does imply uniqueness in the whole range of existence (by applying the result to a sequence of points along the geodesic), given the initial point and direction.

Let $P \in M$, $v \in T_P M$, and suppose that the geodesic with direction v at P can be defined for $|t| \leq 1$. Then we write $\exp(P, v)$ for the point at $|t| = 1$ on the geodesic, and call exp the *exponential map*. We also define the map Exp from a subset of $\mathbb{T}(M)$ to $M \times M$ by $\mathrm{Exp}(P, v) = (P, \exp(P, v))$. We have shown that these maps are defined on a neighbourhood $\mathcal{V}$ of $\mathbb{T}^0(M)$ in $\mathbb{T}(M)$.

A submanifold $V \subset M$ is called *totally geodesic* if each geodesic in M tangent to V is contained in V. Thus a one-dimensional submanifold is totally geodesic if and only if it is a geodesic.

We now obtain further properties of the exponential map.

Proposition 2.2.3 *The Jacobian determinant of* Exp *is non-zero on* $\mathbb{T}^0(M)$.

Proof For $P \in M$, let $\varphi : U \to \mathbb{R}^m$ be a coordinate neighbourhood, and choose $x_1, \ldots, x_m$ as coordinates in M, $dx_1, \ldots, dx_m$ as coordinates in the fibres $T_P M$; write the latter as $v_1, \ldots, v_m$, and write coordinates in $M \times M$ as $x_1, \ldots, x_m$, $z_1, \ldots, z_m$. Then we have $\mathrm{Exp}(x, v) = (x, z)$, so it remains to compute the partial derivatives of the z_i at 0. Now z is the point at $t = 1$ on the solution of the equation $\frac{dz}{dt} = y$ with initial condition $z = x$, $y = v_0$, i.e. at the point t_0 on the solution with initial condition $z = x$, $y = v_0/t_0 = v$. Hence

$$z = x + t_0 v + \text{smaller terms, where } t_0 \text{ is small, } v \text{ fixed,}$$

and so to find $\frac{\partial z_i}{\partial v_j}$, set $(v_0)_i = t_0 \delta_{ij}$; then

$$\frac{\partial z_i}{\partial v_j} = \left.\frac{\partial z_i(v_0)}{\partial t_0}\right|_{t_0=0} = \delta_{ij}.$$

This proves the result: for later reference note also that $\frac{\partial z_i}{\partial x_j} = \delta_{ij}$. □

It follows from Proposition 2.2.3 and the Inverse Function Theorem 1.2.5 that $\mathbb{T}^0(M)$ has a neighbourhood $\mathcal{V}'$ in $\mathbb{T}(M)$ on which Exp is defined, and is a local diffeomorphism. It now follows using Corollary A.2.6 that $\mathbb{T}^0(M)$ has a neighbourhood $\mathcal{V}''$ in $\mathbb{T}(M)$ on which Exp is defined, and is a diffeomorphism.

We have an even sharper statement.

Theorem 2.2.4 *There is a neighbourhood W of $\Delta(M)$ in $M \times M$ such that if $(x, y) \in W$, there is a unique geodesic from x to y of length $\rho(x, y)$. Hence* Exp *defines a diffeomorphism of* $\mathrm{Exp}^{-1}(W)$ *onto W.*

Proof For each $P \in M$, it follows from the above that we can find a neighbourhood V_P of P such that Exp^{-1} defines a diffeomorphism of $V_P \times V_P$ on a neighbourhood of $\mathbb{T}^0(V_P)$. Then if U_P is a sufficiently small neighbourhood of P, each pair of points in U_P is joined by a unique geodesic lying in U_P, and (as in the proof of Theorem 2.1.1) each geodesic going outside U_P is longer. Thus this geodesic gives a minimum length for curves in U_P joining the two points. (In the technical language of Calculus of Variations, the metric is positive definite, the problem is regular, and we have constructed a semi-field of extremals, passing through a point and covering a neighbourhood.)

The geodesic gives the global minimum, which we defined as the distance $\rho(x, y)$. Thus Exp^{-1} is a diffeomorphism on $U_P \times U_P$: we take W as the union of such neighbourhoods. □

This has the following useful application.

Corollary 2.2.5 *There exist a neighbourhood W of $\Delta(M)$ in $M \times M$ and a C^∞ map $H : W \times [0, 1] \to M$ such that for each $(P, Q) \in W$, $H(P, Q, 0) = P$ and $H(P, Q, 1 - t) = H(Q, P, t)$.*

Proof Take W as given by the theorem. Then for each $(P, Q) \in W$ there is a unique geodesic $g_{P,Q} : [0, \rho(P, Q)] \to M$ with $g_{P,Q}(0) = P$ and $g_{P,Q}(1) = Q$. We can thus take $H(P, Q, t) = g_{P,Q}(t.\rho(P, Q))$. □

We will need a variant of this below (for Proposition 6.4.4).

Proposition 2.2.6 *For M a smooth manifold, the map $e_M : \mathbb{T}(M) \to M \times M$ given by $e_M(\xi) = (exp(\xi), exp(-\xi))$ is a local diffeomorphism along $\Delta(M)$ and there exist neighbourhoods A_M of $\mathbb{T}^0(M)$ in $\mathbb{T}(M)$ and O_M of $\Delta(M)$ in $M \times M$ such that e_M gives a diffeomorphism of A_M on O_M.*

For it follows from the proof of Proposition 2.2.3 that, in the natural local coordinates, the differential of e_M takes the form $(x, v) \mapsto (x + v, x - v)$, so is an isomorphism. The conclusion now follows as above.

In the region where geodesics are unique, the distance function also has the expected properties.

Proposition 2.2.7 *On the set W of Theorem 2.2.4, the square $\rho(x, y)^2$ of the distance is a smooth function.*

Proof In view of Theorem 2.2.4, it suffices to show that taking the square of the length of the geodesic defines a smooth function on a neighbourhood of $\mathbb{T}^0(M)$ in $\mathbb{T}(M)$. But this function is just the square of the length of the tangent vector in question, so is a smooth function since the Riemannian structure is smooth. □

We recall that a metric space is complete if each Cauchy sequence of points converges to a limit point, or equivalently, if each bounded closed subset is compact. With this concept, we can give the global forms of the above theorems.

Theorem 2.2.8 *M is complete if and only if geodesics may be indefinitely produced, i.e. if* exp *and* Exp *are definable on $\mathbb{T}(M)$. Any two points in a complete manifold may be joined by geodesics: the length of at least one such is the distance between them.*

Proof Suppose first M is complete, and $p(t)$ a geodesic which exists only for $t < k$. Then the points $p(t - \frac{1}{n})$ form a Cauchy sequence: since M is complete, these have a limit point P. But by Theorem 2.2.2, P has a compact neighbourhood K such that any geodesic within K may be produced a distance ε. This gives a contradiction.

Now suppose exp globally definable, but that there are pairs of points (P, Q) not joined by a geodesic of length $\rho(P, Q)$. Let r be the greatest lower bound of the distance of such points Q from P (by Theorem 2.2.4, $r > 0$), let $K_1 = \{v \in T_P M \mid \|v\| \leq r\}$, and let $K = \exp(K_1)$. Then K_1 is compact, hence so is K, by definition of r, K contains all points at distance less than r from P. Choose $2\varepsilon < r$ as the number ε in Theorem 2.2.2, and choose Q such that $\rho(P, Q) = r_0 < r + \varepsilon$, but P and Q are not joined by a geodesic of length $\rho(P, Q)$. Now let P_i be a smooth path from P to Q of length at most $r_0 + 1/i$, and let R_i be the point on it at distance $r - \varepsilon$ from P. The R_i lie in the compact set K; let R be a cluster point. Then

$$\rho(P, R) \leq \limsup \rho(P, R_i) = r - \varepsilon,$$
$$\rho(R, Q) \leq \limsup \rho(R_i, Q) = r_0 - r + \varepsilon,$$

so by the triangle inequality we have

$$\rho(P, R) = r - \varepsilon, \qquad \rho(R, Q) = r_0 - r + \varepsilon.$$

By the definition of r, ε; P can be joined to R by a geodesic of length $r - \varepsilon$; R to Q by on of length $r_0 - r + \varepsilon$. If these met at an angle at Q, we could construct a shorter path by rounding the corner in a neighbourhood of Q. Hence

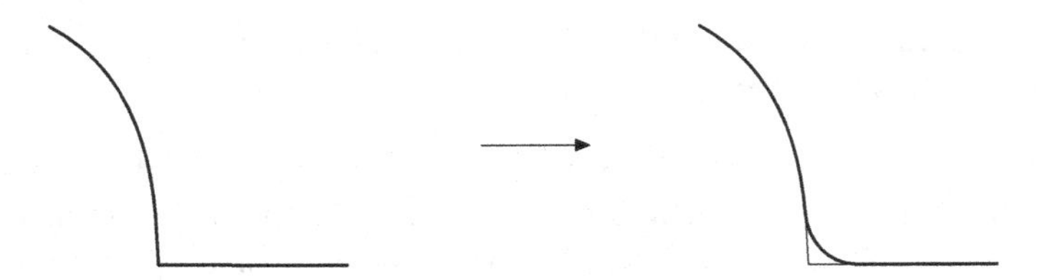

Figure 2.1 Rounding the corner of a path

they have the same direction at Q, so by the uniqueness theorem form part of the same geodesic. Thus P is joined to Q by a geodesic of length $\rho(R, Q)$: a contradiction. The idea of this proof is sketched in Figure 2.1.

Finally, suppose $\exp(T_P M) = M$. Then a bounded set lies within a finite distance from P, so is contained in the image of a closed and bounded, hence compact, subset of $T_P M$. But the image of this set is also compact, so it follows that M is complete. □

Theorem 2.2.9 *Any connected manifold has a Riemannian metric in which it is complete.*

Proof We make a slight refinement of the proof of Theorem 1.3.1, asserting the existence of Riemannian structures. Let $\varphi_\alpha : U_\alpha \to \mathring{D}^m(3)$ be the coordinate neighbourhoods constructed in Theorem 1.1.4, and define $\Phi_\alpha \in \mathcal{F}_i$ by

$$\Phi_\alpha(P) = \begin{cases} Bp(2\frac{1}{2} - \|x\|) & \text{if } P \in U_\alpha, \varphi_\alpha(P) = x \\ 0 & \text{if } P \notin U_\alpha. \end{cases}$$

Then write $ds^2 = \sum \Phi_\alpha(\sum dx_i^2) \circ \varphi_\alpha$. As in the earlier proof, we see that this is a metric. In $\varphi_\alpha^{-1}(\mathring{D}^m(1\frac{1}{2}))$, it is greater than or equal to the Euclidean metric, so the set of points at distance $\leq \frac{1}{3}$ from $\varphi_\alpha^{-1}(D^m)$ is a closed subset of $\varphi_\alpha^{-1}(D^m(2))$, so is compact. As in Theorem 2.2.8, it follows that all geodesics from a point of $\varphi_\alpha^{-1}(D^m)$, and hence from any point of M, may be produced a distance at least $\frac{1}{3}$ from any point. Thus they can all be produced indefinitely. □

Corollary 2.2.10 *(i) For any smooth manifold V, there is a proper map $V \to \mathbb{R}_+$.*

(ii) If M is non-compact, there is a proper map $V \to M$.

Proof (i) Choose a complete Riemannian metric on V; then for any $P_0 \in V$, the distance from P_0 is a proper map $\rho(P_0, -) : V \to \mathbb{R}_+$. For we saw above that the preimage of any set $[0, K]$ is compact. The square $\rho(P_0, -)^2$ is also proper, and is smooth.

Since the composite of two proper maps is proper, (ii) will follow if we can construct a proper map $\mathbb{R}_+ \to M$. Choose a non-compact component M_0 of

M and a point $Q_0 \in M_0$. Suppose inductively chosen $Q_i \in M_i$: then remove $\{P \in M_i \mid \rho(P, Q_0) < i\}$ from M_i, let M_{i+1} be a non-compact component of the complement, and choose any $Q_{i+1} \in M_{i+1}$.

Since Q_i, Q_{i+1} lie in the connected set M_i, they can be joined by a path $[i, i+1] \to M_i$. Joining all these paths gives a map $\varphi : \mathbb{R}_+ = [0, \infty) \to M$. Since, for any $P \in M_i$, $\rho(P, Q_0) \geq i - 1$, the map φ is proper. □

2.3 Tubular neighbourhoods

We will now apply the results of §2.2 in the context of a submanifold V^v of M^m. Then we proceed to consider boundaries.

Proposition 2.3.1 *The Jacobian determinant of* $\exp : \mathbb{N}(M/V) \to M$ *on* $\mathbb{T}^0(V)$ *is non-zero.*

Proof Let $P \in V$, and let $\varphi : U \to \mathbb{R}^n$ be a coordinate neighbourhood of P in M such that $U \cap V = \varphi^{-1}(\mathbb{R}^m)$. Then if $x_1, \ldots, x_n$ are coordinates in $\mathbb{R}^n$, we can take as local coordinates in $\mathbb{N}(M/V)$ $x_1, \ldots, x_m$ (coordinates in V) and $v_{m+1}, \ldots, v_n$ (coordinates in the fibre) where $v_i = dx_i$. Now refer back to Proposition 2.2.3, where we showed that if $\exp(x, v) = z$, then $\frac{\partial z_i}{\partial x_j} = \frac{\partial z_i}{\partial v_j} = \delta_{ij}$ so that with respect to our coordinates, the Jacobian matrix is the unit matrix, so its determinant is non-zero. □

Theorem 2.3.2 *Let V be a submanifold of M. Then*

(i) the map $\exp : \mathbb{N}(M/V) \to M$ *is a local diffeomorphism at* $\mathbb{T}^0(V)$,

(ii) there is a neighbourhood of $\mathbb{T}^0(V)$ *in* $\mathbb{N}(M/V)$ *on which* exp *is a diffeomorphism to a neighbourhood U of V in M,*

(iii) V has a neighbourhood U' in M such that each point P of U is joined to V by a unique geodesic of length $\rho(P, V)$; this meets V orthogonally.

Proof (i) follows from Proposition 2.3.1 and the Inverse Function Theorem 1.2.5.

(ii) follows from this by applying Corollary A.2.6.

(iii) Let $Q \in V$, and let $U_1 \subset U_0$ be neighbourhoods of Q in M as in the proof of Theorem 2.2.4: any two points in U_0 are joined by a unique geodesic of minimal length, and the minimal geodesic joining two points of U_1 lies in U_0. We may suppose $\bar{U}_0$ compact.

For $P \in U_1$, let r_P be the greatest lower bound of distances of P from points of V. If we have points $Q_i \in V$ with $\rho(P, Q_i) < \frac{1}{i}$, then for $i > D^{-1}$ we have $Q_i \in U_0$, and since $\bar{U}_0$ is compact, the points Q_i have a cluster point Q; since V is closed, we have $Q \in V$, and now $\rho(P, Q) = r_P$. By the above choice of U_0, P and Q are joined by a unique geodesic of minimal length. This meets V orthogonally for if not, by a small modification near Q, we could make it shorter (take a path orthogonal to V, and smooth off the corner), giving a shorter path from P to V. Hence there is a point R' of $\mathbb{N}(M/V)$ lying over Q with $\exp(R') = P$.

We may now take U' as the union of the U_1. □

Taking the intersection $U \cap U'$ gives a neighbourhood of V on which both exp is a diffeomorphism and the geodesics give shortest distances from V.

For V^v a closed submanifold of a smooth manifold M^m, a *tubular neighbourhood* of V in M consists of a bundle B over V with fibre the disc D^{n-m} and an embedding $\psi : B \to M$ (as submanifold with boundary) extending the map taking the centre of each disc to the corresponding point of V.

As with coordinate neighbourhoods, the actual neighbourhood $\psi(B)$ is the more geometrical concept; but the mapping ψ is more convenient to work with. A tubular neighbourhood is pictured in Figure 2.2.

For any tubular neighbourhood, the map ψ induces an isomorphism of the normal bundle of V in M with that in B, and hence with the vector bundle associated to B. If M^m has a Riemannian structure, the normal bundle $\mathbb{N}(M/V)$ has group O_{m-v}. We may then take B as the associated disc bundle, consisting of vectors of $\mathbb{N}(M/V)$ of at most unit length.

Figure 2.2 Tubular neighbourhood in a manifold and in one with boundary

Theorem 2.3.3 *For any submanifold V of a smooth manifold M, there exists a tubular neighbourhood of V in M.*

Proof Choose a Riemannian metric on M. Let W be a neighbourhood of $\mathbb{T}^0(V)$ in $\mathbb{N}(M/V)$ mapped diffeomorphically by exp: the existence of such W is guaranteed by Theorem 2.3.2. Let f be a positive continuous function on V such that vectors in $N_P(M/V)$ of length less than $f(P)$, are contained in W: the existence of such f follows from Lemma A.2.4 (i). By Proposition 1.1.7, we can find a positive smooth function g on V such that $0 < g(P) < f(P)$ for all $P \in V$. We now define a diffeomorphism ψ. For each $P \in V$, $v \in N_P(M/V)$, set

$$\psi(P, V) = \exp(P, g(P)v).$$

Multiplication by $g(P)$ in the fibre is possible since $g(P) \neq 0$, and for $\|v\| \leq 1$ we have $|g(P)v| \leq g(P) < f(P)$, so $(P, g(P)v) \in W$. □

We will extend this result to the case of manifolds with boundary, but need first to develop further ideas.

We now combine Whitney's embedding theorem with the existence of tubular neighbourhoods to give a general method of constructing maps into smooth manifolds. We illustrate by showing the existence of smooth approximations, extending Lemma 1.1.7.

Let V be a compact manifold. By Theorem 1.2.11, there exists a smooth embedding $i : V \to \mathbb{R}^N$ for some N. By Theorem 2.3.3 there exist a disc bundle $\pi : W^N \to V$ and a smooth embedding $\psi : W \to \mathbb{R}^N$, extending i, and whose image is a neighbourhood U of $i(V)$. Further, we can choose the discs to have radius ε; U is then a ε-neighbourhood of $i(V)$. We have a retraction $\phi := \pi \circ \psi^{-1} : U \to V$; for each $x \in V$, the preimage $\phi^{-1}(x)$ is a disc of radius ε.

Proposition 2.3.4 *Let M and V be smooth manifolds with $V \subset \mathbb{R}^N$ compact.*

(i) For any continuous $f : M \to V$ and any $\varepsilon > 0$ there exists a smooth $h : M \to V$ with $\|h(x) - f(x)\| < \varepsilon$ for every $x \in M$.

(ii) If moreover F is a closed subset of M such that f is smooth on some open set $U \supset F$, we can find h such that also $h = f$ on a neighbourhood of F.

Proof Choose a tubular neighbourhood of V in $\mathbb{R}^N$ as above. Applying Proposition 1.1.7 to each component of $M \xrightarrow{f} V \subset \mathbb{R}^N$ gives a smooth map $h : M \to \mathbb{R}^N$ within distance ε of f, and hence with image contained in U. Thus $\phi \circ h$ gives a map $M \to V$, and since ϕ moves each point within a disc of radius $< \varepsilon$, h is within ε of f.

The same argument, but using (iii) of Proposition 1.1.7, gives (ii). □

For N a smooth manifold with boundary, the discussion of geodesics at non-boundary points is the same as before. At a boundary point P, we see from the differential equations that local geodesics can be constructed for all inward-pointing tangent vectors and for no outward-pointing ones. There are several possibilities for those tangent to the boundary; as examples, the reader may consider D^2 and the closure of $\mathbb{R}^2 \setminus D^2$, each with the metric induced from $\mathbb{R}^2$.

A Riemannian metric on M is *adapted to the boundary* if ∂M is totally geodesic.

Lemma 2.3.5 *Let M^m have a Riemannian metric. Then the product metric for $M \times \mathbb{R}^1_+$ is adapted to the boundary.*

Proof Let $x_1, \ldots, x_m$ be local coordinates in M, and x_0 the coordinate in $\mathbb{R}^1_+$. Then for the metric $g_{i,j}$ we have $g_{0j} = \delta_{0j}$. Hence one of the defining equations for geodesics is simply $d^2x_0/dt^2 = 0$. Thus if initially $x_0 = dx_0/dt = 0$, we have $x_0 = 0$ all along the geodesic, which thus stays in $M \times \{0\}$. □

A similar argument gives the following.

Lemma 2.3.6 *If $V \subset M$ is a submanifold whose normal bundle is trivial, then M has a Riemannian metric in which the submanifold V is totally geodesic.*

Proof It follows from Theorem 2.3.3 that V has a neighbourhood in M diffeomorphic to $V \times \mathbb{R}^c$, where c is the codimension of V in M. We may choose any metric on V and then take the product metric on $V \times \mathbb{R}^c$: in any coordinate neighbourhood of V with metric $ds^2 = \sum g_{i,j}dx_idx_j$ this is given by $ds'^2 = \sum g_{i,j}dx_idx_j + \sum dy_k^2$. A short calculation shows that any geodesic initially tangent to $V \times \{0\}$ remains in this submanifold.

As in the proof of Theorem 1.3.1, we can now construct a metric on M which agrees with this metric on some neighbourhood of V in M. The result follows. □

Proposition 2.3.7 *(i) Every manifold M^m with boundary has a Riemannian metric adapted to the boundary.*

(ii) Given a submanifold V^v of M^m, there is a metric on M such that V meets ∂M orthogonally, and the restriction of the metric to V is adapted to the boundary.

Proof (i) By Theorem 1.5.5, ∂M has a collar neighbourhood $\psi : \partial M \times I \to M$. Let φ be a metric on M, φ' the product of some metric on ∂M with the standard metric of I. We define a metric φ'' by

$$\varphi'' = \begin{cases} \varphi & \text{outside the image of } \psi \\ \varphi' + (\varphi - \varphi')Bp(3t-1) & \text{at } \psi(P, t). \end{cases}$$

The latter agrees with φ in a neighbourhood of $t = 1$, so is smooth everywhere; it is a Riemannian structure, as a positive linear combination of positive form is another, and it agrees with φ' near $t = 0$, so by Lemma 2.3.5, it is adapted to ∂M.

(ii) By Proposition 1.5.6(ii), we may suppose that the restriction to $\partial V \times I$ of the collar neighbourhood of ∂M gives a collar neighbourhood for ∂V. Then the metric constructed above has both the desired properties. □

The definition of tubular neighbourhood of a closed submanifold V^v of a manifold M^m with boundary is the same as before: we require a bundle B over

V with fibre the disc D^{n-m} and an embedding $\psi : B \to M$ (as submanifold with boundary) extending the map taking the centre of each disc to the corresponding point of V.

If $\pi : B \to V$ is the projection of a disc bundle, Σ the boundary sphere-bundle of B, and $C = \pi^{-1}(\partial V)$, then B has the structure of a smooth manifold with corner, and $\angle B = \Sigma \cap C$ separates ∂B into two parts, with closures Σ and C. It follows that if (B, ψ) is a tubular neighbourhood of V, $\psi(C) = \partial M \cap \psi(B)$.

Theorem 2.3.8 *If M is a manifold with boundary, V a submanifold, then there exists a tubular neighbourhood of V in M.*

Proof By Proposition 2.3.7 (ii), we can choose a Riemannian metric for M, adapted to the boundary, in which V meets ∂M orthogonally. As in the proof of Theorem 2.3.3, we consider the exponential map of the normal bundle $\mathbb{N}(M/V)$. We need to show that this is well defined. The crucial point is that since the metric is adapted to the boundary, and the vectors in C are normal to V and hence tangent to ∂M, integrating them gives curves in ∂M and hence, at least locally, a map $C \to \partial M$. The previous argument shows that this map is a local diffeomorphism.

The arguments needed to go from having a local diffeomorphism to the result are the same as those for Theorem 2.3.3. □

2.4 Diffeotopy extension theorems

Let V^v, M^m be smooth manifolds, possibly with boundary. A *diffeotopy* of V in M is an embedding $h : V \times I \to M \times I$ which is *level-preserving*, i.e. we can write

$$h(x, t) = (h_t(x), t) \qquad m \in V, t \in I.$$

It follows that each h_t is also an embedding. We also say that h is a diffeotopy between h_0 and h_1.

h is called *normalised* if it extends to a level-preserving embedding $h : V \times \mathbb{R} \to M \times \mathbb{R}$ such that $h_t = h_0$ when $t \leq 0$, and $h_t = h_1$ when $t \geq 1$. If h is any diffeotopy, the map $H : V \times \mathbb{R} \to M \times \mathbb{R}$ given by $H(m, t) = (h_{Bp(t)}(m), t)$ is a normalised diffeotopy between h_0 and h_1.

A *diffeotopy* of M is a diffeomorphism k of $M \times I$ which is level-preserving, thus in particular it is a diffeotopy of M in M. The diffeotopy k of M *covers* the diffeotopy h of V in M if, for all $x \in V, t \in I$, $k_t(h_0(x)) = h_t(x)$. A diffeotopy covered by a diffeotopy of M is called an *ambient diffeotopy*.

Lemma 2.4.1 *Diffeotopy is an equivalence relation.*

Proof The definition $h(x,t) = (h_0(x), t)$ gives a diffeotopy between h_0 and itself. If h' gives one between h_0 and h_1, then h'', where $h''(x,t) = h'(x, 1-t)$ gives a diffeotopy between h_1 and h_0. Finally, let h', h'' be normalised diffeotopies between h_0 and h_1 and h_1 and h_2. Then set

$$h_t'''(x) = \begin{cases} h'_{3t}(x) & \text{if } t \leq 1/2 \\ h''_{3t-2}(x) & \text{if } t \geq 1/2; \end{cases}$$

this is a smooth embedding, since h' and h'' are so, and we have $h_t''' = h_1$ for $\frac{1}{3} \leq t \leq \frac{2}{3}$, so that the two parts of the definition fit smoothly. □

One of the basic problems in differential topology is to determine the set of equivalence classes. We will accomplish this in some cases in Chapter 6.

The *support* of a diffeomorphism h of a smooth manifold M is the closure of the set of points P with $h(P) \neq P$.

The support of a diffeotopy h of V in M is the closure of the set of points $P \in V$ such that $h_t(P)$ is not independent of t.

Theorem 2.4.2 (Diffeotopy Extension Theorem) *Let V, M be smooth manifolds, perhaps with boundary, and let $h : V \times \mathbb{R} \to M \times \mathbb{R}$ be a diffeotopy of V in M, whose support K is compact, and contained in $\mathring{M}$. Then there is a diffeotopy k of M, whose support is compact and contained in $\mathring{M}$, and which covers h; in particular, h is ambient.*

Proof Since K is contained in $\mathring{M}$, we can ignore the boundary of M, and suppose simply that M is a smooth manifold, for if the result is proved in this case, the diffeotopy k of M which we obtain, having compact support, equals the identity on a neighbourhood of $\partial M \times \mathbb{R}$, and can therefore be extended to the boundary as the identity.

Let k be a diffeotopy of $M \times \mathbb{R}$. Then k defines a vector field on $M \times \mathbb{R}$ as follows. Write ∂_t for the vector field which projects to 0 on M and to $\partial/\partial t$ on $\mathbb{R}$, and define a vector field on $M \times \mathbb{R}$ by $\xi_k := dk(\partial_t)$. Since k is level-preserving, the projection of ξ_k on the second factor is still $\partial/\partial t$. Also, if k has compact support, $\xi_k = \partial_t$ except at some points of a compact set.

Conversely, suppose given a vector field ξ whose projection on $\mathbb{R}$ is $\partial/\partial t$. If ξ is complete, it gives rise to a 1-parameter group (φ_t) of diffeomorphisms of $M \times \mathbb{R}$, and hence to the diffeotopy given by $k(P,t) = (\varphi_t(P), t)$. Moreover, the local uniqueness clause in Theorem 1.4.2 implies that if k gives rise to ξ, then we recover the original k.

Since by Proposition 1.4.4 (ii) and (iii), the vector field ∂_t on $M \times \mathbb{R}$ is complete, it follows by (iv) that if $\xi = \partial_t$ except on a compact set, then ξ is complete.

We conclude that to construct the diffeotopy, it is sufficient to construct the vector field ξ. By the above argument, we see that the necessary and sufficient condition that k covers h is that on $h(V \times \mathbb{R})$, we have $\xi = dh(\partial/\partial t)$. Thus the problem is reduced to the construction of a vector field ξ on $M \times \mathbb{R}$ satisfying:

(i) $\xi = \partial_t$ outside a compact set,

(ii) the projection of ξ on $\mathbb{R}$ is everywhere $\partial/\partial t$,

(iii) on $h(V \times \mathbb{R})$, $\xi = dh(\partial/\partial t)$.

We assert that if we can do this in a neighbourhood of each point of $h(V \times \mathbb{R})$, ξ can be constructed. For such neighbourhoods, together with the complement U_0 of $h(V \times \mathbb{R})$, form an open covering of $M \times \mathbb{R}$. By Theorem 1.1.5, there is a smooth partition $\{\Psi_\alpha\}$ of unity strictly subordinate to this covering. If ξ_α is a function on the support U_α of Ψ_α which satisfies conditions (i) – (iii), the function $\xi := \sum_\alpha \xi_\alpha \Psi_\alpha$ (where $\xi_0 := \partial_t$) will satisfy all the conditions.

Now $h(V \times \mathbb{R})$ is a submanifold of $M \times \mathbb{R}$, hence in a neighbourhood of any point of it we can find a coordinate neighbourhood $\varphi : U \to \mathbb{R}^{m+1}$ with $U \cap \operatorname{Im} h = \varphi^{-1}(\mathbb{R}^{v+1})$; say for simplicity that the image of U is $\mathring{D}^{m+1}$. Then $d\varphi(dh(\partial/\partial t)) = \sum a_i \partial/\partial x_i$ in $\mathring{D}^{v+1}$; we define ξ by taking the same formula in $\mathring{D}^{m+1}$ (i.e. by taking the a_i independent of the last $m - v$ coordinates).

In the case of boundaries, the a_i are only defined on $\mathring{D}^{v+1}_+$. But by Whitney's Extension Theorem 1.5.1, they can be extended to smooth functions on $\mathring{D}^{v+1}$, and then extended to $\mathring{D}^{m+1}$ as above. This completes the proof of the result. □

Corollary 2.4.3 *If M is a smooth manifold, V a compact submanifold (perhaps with boundary), then any diffeotopy of the inclusion $i : V \subset M$ is an ambient diffeotopy.*

Corollary 2.4.4 *If M is a smooth manifold with boundary, any diffeotopy of a compact submanifold (perhaps with boundary) of $\mathring{M}$ is covered by a diffeotopy of M.*

Proof By the theorem, it is covered by a diffeotopy of $\mathring{M}$ with compact support. Thus ∂M has a neighbourhood in $\mathring{M}$ fixed by the diffeotopy, which can thus be extended to M, defining it to be fixed on ∂M. □

Proposition 2.4.5 *Any diffeotopy of ∂M is covered by a diffeotopy of M.*

Proof We shall suppose the diffeotopy h_t of ∂M normalised so that $h_t = 1$ for $t \le \frac{1}{3}$ and $h_t = h_1$ for $t \ge \frac{2}{3}$. Let $\psi : \partial M \times I \to M$ be a collar neighbourhood of ∂M in M (such exist by Theorem 1.5.5). Then we define a covering diffeotopy

k_t of M by

$$k_t(Q) = \begin{cases} Q & \text{if } Q \notin \mathrm{Im}(\psi), \\ Q & \text{if } Q = \psi(P, s),\ s \geq t, \\ \psi(h_{t-s}(P), s) & \text{if } Q = \psi(P, s),\ s \leq t. \end{cases}$$

Thus for $s = 0$, k_t agrees with h_t, and for $s \geq \frac{2}{3}$, $k_t(P) = P$, so that k is everywhere smooth, and does cover h. □

Theorem 2.4.6 *Let M be a manifold with boundary, V a submanifold (perhaps with boundary). Any diffeotopy of V in M with compact support is covered by a diffeotopy of M with compact support.*

Proof Following the proof of Theorem 2.4.2, we see that it only remains to show that we can construct ξ in a neighbourhood of each point of $h(V \times \mathbb{R})$. In this case, in a neighbourhood of any point of $h(V \times \mathbb{R})$ we can find a coordinate neighbourhood $\varphi : U \to \mathbb{R}^{m+1}$ with $U \cap \mathrm{Im}h = \varphi^{-1}(\mathbb{R}_+^{v+1})$. By Theorem 1.5.1 we can write $d\varphi(dh(\partial/\partial t)) = \sum a_i \partial/\partial x_i$ in $\mathring{D}^{v+1}$ with the a_i smooth in $\mathbb{R}^{v+1}$ and define ξ by taking the same formula in $\mathbb{R}^{m+1}$. □

We shall need one or two further kinds of diffeotopy extension, when we come to consider corners, but feel that by now proofs may be left to the reader. We mention one immediate application of our results.

Proposition 2.4.7 *Let M^m be a manifold (perhaps with boundary), V^v a compact submanifold with boundary. Then there is a submanifold U^v of M^m containing V^v.*

Proof First suppose that M has no boundary. Let $\varphi : \partial V \times I \to V$ be a tubular neighbourhood of ∂V in V. We define a diffeotopy of V by

$$\begin{cases} h_t(P) = P & P \notin \mathrm{Im}\,\varphi \\ h_t\varphi(P, u) = \varphi(P, f(t, u)) & \end{cases},$$

where f is chosen with

$f(t, u) = u$ for $u > 1 - \varepsilon$,
$f(0, u) = u$,
$f(t, 0) > 0$ for $0 < \varepsilon$,

and $\partial f/\partial u > 0$ everywhere; so that the diffeotopy 'pushes' the boundary a little way into V: for example, we can take $f(t, u) = u + Bp(t - u)$ provided $t \leq k$, where in this range $Bp'(t) < 1$. Now h_t is a diffeotopy, hence (V being compact) is ambient, and so covered by H_t, say, $h_k(V) \subset \mathring{V}$. We can thus take $U = H_k^{-1}(\mathring{V})$.

If M is bounded, we argue similarly, using that part of the boundary of V not contained in ∂V. □

This result has the effect that to describe a neighbourhood of V in M, we can use tubular neighbourhoods of U; tubes round V do not give neighbourhoods.

2.5 Tubular neighbourhood theorem

We shall now use our results on diffeotopy extension to complete the discussion in §2.3 of tubular neighbourhoods by showing that these are, essentially, unique.

We recall the definition. If B is an $(m - v)$-disc bundle over V, with group O_{m-v}, and central cross-section B_0, then a tubular neighbourhood of V in M is an embedding $\varphi : B \to M$, as submanifold with boundary, extending the projection of B_0 on V.

We say that two tubular neighbourhoods $\varphi : B \to M$ and $\varphi' : B' \to M$ are *equivalent* if there is a bundle map $\chi : B \to B'$ over the identity map of V, and an ambient diffeotopy of φ on $\varphi'_0\chi$ which is fixed on B_0.

Our object is to show that any two tubular neighbourhoods are equivalent. Since we shall use the result of §2.4 we shall have to assume that V is compact. One might expect that this assumption was unnecessary; however, it cannot be omitted, as the example of Figure 2.3 illustrates.

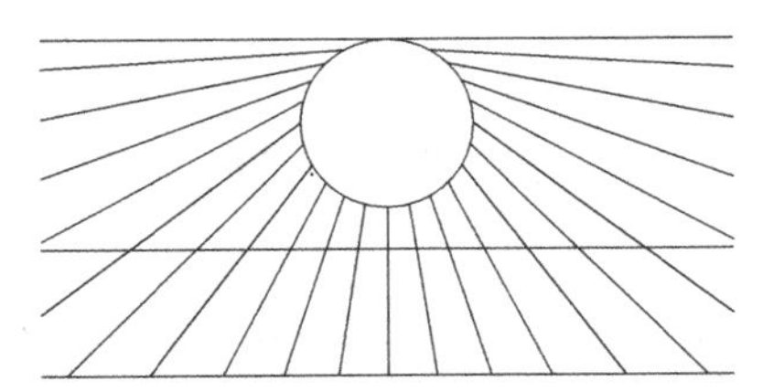

Figure 2.3 Example of a bad tubular neighbourhood

In the figure, T is the set defined by $-3 \leq y < 3$ and $x^2 + (y - 2)^2 \geq 1$, and the projection of T on $\mathbb{R}^1$ is defined by straight lines through $(0, 3)$. This gives a tubular neighbourhood of $\mathbb{R}^1$ in $\mathbb{R}^2_+$, which is not a closed subset, so is not equivalent to a standard one.

The same example thus also shows the necessity of the compactness hypothesis in Theorem 2.4.2.

Let $\varphi : B \to M$ be a tubular neighbourhood for V in M. We consider the bundle E associated to B but with fibre $\mathbb{R}^{m-v}$. Then B is a submanifold with boundary of E. For the tubular neighbourhoods of §2.3, E is simply the normal bundle $\mathbb{N}(M/V)$.

We say that an embedding $\bar{\varphi} : E \to M$ as open submanifold, extending the projection of E_0 on V, is a *open tubular neighbourhood* of V in M.

Lemma 2.5.1 *Any tubular neighbourhood $\varphi : B \to M$ can be extended to an open tubular neighbourhood $\bar{\varphi} : E \to M$.*

Remember that we are assuming that V is compact. We use the same idea as for Proposition 2.4.7.

Proof We can define a diffeotopy of φ as follows. Recall that over each neighbourhood U in V, B is a product of U with a vector space; in the sequel, we permit ourselves to form sums and products by scalars in these vector spaces, using the standard notation. Then our diffeotopy is $\varphi_t(x, v) = \varphi(x, tv)$ for $\frac{1}{2} \leq t \leq 1$ (where $x \in V$, $v \in D^{m-v}$). Since V, and so also B, is compact, the diffeotopy is ambient: say it is covered by the diffeotopy k_t of M. But $\varphi_{1/2}$ can be extended to a open tubular neighbourhood, for example, by the map

$$\bar{\bar{\varphi}}(x, v) = \varphi\left(x, \frac{\gamma(\|v\|)}{\|v\|} \cdot v\right),$$

where γ is smooth, $\gamma(t) = \frac{1}{2}t$ for $0 \leq t \leq 1$, $\gamma'(t) > 0$, and $\gamma(t) < 1$. We can now define $\bar{\varphi} = k_{1/2}^{-1} \circ \bar{\bar{\varphi}}$.

A suitable γ can be constructed by using bump functions, for example, we may take

$$\gamma(t) = \frac{1}{3} \int_0^t \{1 + (e^{-x} - 1)Bp(x - 1)\} dx.$$

□

Lemma 2.5.2 *Let $\bar{\varphi} : E \to M$, $\bar{\varphi}' : E' \to M$ be open tubular neighbourhoods of V in M such that* $\operatorname{Im} \bar{\varphi} \subset \operatorname{Im} \bar{\varphi}'$. *Then for some bundle map $\bar{\chi} : E \to E'$, there is a diffeotopy of $\bar{\varphi}$ on $\bar{\varphi}' \circ \bar{\chi}$ which is fixed on B_0.*

Proof Let $j = \bar{\varphi}'^{-1} \circ \bar{\varphi} : E \to E'$, then j is an embedding. Consider the mappings j_t given by $j_t(e) = t^{-1} j(te)$ for $0 < t \leq 1$, $e \in E$; where the multiplications by t^{-1}, t are again scalar multiplications in the fibre. Clearly $j_1 = j$; we shall show that the definition of j_t can be extended to $t = 0$, and that j_0 can be taken as $\bar{\chi}$: $\bar{\varphi}' \circ j_t$ will then give the required diffeotopy of $\bar{\varphi} = \bar{\varphi}' \circ j$ on $\bar{\varphi}'\bar{\chi}$; it is fixed on B_0.

Take local coordinates $x = (x_1, \ldots, x_v)$ in V, and let y, z be Euclidean coordinates in the fibres of E, E'. Then setting $j(x, y) = (\alpha(x, y), \beta(x, y))$ we have

$$j_t(x, y) = (\alpha(x, ty), t^{-1}\beta(x, ty)).$$

But j carries the zero cross-section of E onto that of E', so

$$\alpha(x, 0) = x, \qquad \beta(x, 0) = 0.$$

Now by Lemma 1.2.3, applied to β (regarded as a function of y with x as a parameter), there are smooth functions β_i with $\beta(x, y) = \sum y_i\beta_i(x, y)$. Then $t^{-1}\beta(x, ty) = \sum y_i\beta_i(x, ty)$, so we can write j_t in the form

$$j_t(x, y) = (\alpha(x, ty), \sum y_i\beta_i(x, ty)),$$

where the left-hand side is a smooth function also at $t = 0$. This shows that we have a smooth map $J : E \times I \to E' \times I$ defined by the j_t; to have a diffeotopy, we must check that the Jacobian is everywhere non-zero. This follows for $t \neq 0$, since j is a diffeomorphic embedding, and multiplication by t or t^{-1} gives a diffeomorphism. Now

$$j_0(x, y) = \left(x, \sum y_i\beta_i(x, 0)\right) = \left(x, \sum y_i \left.\frac{\partial\beta}{\partial y_i}\right|_{y=0}\right)$$

induces a linear map of each fibre, with matrix $(\partial\beta_j/\partial y_i) = (\partial z_j/\partial y_i)$ which is also the matrix of partial derivatives of j on B_0. Since j_0 is an embedding, this is non-zero. It follows that j_0 is a GL_{m-v}-bundle map, hence a diffeomorphism. We can thus take $\bar{\chi} = j_0$. We have also verified by the same token that J is a diffeotopy. □

Corollary 2.5.3 *The result holds also without the assumption* $\operatorname{Im}\bar{\varphi} \subset \operatorname{Im}\bar{\varphi}'$.

Proof For $\operatorname{Im}\bar{\varphi} \cap \operatorname{Im}\bar{\varphi}'$ is a neighbourhood of V, which thus has a tubular neighbourhood, hence also a open one $\bar{\varphi}''$, with $\operatorname{Im}\bar{\varphi}'' \subset \operatorname{Im}\bar{\varphi} \cap \operatorname{Im}\bar{\varphi}'$. Then there are bundle maps modulo which $\bar{\varphi}''$ is diffeotopic both to $\bar{\varphi}$ and $\bar{\varphi}'$, whence the result follows. □

Lemma 2.5.4 *Let $\bar{\varphi} : E \to M$, $\bar{\varphi}' : E' \to M$ be open tubular neighbourhoods of V in M where the bundles E, E' have group O_{m-v}. Then the conclusion of Lemma 2.5.2 holds, with $\bar{\chi}$ an O_{m-v}-bundle map.*

Proof It suffices to show that any $\psi : E \to E'$ which is a $GL_{m-v}(\mathbb{R})$-bundle map is diffeotopic to an O_{m-v}-bundle map. As above, in coordinates, ψ is given by

$$\psi(x, y) = (x, z) \quad \text{where} \quad z_i = \sum a_{ij}(x)y_j.$$

Now since the group is the orthogonal group, we can speak of the length of a vector in the fibre (compare §1.2). By the Gram–Schmidt orthogonalisation process, take the vectors b_i with components a_{ij}, and write $b_i = \sum_{j=1}^{i} \lambda_{ij}e_j$, where the e_i are orthonormal, and each $\lambda_{ij} > 0$. If e_i has components e_{ij}, consider now the diffeotopy

$$k_t(x, y) = (x, z_t), \quad \text{where} \quad (z_t)_i = \sum_{j,k}(t\lambda_{ij} + (1-t)\delta_{ij})e_{jk}y_k.$$

That this is a diffeotopy follows as no matrix $(t\lambda_{ij} + (1-t)\delta_{ij})$ is singular (for the matrix is triangular, with non-zero diagonal terms); k_1 is the given map ψ, and k_0 takes one orthonormal base to another, so is an O_{m-v}-bundle map. □

Theorem 2.5.5 (Tubular Neighbourhood Theorem) *Let M^m be a smooth manifold and V^v a compact submanifold. Then any two tubular neighbourhoods of V in M are equivalent.*

Proof Let $\varphi : B \to M$, $\varphi' : B' \to M$ be tubular neighbourhoods of V in M. By Lemma 2.5.1, φ and φ' extend to open tubular neighbourhoods $\bar{\varphi}$, $\bar{\varphi}'$. By Corollary 2.5.3, there is a bundle map $\bar{\chi} : E \to E'$ such that there is a diffeotopy of $\bar{\varphi}$ on $\bar{\varphi}' \circ \bar{\chi}$, fixed on B_0. By Lemma 2.5.4, we may take $\bar{\chi}$ as an O_{m-v}-bundle map. Then $\bar{\chi}$ maps B into B', and so we can take χ as its restriction. It follows that χ is a bundle isomorphism. Also, by Theorem 2.4.2, the diffeotopy we have constructed is in fact ambient. □

As a first corollary, we obtain a useful little result.

Theorem 2.5.6 (Disc Theorem) *Let M be a connected manifold (perhaps with boundary), $f_1, f_2 : D^m \to M^m$ embeddings as submanifold with boundary. Then f_1 and f_2 are ambient diffeotopic unless M is oriented and f_1, f_2 have opposite orientations.*

Proof Let $P_i = f_i(O) \quad (i = 1, 2)$. Since $\mathring{M}$ is connected, there is a smooth path connecting P_1 and P_2 in $\mathring{M}$, i.e. a diffeotopy of P_1 and P_2, considered as submanifolds of zero dimension. By the diffeotopy extension theorem, there is an ambient diffeotopy. Hence we may suppose $P_1 = P_2 = P$. Now f_1, f_2 are tubular neighbourhoods of P, so by Theorem 2.5.5, there is an orthogonal transformation χ of D^m, such that f_1 and $f_2 \circ \chi$ are ambient diffeotopic.

Now if $\chi \in SO_m$, then f_2 is diffeotopic, so also ambient diffeotopic to $f_2 \circ \chi$, so the result follows. If not, and M is orientable, we have the case excluded by the theorem. If M is non-orientable, there is an orientation-reversing smooth path (see the discussion after the definition of orientability), and if we take P on an ambient diffeotopy round such a path, the sign of the determinant of χ will change. □

We shall use numerous extensions of Theorem 2.5.5 in the sequel; let us indicate one or two briefly here. The definition of equivalence remains the same.

Proposition 2.5.7 *Any two collar neighbourhoods of ∂M in M are equivalent if ∂M is compact.*

Proof The proof follows the same pattern. The analogues of Lemma 2.5.1 and Lemma 2.5.2 follow as before. In Lemma 2.5.4, note only that our group is not $GL_1(\mathbb{R})$ or O_1, but simply $GL_1^+(\mathbb{R})$ or SO_1 – the trivial group. This makes for a slight simplification in the argument. □

Proposition 2.5.8 *The result of Theorem 2.5.5 holds also if M has a boundary.*

We note that in proving uniqueness of tubular neighbourhoods, in contrast to the case where we had to prove existence, no extra difficulties arise in the case where we have boundaries.

We now present an alternative approach to the existence of tubular neighbourhoods which, while less immediate than the use of the exponential map, is more flexible for generalisations.

We begin with notation. For $\pi : E \to B$ the projection map of a vector bundle, we identify B with the zero cross-section (the zero vectors in the fibres). The map π induces $\pi_* : \mathbb{T}(E) \to \mathbb{T}(B)$, hence for each $e \in E$ a linear map $T_eE \to T_eB$. Vectors in the kernel are called *vertical tangent vectors* of E, and we write $\mathbb{T}^v(E)$ for the bundle of vertical tangent vectors.

Define a *partial tubular neighbourhood* of a submanifold V of M to consist of a neighbourhood U of V in the normal bundle $\mathbb{N}(M/V)$ together with a map $\psi : U \to M$ such that, for each $v \in V$, $\psi(v) = v$ and the composite

$$N_v(M/V) = T_v^v(\mathbb{N}(M/V)) \xrightarrow{d\psi} T_v(M) \to N_v(M/V)$$

is the identity. We will construct partial tubular neighbourhoods by piecing together ones constructed over coordinate neighbourhoods in V. The definition implies that at each $v \in V$ the map $d\psi : T_vU \to T_uM$ is the identity on the common subspace T_vV and an isomorphism on the quotient, hence by the Inverse Function Theorem 1.2.5 that ψ is a local diffeomorphism. Thus if V is a closed submanifold with a partial tubular neighbourhood in M, it follows from Corollary A.2.6 that V has a neighbourhood U' in U such that $\psi \mid U'$ is an embedding; and so, by the same argument as in Theorem 2.3.3, that V has a tubular neighbourhood in M.

Proposition 2.5.9 *Any submanifold V of a manifold M has a partial tubular neighbourhood.*

Proof Since V is a submanifold, at any point $P \in V$ there is a chart $\phi : (U_P, U_P \cap V) \to (\mathbb{R}^m, \mathbb{R}^v)$. Identifying $\mathbb{R}^m \cong \mathbb{R}^v \times \mathbb{R}^{m-v}$ gives a partial tubular neighbourhood for $U_P \cap V$.

These open sets $U_P \cap V$ form an open cover $\{V_a\}$ of V; by Theorem 1.1.5, there is a smooth partition $\{\eta_a\}$ of unity strictly subordinate to it. Write W_a for the closure of the support of η_a: thus W_a is closed, $W_a \subset V_a$, and the W_a cover V. We will construct a partial tubular neighbourhood over V by extending over a neighbourhood of one set W_a at a time. Arguing as in Proposition 1.1.7, we can construct a smooth function ε_a on V, vanishing outside V_a, taking the value 2 on a neighbourhood of W_a, and with all values in [0,2].

First consider two open subsets V_a, V_b of V and partial tubular neighbourhoods $\psi_a : U_a \to M$, $\psi_b : U_b \to M$. Since each of the images is a neighbourhood of $V_a \cap V_b$ in M, the composite $\phi_{ab} := \psi_b^{-1} \circ \psi_a$ is defined on a neighbourhood X of the zero cross-section of $V_a \cap V_b$ in $\mathbb{N}(M/V)$. Using a trivialisation of $\mathbb{N}(M/V)$ over V_b, we can write ϕ_{ab}, which is a partial map of $\mathbb{R}^k \times (V_a \cap V_b)$ to itself, as $\phi_{ab}(x, y) = (\{f_i(x, y)\}, g(x, y))$ in a neighbourhood of $\{0\} \times (V_a \cap V_b)$. Since ϕ_{ab} preserves the zero cross-section, each $f_i(x, y)$ vanishes when $x = 0$. Hence by Lemma 1.2.3, we can write $f_i(x, y) = \sum_k x_k f_{ik}(x, y)$. Define a deformation by

$$\Phi^t(x, y) := \left(\left\{ \sum_k x_k f_{ik}(tx, y) \right\}, g(tx, y) \right).$$

As in the proof of Lemma 2.5.2, this is well defined and smooth for a range including $t = 0$. By definition, $\Phi^1 = \phi_{ab} = \psi_b^{-1} \circ \psi_a$. It follows from the definition of partial tubular neighbourhood that Φ^0 reduces to the identity.

Define $\epsilon : V_a \cap V_b \to I$ by $\epsilon(z) = Bp(\frac{1}{2}(1 + \varepsilon_a(z) - \varepsilon_b(z)))$; thus $\epsilon = 1$ if $\varepsilon_a - \varepsilon_b \geq 1$ and $\epsilon = 0$ where $\varepsilon_a - \varepsilon_b \leq -1$. Now define ψ_{ab} by

$$\psi_{ab}(z) = \begin{cases} \psi_a(z) & \text{if } z \in W'_a \setminus X_b, \\ \psi_b(\Phi^{\epsilon(z)}(z)) & \text{if } z \in V_a \cap V_b, \\ \psi_b(z) & \text{if} z \in W'_b \setminus X_a. \end{cases}$$

Each formula defines a smooth map on an open set.

On the overlap $W'_a \cap (V_b \setminus X_b)$ we have $\varepsilon_a(z) = 2$, $\varepsilon_b(z) \leq 1$, so $\epsilon(z) = 1$ and the first two formulae agree. Similarly on $W'_b \cap (V_a \setminus X_a)$ we have $\epsilon(z) = 0$, and the latter two formulae agree. Hence ψ_{ab} is defined and smooth on $W'_a \cup W'_b$.

It remains to check the derivative along the zero section. This reduces to checking the x-derivative at $x = 0$ of $\sum_k x_k f_{ik}(tx, y)$, which indeed reduces to the identity.

By Theorem 1.1.4 we may suppose the covering $\{V_a\}$ locally finite and hence countable, so label the pairs by $n \in \mathbb{N}$. We now construct a partial tubular neighbourhood over V by extending over one set at a time. Suppose a partial tubular neighbourhood constructed on a neighbourhood X of $\bigcup_{r=1}^{n-1} W_r$. Then the construction in the first part of the proof yields a partial tubular neighbourhood on a neighbourhood of $\bigcup_{r=1}^{n} W_r$; moreover, the alteration takes part only inside V_n. Since the covering is locally finite, each point of V has a neighbourhood which is only affected by a finite number of steps of the construction, so the sequence of maps converges, being ultimately constant on a neighbourhood of any given point. The limit gives the desired partial tubular neighbourhood of V. □

Proposition 2.5.10 *Given submanifolds $V \subset W \subset M$, there exists a partial tubular neighbourhood $\psi : U \to M$ of V in M such that the restriction of ψ to $U \cap \mathbb{N}(W/V)$ is a partial tubular neighbourhood of V in W. Hence if V is closed, there exists a tubular neighbourhood $\psi : \mathbb{N}(M/V) \to M$ of V in M whose restriction to $\mathbb{N}(W/V)$ is a tubular neighbourhood of V in W.*

Proof The proof of Proposition 2.5.9 goes through with the only change being the requirement on each of the partial tubular neighbourhoods of compatibility with W. As before, the existence of a tubular neighbourhood follows from that of a partial tubular neighbourhood. □

This rather weak relative form of Theorem 2.3.3 will be used in §6.3.

Clearly the argument adapts to further cases such as $V \subset W_1 \subset W_2 \subset M$ or to having two submanifolds W_1 and W_2 of M such that at each $v \in V$ there is a chart with each of the W_i mapping to a coordinate subspace of $\mathbb{R}^m$. Let us make one such result explicit.

Lemma 2.5.11 *Let $V^v \to M^m$ be an embedding of connected oriented manifolds. Then there exist orientation preserving embeddings $\phi : (D^m, D^v) \to (M, V)$, and any two such are isotopic.*

2.6 Corners and straightening

We recall that M^m is a manifold with corner if it has an atlas, with charts mapping to open sets in $\mathbb{R}^m_{++}$, and that the corner $\angle M$ is the set of points mapping to $\mathbb{R}^{m-2}$. At such a point $\angle M$ has two sides in ∂M: one corresponding locally to $x_1 = 0$, the other to $x_2 = 0$. Globally, the two sides define a double covering of $\angle M$, and we say that the corner is two-sided if this covering is trivial.

Lemma 2.6.1 *If $\angle M$ is two-sided, there is a smooth embedding $h : \angle M \times I^2 \to M$ with $h(x, 0, 0) = x$ for each $x \in \angle M$ and $h^{-1}(\partial M) = (I \times \{0\}) \cup (\{0\} \times I)$. Moreover, h is unique up to diffeotopy.*

Proof Since we are only interested in a neighbourhood of $\angle M$, we can delete from ∂M the complement of a neighbourhood of $\angle M$, and thus suppose that ∂M consists of such a neighbourhood and hence, since $\angle M$ is two-sided, can be split into two components, $\partial_1 M$ and $\partial_2 M$, each with boundary $\angle M$.

Let $h_1 : \angle M \times I \to \partial_1 M$ be a collar neighbourhood of $\angle M$ in $\partial_1 M$.

By Proposition 1.5.6, there is a smooth embedding $h_2 : (\partial_1 M) \times I \to M$ giving a neighbourhood of $\partial_1 M$. We may now set $h(x, t_1, t_2) = h_2(h_1(x, t_1), t_2)$ for $x \in \angle M$ and $t_1, t_2 \in I$. Uniqueness up to diffeotopy follows from the corresponding result for collars. □

We can call the map we have constructed a *bicollar neighbourhood* of $\angle M$.

Define $\partial^s M$ by cutting ∂M along $\angle M$. By the arguments of Proposition 1.5.6, we can find a map $\partial^s M \times I \to M$ which is an embedding except that a bicollar neighbourhood is covered twice: call the image a *semicollar* of ∂M. Both a bicollar and a semicollar are pictured in Figure 2.4.

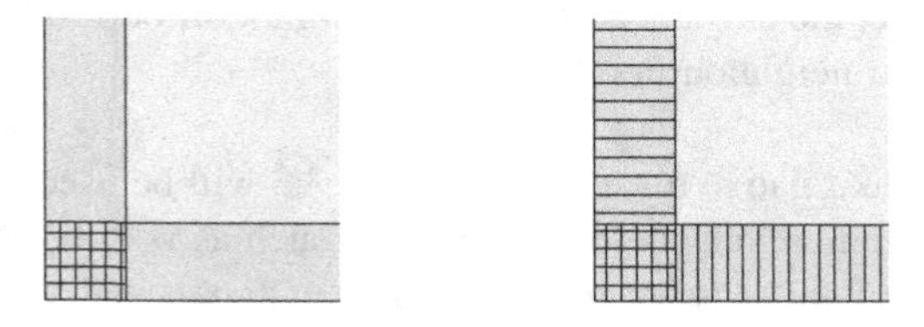

Figure 2.4 A bicollar and a semicollar

Proposition 2.6.2 *If M is a manifold with corner, there exist a manifold with boundary N and a homeomorphism $h : M \to N$ which is a diffeomorphism except on $\angle M$. Moreover, there is a construction of such an N which gives a result unique up to diffeomorphism.*

Proof Our construction is as follows. N will be M itself, with a different differential structure, defined by a new set of coordinate neighbourhoods. At points of $M \setminus \angle M$, the differential structure and coordinate neighbourhoods are unchanged. Let $h : \angle M \times I^2 \to M$ be a bicollar neighbourhood as above. Then a coordinate neighbourhood for $\angle M$, with coordinates $x_3, \ldots, x_m$ determines one for the neighbourhood with additional coordinates t_1, t_2.

We define N by the same mapping, but followed by taking the new coordinates as $(z_1, z_2) = (t_1^2 - t_2^2, 2t_1t_2)$. Since $t_1 + it_2$ lies in the first quadrant of the complex plane $\mathbb{C}$, $z_1 + iz_2 = (t_1 + it_2)^2$ lies in the upper half-plane $z_2 \geq 0$. We thus have the structure of smooth manifold with boundary. Uniqueness up to diffeomorphism follows from the uniqueness in Lemma 2.6.1. □

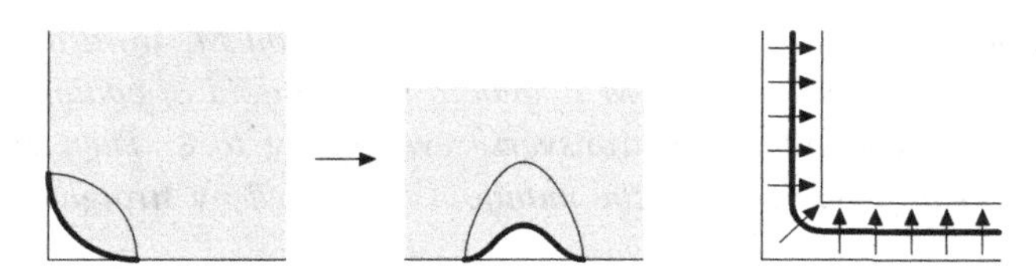

Figure 2.5 Rounding a corner

The resulting manifold N is said to be derived from M by *straightening the corner*.

We have discussed straightening corners, but may also consider the converse process, the introduction of corners. Given a manifold with boundary N, and a submanifold L of ∂N of codimension 1, we can construct a tubular neighbourhood of L in N, and redefine the differentiable structure, again using the change of coordinates $(z_1, z_2) = (t_1^2 - t_2^2, 2t_1t_2)$ in $\mathbb{R}^2$, to introduce a corner along L. The resulting M is unique up to diffeomorphism.

Since we have just reversed the above procedure, if we straighten the corner, we return to a manifold diffeomorphic to N. The procedure is roughly illustrated in Figure 2.5.

Lemma 2.6.3 *If L is a submanifold of ∂N of codimension 1, we can introduce a corner on L in an essentially unique way. If we straighten it, we recover L.*

While the above method of straightening is satisfactory, it is desirable to have alternative constructions, and be able to recognise when they give the same result.

We begin with the picture in the case when M has no corner. We can take a smooth vector field ξ on M, inward pointing at the boundary, and integrate to construct a collar neighbourhood $\varphi : \partial M \times I \to M$. A smooth submanifold $L \subset M$ of codimension 1, contained in the collar neighbourhood, and transverse everywhere to ξ, can be identified with the graph of a smooth map $\partial M \to I$. If L lies in the interior of the collar, it separates the collar into two pieces; it follows from Theorem 1.5.4 (taking the function $f = t$ as the projection on I and the vector field as $\partial/\partial t$) that each is diffeomorphic to $\partial M \times I$. Now L separates M into an outer part lying between L and ∂M, and hence diffeomorphic to $\partial M \times I$ and an inner part L^*; it follows from Lemma 2.7.2 below that the inner part is diffeomorphic to M.

We say that a smooth vector field $\sum_1^n a_i \frac{\partial}{\partial x_i}$ is inward pointing in $\mathbb{R}^n_{++}$ if $a_1 > 0$ on $x_1 = 0$ and $a_2 > 0$ on $x_2 = 0$. This definition is intrinsic, so passes to manifolds with corners.

Proposition 2.6.4 *Let ξ be a smooth vector field on M, inward pointing at boundaries and corners, $L \subset M$ a smooth submanifold of codimension 1, contained in a semicollar, and transverse everywhere to ξ. Then the inner region of L is diffeomorphic to the manifold N defined by straightening the corner.*

Proof The manifold N is obtained by applying the above change of coordinates $(z_1, z_2) = (t_1^2 - t_2^2, 2t_1t_2)$ at the corner. The image of the vector field ξ is not smooth at points corresponding to the corner, so we argue as follows.

In N we have a collar neighbourhood of $z_2 = 0$ given locally by (z_1, t). It contains a smooth submanifold L_ε given locally by $z_2 = \varepsilon$. The region between L_ε and ∂N is a collar, and the inner region for L_ε is diffeomorphic to N.

The region $0 \leq z_2 \leq \varepsilon$ in N becomes $0 \leq t_1, t_2$ and $2t_1t_2 \leq \varepsilon$ in M. The boundary L_ε given by $2t_1t_2 = \varepsilon$ is transverse to ξ, for we have $\sum_i a_i \frac{\partial}{\partial t_i}(2t_1t_2) = 2a_1t_2 + 2a_2t_1 > 0$ since, at least for ε small enough, we have $a_1, a_2, t_1, t_2 > 0$.

If L is any other submanifold transverse to ξ, there is an L' contained in the collar region, transverse to ξ, and disjoint from both L and L_ε. Hence the inner regions for L, L' and L_ε are all diffeomorphic. □

Once we have a semicollar, we can regard a neighbourhood of $\angle M$ as the product of $\angle M$ with the region $1 > y > |x|$ in $\mathbb{R}^2$ and then construct L as the graph of a function $\mu(x)$ defined by smoothing $|x|$. An example of such a function can be constructed as follows.

The function $\varepsilon^{-1}Bp(1 - \frac{|x|}{\varepsilon})$ is smooth, non-negative, vanishes unless $|x| < \varepsilon$, and has $\int_{-\infty}^{\infty} \delta_\varepsilon(x)dx = 1$. Now set $\mu(x) := \int_{-\infty}^{\infty} |y|\delta_\varepsilon(x-y)dy$. Then $\mu(x)$ is smooth, even, $\mu(x) = |x|$ if $|x| \geq \varepsilon$, and $\mu(x)$ is strictly increasing for $x > 0$.

Corollary 2.6.5 *D^{r+s} is derived from $D^r \times D^s$ by straightening the corner.*

Proof We can take the vector field in $D^r \times D^s$ to be the radial vector field $\sum_1^{r+s} x_i \frac{\partial}{\partial x_i}$, which is indeed inward pointing at the corner. We can then take S^{r+s-1} as the above L. □

We have given details for rounding corners in the simplest case. It is not possible to approximate any submanifold (not even any locally tame one) of a smooth manifold by a smooth submanifold, but the technique of rounding corners can be extended to the boundary of a submanifold of zero codimension: we have already mentioned the existence of smooth regular neighbourhoods.

2.7 Cutting and glueing

Let $M_i (i = 1, 2)$ be manifolds with boundary, $\partial M_i = Q_i$, and suppose given a diffeomorphism $h : Q_1 \to Q_2$. Take the disjoint union $M_1 \cup M_2$, and identify points corresponding under h to give a topological space N, and an identification map $\pi : M_1 \cup M_2 \to N$. Choose collar neighbourhoods $\varphi_i : Q_i \times I \to M_i$, and define a map $\varphi_i : Q_1 \times D^1 \to N$ by

$$\varphi(q, t) = \begin{cases} \pi\varphi_1(q, t) & \text{if } t \geq 0 \\ \pi\varphi_2(h(q), t) & \text{if } t \leq 0; \end{cases}$$

these agree on $t = 0$ since Q_1 and Q_2 were identified using h. Then φ is injective; in fact, it is an embedding. Define a function f on N to be smooth provided $f \circ \pi$ is a smooth function on $M_1 \cup M_2$ and $f \circ \varphi$ a smooth function on $Q_1 \times D^1$. The axioms defining a smooth manifold are now satisfied: coordinate neighbourhoods in M_1, $Q_1 \times D^1$, and in M_2 give rise to coordinate neighbourhoods in N, and where these overlap, they agree.

We have not made full use of the assumption $\partial M_i = Q_i$, and none of the above argument is affected if ∂M_i is the disjoint union of a certain set of components, and Q_i the union of a subset of these components. In this case, the remaining boundary components form the boundary of N.

More generally, suppose given manifolds M_1, M_2 with corner, smooth parts Q_i of ∂M_i, and a diffeomorphism $h : Q_1 \to Q_2$. Then by Proposition 1.5.6 (i) we have collar neighbourhoods of each Q_i, and the same definition now applies.

We say that N is obtained by *glueing* M_1 to M_2 by h (or, along Q_1).

Lemma 2.7.1 *The manifold defined by glueing M_1 to M_2 by h is determined up to diffeomorphism.*

Proof The only arbitrary element in the definition was the choice of collar neighbourhoods of the Q_i. The result follows since these are unique up to diffeotopy. □

The manifold obtained by glueing M to itself via the identity map $\partial M \to \partial M$ is said to be obtained by *doubling* M, and denoted $D(M)$.

Another simple but useful case is the following.

Lemma 2.7.2 *The result of glueing M to $\partial M \times I$ by the map $h : \partial M \to \partial M \times \{0\}$ given by $h(x) = (x, 0)$ is diffeomorphic to M.*

Proof Let $k : \partial M \times I \to M$ be a collar neighbourhood of ∂M. Define $p : M \cup (\partial M \times I) \to M$ by:

p is the identity on $M \setminus Im(k)$,

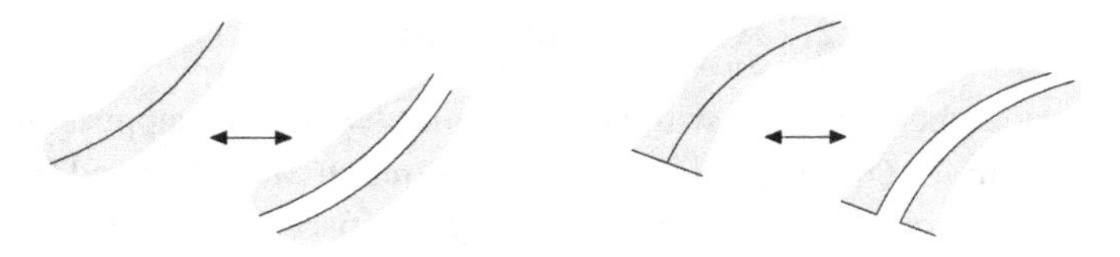

Figure 2.6 Cutting and glueing

$p(k(x,t)) = k(x, \alpha(t))$ for $x \in \partial M,\ t \in I$,
$p(x,t) = k(x, \frac{1}{2}(1-t))\ x \in \partial M,\ t \in I$.
This induces a bijection between the manifold obtained by glueing and M provided that $\alpha(t)$ increases from $\frac{1}{2}$ to 1 as t increases from 0 to 1. To make it a diffeomorphism it will suffice if also $\alpha(t) = \frac{1}{2}(1+t)$ for $t < \varepsilon$ and $\alpha(t) = t$ for $t > 1-\varepsilon$, for some small ε, for example, take $\alpha(t) = \frac{1}{2}\{(t+1) + (t-1)$ $Bp(3t-1)\}$. □

Glueing, and its inverse operation cutting, are both illustrated in Figure 2.6. Now let Q^{n-1} be a submanifold of N^n, with inclusion map $i : Q \to N$. For each point $P \in Q$, $di(T_P Q)$ is a subspace of $T_P N$ of unit codimension, and so separates this real vector space into two components. We define a manifold M as follows. Its points are those of $N \setminus Q$, together with two points for each point P of Q, one associated with each complementary component of $di(T_P Q)$ in $T_P N$ or, as we shall say, *side* of Q in N. There is thus a natural projection $\pi : M \to N$. We take for coordinate neighbourhoods in M those induced by π from coordinate neighbourhoods in $N \setminus Q$; in addition, for each coordinate neighbourhood $f : U \to \mathbb{R}^n$ with $f^{-1}(\mathbb{R}^{n-1}) = U \cap Q$ two coordinate neighbourhoods in M; induced by π from the restriction of f to the inverse images of $\mathbb{R}^n_+$ and $\mathbb{R}^n_-$ (in the latter case, we must change the sign of the first coordinate to obtain a coordinate neighbourhood of standard type). Here, of course, the points of N corresponding to a certain side of Q in N are mapped by the coordinate neighbourhood for the corresponding side of $\mathbb{R}^{n-1}$ in $\mathbb{R}^n$; since df is nonsingular, it preserves the distinction between sides.

We say that the resulting manifold M is obtained by *cutting* N along Q.

The same definition can be given more generally in the case when N has a boundary and Q is a submanifold of codimension 1 (so $\partial Q = Q \cap \partial N$): in this case the points corresponding to ∂Q form the corner $\angle M$; this divides ∂M into two parts: a part $\partial_1 M$ obtained by cutting ∂N along ∂Q and a part $\partial_2 M$ which is a double covering of Q. The double covering is given by the two sides of Q, or equivalently by the normal bundle, which we can take to have fibre D^1 with boundary giving the two points.

For example, if $N \setminus Q$ has just two components, with closures M_1 and M_2 so that $\partial M_1 = Q = \partial M_2$, then cutting N along Q yields the disjoint union of M_1 and M_2.

Proposition 2.7.3 *If N is defined by glueing M_1 to M_2 along Q_1, and we cut N along $\pi(Q_1)$, we recover M_1 and M_2. Conversely, if N^n and its submanifold Q^{n-1} are connected, Q separates N with parts M_1 and M_2 and we glue M_1 to M_2 along Q, we recover N.*

Proof The first part is immediate from the definition of glueing. For the converse, if the above conditions are satisfied, we obtain M_1 and M_2. Now if $\varphi : Q \times D^1 \to N$ is a tubular neighbourhood of Q in N, φ defines by restriction collar neighbourhoods of Q in M_1, M_2. If these are used in the glueing process, we recover N. The second part of the result now follows from Lemma 2.7.1. □

There are alternative definitions of cutting, which yield the same result up to diffeomorphism. One is to let ρ be a complete metric on N, and define M as the metric completion of $N \setminus Q$.

We can also define a manifold M' by deleting from N the interior of a tubular neighbourhood of Q. We see directly that this is obtained from the manifold M obtained by cutting N along Q by removing the interior of a collar neighbourhood of the boundary, hence by Lemma 2.7.2 is diffeomorphic to M.

We have seen that cutting and glueing are inverse operations, but cutting as defined above is more general than the inverse of glueing as it includes the case when the normal covering of Q in N is non-trivial. However we can also define glueing more generally: let Q_1, Q_2 be smooth parts of ∂M, not necessarily disjoint, and $h : Q_1 \to Q_2$ a diffeomorphism. The definition of glueing along h is now, as above, the quotient of M by identifying along h, with smooth structure defined using a choice of collar neighbourhoods of the Q_i. We see easily that this remains inverse to the cutting operation.

An important application of glueing is the following. Let M_1^m, M_2^m be connected smooth manifolds, $f_i : D^m \to M_i^m$ embeddings. Delete the interiors of the images of the f_i, and glue the result along the boundary $f_i(S^{m-1})$ by $f_2 f_1^{-1}$. Since removing a disc does not disconnect M if $m > 1$, the result is connected: it is called the *connected sum*, and written $M_1 \# M_2$. The construction is pictured in Figure 2.7.

Theorem 2.7.4 *$M_1 \# M_2$ is determined up to diffeomorphism by summands, unless these are both orientable, when there are two determinations.*

Proof By the Disc Theorem 2.5.6, the embeddings f_i are unique up to ambient diffeotopy and a possible change of orientation. By Lemma 2.7.1 the result

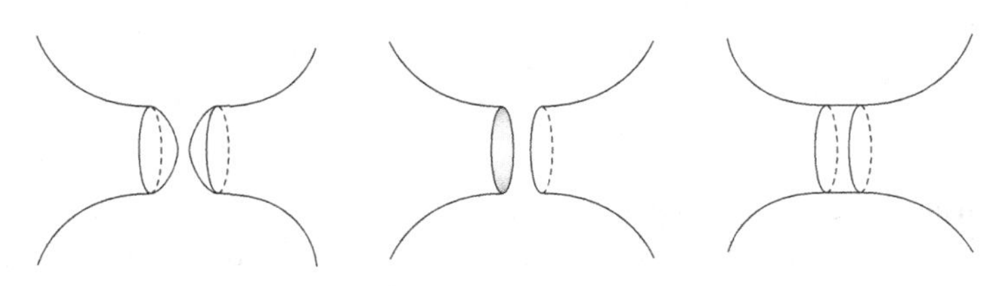

Figure 2.7 The connected sum

of glueing, given f_1 and f_2, is unique up to diffeomorphism. Hence the result follows, except for considerations of orientation. Now if f_1,f_2 are replaced by $f_1 \circ r$, $f_2 \circ r$, where r is a reflection, the connected sum is unaltered. If neither M_i is orientable, the result is trivial; if only M_2 is orientable, using the above possibility of simultaneous reversal, uniqueness again follows. If both are orientable, the result has two possible cases. □

To make the result precise in the orientable case, we suppose the M_i both oriented, and that one of the f_i preserves, the other reverses orientation. The result is then again unique, and has a canonical orientation inducing the given ones of the M_i.

The connected sum is also defined for manifolds with boundaries and corners; we simply suppose that the f_i map into the interior. However, in this case we also have a different sum operation. Let $f_i : D^{m-1} \to \partial M_i^m$ be an embedding. Introduce a corner along $f_i(S^{m-2})$. We may now glue the $f_i(D^{m-1})$ together by $f_2 f_1^{-1}$. The result is called a *boundary sum* $M_1 + M_2$ of M_1 and M_2.

Proposition 2.7.5 *If M_1^m, M_2^m are connected manifolds with connected boundaries, $M_1 + M_2$ is determined up to diffeomorphism by M_1 and M_2 unless ∂M_1 and ∂M_2 are both orientable, when there are two sums.*

Proof This follows by the Disc Theorem exactly as for Theorem 2.7.4. □

We conclude by summing up the simple properties of those operations.

Proposition 2.7.6 *Both operations are commutative and associative, with units: $M^m \# S^m \cong M^m$, $M^m + D^m \cong M^m$. We have $\partial(M_1 + M_2) = \partial M_1 \# \partial M_2$.*

Proof Commutativity and associativity are immediate. To form $M^m \# S^m$ we simply delete one disc from M^m, and replace it by another disc.

The second result may be seen as follows. D^m is obtained from $D^{m-1} \times I$ by straightening the corner. Derive N from M by introducing a corner along $f(S^{m-2})$ as above; then glueing on $D^{m-1} \times I$ does not affect N other than by a diffeomorphism by Lemma 2.7.2. The result follows by straightening the corner.

The last part is merely an observation of what happens to the boundary for the sum operation. □

2.8 Notes on Chapter 2

§2.1 I have proved a little more than I need at this point, but the existence of a neighbourhood of $\Delta(M)$ of pairs joined by minimal geodesics allows us to go further and define a continuous family of paths joining nearby points.

§2.2 These (classical) results on geodesics could be taken as the jumping off point for further results in differential geometry. Another treatment of this material is given in Milnor [98, II].

§2.3 It seems that tubular neighbourhoods, along with fibre bundles, were first introduced by Whitney [174].

§2.4 Our results are restricted to the case of diffeotopies of compact support. This restriction is necessary; otherwise we have counterexamples; but it may be possible to improve the result. The result was first proved by Thom [152], with a sharper version obtained independently by Cerf [36] and Palais [118].

§2.5 The tubular neighbourhood theorem was first proved by Milnor in lectures at Princeton University in 1961; an equivalent result was obtained in [36].

The construction of tubular neighbourhoods by local piecing together of partial tubular neighbourhoods is the method adopted by Cerf [36] and Lang [83]; it gives a proof of Theorem 2.5.10 without using the clumsy hypothesis that the normal bundle is trivial.

§2.6, §2.7, For a corner which is not two-sided, there is an analogue of a bicollar neighbourhood which is an embedding of a bundle over $\angle M$ with fibre $I \times I$ and group $\mathbb{Z}_2$ interchanging the components.

The disc theorem justifies the definition of connected sum. This seems to be due to Milnor, in the context of homotopy spheres.

Both these sections are designed for use in Chapter 5 for the theory of handle decompositions.

3
Differentiable group actions

We begin by recalling the definitions of Lie groups, of group actions, and of smooth actions, and establish some elementary properties.

Although the centre of our interest is in actions of compact (including finite) groups, the geometrical properties extend to all proper group actions. A key step is the notion of slice. We establish the existence of slices for arbitrary proper actions. This leads at once to a local model for a proper smooth actions, which is the basis for all the subsequent results.

We show that the development of basic results in §1.1 can be parallelled in the group action situation: we have covers by coordinate neighbourhoods, partitions of unity, an approximation lemma, and invariant Riemannian metrics. There is also a theorem on the existence of an equivariant embedding in Euclidean space (with an orthogonal action), which applies when the group is compact.

We continue by defining orbit types, and the stratification of the manifold by orbit types. This stratification is locally finite and smoothly locally trivial. One consequence is that if the manifold is connected, one orbit type is dense and open: orbits of this type are called principal orbits. We give a model for a neighbourhood of a stratum, and proceed to an analysis of the case with two strata.

We conclude with examples.

3.1 Lie groups

We recall from §1.3 that a *Lie group* is a smooth manifold G, which is also a group, such that the group operations $g \mapsto g^{-1}$, $(g, h) \mapsto gh$ are smooth maps $G \to G, G \times G \to G$.

Important examples are the general linear groups $GL_m(\mathbb{R})$ and $GL_m(\mathbb{C})$ of nonsingular $m \times m$ real, respectively complex, matrices, which are open submanifolds of the vector space of all matrices. We also use the notation $GL(V)$ for the group of linear endomorphisms of the vector space V.

A *Lie subgroup* is a smooth submanifold which is also a subgroup. Any subgroup of a Lie group G which is a closed subset is a Lie subgroup. This result is not trivial: a proof is given, for example, in [146, Theorem 4.1] or in [148, §3.1].

Not every subgroup of a Lie group is a closed subset: a simple example is the additive subgroup $\mathbb{Q}$ of $\mathbb{R}$. However, the closure of any subgroup is also a subgroup, hence is a Lie subgroup. If H is a Lie subgroup of G, as H is locally closed, it is open in its closure and hence by homogeneity is equal to its closure.

Among the Lie subgroups of $GL_m(\mathbb{R})$ are the group $GL_m^+(\mathbb{R})$ of matrices with positive determinant, the group $SL_m(\mathbb{R})$ of matrices of determinant 1, the orthogonal group O_m (orthogonal matrices can be characterised by the equation $AA^t = I$), and $SO_m = SL_m(\mathbb{R}) \cap O_m$. Lie subgroups of $GL_m(\mathbb{C})$ include $SL_m(\mathbb{C})$, U_m (here we have $A\overline{A}^t = I$) and SU_m. Further important examples are the spinor groups $Spin_m$ (the double covering group of SO_m), and the symplectic group Sp_m, defined like U_m, but using the algebra $\mathbb{H}$ of quaternions. We identify SO_2 with the multiplicative group S^1 of complex numbers of modulus 1, and SU_2 (also $Spin_3$ and Sp_1) with the multiplicative group S^3 of unit quaternions.

There is a general classification of compact Lie groups, which has its origin in the work of Lie and Killing: a convenient recent account is given in [125] (see Theorem 10.7.2.4). Any connected compact Lie group G has a finite covering group which is a direct product of copies of groups of the type S^1, SU_m, $Spin_m$, Sp_m and five other groups denoted G_2, F_4, E_6, E_7, and E_8. We will not use this in this book, but it opens the way to enumerations of groups and group actions satisfying prescribed conditions.

For G a Lie group and $g \in G$, the map $\rho_g : G \to G$ defined by $\rho_g(x) = xg$ is a diffeomorphism, with inverse $\rho_{g^{-1}}$; it is called *right translation* by g. Left translation λ_g is defined similarly.

If G is a group and H a subgroup, we write G/H for the set of right cosets $\{gH \mid g \in G\}$ and $\pi : G \to G/H$ for the natural projection given by $\pi(g) = gH$. We also have left cosets $Hg := \{hg \mid h \in H\}$ and the coset space $H\backslash G := \{Hg \mid g \in G\}$.

If G is a Lie group and H a Lie subgroup, the coset space G/H (with the quotient topology) has a natural structure as a smooth manifold. For at any $g \in G$, choose a chart $\varphi : U \to \mathbb{R}^{p+q}$ such that the submanifold $U \cap gH$ corresponds to the subspace $\mathbb{R}^p$; then take the composite $\mathbb{R}^q \subset \mathbb{R}^{p+q} \to U \subset G \to G/H$ as a chart at $gH \in G/H$. It is easy to see that transformations between overlapping charts are smooth, using as guideline the fact that a function $f : G/H \to \mathbb{R}$ is smooth if and only if $f \circ \pi \in \mathcal{F}_G$ is smooth. A similar argument shows that the projection $G \to G/H$ is that of a smooth fibre bundle. More generally, if we have two Lie subgroups $H_1 \subset H_2 \subset G$, the projection $G/H_2 \to G/H_1$ is that of a fibre bundle, with fibre H_2/H_1.

If G is a Lie group, the tangent space $T_1(G)$ at the identity has the structure of a Lie algebra. We will not use this in this book.

If H is a Lie subgroup of G we may choose an additive complement Y to $T_1(H)$ in $T_1(G)$. Then the differential at $(0, 1)$ of the map $Y \times H \to G$ given by $(y, h) \mapsto exp(y)h$ is an isomorphism (here we may use any Riemannian metric on G to define the exponential map), so by Theorem 1.2.5 the map is a local diffeomorphism. We can thus choose open neighbourhoods U of 1 in $exp(Y)$ and V of 1 in H such that $U \times V \to G$ is an embedding. We will call U a *local section* of H in G.

Lemma 3.1.1 *There exist local sections U_1 such that the map $\mu : U_1 \times H \to G$ is an embedding.*

Proof The fact that the differential of μ at any $(u, h) \in U \times H$ is bijective follows since this holds at $(u, 1)$ by hypothesis, and (right) translation by h is a diffeomorphism. It follows (as in the proof of Theorem 2.3.2) from Corollary A.2.6 that there is a neighbourhood of $1 \times H$ such that the restriction of μ to it is an embedding and by Lemma A.2.4 that for ε small enough if U_1 is the ε-neighbourhood of 1 in U, the restriction of μ to $U_1 \times H$ is an embedding. □

It follows that the induced map $U_1 \to G/H$ is an embedding. We can also argue similarly for $H \times U \to G$ and $U \to H\backslash G$.

If G is a Lie group, the connected component of the identity is a subgroup G_0, as if $x(t)$ is a path from 1 to $g \in G$ and $y(t)$ a path from 1 to h, then $x(t)^{-1}$ gives a path from 1 to g^{-1} and $x(t)y(t)$ a path from 1 to gh. As G is a manifold, G_0 is an open subset. Any open subgroup G^* of G is closed, since its complement is a union of cosets of G^*, each open, hence is open. Now if $p : I \to G$ is any path, $p^{-1}(G^*)$ is open and closed in I, hence is either I or the empty set. Thus G^* contains all paths from $1 \in G$, hence contains G_0.

If N is a neighbourhood of 1 in G, then the subgroup G^* of G generated by N contains an open neighbourhood of 1, so by homogeneity is open, so $G_0 \subset G^*$. Thus if $N \subset G_0$ we also have $G^* \subset G_0$, so the two coincide.

The point G_0/G_0 of G/G_0 is open; since G acts by homeomorphisms, all points, hence all subsets, are open, so the coset space G/G_0 has the discrete topology. If G is compact, then G/G_0 also is compact, so is finite.

Proposition 3.1.2 *Let G be a compact Lie group and let $H \neq G$ be a Lie subgroup. Then either*
$\dim H < \dim G$ *or*
$\dim H = \dim G$, *and H has fewer components than G.*

Proof Since $H \subset G$ is a submanifold, we have $\dim H \leqq \dim G$.
Suppose $\dim H = \dim G$. Then H contains a neighbourhood of 1 in G, so H contains G_0. As H is a proper subgroup of G, H/G_0 is a proper subgroup of G/G_0, which is compact and discrete, hence finite. But the components of G are the cosets of G_0. □

If G is a compact topological group, there is an averaging operator on the space $C^0(G)$ of continuous functions on G: it is the unique linear map $\int_G : C^0(G) \to \mathbb{R}$ such that
(i) if $g.f : G \to \mathbb{R}$ is defined by $g.f(x) := f(gx)$, then $\int_G(g.f) = \int_G(f)$,
(ii) if f_c is given by $f_c(g) = c$ for all $g \in G$, then $\int_G(f_c) = c$,
(iii) if $f(g) \geq 0$ for all $g \in G$, then $\int_G(f) \geq 0$.
For the reader familiar with integration theory, we can give a quick account as follows. The bundle of differential n-forms on a smooth manifold M is defined to be the nth exterior power $\Lambda^n T_1^\vee M$. If M has dimension n, then for any section ω of this bundle with compact support we can integrate ω over M: the result is denoted $\int_M \omega$.
If G is a Lie group of dimension n, we choose a form ω_0 at the identity $1 \in G$ to be any element of the exterior power $\Lambda^n T_1^\vee G$. Now for any $g \in G$, left translation by g gives a diffeomorphism of G taking 1 to G and hence ω_0 to an n-form at $g \in G$; assembling these gives an n-form ω' on G invariant under left translations by elements $g \in G$. For any (smooth or even just continuous) function f of compact support on G we can now form the integral $\int_G f\omega'$.
In the case when G is compact we can now define $\int_G f := \int_G f\omega / \int_G \omega$; properties (i)–(iii) follow easily. We will not give the proof of uniqueness; however from uniqueness follows that if $f.g' : G \to \mathbb{R}$ is defined by $f.g'(x) := f(xg')$, then $\int_G(f.g') = \int_G(f)$. For since the averaging operator is unique, it suffices to show that $f \mapsto \int_G(f.g')$ satisfies (i)–(iii). But these follow from the same results for $\int_G$ by substitution. It follows similarly that if we define f^* by $f^*(g) = f(g^{-1})$, then $\int_G f^* = \int_G f$.

A proof of existence of an averaging operator for arbitrary compact groups (due to Haar) may be found in [68], also a theory of (left invariant) integration for any locally compact topological group.

3.2 Smooth actions

A (left) *action* of a group G on a set X is a map $\phi : G \times X \to X$ such that $\phi(1, x) = x$ for all $x \in X$ and $\phi(g, \phi(h, x)) = \phi(gh, x)$ for all $x \in X$ and $g, h \in G$. If the action is understood, it is frequently denoted by a dot: thus $\phi(g, x)$ becomes $g.x$; we will call X a G-space. If G is a Lie group, X a smooth manifold, and ϕ a smooth map, we speak of a *smooth action* and a smooth G-space.

If we have an action of a group G on a set X, and $x \in X$, the *isotropy group* of x is defined to be $G_x := \{g \in G \mid g.x = x\}$. It follows from the definition of group action that this is a subgroup of G; it is also sometimes called the stabiliser of x. The *orbit* of x is defined to be $\{g.x \mid g \in G\}$, and is denoted $G.x$. The action induces a bijection $G/G_x \to G.x$ since

$$g.x = h.x \Leftrightarrow h^{-1}g.x = x \Leftrightarrow h^{-1}g \in G_x \Leftrightarrow hG_x = gG_x.$$

Equivalently, the map $Op_x : G \to X$ defined by $Op_x(g) := g.x$ induces an injection of G/G_x into X.

The set of orbits of a left group action is denoted $G\backslash X$; in the case of continuous, in particular smooth actions, we give $G\backslash X$ the quotient topology and call it the *orbit space*. Even for a smooth action, this is only rarely a manifold.

For a smooth action, any isotropy group is a closed subgroup of G, hence is a Lie subgroup. A sufficient condition for the injection $G/G_x \to X$ to be a smooth embedding will be given in the next section.

A point $x \in X$ is *fixed* under G if $g.x = x$ for all $g \in G$, i.e. if $G_x = G$. The *fixed set* of the action is the set of all fixed points, and is denoted X^G. At the opposite extreme to the fixed set, an action is called *free* if $g.x = x$ implies $g = 1$: thus $\{1\}$ is the only isotropy group. The action is *semi-free* if the only isotropy groups are $\{1\}$ and G.

A subset $Y \subseteq X$ is *invariant* under G if $g.y \in Y$ for all $g \in G$ and $y \in Y$.

Given two actions $\phi : G \times X \to X$ and $\psi : G \times Y \to Y$, a map $f : X \to Y$ is *equivariant* (more precisely, G-equivariant) if, for all $g \in G$ and $x \in X$, we have $f(\phi(g, x)) = \psi(g, f(x))$.

Given a subgroup H of G and an action of H on X, we define $G \times_H X$ to be the quotient of $G \times X$ by the relation $(gh, x) \sim (g, hx)$ for all $g \in G$, $h \in H$ and $x \in X$: this is an equivalence relation since H is a subgroup. We denote the equivalence class containing (g, x) by $[g, x]$. Setting $g'.[g, x] := [g'g, x]$ defines an action of G on $G \times_H X$.

Lemma 3.2.1 *The isotropy group of $[g, z] \in G \times_H X$ is gH_zg^{-1}.*

For if $[g, z] \in G \times_H X$, we have $g'.[g, z] = [g'g, z]$, and this is equal to $[g, z]$ if and only if, for some $h \in H$, $g'g = gh$ and $h^{-1}.y = y$; so $h.y = y$ and $g' = ghg^{-1}$.

If these groups and actions are smooth, then by Lemma 3.1.1 we can pick a local section Y such that $Y \times H \to G$ is a diffeomorphism onto an open set. It follows that $Y \times X \to G \times_H X$ is a diffeomorphism onto an open set.

We will give many examples of smooth group actions at the end of this chapter, but offer two here.

If H is a subgroup of G, G acts on G/H by left translations: $g.g'H := gg'H$. If G is a Lie group and H a Lie subgroup, this action is smooth.

The group $GL(V)$ acts on the vector space V: for example, $GL_m(\mathbb{R})$ acts on the space $\mathbb{R}^m$ of column vectors by matrix multiplication. If G is any group and $f : G \to GL(V)$ a homomorphism, there is an induced action of G on V by linear maps; we refer to V as a linear G-space. The action is called a linear representation of G.

A classical theorem, known as the Peter–Weyl Theorem, states that for any compact group G there exist a (finite dimensional) real vector space V and an injective continuous homomorphism $G \to GL(V)$. Moreover, the function algebra $L^2(G)$ is a direct sum of finite dimensional invariant subspaces so, for example, any smooth function on G can be approximated by functions of the form $g \mapsto \ell(g.x)$, where $x \in V$ for some linear G-space V and $\ell : V \to \mathbb{R}$ is linear.

Lemma 3.2.2 *For any continuous linear action of a compact Lie group H on a vector space V, there is an inner product on V invariant under H.*

Proof Choose an inner product $V \times V \to \mathbb{R}$, and denote it $\langle x, y\rangle$. Define $\langle x, y\rangle_H := \int_H \langle g.x, g.y\rangle$. This is linear in each of x and y, and invariant in the sense that $\langle g.x, g.y\rangle_H = \langle x, y\rangle_H$ for all $g \in H$ and $x, y \in V$. Moreover we have $\langle x, x\rangle_H > 0$ if $x \neq 0$, so $\langle *, *\rangle_H$ is an inner product. □

The image of H in $GL(V)$ is a subgroup of the orthogonal group of V with respect to this product. Since any two inner products on V are equivalent under the general linear group $GL(V)$, it follows that any compact subgroup of $GL(V)$ is conjugate to a subgroup of $O(V)$. Extending Lemma 3.2.2, we have

Proposition 3.2.3 *For any smooth action of a compact Lie group H on a smooth manifold M, there is a Riemannian metric on M invariant under H.*

Proof Choose any Riemannian metric on M: we can regard it as a collection of inner products on all the tangent spaces T_xM. The action of $g \in H$ takes $g^{-1}.x$ to x and gives an isomorphism of $T_{g^{-1}.x}M$ on T_xM, so transporting the given inner product $\langle , \rangle$ on $T_{g^{-1}.x}M$ gives an inner product $\langle , \rangle_g$ on T_xM. Integrating over H as above gives a new family of scalar products giving a Riemannian metric invariant under H. □

Since the exponential map $Exp : \mathbb{T}(M) \to M$ was directly constructed from the metric, it follows that if we have a G-invariant metric on M, the corresponding exponential map is G-equivariant.

Corollary 3.2.4 *The fixed set M^H of a smooth action of a compact Lie group H on a smooth manifold M is a smooth submanifold of M.*

Proof By the Proposition, we can choose an H-invariant Riemannian metric on M. Let $x \in M^H$ be a fixed point, then the exponential map $T_xM \to M$ is a local diffeomorphism and is H-equivariant. Since H acts orthogonally on T_xM, the fixed set $(T_xM)^H$ is a linear subspace, and so a smooth submanifold. The result follows. □

3.3 Proper actions and slices

The main geometrical results about smooth group actions depend on compactness. The theory is usually written in terms of actions of a compact group G, but with a little effort, the results extend to arbitrary Lie groups, provided the action satisfies the following key condition.

An action $\phi : G \times X \to X$ is said to be *proper* if the map

$$(\phi, i) : G \times X \to X \times X$$

given by $(g, x) \mapsto (g.x, x)$ is a proper map.

Proposition 3.3.1 *Let $\phi : G \times X \to X$ be a proper group action and $x \in X$. Then*

(i) the isotropy group G_x is compact;
(ii) the map $Op_x : G \to X$ is proper;
(iii) the orbit $G.x$ is a closed subset of X;
(iv) the induced map $G/G_x \to G.x$ is a homeomorphism;
(v) for any compact subsets $K, L \subseteq X$, $\{g \in G \mid g.K \cap L \neq \emptyset\}$ is compact;
(vi) the orbit space $G\backslash X$ is Hausdorff.

A fuller discussion is given in §A.3. The above result is contained in Propositions A.3.1 and A.3.3.

By Lemma A.3.2(i), a smooth group action with G compact is always proper. So is the action on G by a Lie subgroup H by left translation. More generally, by (ii) of the Lemma, given two Lie subgroups H, K of G with K compact, the natural action of H on the coset space G/K is proper.

To illustrate the importance of properness, we give examples where the condition fails, and the geometrical picture is very different from what we obtain below in the proper case.

First, we can consider $\mathbb{Q}$ as a discrete group and let it act additively on $\mathbb{R}$.

Second, take G as $\mathbb{R}$, $M := \mathbb{R}^2/\mathbb{Z}^2$ and let $\alpha \in \mathbb{R}$ be irrational: define an action by $\phi(t, [x, y]) := [x + t, y + \alpha t]$.

In these two cases, all isotropy groups are trivial but all orbits are dense in M. In general, a smooth action of $\mathbb{R}$ on M (also called a dynamical system) defines a vector field on M, and we saw in Theorem 1.4.2 that conversely any vector field defines a flow and subject to a completeness condition (see, for example, Proposition 1.4.4) gives a group action.

For a third example take $M = \mathbb{R}$ and $\frac{d\theta}{dt} = \sin\theta$ (which is certainly bounded). The fixed set of this action is the set of θ with $\sin\theta = 0$, so consists of integer multiples of π.

Theorem 3.3.2 (The Rank Theorem) *Let $f : \mathbb{R}^m \rightsquigarrow \mathbb{R}^n$ be a smooth map defined on a neighbourhood A of $a \in \mathbb{R}^m$ such that, for all $x \in A$, df_x has rank p, for some fixed $p > 0$. Then there exist open neighbourhoods $U \subset A$ of a, $V \supset f(U)$ of $f(a)$, and diffeomorphisms $u : U \to (\mathring{D}^1)^m$, $v : V \to (\mathring{D}^1)^n$ such that $f|U = v^{-1} \circ \pi \circ u$, where $\pi(x_1, \cdots, x_m) = (x_1, \cdots, x_p, 0, \cdots, 0)$.*

We regard this as an extension of Theorem 1.2.5 and, as for that result, proofs can be found in [40] and [52]. As for Theorem 1.2.5, the given statement refers only to a neighbourhood of a point in $\mathbb{R}^m$, but the result translates at once to one valid for any manifold.

Theorem 3.3.3 *For any smooth action of G on M and any $x \in M$, the induced map $j : G/G_x \to M$ is a smooth immersion with image $G.x$.*

If the action is proper, j is an embedding as a closed submanifold.

Proof We first apply Theorem 3.3.2 to the map $Op_x : G \to M$. We claim that it follows from the group action property that this map has the same rank at all points. For left translation ℓ_g by $g \in G$ is a diffeomorphism of G taking a neighbourhood of $1 \in G$ to a neighbourhood of g. The action of g is a diffeomorphism r_g of M taking x to $g.x$. The diagram

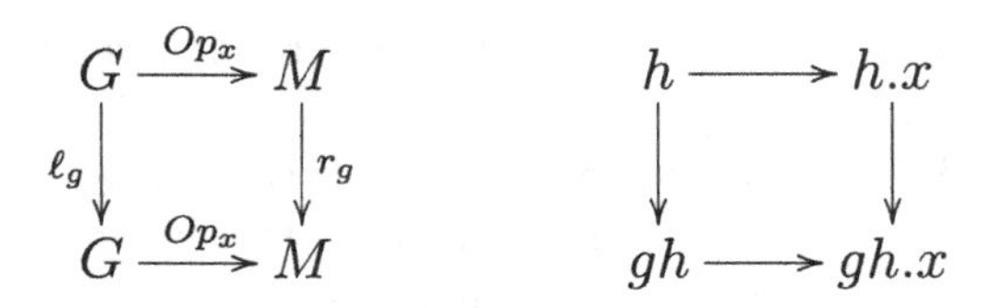

is commutative and the vertical maps are diffeomorphisms. Taking tangent spaces thus gives a commutative diagram with the vertical maps linear isomorphisms. Thus indeed dOp_x has the same rank at 1 and at g.

It follows from the rank theorem that the map Op_x is locally trivial. Hence the rank of dOp_x is equal to the dimension of the image, namely of the orbit $G.x$; and the rank of the kernel is equal to the dimension of the fibre, which is the isotropy group G_x. Thus the induced map $G/G_x \to G.x$ is an immersion.

By Proposition 3.3.1 (ii), if the action is proper, the map $j : G/G_x \to M$ is proper. It follows from Proposition 1.2.10 that j is an embedding as a closed submanifold. □

Although the basic idea of taking a slice is simple, the following definition is important; the existence of slices is key to the structure results that follow.

Given a smooth action of the G on M, and a closed subgroup H of G, a smooth H*-slice* to the action is a smoothly embedded submanifold V of M such that

(S1) For all $y \in V$, $T_yM = T_y(G.y) + T_xV$.

(S2) V is H-invariant.

(S3) If $s \in V$, $g \in G$ and $g.s \in V$, then $g \in H$.

The definition includes the case when V is a submanifold with boundary.

Theorem 3.3.4 *For any proper smooth action of G on M and any $x \in M$ there exists a smooth G_x-slice V to the action with $x \in V$.*

Proof Since the action is proper, the isotropy group G_x is compact; write $H := G_x$. By Proposition 3.2.3 we can choose a Riemannian metric of M invariant under H. Then $T_x(G.x)$ is a subspace of T_xM; write E for its orthogonal complement. Then E is also invariant under the induced action of H on T_xM.

Since the metric is H-invariant, the exponential map of M is H-equivariant. Denote by D_a, $\mathring{D}_a$ the closed and open discs of radius a in E, and write $V_a := \exp(D_a)$ and $\mathring{V}_a = \exp(\mathring{D}_a)$. As in the construction of tubular neighbourhoods, if a is small enough, the restriction of the exponential map to D_a is an embedding. We will show that for b small enough V_b, hence also $\mathring{V}_b$, is a smooth H-slice.

It follows from the construction that for any $b \leq a$, V_b is a smoothly embedded disc and is H-invariant, so satisfies (S2).

Now choose a local section U to H in G: then UH is an open neighbourhood of H in G, so its complement is closed. Since by Proposition 3.3.1(ii), Op_x is a closed map, $(G \setminus UH).x$ is a closed set. It does not contain x, so is at a positive distance 2ε from x.

We also have $T_xV_a = E$, so $T_xM = T_x(G.x) \oplus T_xV_a$. The action induces a smooth map $G \times V_a \to M$ whose differential is surjective at $(1, x)$. Hence it is also surjective on some neighbourhood of this point. If b is small enough, this neighbourhood contains $1 \times V_b$, thus V_b satisfies (S1).

Since $T_1U \oplus T_1H = T_1G$, and it follows from the Rank Theorem that $T_1G/T_1H \cong T_x(G.x)$, the map $T_1U \to T_x(G.x)$ is an isomorphism. Thus the map $U \times V_a \to M$ induces an isomorphism $T_1U \oplus T_xV_a \to T_xM$ of tangent spaces, so induces a diffeomorphism of some neighbourhood; shrinking U if necessary, and taking b small enough, we may suppose this neighbourhood contains $U \times V_b$. Then $u \neq 1 \in U$ and $y \in V_b$ implies $u.y \notin V_b$.

Since the action is proper and V_a is compact, it follows from Proposition 3.3.1(v) that $K := \{g \in G : V_a g \cap V_a \neq \varnothing\}$ is compact; note that $H \subseteq K$. It follows from Lemma A.2.1 that for any ε we can find δ such that if $s \in V_a$, $g \in K$ and $\rho(s, x) < \delta$ we have $\rho(g.s, g.x) < \varepsilon$.

Now if $s \in V_b$ and $g.s \in V_b$, then

$$\rho(x, g.x) \leq \rho(x, g.s) + \rho(g.s, g.x) \leq b + \varepsilon < 2\varepsilon.$$

Hence $g \notin G \setminus UH$. i.e. $g \in UH$: say $g = uh$. Then $h.s \in V_b$ and $u.(h.s) \in V_b$. It now follows from the above that $u = 1$, so indeed $g = h \in H$. Thus V_b also satisfies (S3). □

We now derive a local model giving a description of the neighbourhood of an orbit in a proper group action.

Theorem 3.3.5 *Let V be an H-slice at x to a smooth proper action of G on M, with $H = G_x$. Then the action induces a smooth map $j : G \times_H V \to M$ giving an equivariant diffeomorphism onto a neighbourhood Y of $G.x$ in M.*

If V is a closed disc, this gives a tubular neighbourhood of $G.x$ in M.

Proof By (S2) V is H-invariant, so $G \times_H V$ is defined. The action ϕ now induces a smooth equivariant map j, and it follows from (S3) that j is injective and from (S1) that j is a submersion, hence a diffeomorphism.

We recall that a tubular neighbourhood of a (closed) submanifold F in M is defined to consist of a bundle B over F with fibre a disc and an embedding $\psi : B \to M$ (as submanifold with boundary) extending the map taking the centre of each disc to the corresponding point of V. Here we take $F = G.x$ and $B = G \times_H V$. A projection $G \times_H V \to G/H \cong G.x$ is given by $[g, s] \to gH \to g.x$:

we see at once that this is well defined, and its fibre is V. A local section U for H in G induces a local trivialisation. □

For a smooth proper group action of G on M, by Theorem 3.3.5 there is a smooth map $j : G \times_H V \to M$ giving an equivariant diffeomorphism onto a G-invariant neighbourhood Y of $G.x$ in M. We constructed V as a metric disc in the orthogonal complement E of $T_x(G.x)$ in T_xM. Moreover, since we have a diffeomorphism of $\mathring{D}^m$ on $\mathbb{R}^m$ which is invariant under rotations, we may also replace V by E itself, and have an equivariant diffeomorphism of Y with $G \times_H E$. Here E is a real vector space on which H acts orthogonally. This choice gives a convenient local model, which we use for further analysis below.

3.4 Properties of proper actions

From now on, we suppose M a smooth proper G-space. By Proposition A.3.4, the quotient space $G\backslash M$ is Hausdorff, locally compact, and a countable union of compact sets. We can now parallel the development in §1.1.

First, we can apply Proposition 1.1.3 to express $G\backslash M$ as $\bigcup_n C_n$, where we have compact subsets C_n and open subsets $B_{n+\frac{1}{2}}$ such that for all $n \geq 1$, $C_n \subset B_{n+\frac{1}{2}} \subset C_{n+1}$.

The map $j : G \times_H V \to M$ of Theorem 3.3.5 induces a homeomorphism of $G\backslash(G \times_H V)$ onto a neighbourhood of the image $[x]$ of x in $G\backslash M$. We will regard such a map as a coordinate neighbourhood[1] for $G\backslash M$. Observe that $G\backslash(G \times_H V) \cong H\backslash V$, so this neighbourhood is a quotient of V. We will use the term 'nice neighbourhood' in the case when V is a disc D (we suppress the affix giving the dimension of the disc, which depends on the slice, and will be clear from the context). We think of this as a map $\overline{j} : H\backslash D \to G\backslash M$ coming from a map $j : G \times_H D \to M$.

Theorem 3.4.1 *We can find a set of nice coordinate neighbourhoods $\varphi_\alpha : \mathring{D}(3) \to G\backslash M$, with images denoted U_α, such that*

(i) The sets $\varphi_\alpha(\mathring{D})$ cover $G\backslash M$.

(ii) Each $P \in G\backslash M$ has a neighbourhood which meets only a finite number of sets U_α, i.e. the U_α are locally finite.

Moreover, the covering $\{U_\alpha\}$ may be chosen to refine any given covering of $G\backslash M$.

The proof of Theorem 1.1.4 goes through here with essentially no change.

[1] This differs from the notation of 1.1, where the map went from a neighbourhood in the manifold to one in Euclidean space.

It follows that the quotient space $G\backslash M$ is locally modelled by quotients $H\backslash D^k$ of discs D^k by compact subgroups H of O_k: although this is not smooth, it is a topological space with very good properties (triangulable, semi-algebraic, etc.).

We define a function $f : G\backslash M \to \mathbb{R}$ to be smooth: if the composite function $f \circ p : M \to \mathbb{R}$ is smooth.

Theorem 3.4.2 *For any covering $\mathcal{V}$ of M by G-invariant open sets, there is a smooth partition of unity by invariant functions strictly subordinate to it.*

Proof This is an analogue of Theorem 1.1.5, and the proof of the earlier result carries over with only minor change. The images of the elements of $\mathcal{V}$ define an open covering $\mathcal{U}$ of $G\backslash M$. By Theorem 3.4.1 there is a locally finite refinement of $\mathcal{U}$ by a set of coordinate neighbourhoods $\varphi_\alpha : \mathring{D}^k(3) \to G\backslash M$ such that the $\varphi_\alpha(\mathring{D}^k)$ cover $G\backslash M$. As in the earlier proof, we use these to construct smooth functions Ψ_α on $G\backslash M$, with Ψ_α supported on the image of φ_α, such that for each $P \in G\backslash M$, there is an α with $\Psi_\alpha(P) = 1$, and that each $P \in G\backslash M$ has a neighbourhood on which all but a finite number of functions Ψ_α vanish. Hence $\Sigma(P) := \sum_\alpha \Psi_\alpha(P)$ is defined, and is everywhere smooth. Thus the functions $\psi_\alpha(P) = \Psi_\alpha(P)/\Sigma(P)$ give a partition of unity; by construction it is strictly subordinate to $\mathcal{U}$. Now the functions $\psi_\alpha \circ p$ are smooth invariant functions on M giving the desired partition of unity. □

Next we have an equivariant version of Proposition 1.1.7.

Proposition 3.4.3 *(i) Let f be a continuous positive invariant function on M. Then we can find a smooth invariant function g, with $0 < g(P) < f(P)$ for all $P \in M$.*

(ii) For any continuous invariant function f on M and any $\varepsilon > 0$ there exists a smooth invariant function h on M with $|h(x) - f(x)| < \varepsilon$ for every $x \in M$.

(iii) If $f : M \to \mathbb{R}$ is continuous and invariant, $\varepsilon > 0$, and F is a closed invariant subset of M such that f is smooth on some open invariant set $U \supset F$, we can find h such that also $h = f$ on an invariant neighbourhood of F.

We can carry over the whole proof of the earlier result: it suffices to work throughout in $G\backslash M$ rather than in M.

For actions of a compact group, it is shown in Proposition A.3.5 that any neighbourhood of an invariant set contains an invariant neighbourhood. This is *not* true in general for proper actions. For an example, consider the translation action of $\mathbb{R} \times \{0\}$ on $\mathbb{R}^2$. The subset $\mathbb{R} \times \{0\}$ is invariant, and the set $\{(x, y) \mid |xy| \leq 1\}$ is a neighbourhood, but any invariant neighbourhood contains $\{(x, y) \mid |y| \leq \varepsilon\}$ for some $\varepsilon > 0$.

We turn to the existence of an invariant metric. First we consider Riemannian metrics on G itself. A positive definite scalar product on the tangent space T_1G at the identity gives rise under left translation λ_g to a scalar product on T_gG; collecting these for all $g \in G$ gives a Riemannian structure on G, invariant under left translation by elements of G.

Inner automorphism $x \to g^{-1}xg$ by $g \in G$ is a diffeomorphism of G fixing the identity, so induces a linear automorphism of T_1G. Collecting these for all $g \in G$ gives a homomorphism $ad_G : G \to GL(T_1G)$. If H is a compact subgroup of G, we know by Lemma 3.2.2 that there is an inner product on T_1G invariant under H. If we begin with such an inner product, it follows that the Riemannian metric on G is also invariant under right translation by elements of H.

Theorem 3.4.4 *A smooth proper G-manifold M has a G-invariant Riemannian structure.*

Proof This is an analogue of Theorem 1.3.1, and the proof is again modelled on the previous one. As there, we begin with a cover by charts $\varphi_\alpha : \mathring{D}^k(3) \to G\backslash M$, associated to maps $j_\alpha : G \times_H \mathring{D}^k(3) \to M$, and a strictly subordinate partition ψ_α of unity.

We next construct a G-invariant metric on $Y := G \times_H E$. Since H is compact we can, as in Proposition 3.2.3, find an H-invariant Riemannian structure on the restriction of $\mathbb{T}(Y)$ to $E = H \times_H E$ (an explicit construction can be given using an H-invariant inner product on E, and a Riemannian metric on G). The action of G gives a unique G-invariant Riemannian metric on $\mathbb{T}(Y)$ extending this structure over E.

Pulling back this metric by j_α gives a metric m_α on $j_\alpha(G \times_H \mathring{D}^k(3))$. Then $\psi_\alpha m_\alpha$ extends to an invariant section over M of the Riemannian bundle which is supported in $j_\alpha(G \times_H D^k(2))$. Now consider $\sum_\alpha \psi_\alpha m_\alpha$. Since the U_α are locally finite, the sum is defined; since the partition was strictly subordinate to the cover, the sum is smooth. Since a linear combination of positive definite quadratic forms is again positive definite, the sum is everywhere positive definite. Thus it defines an invariant Riemannian structure on M^m. □

I expect that the existence of a complete invariant metric can be established, but have not found a proof.

Under some restrictions, one can prove the existence of equivariant embeddings in Euclidean space. We first need a couple of results about linear actions.

Lemma 3.4.5 *Let G be a compact Lie group, H a Lie subgroup. Then*

(i) if V is a linear H-space, there exist a linear G-space W and an H-equivariant linear embedding $V \to W$;

(ii) there exist a linear G-space U and $u \in U$ *with* $G_u = H$.

We omit the proofs, which can both be deduced from the Peter–Weyl Theorem. For (i) we consider the vector bundle over G/H with fibre V. The space of L^2 sections is an infinite-dimensional linear G-space, and one needs to extract a finite dimensional subspace. For (ii) one similarly begins with the action of G on the space of functions on G/H (see [20, p. 105]).

Theorem 3.4.6 *For any smooth action of a compact group G on a compact manifold M, there exist a linear G-space E and a G-equivariant embedding $M \to E$.*

Proof By Theorem 3.4.1, we can cover $G\backslash M$ by a finite set of nice coordinate neighbourhoods $U_\alpha = j_\alpha(G \times_{H_\alpha} \mathring{D}_\alpha(3))$ coming from maps $\varphi_\alpha : \mathring{D}_\alpha(3) \to G\backslash M$, where $\mathring{D}_\alpha(3)$ is the disc of radius 3 in the H_α-space E_α. We define a smooth map $\Phi_\alpha : G\backslash M \to \mathbb{R}$ by $\Phi_\alpha(\varphi_\alpha(x)) = Bp(2 - \|x\|)$ for $x \in \mathring{D}_\alpha(3)$, $\Phi_\alpha(P) = 0$ otherwise.

By Lemma 3.4.5 (i) we can choose an H_α-linear embedding $f_\alpha : E_\alpha \to W_\alpha$ with W_α a linear G-space. By (ii) of the Lemma we can choose a linear G-space U_α and $u_\alpha \in U_\alpha$ with $G_{u_\alpha} = H_\alpha$. Now define $\phi_\alpha : G \times E_\alpha \to W_\alpha \oplus U_\alpha$ by $\phi_\alpha(g, s) = (g.f_\alpha(s), g.u_\alpha)$. Then for $h \in H_\alpha$, $\phi_\alpha(gh, s) = (gh.f_\alpha(s), gh.u_\alpha)$; since $G_{u_\alpha} = H_\alpha$, $h.u_\alpha = u_\alpha$ so $gh.u_\alpha = g.u_\alpha$; since f is H_α-equivariant, $gh.f_\alpha(s) = g.h.f_\alpha(s) = g.f_\alpha(h.s)$. Thus $\phi_\alpha(gh, s) = (g.f_\alpha(h.s), g.u_\alpha) = \phi_\alpha(g, h.s)$, so ϕ_α factors through $\psi_\alpha : G \times_{H_\alpha} E_\alpha \to W_\alpha \oplus U_\alpha$. By construction, ψ_α is a G-equivariant map.

The map ψ_α is injective since as $G_{u_\alpha} = H_\alpha$, $g.u_\alpha = g'.u_\alpha$ implies $g' = gh$ for some $h \in H_\alpha$; thus if $\phi_\alpha(g, s) = \phi_\alpha(g', s')$ then $\phi_\alpha(g, s) = \phi_\alpha(g, hs')$ so $g.f_\alpha(s) = g.f_\alpha(hs')$ and as f is injective, $s = hs'$. A corresponding argument on tangent spaces proves ψ_α a smooth embedding.

Now define $\Psi_\alpha : M \to W_\alpha \oplus U_\alpha \oplus \mathbb{R}$ by $\Psi_\alpha(Q) = (\Phi_\alpha(p(Q))\psi_\alpha([g, s]), \Phi_\alpha(p(Q)))$ if $Q = j_\alpha([g, s])$ with $s \in \mathring{D}_\alpha(3)$, and $\Psi_\alpha(Q) = 0$ otherwise. Since ψ_α is G-equivariant, so is this, where G acts trivially on $\mathbb{R}$. In view of the definition of Φ_α, Ψ_α is a smooth map. It now follows exactly as in the proof of Theorem 1.2.11 that the product map

$$\prod_\alpha \Psi_\alpha : M \to \bigoplus_\alpha (W_\alpha \oplus U_\alpha \oplus \mathbb{R})$$

is a smooth embedding. □

3.5 Orbit types

If we denote by ρ the (orthogonal) representation of H on E, then by Theorem 3.3.5 the structure of M in a neighbourhood of the orbit is determined by the pair $(H \subseteq G, \rho)$. In turn, this pair is determined by the action and the point $x \in M$. If we replace x by another point $g.x$ on the same orbit, $H = G_x$ is replaced by $G' = G_{g.x} = gHg^{-1}$ and ρ by an action ρ' of G' on E', where there is an isomorphism $\lambda : E \to E'$ with $\lambda(\rho(h).e) = \rho'(ghg^{-1}).\lambda(e)$ for all $h \in H,\ e \in E$. We will call two such pairs equivalent: then the equivalence class of the pair (H, ρ) depends only on the orbit $G.x$. We call it the *orbit type* of the orbit.

We may also define the *weak orbit type* of an orbit $G.x$ to be the conjugacy class of the isotropy group G_x of x. Since $G_{g.x} = g^{-1}G_x g$, this too is determined by the orbit. Two orbits have the same weak orbit type if and only if there is an equivariant bijection between them. Write $M^{\langle H \rangle} := \{x \in X \mid G_x = H\}$ for the set of points with isotropy group H. For M a proper smooth G-space, Theorem 3.3.5 describes the neighbourhood of an orbit as $Y = j(G \times_H E)$.

Lemma 3.5.1 *In the notation of Theorem 3.3.5,*

$$Y^H = Y^{\langle H \rangle} = j(N_G(H) \times_H E^H) = j((N_G(H)/H) \times E^H).$$

Thus $M^{\langle H \rangle}$ is an open submanifold of M^H.

Proof By Lemma 3.2.1, the isotropy group of $[g, z]$ is $gH_z g^{-1}$. For this to be conjugate to H, we need $H_z = H$, so $z \in E^H$; otherwise the isotropy group is strictly smaller (in the sense of Proposition 3.1.2). The calculation follows. □

The manifold $M^{\langle H \rangle}$ is not in general closed; nor need it be dense in M^H: if H does not itself occur as an isotropy group, the open subset $M^{\langle H \rangle}$ of M^H will be empty.

Different components of M^H, or of $M^{\langle H \rangle}$, may well have different dimensions. A simple example is given by the action of $\mathbb{Z}_2$ on the projective plane $P^2(\mathbb{R})$ defined by $T.(x_0 : x_1 : x_2) = (-x_0 : x_1 : x_2)$. The fixed point set consists of the point $(1 : 0 : 0)$ and the projective line $x_0 = 0$. Thus it is not convenient to partition M according to weak orbit type, and we focus on the study of orbit types.

Having the same orbit type is an equivalence relation on orbits, which we use to define partitions of $G\backslash M$ and of M. We will study these partitions, and begin with a key finiteness result.

Theorem 3.5.2 *Let φ be a proper smooth action of G on M. Then M has locally a finite number of orbit types.*

Proof We prove the result by induction on the dimension of M. If M is 0-dimensional, for each $x \in M$, the point $\{x\}$ is a neighbourhood of x and contains just one orbit type. So the assertion holds in this case. Now suppose M of dimension m and the result proved for manifolds of dimension $k < m$.

Let $x \in M$, set $H = G_x$ and let V be an H-slice at x. By Theorem 3.3.5, $G.x$ has an invariant neighbourhood diffeomorphic to $G \times_H V$. Every orbit in this neighbourhood meets V, so it is sufficient to show that there are only a finite number of orbit types in V.

We may also suppose $D^k \cong V \subset E$ a disc, and the action of H on E linear. All points on the same open radius have the same isotropy group, so the different orbit types occur at 0 and on the boundary of D^k, which is a sphere S^{k-1}, for some $k \leqq m$. By the inductive hypothesis, there are only a finite number of orbit types on S^{k-1}; and there is just one orbit type at 0. Thus $D^k \cong V$ has a finite number of orbit types. □

Let τ denote an orbit type, and write M^τ for the union of orbits of type τ.

Proposition 3.5.3 *M^τ is a smooth submanifold of M.*

Proof Let $x \in M^\tau$, and consider the neighbourhood $j(G \times_H V)$ of x constructed in Theorem 3.3.5. By Lemma 3.5.1 the points with the same weak orbit type as x in this neighbourhood form $j(G \times_H V^H)$, which is isomorphic to $(G/H) \times V^H$, and hence smooth. For $v \in V^H$, the translation in E by v is H-equivariant and takes a neighbourhood of O to one of v; thus we have the same orbit type. It follows that the set of points of orbit type τ in $j(G \times_H V)$ is also $j(G \times_H V^H)$. Thus M^τ is a smooth neighbourhood of each of its points. □

It follows from this proof that the orbit type is locally constant along $M^{\langle H \rangle}$, hence also along the space $G.M^{\langle H \rangle}$ of points of the same weak orbit type: thus is constant on each connected component of this set.

A *stratification* of a manifold is a locally finite partition into smooth submanifolds. The preceding two results show that given a smooth proper action of G on M, the partition by orbit types is a stratification. We next show that this partition has a local triviality property.

For a stratification to be used geometrically one usually imposes some condition on the way strata fit together; in particular on the behaviour of a bigger stratum near a smaller one. The strongest such condition is local triviality. We say a stratification $\mathcal{S} = \{S_\alpha\}$ of M is *locally trivial* if at each point $x \in M$ there is a neighbourhood W of x in M and a diffeomorphism $\phi : W \to A \times B$ with A, B smooth manifolds such that if S_α is the stratum containing x, $\phi(S_\alpha \cap W) = A \times \{x_0\}$ for some $x_0 \in B$ and for any other stratum S_β, $\phi(S_\beta \cap W) = A \times B_\beta$ for some smooth submanifold B_β of B.

Theorem 3.5.4 *The stratification of M by orbit types is locally trivial.*

Proof Again we use the model given by Theorem 3.3.5, and work in a neighbourhood $Y = j(G \times_H E)$ of $G.x$ in M. The orbit type α of the point $[g, y]$ is determined by that of y under the action of H on V. Now split E as a direct sum $E^H \oplus E^\alpha$, where E^α is the orthogonal complement to E^H in E. Then the orbit type of y under H depends only on the component of y in E^α. The result follows, taking E^α as the B in the above definition. □

We defined above the stratification of M by orbit types: the strata M^α are smooth submanifolds of M, and are locally finite. Thus at least one must have the same dimension as M. More precisely,

Theorem 3.5.5 (Principal Orbit Theorem) *For any smooth proper group action on a connected manifold M, there is one orbit type stratum which is open and dense in M.*

Proof We use the model given by Theorem 3.3.5: an invariant neighbourhood of a point x with $H = G_x$ is equivariantly diffeomorphic to $G \times_H E$ for some H-vector space E. The set M^α of points with the same orbit type α as x locally form $G \times_H E^H$. Thus if $\dim E^\alpha \geq 2$, M^α has codimension at least 2 and does not separate M. Now consider the case $\dim E^\alpha = 1$. Since the action of H on E^α is orthogonal and non-trivial, there is a subgroup H^+ of index 2 which acts trivially, and H acts by reflection. In this case M^α does locally separate M, but points on opposite sides lie on the same orbit. So here also $G\backslash M^\alpha$ does not separate $G\backslash M$, so the complement of the union of the $G\backslash M^\alpha$ with $\dim E^\alpha > 0$ is connected, and so is a single orbit type stratum. □

There is a natural partial order on the set of orbit types which is defined as follows. An orbit type α determines (up to equivalence) a subgroup H_α of G and a linear H_α-space E_α. Then a neighbourhood of an orbit of this type is equivariantly diffeomorphic to $N_\alpha := G \times_{H_\alpha} E_\alpha$. If β is an orbit type occurring in N_α, we write $\beta \prec \alpha$.

Lemma 3.5.6 *The relation $\prec$ is a partial order. If $\beta \prec \alpha$ then $M^\alpha \subset \bar{M}^\beta$. For any α there are only finitely many types β with $\beta \prec \alpha$. For M connected, the principal orbit type of M is the least α with $M^\alpha \neq \emptyset$.*

Proof It follows from the definition that if $\beta \prec \alpha$ there is an equivariant embedding of N^β in N^α, so the relation is transitive. Moreover, by Proposition 3.1.2 H^β has either a lower dimension than H^α or the same dimension and fewer components, so the relation is antisymmetric.

The second clause also follows from the definition; the third from Theorem 3.5.2; and the fourth is the definition of 'principal'. □

There is scope for confusion here: if $\beta \prec \alpha$ then H^β is 'smaller' than H^α but the stratum M^β is 'larger' than M^α. If A is a set of orbit types we say that A is *closed* if $\alpha \in A$ and $\beta \prec \alpha$ imply $\beta \in A$: thus the set of orbit types with M^α non-empty is always closed. If A is a closed set of orbit types, then $\bigcup_{\alpha \in A} M^\alpha$ is an open subset of M. We observe that if α and β are distinct orbit types with the same class of isotropy groups $H^\alpha = H^\beta$, then neither precedes the other.

We return to the problem of equivariant embedding in a linear G-space L. We see from Theorem 3.5.2 that there are only finitely many orbit types for the action of G on L (they all appear in any neighbourhood of the origin), and it follows that there are only finitely many orbit types on any G-submanifold of L: thus the hypothesis in Theorem 3.4.6 that M be compact cannot simply be removed. However we do have

Theorem 3.5.7 *For any smooth action of a compact group G on M with only finitely many orbit types, there exist a linear G-space E and a G-equivariant embedding $M \to E$.*

Proof Write T for the set of orbit types α. As before, we cover $G\backslash M$ by a set of nice coordinate neighbourhoods $\{U_i \mid i \in I\}$, so there is a map $a : I \to T$ and for each $i \in I$, $U_i = j_i(G \times_{H_{a(i)}} D_{a(i)})$ coming from a map $\varphi_i : D_{a(i)} \to G\backslash M$, where $D_{a(i)}$ is a disc in the $H_{a(i)}$-space $E_{a(i)}$. Define $\Phi_i : G\backslash M \to \mathbb{R}$ by $\Phi_i(\varphi_i(P)) = Bp(2 - |P|)$ for $P \in \mathring{D}_\alpha(3)$, $\Phi_\alpha(Q) = 0$ otherwise.

In the former proof, for each α we chose an H_α-linear embedding $f_\alpha : E_\alpha \to W_\alpha$ with W_α a linear G-space, and a linear G-space U_α and $u_\alpha \in U_\alpha$ with $G_{u_\alpha} = H_\alpha$, and then formed the G-equivariant embedding $\psi_\alpha : G \times_{H_\alpha} E_\alpha \to W_\alpha \oplus U_\alpha$. Here we need to separate the different φ_i for the different i with the same $a(i) = \alpha$; the difficulty is that these neighbourhoods U_i overlap.

The images $\varphi_i(D_\alpha \cap E_\alpha^{H_\alpha})$ with $a(i) = \alpha$ give an open covering of B^α. Since B^α is finite dimensional, it follows from Proposition A.2.9 that this covering has a finite dimensional refinement. More precisely, there exist an open covering $\{S_j \mid j \in J\}$ of B^α, with each S_j contained in $\varphi_i(D_\alpha \cap E_\alpha^{H_\alpha})$ for some $i(j)$, and a map $d : J \to \{0, \dots, N\}$ such that if $d(j) \neq d(j')$, then $\bar{S}_j \cap \bar{S}_{j'} = \emptyset$. Choose an open set C_j in D_α such that $C_j \cap E_\alpha^{H_\alpha} = \varphi_i^{-1}(S_j)$; then by shrinking the C_j if necessary, we may suppose that if $d(j) \neq d(j')$, then also $\overline{\varphi_{i(j)}(C_j)} \cap \overline{\varphi_{i(j')}(C_{j'})} = \emptyset$.

Now for each r with $0 \leq r \leq N$ we define a map $F_{\alpha,r} : E_\alpha \times d^{-1}(r) \to W_\alpha \times \mathbb{R}$ as follows. Choose an injective map $n : d^{-1}(r) \to \mathbb{Z}$ and set

$$F_{\alpha,r}(x, s) := (f_\alpha(x), \Phi_\alpha(x) + 3n(s)).$$

Since the f_α are injective and the values of Φ_α lie in $[0, 1]$, this is injective; since Φ_α is invariant, this is equivariant (where G acts trivially on $\mathbb{R}$). As above we can now form a G-equivariant embedding $\Psi_{\alpha,r} : G \times_{H_\alpha} (\bigcup_{d(j)=r}(C_j \times \{j\})) \to W_\alpha \oplus U_\alpha \oplus \mathbb{R}$.

Since we now have a finite set of embeddings, we can piece them together as before. □

There is also an equivariant embedding theorem when G is not compact. It is clear that some restriction on G is needed, as there exist Lie groups with no faithful finite dimensional linear G-space. A notorious example is the so-called Weil–Heisenberg group, which can be considered as the group of matrices $\begin{pmatrix} 1 & a & b \\ 0 & 1 & c \\ 0 & 0 & 1 \end{pmatrix}$ with $a,\ c \in \mathbb{R}$ and $b \in \mathbb{R}/\mathbb{Z}$.

Palais gives a result in [119] using only the hypothesis that there exists a faithful finite dimensional linear G-space.

If A^α is a stratum of the orbit type stratification of M, the quotient $B^\alpha := G\backslash A^\alpha$ is a smooth manifold. We next give a model for the action of G in a neighbourhood of A^α.

Theorem 3.5.8 *A neighbourhood of A^α in M is equivariantly diffeomorphic to a bundle over B^α with fibre $G \times_H E^\alpha$.*

Proof By Theorem 3.4.4 since the action is proper we can choose a G-invariant metric for M: this induces metrics on the submanifold A^α and, by Proposition A.3.6, on $G\backslash M$; it also induces a reduction of the structure group of the normal bundle N^α to the orthogonal group.

By Proposition 2.3.1, the exponential map e^α for the normal bundle $\mathbb{N}(M/A^\alpha)$ has non-zero Jacobian along the zero cross-section of N^α, so is a local diffeomorphism at A^α; since the metric is invariant, e^α is equivariant. We now follow the proof of Theorem 2.3.3: we know some neighbourhood of the zero cross-section A^α is embedded, but need an invariant one.

In the model given by Theorem 3.3.5, we can choose the slice at x as the image of the normal space by the exponential map; by equivariance, the same holds at each point on the orbit $G.x$. In the model $G \times_H (E^H \oplus E^\alpha)$, we can identify A^α with $G \times_H E^H$; by Lemma 3.5.1 the normal space at x to A^α can be identified with E^α, the normal bundle N^α is identified with the projection with kernel E^α; and e^α is represented by the identity map.

Now factor out G: $e^\alpha : N^\alpha \to M$ yields $G\backslash e^\alpha : G\backslash N^\alpha \to G\backslash M$. Near the zero cross-section $G\backslash A^\alpha = B^\alpha$ this too is represented by the identity map (of $E^H \times H\backslash E^\alpha$); thus it is a local homeomorphism. It follows from Corollary A.2.6 that there is a neighbourhood W of B^α on which $G\backslash e^\alpha$ is an embedding. Hence also the restriction of e^α to $Z := q^{-1}(W)$ is an embedding.

We have a function $\Phi : N^\alpha \to \mathbb{R}$ measuring the length of the normal vector. Define $f : A^\alpha \to \mathbb{R}$ by

$$f(x) = inf\{\rho(x, y) + \Phi(z) \mid z \in (N^\alpha \setminus Z), \pi(z) = y\}.$$

Since each $x \in A^\alpha$ has a neighbourhood disjoint from Z, we have $f(x) > 0$. It follows from the definition that $|f(x) - f(y)| \leq \rho(x, y)$, so f is continuous; and clearly f is invariant. By Proposition 3.4.3 there is a positive smooth invariant function F on A^α with $F(x) \leq f(x)$ for all x. The proof is completed, as for Theorem 2.3.3, by writing down a diffeomorphism of the bundle with fibre the unit disc to the submanifold $\Phi(z) \leq F(\pi(z))$. □

3.6 Actions with few orbit types

We can decompose a G-manifold M into orbit types and then build it up piece by piece. We begin with a stratum A^α of least dimension: this is a compact smooth manifold, and has a neighbourhood N^α given by a bundle over A^α with fibre E^α. The next piece A^β overlaps this bundle; the details are made precise by the local structure theorem. We now explore how M is built up in the case when there are at most two strata.

For the principal orbit type α we have $E^\alpha = 0$, A^α is open in M, and is equivariantly diffeomorphic to a bundle over B^α with group G and fibre G/H^α.

If there is only one stratum, it is necessarily principal: the orbit map $M \to G\backslash M$ is a fibration with fibre G/H. To regard this as a bundle, first consider the submanifold $M^{\langle H\rangle} = M^H$ of points with isotropy subgroup equal to H. This meets all orbits, and $g.M^H$ is equal to M^H if $g^{-1}Hg = H$, and is disjoint from M^H otherwise. The elements $g \in G$ satisfying $g^{-1}Hg = H$ form a subgroup of G, called the *normaliser* of H in G and denoted $N_G(H)$. The action of $N_G(H)$ on M^H factors through $N_G(H)/H$ (since H acts trivially here). We thus see that $N_G(H)/H$ acts freely on M^H and the quotient is just $G\backslash M$, so we have a principal bundle.

If in particular M is a sphere, we have a fibration of a sphere. The possibilities for fibrations of spheres are strictly limited: the standard examples are the Hopf fibrations $S^1 \to S^{2n-1} \to P^{n-1}(\mathbb{C})$, $S^3 \to S^{4n-1} \to P^{n-1}(\mathbb{H})$, and $S^7 \to S^{15} \to S^8$. It follows from a result of Browder [29] that for any non-trivial fibration of a sphere with connected fibres, the fibre is homotopy equivalent to S^1, S^3, or S^7. In the case of manifolds it follows from the generalised Poincaré conjecture (see §5.6 and discussion following) that the fibre is homeomorphic to a sphere and, except perhaps for S^7, diffeomorphic.

In the present situation we can be even more precise.

Theorem 3.6.1 *If H is a non-trivial compact connected Lie group, acting on S^n $(n \geq 2)$ with just one weak orbit type, then either (a) the action is transitive or (b) H has rank 1 and the action is free.*

We refer to Borel [20, p 185] for the proof which, after several preliminaries, is homological in nature, so Borel's result is stated in more general terms.

It was shown by Poncet [122] that the only faithful transitive actions on spheres are the classical actions of SO_n and O_n on S^{n-1}, U_n and SU_n on S^{2n-1}, and Sp_n on S^{4n-1}; also three exceptional cases $S^6 = G_2/SU_3$, $S^7 = Spin_7/G_2$, and $S^{15} = Spin_9/Spin_7$.

Now consider groups H acting freely on spheres; first suppose H finite. Then (see [35, Chapter XII]) H has periodic cohomology, and hence all Sylow subgroups of G are cyclic or generalised quaternionic. The classification up to isomorphism of such groups is known: see [170], which also gives the latest known results about the classification of these actions.

In particular, $\mathbb{Z}_2 \oplus \mathbb{Z}_2$ cannot act freely on a sphere, hence neither can a torus $S^1 \times S^1$. Thus if H acts freely, it has rank at most 1. The only connected groups of rank 1 are S^1, S^3, and SO_3, and SO_3 has a subgroup isomorphic to $\mathbb{Z}_2 \oplus \mathbb{Z}_2$, so is excluded.

If $H \neq H_0 = S^1$ and $g \in H \setminus H_0$, conjugation of H_0 by g is an automorphism, hence is either the identity or the map $x \to x^{-1}$. If g centralises H_0, the subgroup $\langle H_0, g\rangle$ is isomorphic to a direct sum $S^1 \oplus \mathbb{Z}_k$ for some k, hence contains a subgroup $\mathbb{Z}_k \oplus \mathbb{Z}_k$; hence this case does not occur. Thus H/H_0 has order 2 and H is isomorphic to the subgroup $S^1 \cup jS^1$ of S^3. (This group can also be identified with the group Pin_2 of [15].)

If $H_0 = S^3$ and $g \in H \setminus H_0$, $g^{-1}S^1 g$ is a circle subgroup of S^3, hence conjugate in S^3 to S^1, so for some $h \in S^3$ gh normalises S^1. Arguing as above now yields a contradiction.

There are many free actions of S^1 on spheres; the classification is described, for example, in §14C of my surgery book [167]; a similar analysis holds for actions of S^3. The same methods could be applied to the $S^1 \cup jS^1$ case, but to the author's knowledge this has not been attempted.

We next consider the case of just two orbit types α (principal) and β. Choose $x \in M^\beta$ and set $H := G_x$ $(= H^\beta)$. By Theorem 3.3.5, a neighbourhood of $G.x$ is equivariantly diffeomorphic to $G \times_H E$, where H acts orthogonally on E $(= E^\beta)$ and the only fixed point is the origin. Thus there is only one orbit type for the action of H on the unit sphere S^{k-1} in E (where we choose an isomorphism of E with $\mathbb{R}^k$) and we can apply the classification just discussed; so by Theorem 3.6.1, either (a) H acts transitively on S^{k-1} or (b) H has rank at most 1.

In the present situation these H-spaces are the restrictions to S^{k-1} of linear H-spaces, so the list of cases is shorter. For (b) if H is finite a complete list of fixed point free representations (and of groups) was given by Wolf [182] (the list is repeated in a simpler notation in [170]). For $H = S^1$ any fixed point free representation is isomorphic (over $\mathbb{R}$) to the action on $\mathbb{C}^n$ for some n; and for S^3 and $S^1 \cup jS^1$ to $\mathbb{H}^n$.

By Theorem 3.5.8, a neighbourhood $N(M^\beta)$ of M^β in M is equivariantly diffeomorphic to a bundle over B^β with fibre $G \times_H E^\beta$: here M^β itself corresponds to choosing $0 \in E^\beta$. Choose $y \in M^\alpha$ to lie in the fibre over x corresponding to a point in $E^\beta \setminus \{0\}$, and set $K := G_y$ $(= H^\alpha)$.

The isomorphism $E^\beta \to \mathbb{R}^k$ induces $E^\beta \setminus \{0\} \cong S^{k-1} \times]0, \infty[$. Thus we can identify $M^\alpha \cap N(M^\beta)$ with the bundle over B^β with fibre $(G \times_H S^{k-1}) \times]0, \infty[$. Factoring out G gives an identification of $B^\alpha \cap N(B^\beta)$ with the bundle over B^β with fibre $(H \backslash S^{k-1}) \times]0, \infty[$: note that this projection indeed has fibre G/K. Now B is the union of B^α and $N(B^\beta)$ modulo this identification on the intersection. Correspondingly, M is the union of M^α and $N(M^\beta)$ modulo an identification on the intersection of bundles with fibre G/K over the above. In principle, this reduces the classification problem to a problem about manifolds (with no group action) and bundles over them.

In case (a), H acts transitively on S^{k-1}: here $B = G \backslash M$ is a smooth manifold with boundary: B^β is the boundary and B^α its complement; the identification takes place over a collar neighbourhood of the boundary. This necessarily occurs if a principal orbit has codimension 2. Here some classifications have been effectively done. If also $M = \mathbb{R}^m$, it was shown by Borel (see [20, XIV]) that G has a fixed point P, so the whole action is modelled by the induced linear action on the tangent space at P.

Interesting examples were given by Bredon [25]. Begin with the linear action of SO_n on $\mathbb{R}^n \oplus \mathbb{R}^n$. Then (see example (vb) below) there are just two orbit types; the isotropy subgroups are SO_{n-1} and SO_{n-2}. Next restrict to $D^n \times S^{n-1}$. For $x \in S^{n-1}$ define $\theta_x \in O_n$ to be the reflection in the radius through x. Then the map of $S^{n-1} \times S^{n-1}$ given by $\psi_k(x, y) := ((\theta_x\theta_y)^k x, (\theta_x\theta_y)^k y)$ is a diffeomorphism equivariant for the action of SO_n; it acts on $H_{n-1}(S^{n-1} \times S^{n-1})$ by the identity (if n is odd) and by the matrix $\begin{pmatrix} 2k+1 & 2k \\ 2k & 1-2k \end{pmatrix}$ (if n is even). Now glue two copies of $D^n \times S^{n-1}$ together using the diffeomorphism ψ_k. We obtain a closed manifold M with an action of SO_n; it still has just the two orbit types. If n is odd, M has the homology of S^{2n-1}; if n is even, $H_{n-1}(M) \cong \mathbb{Z}_{2k+1}$. For $n = 3$ this coincides with the manifold denoted M_{2k+1} in §7.8.

In [26], Bredon goes on to give a classification of actions of compact Lie groups G on manifolds with the homology of S^m and just two orbit types, one with orbits of codimension 2, and one with orbits of lower dimension. He proves that $m = 2n - 1$ is odd and either $G = SO_n$ with one of the above actions (so if n is even we have $k = 0$); or we have the action restricted to the subgroup $Spin_7$ of SO_8 or the subgroup G_2 of SO_7. We do not give the proof: a large part of it is devoted to identifying the possibilities for the group G and the isotropy subgroups H^α and H^β.

3.7 Examples of smooth proper group actions

Most of the following examples are linear actions; each of these induces also an action on the unit sphere in the vector space, also one on the corresponding projective space. Write, for $n \in \mathbb{N}$, $\zeta_n := e^{2i\pi/n}$.

(ia) The symmetric group $\mathfrak{S}_n$ acts on $\mathbb{R}^n$ by permutation of the coordinates. For each partition $\lambda : n = \lambda_1 + \lambda_2 + \ldots + \lambda_r$ (with $\lambda_1 \leq \lambda_2 \ldots$) there is an orbit type with isotropy subgroup $\prod_i \mathfrak{S}_{\lambda_i}$. The orbit type containing $(x_1, \ldots, x_n)$ is given by the partition defined by $i \sim j \Leftrightarrow x_i = x_j$. For a principal orbit, the x_i are distinct; each $\lambda_i = 1$; and the isotropy group is trivial.

The orbit space $\mathbb{R}^n/\mathfrak{S}_n$ can be identified with the subset $x_1 \leq x_2 \leq \ldots \leq x_n$.

(ib) If we replace $\mathbb{R}^n$ by $\mathbb{C}^n$ in example (i), the description of orbit types is the same, but now the orbit space $\mathbb{C}^n/\mathfrak{S}_n$ is isomorphic (using elementary symmetric functions) with $\mathbb{C}^n$.

(ic) The orthogonal group O_n acts on the space of symmetric $n \times n$ matrices by $P.A := PAP^t$ (where the affix t denotes transpose). Each orbit contains a diagonal matrix; to calculate the isotropy group we partition the eigenvalues (as above) into sets of equal ones: say this gives $n = \sum \lambda_i$. Then the isotropy group is (conjugate to) $\prod_i O_{\lambda_i}$. Principal orbits occur where all eigenvalues are distinct: here the isotropy group is $O_1^n = \{\pm 1\}^n$. The orbit space is as in (i) the simplicial cone $x_1 \leq x_2 \leq \ldots \leq x_n$. In this example, we can interpret the corresponding projective space as the space of (central) quadrics.

(id) A similar example is the action of the unitary group U_n on the set of self-adjoint $n \times n$ matrices over $\mathbb{C}$. Here the eigenvalues can be any non-zero complex numbers; the orbit space is all of $\mathbb{C}^n$.

(ie) The unitary group U_n acts on itself by conjugation: $x.y = xyx^{-1}$. As any unitary matrix is conjugate to a diagonal matrix, we again have a similar situation: here the eigenvalues satisfy $|\lambda| = 1$.

(iia) The circle group S^1 acts on the sphere S^2 by rotations, say $e^{i\theta}.(x, y, z) = (x\cos\theta + y\sin\theta, y\cos\theta - x\sin\theta, z)$. We have two fixed points at the poles $(0, 0, \pm 1)$, and the remaining orbits are principal, with trivial isotropy group. We can identify the orbit space with $[-1, 1]$ and $q : S^2 \to S^1\backslash S^2$ with $z : S^2 \to [-1, 1]$.

(iib) The group S^3 acts on itself by conjugation. The isotropy group of ± 1 is S^3; of other points in S^1 is S^1 and at other points is conjugate to S^1. The orbit space is $[-1, 1]$.

(iii) For any sequence $\mathbf{a} = (a_0, a_1, \ldots, a_n)$ of integers, the circle group $S^1 := \{t \in \mathbb{C} \mid |t| = 1\}$ acts on $\mathbb{C}^{n+1}$ by $t.(z_0, z_1, \ldots, z_n) = (t^{a_0}z_0, t^{a_1}z, \ldots, t^{a_n}z_n)$; the induced action on $P^n(\mathbb{C})$ is thus $t.(z_0 : z_1 : \ldots : z_n) = (t^{a_0}z_0 : t^{a_1}z_1 : \ldots : t^{a_n}z_n)$.

A point z is fixed under $t \in S^1$ if and only if, for all values of i with $z_i \neq 0$, the corresponding t^{a_i} are equal. Thus if t has multiplicative order r, we need the a_i for these i to be congruent mod r to each other; and, for the isotropy subgroup to have order r, no more. The isotropy action is then given by the $t^{a_j - a_i}$ for the j with $a_j \neq a_i$.

(iva) The quaternion group of order $4n$ has a presentation $\{t, u \mid t^{2n} = 1,\ u^2 = t^n,\ u^{-1}tu = t^{-1}\}$. There is a semi-free action on $\mathbb{C}^2$ with $t.(x_1, x_2) = (\zeta_{2n}x_1, \zeta_{2n}^{-1}x_2)$, $u.(x_1, x_2) = (x_2, -x_1)$. The ring of invariants is generated by $Y = x_1^{2n} + x_2^{2n}$, $Z = x_1^2x_2^2$ and $W = x_1x_2(x_1^{2n} - x_2^{2n})$; these have the unique syzygy $Y^2Z - W^2 = 4Z^{n+1}$.

(ivb) Let $G = \{u, v \mid u^7 = v^3 = 1, v^{-1}uv = u^2\}$. The subgroup $U = \langle u \rangle$ has a 1-dimensional representation $u \to \zeta_7$. The induced representation of G takes u to the diagonal matrix $(\zeta_7, \zeta_7^2, \zeta_7^4)$ and v to the matrix which cyclically permutes the coordinates. Thus v fixes the line $x_1 = x_2 = x_3$.

(ivc) Let $G = \{u, v \mid u^7 = v^9 = 1, v^{-1}uv = u^2\}$. The subgroup $U = \langle u, v^3 \rangle$ is cyclic and has a 1-dimensional representation $u \to \zeta_7$, $v^3 \to \zeta_3$. In this case, the induced representation of G on $\mathbb{C}^3$ is semi-free, and we have a free action of G on the unit sphere S^5.

(va) Consider the natural action of $SO_n \subset SO_{n+r}$ on $S^{n+r-1} \subset \mathbb{R}^n \times \mathbb{R}^r$. The isotropy subgroup of (x, y) is trivial, and the orbit an $(n-1)$-sphere unless $x = 0$, when we have a fixed point, so the action is semi-free. The orbit space is homeomorphic to D^r.

(vb) The diagonal subgroup $SO_n \subset SO_n \times SO_n$ acts on $S^{2n-1} \subset \mathbb{R}^n \times \mathbb{R}^n$. For $(x, y) \in S^{2n-1}$, if x and y are independent we have a principal orbit; the isotropy subgroup is (conjugate to) SO_{n-2} and the orbit a Stiefel manifold $V_{n,2}$. If x and y are linearly dependent, the isotropy subgroup is SO_{n-1} and the orbit S^{n-1}. The orbit space is homeomorphic to D^2.

(via) The group $SL_2(\mathbb{R})$ acts on the upper half-plane $\mathcal{H}^2 = \{z \in \mathbb{C} \mid \mathrm{Im} z > 0\}$ by

$$\begin{pmatrix} a & b \\ c & d \end{pmatrix} \cdot z = \frac{az + b}{cz + d}.$$

This action is transitive, and the isotropy subgroup of i is the rotation group SO_2: thus we have a diffeomorphism of $SO_2 \backslash SL_2(\mathbb{R})$ on $\mathcal{H}^2$ and the action is proper. The action is not effective: $-I$ acts trivially, so the action factors through $PSL_2(\mathbb{R})$.

(vib) The restriction of the action in (via) to an action of $SL_2(\mathbb{Z})$ is thus also proper. There are only two non-principal orbits for this action: they are the orbits of i, with isotropy group of order 4, and of a cube root ζ_3 of 1, with isotropy group of order 6. The orbit space is usually identified with a sphere S^2 with one point deleted (puncture).

3.8 Notes on Chapter 3

§3.1 and §3.2 contain little more than basic definitions and terminology.

There are many introductory books on these subjects: for Lie groups: [37], for example, has an algebraic approach; and [6] gives an excellent account for topologists.

A good general reference for (compact) differentiable group actions is [27]. An early account is in [20], which is a good source for early references.

§3.3: Although slices in the sense of a submanifold transverse to an orbit had appeared long before, the use of 'slice' in the precise sense needed here perhaps appeared first in Montgomery and Yang [104], where existence is proved for actions of compact groups; for proper actions the result is due to Palais [119].

The concept of proper group action developed from special cases and seems to have been first formalised about 1960. It appears in the later revisions of Bourbaki (not yet in [24]): the first reference I have is [119]. (The volume [20] only considers actions of compact groups.)

We commented in §1.6 that (M4) was equivalent to various other conditions. A similar situation exists here. It is shown in Proposition A.3.1 that the action $\phi : G \times X \to X$ is proper if and only if

(i) the map $(\phi, \pi) : G \times X \to X \times X$ (where π denotes the projection) is a proper map;

(ii) (ϕ, π) is closed and all isotropy groups G_x are compact;

(iii) for any compact subsets $K, L \subseteq X$, $T_{K,L} := \{g \in G \,|\, g.K \cap L \neq \emptyset\}$ is compact;

further equivalent conditions are mentioned in Proposition 3.3.1:

(iv) for any compact subsets $K, L \subseteq X$, $\{g \in G \,|\, g.K \cap L \neq \emptyset\}$ is compact;

(v) the orbit space $G\backslash X$ is Hausdorff.

§3.4 Most of the results in this section are fairly easy for actions of compact groups; the extension to proper actions is again in [119], though his emphasis is on continuous actions on metric spaces.

§3.5 Several results on weak orbit type appear in [20]. The Principal Orbit Theorem is due to Montgomery and Yang [105]. However, orbit types in our sense are what is required in the study of cobordism of group actions. The Atiyah–Singer fixed point theorem gives formulae expressed in terms of sums where the character of the representation of H on E plays a role. The local finiteness theorem is due to Mostow [112]. The earlier literature does not explicitly mention the stratification.

There was an explosion of papers on group actions in the 1960s: see, for example, the conference proceedings [110].

4

General position and transversality

We open our discussion of the deeper properties of smooth manifolds with Whitney's embedding theorem for two reasons. The first is historical: smooth manifolds were originally considered as submanifolds of Euclidean spaces, and this theorem reconciled this approach with the abstract form of definition which we prefer. Secondly, the proof is quite simple, and opens the way to our later discussion of the general transversality theorem.

In Chapter 5 we will give a method for describing compact manifolds up to diffeomorphism. The method consists in defining a smooth function $f : M^m \to \mathbb{R}$; and then we can regard M as 'filtered' by the subset $f^{-1}(-\infty, a]$ as a increases. In order to carry out this process in detail, it is necessary to suppose f non-degenerate. Thus we next give a direct proof of the existence of non-degenerate functions.

We proceed to techniques for moving a smooth map into 'general position'. The language of jet spaces, which is basic to the study of singularities of smooth maps, is introduced in §4.4. Jets are also used to define topologies on function space (we give some proofs of properties of these topologies in §A.4).

The fundamental technical general position result is the transversality theorem, which is stated and proved in §4.5, and extended in the following section to multitransversality, to deal with the interaction of two maps with a common target. The development of transversality as a tool is due to Thom [150]; the very flexible formulation of multitransversality is due to Mather [88].

The main theorems include 'general position' results which we will often use in later chapters. In particular, a map $f : V^v \to M^m$ may be supposed an embedding if $m > 2v$ (or an immersion if $m = 2v$); it may be deformed to avoid any subset of M of dimension $< (m - v)$, and to be transverse to any given submanifold of M.

However the results allow a much wider range of application: for example, dealing with transversality to submanifolds of jet space rather than just of M; and establishing that the set of smooth maps satisfying such conditions is open and dense in function space. We thus spend some time in §4.7 applying the main results to describe the singularities of a dense open set of maps when the target dimension is either small (≤ 2) or large ($\geq \frac{3}{2}m$). The main results also lead to local normal forms for smooth maps, and in §4.8 we obtain these in the same cases. The details here are somewhat technical, and the reader may prefer to pass over them and just read the statements of the theorems to get a feel for what can be proved.

4.1 Nul sets

We say that a subset A of $\mathbb{R}^n$ is *nul* if for each $\varepsilon > 0$, A can be enclosed in a countable union of discs of total volume (i.e. the sum of the volumes) $< \varepsilon$. The useful terminology 'nul' is now out of fashion; it is equivalent to saying that A has Lebesgue measure zero.

It is trivial that a countable union of nul sets is nul; also that a nul set has no interior: its complement is everywhere dense.

Lemma 4.1.1 *Suppose U open in $\mathbb{R}^n$, $f : U \to R^n$ smooth, and $A \subset U$ nul. Then $f(A)$ is nul.*

Proof Let K be a compact subset of U. Then in K the partial derivatives of f of first order are bounded, so infinitesimal lengths are multiplied by a bounded factor: let c be a bound. Then the image of a ball of radius r is contained in a ball of radius cr. If $A \subset K$ is nul, for any $\varepsilon > 0$ it is contained in a number of balls in K of total volume less than ε, so $f(A)$ is contained in a union of balls of total volume less than $c^n\varepsilon$, so is nul.

Now as in Theorem 1.1.4, we may find a countable set of discs $K_i = D^n_{x_i}(2\delta_i)$ contained in U, with the $\mathring{D}^n_{x_i}(\delta_i)$ covering U. As K_i is compact, and $A_i := A \cap K_i$ is nul, $f(A_i)$ is nul. Hence so is the countable union $f(A) = \bigcup_i f(A_i)$. □

We say that a subset A of a smooth manifold N is *nul* if, for each coordinate neighbourhood $\varphi : U \to \mathbb{R}^n$, $\varphi(U \cap A)$ is nul. Since by the lemma, nul sets are preserved by smooth maps, it is sufficient to verify the condition for a set $(U_\alpha, \varphi_\alpha)$ of coordinate neighbourhoods with the U_α covering N.

Corollary 4.1.2 *(i) If $A \subset N^n_1$ is nul, and $f : N^n_1 \to N^n_2$ smooth, $f(A)$ is nul.*
(ii) Suppose U open in $\mathbb{R}^v$, $v < n$, $f : U \to \mathbb{R}^n$ smooth. Then $f(U)$ is nul.
(iii) If $v < n$ and $f : V^v \to N^n$ is smooth, $f(V)$ is nul.

Proof (i) follows at once from Lemma 4.1.1 and the definition. For (ii) define $F : U \times \mathbb{R}^{n-v} \to \mathbb{R}^n$ by $F(x, y) = f(x)$. Then $f(U) = F(U \times O)$, but $U \times O$ is nul in $\mathbb{R}^n$. Similarly for (iii). □

These give the basic properties of nul sets: we now go on to the deeper result which we will need. If $f : V^v \to M^m$ is a smooth map, a point $P \in V$ is a *regular point* of f if $df : T_{g(P)}V \to T_{f(P)}M$ has rank m. Otherwise P is a *critical point*, and $f(P)$ a *critical value* of f.

Theorem 4.1.3 (Sard's Theorem) *Let $f : V^v \to M^m$ be a smooth map. Then the set of critical values of f is nul.*

We give the proof here only for $v \leq m$. For $v > m$, we refer the reader to the original paper of Sard [132] or to Milnor's account [100].

Proof We observe that it is sufficient to consider values in a coordinate neighbourhood of M, and further that, since V is a countable union of coordinate neighbourhoods, we may also restrict attention to a coordinate neighbourhood of V. This reduces the proof to the case $M = \mathbb{R}^m$, V an open subset of $\mathbb{R}^v$. For $v < m$, the result follows by Corollary 4.1.2 (ii).

Now let $m = v$. If P is a critical point, the Jacobian determinant of f vanishes at P, so given δ, we can find a ball containing P with $|J(f)| < \delta$ in the ball. Hence the volume of the image is at most δ times the volume of the original ball, so it can be contained in balls of at most twice this total volume.

If K is a compact submanifold of $\mathbb{R}^v$, A the set of critical points in K, we enclose these in small balls of total volume less than $2\mu(K)$, say. Then $f(A)$ can be enclosed in balls of total volume less than $4\delta\mu(K)$. But δ is arbitrarily small, so $f(A)$ is nul. The set of critical values is a countable union of sets $f(A)$, hence also is nul. □

4.2 Whitney's embedding theorem

The proof of the embedding theorem 1.2.11 is very simple, but the result is rather weak. We shall now obtain a stronger version, with a bound on the dimension of the Euclidean space, and an approximation clause. It is possible by similar methods to give a proof for non-compact manifolds; we defer this extension till Corollary 4.7.8. First remark that the result extends to manifolds with boundary, as if M has boundary, form the double $D(M)$: then any embedding of $D(M)$ restricts to give an embedding of M.

Each non-zero vector in $\mathbb{R}^n$ determines the parallel unit vector from the origin, and hence its end-point, which lies on S^{n-1}. Define $\mathbf{u} : (\mathbb{R}^n \setminus \{0\}) \to S^{n-1}$ by $\mathbf{u}(x) := \frac{x}{\|x\|}$.

Lemma 4.2.1 *Let $f : M^m \to \mathbb{R}^n$ be an embedding. Then the set of points of S^{n-1} parallel to a tangent of M^m is nul if $n \geq 2m + 1$, and the set of those parallel to a chord is nul if $n \geq 2m + 2$.*

Proof Any tangent of M^m is parallel to a unit tangent. Let B be the sub-bundle of $\mathbb{T}(M)$ consisting of unit vectors. Then $df : \mathbb{T}(M) \to \mathbb{T}(\mathbb{R}^n)$ restricts to $df : B \to \mathbb{T}(\mathbb{R}^n)$, and the identification of tangent spaces to $\mathbb{R}^n$ with $\mathbb{R}^n$ defines a smooth map $\mathbb{T} : \mathbb{T}(\mathbb{R}^n) \to \mathbb{R}^n$. Moreover, since B consists of unit vectors, $\mathbb{T} \circ df$ maps B to S^{n-1}. Hence the set of points in S^{n-1} whose vectors are parallel to a tangent of M is the image of B under a smooth map. Since B has dimension $2m - 1$, the first result follows from Corollary 4.1.2 (iii).

For chords we proceed similarly. Let $M \times M$ be the product manifold, $\Delta(M)$ the diagonal, and write $M^{(2)}$ for $M \times M \setminus \Delta(M)$: this is a smooth manifold. Since f is an embedding, any two distinct points have distinct images, so if we define $\Delta_f : M^{(2)} \to \mathbb{R}^n$ by $\Delta_f(P, Q) = f(P) - f(Q)$ (vector subtraction), the image does not contain O. Thus we can define $\delta_f := \mathbf{u} \circ \Delta_f : M^{(2)} \to S^{n-1}$. Again we see that the set of points of S^{n-1} whose vectors parallel to a chord of M is the image under a smooth map; this time of $M^{(2)}$. Since $M^{(2)}$ has dimension $2m$, the result follows as before. □

Theorem 4.2.2 (Whitney's Embedding Theorem) *Let M^m be a smooth compact manifold. Any map of M^m to $\mathbb{R}^{2m+1}$ may be approximated arbitrarily closely by an embedding.*

Since we have not yet discussed topologies for mapping spaces (see §4.4 below), approximation is here to be understood in the sense of pointwise convergence.

Proof Let $f_1 : M^m \to \mathbb{R}^{2m+1}$ be the given map; by Proposition 1.1.7 (applied to each component), we may suppose f_1 a smooth map. By Theorem 4.2.2, we can choose an embedding $f_2 : M^m \to \mathbb{R}^n$ for some n. The product map $f_3 : M^m \to \mathbb{R}^{2m+1+n}$ is an embedding, for since f_2 is an immersion and injective, so is f_3.

By Lemma 4.2.1, the set E of points of S^{2m+n} whose vector is parallel to a tangent or chord is nul, thus its complement is everywhere dense. Choose a point x, close to the unit point on the last axis, and not in E, and project $f_3(M)$ orthogonally in the direction x to $\mathbb{R}^{2m+n}$. The first $2m + 1$ coordinates of the projected map f_4 differ from those of f_3, and hence of f_1, by an amount which can be made arbitrarily small by choice of x.

We claim that f_4 is an embedding. For since x is parallel to no chord of $f_3(M^m)$, no two distinct points of M have the same image under f_4; and since x is parallel to no tangent vector, there is no tangent vector which is mapped to zero by df_4. Thus f_4 is an immersion and injective, hence an embedding.

We may now repeat the projection process a further $(n-1)$ times, obtaining ultimately an embedding in $\mathbb{R}^{2m+1}$ with coordinates differing by arbitrarily little from those of f_1. □

Theorem 4.2.3 *Any map of a compact smooth manifold M^m to $\mathbb{R}^{2m}$ may be approximated by an immersion.*

Proof As for Theorem 4.2.2, we obtain an embedding in $\mathbb{R}^{2m+1}$, and then choose $x \in S^{2m}$, arbitrarily close to the unit point on the last axis, and parallel to no tangent vector (which is possible, as before, using Lemma 4.2.1). Projecting parallel to x, we obtain the desired immersion. □

4.3 Existence of non-degenerate functions

Let f be a smooth function on M, and P a critical point of f, so that $df(T_PM) = 0$. If we take local coordinates with P as origin, we have $f(O) = 0$ and $\partial f/\partial x_i$ vanishes at O for $1 \leq i \leq m$. It is now natural to consider the Hessian matrix $(\partial^2 f/\partial x_i \partial x_j)$ of second derivatives of f at O. We regard the Hessian as a symmetric bilinear form $H(f) : T_PM \times T_PM \to \mathbb{R}$, given in local coordinates by

$$H(f)\left(\sum a_i \frac{\partial}{\partial x_i}, \sum b_i \frac{\partial}{\partial x_i}\right) = \sum a_i b_j \frac{\partial^2 f}{\partial x_i \partial x_j}.$$

We can also formulate an equivalent definition without referring to coordinates: given $u, v \in T_PM$, extend v to a local vector field $\mathbf{v}$ defined (at least) in a neighbourhood of P; then $H(f)(u, v) = u(\mathbf{v}(f))$ is independent of the extension $\mathbf{v}$ of v (since P is a critical point). (Recall here that a tangent vector is a mapping of functions on M to the reals, and a vector field maps functions to functions.)

We say that P is a *degenerate (resp. non-degenerate) critical point* of f if $H(f)$ is a singular (resp. nonsingular) bilinear form. Thus P is singular if and only if the matrix $\frac{\partial^2 f}{\partial x_i \partial x_j}$ is; equivalently, if the rows are linearly dependent, i.e. if for some constants λ_i not all zero we have $\sum_i \lambda_i \frac{\partial^2 f}{\partial x_i \partial x_j} = 0$ for all j.

We call f *non-degenerate* if it has no degenerate critical point. Many authors call such functions 'Morse functions'.

For $i : M \to \mathbb{R}^n$ an embedding, since we identify $\mathbb{T}(\mathbb{R}^n)$ with $\mathbb{R}^n \times \mathbb{R}^n$, we may identify $\mathbb{N}(\mathbb{R}^n/M)$ with the submanifold of $\mathbb{R}^n \times \mathbb{R}^n$ given by

$$\mathbb{N}(\mathbb{R}^n/M) = \{(P, v) : P \in M, v \text{ orthogonal to } di(T_P M)\}.$$

Here the exponential map is given by $\exp(P, v) = P + v$ (vector addition).

In general, if M is a submanifold of the complete Riemannian manifold N, a critical value of $\exp : \mathbb{N}(N/M) \to N$ is called a *focus* of M; if the corresponding critical point is a vector at P, it is a focus of M at P. It follows from Sard's theorem 4.1.3 that the set of foci of M in N is nul.

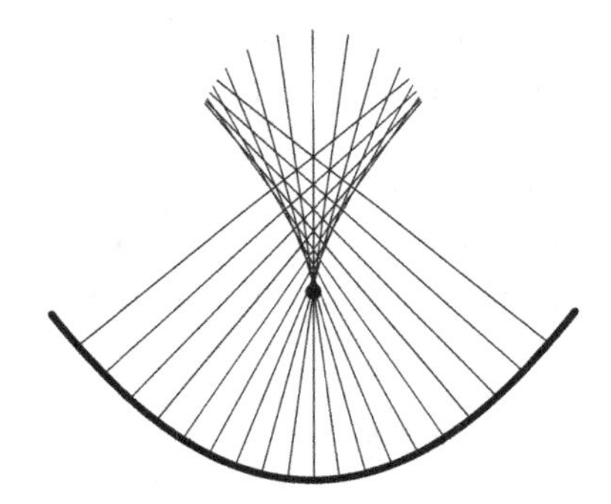

Figure 4.1 A focus

The existence of non-degenerate functions will now follow from the theorem below. Let M be a smooth submanifold of $\mathbb{R}^{m+n}$; for $P \in \mathbb{R}^{m+n}$, define $L_P : M \to \mathbb{R}$ by $L_P(Q) := \|P - Q\|^2$.

Theorem 4.3.1 *L_P has a critical point at $Q \in M$ if and only if the vector $Q - P$ is normal to M at Q. Q is a degenerate critical point if and only if P is a focus of M at Q.*

Proof The first statement is clear. For the second, first suppose M is a curve in $\mathbb{R}^2$. Then a focus must be a point of intersection of consecutive normals, i.e. a centre of curvature. But L_P has a degenerate critical point at Q if and only if $\|P - X\|^2$ is constant to the second order at $X = Q$, i.e. again if and only if P is the centre of curvature of M at Q. The notion of focus of a curve is illustrated in Figure 4.1.

The result holds in general for essentially the same reasons, but for clarity we calculate in convenient coordinates. We may suppose M given in the neighbourhood of Q as the graph B of a map $A : \mathbb{R}^m \to \mathbb{R}^n$ with $A(0) = 0$ and $d_0 A = 0$: thus A has components a_r whose Taylor expansions at 0 begin $a_r = \sum_{i,j=1}^m p_r^{i,j} x_i x_j$, where $p_r^{i,j}$ is symmetric in i and j. Differentiating B with respect to x_i gives a vector α_i whose jth component is $\delta_{i,j}$ (i.e. 1 if $i = j$, 0 if not) and with rth component $\frac{\partial a_r}{\partial x_i}$. These span T_xM, hence a base for N_xM is given by the vectors β_r with ith component $-\frac{\partial a_r}{\partial x_i}$ and sth component $\delta_{r,s}$.

We now have $\exp(x, v) = B(x) + \sum_r v_r\beta_r$. Its derivative with respect to v_r is β_r; the derivative with respect to x_i has the last n coordinates zero. Thus at a singular point of exp there must be a linear relation of the form $\sum \lambda_i \frac{\partial}{\partial x_i} \exp(x, v) = 0$ with the λ_i not all zero. This reduces to $\sum_i \lambda_i \frac{\partial}{\partial x_i}(x_j - \sum_r v_r \frac{\partial a_r}{\partial x_j})$ for each j, so occurs at $x = 0$ if and only if $\lambda_j - 2\sum_{i,r} \lambda_i v_r p_r^{i,j} = 0$ for each j.

On the other hand, the square of the distance of $B(x)$ from a typical point on N_QM, with coordinates $(0, \ldots, 0, c_1, \ldots, c_n)$ is $\sum_1^m x_i^2 + \sum_1^n (c_r - a_r(x))^2$, whose Taylor expansion at 0 is $\sum_1^n c_r^2 + \sum_1^m x_i^2 - 2\sum_{r,i,j} c_r p_r^{i,j} x_i x_j$.

The quadratic form $q(x) := \sum_1^m x_i^2 - 2\sum_{r,i,j} c_r p_r^{i,j} x_i x_j$ is degenerate if and only if, for some λ_i not all zero, the derivative $\sum_i \lambda_i \frac{\partial q}{\partial x_i}$ vanishes identically, i.e. $2\sum_i \lambda_i x_i - 4\sum_{r,i,j} \lambda_i c_r p_r^{i,j} x_j = 0$, i.e. $2\lambda_i - 4\sum_{r,j} \lambda_i c_r p_r^{i,j} = 0$ for each i. This coincides with the previous condition on setting $c_r = v_r$. The result follows. □

Corollary 4.3.2 *Any compact manifold M admits non-degenerate functions.*

Proof By Theorem 4.2.2, M can be imbedded in Euclidean space. By Sard's theorem, the set of foci, which are critical values of a smooth map, is nul. So we can choose $P \notin M$ not a focus, and then by the theorem L_P is a non-degenerate function. □

We remark that compactness is inessential, and also that using the approximation clause in Theorem 4.2.2, we could obtain one here.

If $P \notin M$, we can also replace $L_P = \|P - Q\|^2$ by the distance function $\|P - Q\|$.

4.4 Jet spaces and function spaces

We now introduce the methods for studying smooth mappings in general. We begin by introducing the language for describing a mapping locally, near a point.

Two functions f, g each defined on some neighbourhood of a point x of a topological space X have the same *germ* at x if there is a neighbourhood of x on which they take the same value. The definition applies whether the values are real numbers or lie in any space. We talk of germs, or map-germs at (X, x).

Lemma 4.4.1 *Let $f,\ g : \mathbb{R}^v \to \mathbb{R}^m$ be smooth map-germs at O such that the values of f and all its partial derivatives of orders $\leq r$ agree with those of g at O. Let φ, ψ be diffeomorphisms of $\mathbb{R}^v$, $\mathbb{R}^m$ keeping O fixed. Then the values of $\psi \circ f \circ \varphi$ and all its partial derivatives of orders $\leq r$ agree with those of $\psi \circ g \circ \varphi$ at O.*

Proof The result is an immediate consequence of the chain rules for differentiating a composite: 'a function of a function'. □

For $g, h : (V^v, P) \to M^m$ smooth map-germs, write $g \sim_r h$ at P if, with respect to some local coordinates at P and $g(P)$, we have $g(P) = h(P)$, and all partial derivatives of order $\leq r$ of g and h at P agree. By the lemma, this is independent of the chosen coordinate system. Clearly, $\sim_r$ is an equivalence relation for maps defined on a neighbourhood of P. An equivalence class is called an *r-jet* of maps from V to M at P. The set of all jets of maps of V to M is the *jet space* $J^r(V, M)$.

Each jet is a jet of a smooth map at some $P \in V$, so there is a natural projection $\pi_s : J^r(V, M) \to V$. Similarly (since $r \geq 0$), since two functions g, h with the same r-jet at P have $g(P) = h(P)$, there is another projection $\pi_t : J^r(V, M) \to M$. We call the point $\pi_s(j) \in V$ the *source* of the jet j and the point $\pi_t(j) \in M$ its *target*. The map (π_s, π_t) identifies $J^0(V, M)$ with the product $V \times M$. For any $k \geq r \geq 0$ there is a natural projection $\pi_r^k : J^k(V, M) \to J^r(V, M)$.

In terms of local coordinates $(x_1, \cdots, x_v)$ on V at P, $(y_1, \cdots, y_m)$ on M at Q, since two functions with the same partial derivatives define the same jet, we may take these partial derivatives as coordinates in $J^r(V, M)$. If $\omega = (\omega_1, \cdots, \omega_v)$ is a string of non-negative integers, write

$$x^\omega = (x_1^{\omega_1} \cdots x_v^{\omega_v}), \qquad \partial_\omega = (\partial/\partial x_1)^{\omega_1} \cdots (\partial/\partial x_v)^{\omega_v},$$
$$|\omega| = \omega_1 + \cdots + \omega_v, \qquad \omega! = \omega_1! \cdots \omega_v!.$$

Then if f is a smooth map-germ on (V, P) to M with target Q, its partial derivatives of order $\leq r$ are the numbers $u_j^\omega = \partial_\omega y_j$ $(0 \leq |\omega| \leq r,\ 1 \leq j \leq m)$, and these values determine the r-jet of f at P. We sometimes write y_j for the constant term u_j.

Conversely, given a set of numbers a_j^ω (where the point (a_j) must lie in the prescribed neighbourhood of Q), there exists a corresponding smooth map-germ: we may choose the polynomials $y_j = \sum_{0 \leq |\omega| \leq r} a_j^\omega x^\omega / \omega!$. Hence the set of r-jets j with source P and target Q is isomorphic to a Euclidean space. We can take (x_i, u_j^ω) as local coordinate system in $J^r(V, M)$, and coordinate changes are smooth (they exhibit, again, the chain rule for partial differentials: we shall spare the reader a detailed exhibition of them). We conclude that $J^r(V, M)$ is a smooth manifold, and the projections π_s and π_t are smooth maps.

The above polynomial is called the *polynomial representative* of the r-jet (in the given coordinates). It agrees with the sum of terms of degree $\leq r$ in the Taylor series expansion of f in the given coordinates. We are not concerned here with the question of convergence of this series.

For $f : V \to M$ a smooth map, the equivalence class of f at a point $P \in V$ is an r-jet at P, so f defines a cross-section $j^r f : V \to J^r(V, M)$ to π_s, which is smooth since f (and hence all its partial derivatives) is. Here the restriction to infinitely differentiable maps allows simpler statements: if g is a C^N map (with continuous partial derivatives of order $\leq N$), then $j^r g$ is C^{N-r} for $r \leq N$.

We can calculate the derivative of $j^r f$: the following result will be used explicitly below.

$$dj^1 f\left(\frac{\partial}{\partial x_i}\right) = \frac{\partial}{\partial x_i} + \sum_j u^i_j \frac{\partial}{\partial y_j} + \sum_{j,k} u^{ik}_j \frac{\partial}{\partial u^k_j}. \tag{4.4.2}$$

For $dj^2 f(\partial/\partial x_i)$ we add a further sum $\sum_{j,k,l} u^{ikl}_j \partial/\partial u^{kl}_j$, and so on.

Since $J^0(V, M) \cong V \times M$, $j^0 f$ is just the graph of f. A 1-jet with source P and target Q is determined by these points and a linear map $T_P V \to T_Q M$, and $j^1 f(P) = (P, f(P), df_P)$.

One can also consider jets at more than one point. We define ${}_rJ^k(V, M)$ to be the subset of the r-fold direct product $(J^k(V, M))^r$ consisting of r-tuples $(j_1, \ldots, j_r)$ such that the source points of the j_i are all distinct. We do not insist that the targets are distinct, and indeed we are largely interested in the case when they are not. Extending the notation $M^{(2)}$ of §4.2, write $V^{(r)}$ for the set (the configuration space) of ordered r-tuples of *distinct* points of V. Then the Cartesian power $(j^k f)^r : V^r \to (J^k(V, M))^r$ induces a map ${}_rj^k f : V^{(r)} \to {}_rJ^k(V, M)$. We call ${}_rJ^k(V, M)$ the *multijet space* and ${}_rj^k f$ the *multijet* of f.

We use jets to define topologies on spaces of smooth maps. One standard topology on function spaces is the so-called *compact-open topology*, which we call the C^0 topology. This is the topology on the space $C^0(X, Y)$ of continuous maps $X \to Y$ defined by taking the sets

$$A(K, U) := \{f \mid f(K) \subset U\} \quad \text{with } K \subset X \text{ compact, } U \subset Y \text{ open}$$

as a sub-base of open sets. It can be described as the topology of uniform convergence of f on compact sets.

There is also the *fine topology* (or fine C^0 topology), which we define by taking the

$$B(U) := \{f \mid (1 \times f)(X) \subset U\} \quad \text{with } U \text{ open in } X \times Y$$

as a base of open sets.

For smooth manifolds V^v and M^m, write $C^r(V, M)$ for the set of maps $V \to M$ whose restrictions in any local coordinates have continuous partial derivatives of all orders $\leq r$; in particular, $C^\infty(V, M)$ is the set of smooth maps of V to M. Taking r-jets gives an injective map $j^r : C^r(V, M) \to C^0(V, J^r(V, M))$. The topology on $C^r(V, M)$ induced by regarding it as a subspace of $C^0(V, J^r(V, M))$ with the compact-open topology is called the *C^r topology*, and the topology induced from the fine topology is the *fine C^r topology*.

The inclusion of $C^\infty(V, M)$ in $C^r(V, M)$ induces topologies on it, and we define the *C^∞ topology* to be the union of the C^r topologies, in the sense that a set is open if it is open in one of these topologies. Correspondingly, the fine C^∞ topology, which we christen the *W^∞ topology*, is the union of the fine C^r topologies.

Properties of these topologies are discussed in Appendix A.4. We summarise some key results:

Both topologies on $C^\infty(V, M)$ are completely regular. They agree if V is compact.

With the C^∞ topology, $C^\infty(V, M)$ is a complete metric space. However, a sequence of maps convergent for the W^∞ topology is eventually constant outside a compact set; hence this topology is neither metrisable nor even locally countable.

The space $C^\infty_{pr}(V, M)$ of proper C^∞ maps is open in $C^\infty(V, M)$ in the W^∞ topology.

The composition map $C^\infty(V, M) \times C^\infty(M, N) \to C^\infty(V, N)$ is continuous for the C^∞ topologies; however for the W^∞ topologies this fails unless V is compact: more precisely, for the W^∞ topologies, $C^\infty_{pr}(V, M) \times C^\infty(M, N) \to C^\infty(V, N)$ is continuous, and the map $C^\infty(M, N) \to C^\infty(V, N)$ defined by composition with $f : V \to M$ is continuous if and only if f is proper.

Lemma 4.4.3 *If U is open in $J^k(V, M)$, the set of $f : V \to M$ with $j^k f(V) \subset U$ is open in $C^\infty(V, M)$ in the W^∞ topology. If K is a compact subset of V, the set of $f : V \to M$ with $j^k f(K) \subset U$ is open in $C^\infty(V, M)$ in the C^∞ topology.*

This follows directly from the definitions of the topologies, and explains why we need the W^∞ topology. In particular, since immersions are just the maps whose 1-jet takes values in the open subset of $J^1(V, M)$ with df_P injective, it follows that the set Imm(V, M) of immersions is open in $C^\infty(V, M)$ in the W^∞ topology.

It can be shown (see, for example, [73, 2.1.4]) that the set Emb(V, M) of smooth embeddings is open in $C^\infty(V, M)$ in the W^∞ topology. We will see in Corollary 4.6.4 that the set of injective immersions is open, which will suffice for our purposes. It follows from this using the openness of $C^\infty_{pr}(V, M)$ that the set of closed embeddings is open, and hence taking $V = M$ that the set Diff(M) of diffeomorphisms of M is open.

The following result ties up the notion of approximation in function space with more geometrical notions of equivalence.

Proposition 4.4.4 *If V is a compact manifold and $f : V \to M$ an embedding, there is a neighbourhood $\mathcal{U}$ of f in $C^\infty(V, M)$ such that for any $g \in \mathcal{U}$, g is an embedding and f and g are ambiently diffeotopic.*

Proof Choose a neighbourhood W of $\Delta(M)$ in $M \times M$ and a map $H : W \times [0, 1] \to M$ as in Corollary 2.2.5. Now choose a neighbourhood $\mathcal{U}$ of f such that

(i) for all $g \in \mathcal{V}$ and all $P \in V$, $(f(P), g(P)) \in W$, so we can define a smooth map f_t by $f_t(P) = H(f(P), g(P), t)$,

(ii) with the same notation, for each $t \in [0, 1]$, f_t is a smooth embedding.

Then f_t is a diffeotopy of f to g, and by Theorem 2.4.2, this diffeotopy is ambient. □

A topological space W is said to have the *Baire property*, or to be a Baire space, if the intersection of any countable family of dense open subsets of W is dense. By Baire's Theorem A.4.5, any complete metric space has the Baire property.

In a Baire space, a countable intersection of dense open sets is called a *residual set*. It is not in general open: in examples, to prove openness, further work is required.

Since it has a complete metric, $C^\infty(V, M)$ with the C^∞ topology has the Baire property. The result for the fine C^∞ topology also holds: by Theorem A.4.9, if F is any subspace of $C^\infty(V, M)$ which is closed in the C^∞ topology then F, with either the C^∞ topology or the W^∞ topology, has the Baire property.

From now on, unless explicitly stated otherwise, we use the W^∞ topology on function spaces.

4.5 The transversality theorem

Let V^v, M^m be smooth manifolds, and let N^n be a submanifold of M^m. We say that a smooth map $f : V \to M$ is *transverse* to N if for every $P \in V$ such that $f(P) = Q \in N$, we have $df(T_P V) + T_Q N = T_Q M$. Equivalently, this states that df induces an epimorphism of $T_P V$ on $T_Q M / T_Q N$.

If $\dim V < \operatorname{codim} N$, the map df cannot be surjective: in that case transversality requires $f(V)$ to be disjoint from N.

The following result gives some indication of the geometrical meaning of the condition.

Lemma 4.5.1 *Let $f : V \to M$ be transverse to a submanifold N of M. Then $f^{-1}(N) = W$ is a submanifold of V, whose codimension equals that of N in M. Moreover, $df_P : T_P V \to T_{f(P)} M$ induces an isomorphism of the normal space $N_P(V/W)$ to W in V at P with the normal space $N_{f(P)}(M/N)$ of N in M at $f(P)$.*

Proof Let $P \in V$, $f(P) = Q \in N$, and let N be locally defined at Q by $x_1 = \cdots = x_c = 0$, where the x_i have linearly independent differentials at Q, and $c = \operatorname{codim} N$. Then, by transversality, the functions $x_1 \circ f, \cdots, x_c \circ f$ have linearly independent differentials at P, and their vanishing defines W near P. That W is a smooth submanifold follows using Corollary 1.2.6, as in the proof of Proposition 1.2.10. The same calculation gives the isomorphism of the normal spaces. □

We extend the concept as follows. Let N be a submanifold of $J^r(V, M)$. Then we say that f is transverse to N if $j^r f$ is so.

Lemma 4.5.2 *If K is a closed subset of V, and N a closed submanifold of $J^r(V, M)$, the set of maps which are transverse to N at all points of K is open in $C^\infty(V, M)$ in the W^∞ topology; if K is compact, it is also open in the C^∞ topology.*

Proof The differential of $j^r f$ is determined by the partial derivatives of f of order $\leq (r+1)$, and hence by $j^{r+1} f$. Since the set of linear maps $\mathbb{R}^v \to \mathbb{R}^m$ which *fail* to be transverse to a given subspace of $\mathbb{R}^m$ is defined by the vanishing of some determinants, it is a closed subset. Thus the subset of $J^{r+1}(V, M)$ of jets of maps transverse to N is open. The conclusion now follows from Lemma 4.4.3. □

The transversality theorem states that the set of maps transverse to N is dense. The full proof is somewhat technical, but the following simple idea lies at its heart.

Lemma 4.5.3 *Let N be a submanifold of M, and let $F : V \times U \to M$ be transverse to N (for example, a submersion). Then for a dense set of $u \in U$ the map $f_u : V \to M$ given by $f_u(x) = F(u, x)$ is transverse to N.*

Proof Since F is transverse to N, by Lemma 4.5.1, $W := F^{-1}(N)$ is a submanifold of $V \times U$. Denote by φ the composite $W \subset V \times U \to U$. By Sard's Theorem 4.1.3, the set of critical values of φ is nul, so for a dense set of $u \in U$, u is a regular value of φ. We claim that for such u, f_u is transverse to N.

If u is a regular value of φ and $f_u(P) = Q$ lies in N, then $(P, u) \in W$, so $d\varphi(T_{(P,u)}W) = T_uU$. Thus W meets $V \times \{u\}$ transversely at (P, u). But this implies that f_u is transverse at P to N. □

This leads to a plan for proving the jet transversality result. First define the partial jet map $j_1^rF : V \times U \to J^r(V, M)$ of a family $F : V \times U \to M$ by $j_1^rF(v, u) := j^rf_u(v)$, where $f_u(v) := F(v, u)$. Then seek to embed f in a family $F : V \times U \to M$ such that the partial jet map j_1^rF is a submersion, and hence transverse to N. Then the set of u with f_u transverse to N is dense in U.

It is not so easy to construct such a family directly, but we can do it near a point, and will then be able to obtain the full result using the Baire property. We develop the local results in a lemma.

Lemma 4.5.4 *Let $f : V^v \to M^m$ be a smooth map, $j^rf(P) = Q$. Then we can find:*

a neighbourhood $\mathcal{W}$ of f in $C^\infty(V, M)$,

a coordinate neighbourhood (U_1, φ_1) of P in V,

and a coordinate neighbourhood (U_2, φ_2) of Q in $J^r(V, M)$,

such that for each $g \in \mathcal{W}$ there is a family $G : V \times Y \to M$ with $G_0 = g$, each $g_u \in \mathcal{W}$, and such that the restriction to $U_1 \times Y$ of the partial jet map j_1^rG takes values in U_2 and is a submersion.

Proof Choose coordinate neighbourhoods of P with $\bar{U}_1 \subset U_1'$ and a chart $\varphi_2 : U_2 \to \mathbb{R}^m$ of $f(P)$ in M. Let B be a C^∞ function on V to $[0, 1]$, vanishing outside U_1', and with $B(U_1) = 1$.

Let ε be such that $y \in \mathbb{R}^m$, $\|y\| < \varepsilon$ implies that y is in the image of φ_2. Let $\mathcal{W}_1$ be the set of $g \in C^\infty(V, M)$ such that for all $x \in U_1'$, $\|\varphi_2(f(x))\| < \varepsilon/3$.

Let Y be the set of polynomial maps $y : \mathbb{R}^v \to \mathbb{R}^m$ of degree $\leq r$, and let Y' be the subset such that for $x \in \varphi_1(U_1')$, we have $\|y(x)\| < \varepsilon/3$.

For $g \in \mathcal{W}$ define $G' : U_1' \times Y' \to \mathbb{R}^m$ by $G'(x, y) := g(x) + B(x)y(x)$. Since this takes values y with $\|y\| < \varepsilon$, it lifts under φ_2 to a map $G'' : U_1' \times Y' \to M$. Now define $G : V \times Y' \to M$ by $G(P, y) = G''(P, y)$ if $P \in U_1'$ and $G(P, y) = g(P)$ otherwise.

We claim that $j_1^r G$ restricts to a submersion of $U_1 \times Y'$ to $J^r(V, M)$. For on this subset, G is given in local coordinates by $G(x, y) := g(x) + y(x)$. At $x = 0$, this has Taylor series the sum of those of g and y. But by construction, the tangent space to Y' is Y, essentially the same as the fibre of $\pi_s : J^r(V, M) \to V$, so the derivatives with respect to the y-coordinates span the tangent space to the fibre. Since $j^r g$ is a section of π_s, the derivatives with respect to the x-coordinates span the tangent space to V. Thus the sum is indeed a submersion. The same result holds for points $x \neq 0$ since although the Taylor expansion at x is not the same as at 0, the space of all polynomials of degree $\leq r$ is the same. □

Corollary 4.5.5 *Let $f : V^v \to M^m$ be a smooth map, and let N be a submanifold of $J^r(V, M)$ of codimension p. Let $j^r f(P) = Q \in N$. Then we can find a coordinate neighbourhood U_1 of P in V, a coordinate neighbourhood U_2 of Q in $J^r(V, M)$, and an open neighbourhood $\mathcal{W}$ of f in $C^\infty(V, M)$ such that*

(a) For $g \in \mathcal{W}$, $j^r g(\bar{U}_1) \subset U_2$.

(b) For every $g \in \mathcal{W}$, there are maps h arbitrarily close to g in $C^\infty(V, M)$ such that $j^r h|U_1$ is transverse to N.

Proof Define $\mathcal{W}$ and construct G as above. Since $j_1^r G$ gives a submersion of $U_1 \times Y'$ to $J^r(V, M)$, by Lemma 4.5.3, there exist $y \in Y$ arbitrarily close to 0 such that $j^r g_y \mid U_1$ is transverse to N. □

We can now prove the general theorem.

Theorem 4.5.6 (Transversality Theorem) *Let N be a submanifold of $J^r(V, M)$. The set of maps $f : V \to M$ transverse to N is dense in $C^\infty(V, M)$; if N is closed, it is also open.*

Proof First let K be a compact subset of V. Then K can be covered by a finite number of the neighbourhoods U_1^α of the lemma. The intersection of the corresponding sets $\mathcal{W}_\alpha$ is an open neighbourhood $\mathcal{W}$ of f, and the subset of $\mathcal{W}$ of functions g with $g|U_1^\alpha$ transverse to N is dense in $\mathcal{W}$, by the lemma. Since by Theorem A.4.10 the open set $\mathcal{W}$ has the Baire property, the set of $g \in \mathcal{W}$ with $g|K$ transverse to N is also dense in $\mathcal{W}$. Since this holds for some neighbourhood $\mathcal{W}$ of any f, the set $\mathcal{T}$ of $g \in C^\infty(V, M)$ with $g|K$ transverse to N is dense in $C^\infty(V, M)$. Also, $\mathcal{T}$ is open by Lemma 4.5.2.

Since V may be covered by a countable family of compact sets K, the density result follows since $C^\infty(V, M)$ has the Baire property. Openness is given by Lemma 4.5.2. □

The following addendum is often useful in applications, usually taking $X = \partial V$. For $f : V \to M$ and $X \subset V$ denote by $C^\infty(V, M; f, X)$ the set of $g \in C^\infty(V, M)$ with $g|X = f|X$.

Proposition 4.5.7 *Let N be a submanifold of $J^r(V, M)$, X a closed subset of V, $f : V \to M$ transverse to N along X. The set of maps $g \in C^\infty(V, M; f, X)$ transverse to N is dense in $C^\infty(V, M; f, X)$; if N is closed, it is also open.*

This follows from the same argument on making two changes. First, as well as the sets U_α above, we choose open sets U_β which cover X and are such that f is transverse to N along U_β: we then define $\mathcal{W}_\beta$ to be the open set of g transverse to N along U_β. Secondly, note that by Theorem A.4.9, $C^\infty(V, M; f, X)$ is a Baire space.

The Transversality Theorem is the general tool for proving 'general position' arguments in differential topology, and admits a wide variety of applications. We spend some time giving such examples, beginning with the simplest.

The following easy application seems worth formulating explicitly.

Corollary 4.5.8 *Given two embeddings $f : V \to M$ and $f' : V' \to M$, we can perturb f by an arbitrarily small diffeotopy to a map transverse to f'.*

In general, the set of f satisfying a transversality condition is residual; by further applications of Baire's theorem, we see that the set of f satisfying a finite, or even a countable, number of conditions of the above type is residual, hence dense. Thus given a countable family of submanifolds of various $J^k(V, M)$, the set of maps transverse to all of them is a residual set. Moreover, for those submanifolds of codimension $> v$, we know that transversality means that $j^k f$ avoids these submanifolds. In particular,

Lemma 4.5.9 *Given a finite or countable collection of submanifolds A_α of M, each of dimension $< (m - v)$, the set of maps $f : V^v \to M^m$ with $f(V) \cap \bigcup_\alpha A_\alpha = \emptyset$ is residual in $C^\infty(V, M)$.*

Any embedding $V \to M$ is diffeotopic to one avoiding all the A_α.

The first assertion is an immediate consequence of the theorem, since transversality to A_α implies that the two are disjoint. The second follows by Lemma 4.4.4.

The following was an early application of transversality.

Proposition 4.5.10 *Let $f : V \to M$ be a smooth map, N a submanifold of $J^k(V, M)$, and suppose F closed in V such that $f|F$ is transverse to N, then f can be approximated by g, transverse to N, and with $g|F = f|F$.*

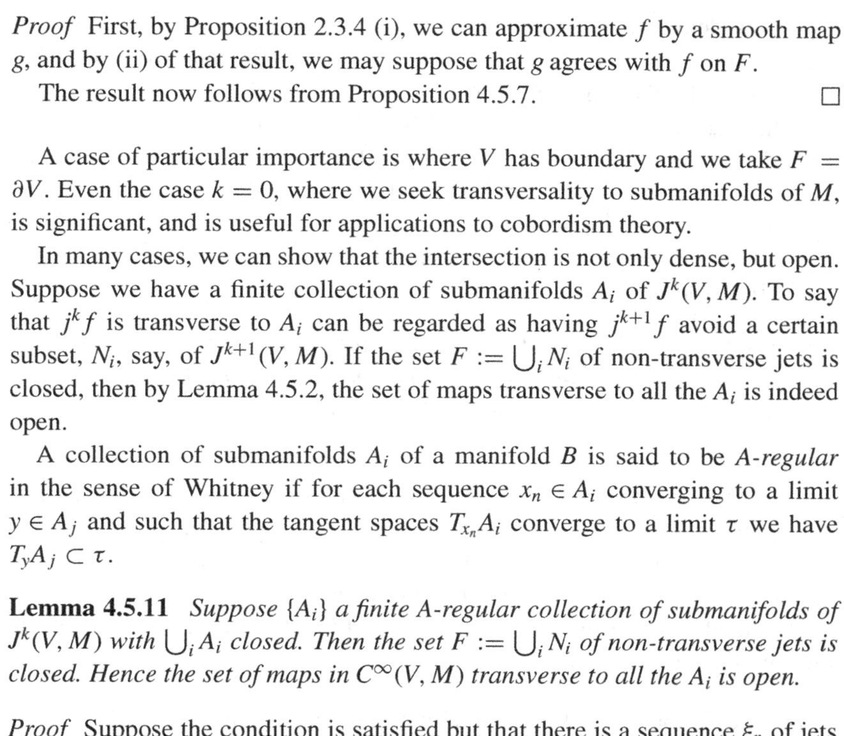

Proof First, by Proposition 2.3.4 (i), we can approximate f by a smooth map g, and by (ii) of that result, we may suppose that g agrees with f on F.

The result now follows from Proposition 4.5.7. □

A case of particular importance is where V has boundary and we take $F = \partial V$. Even the case $k = 0$, where we seek transversality to submanifolds of M, is significant, and is useful for applications to cobordism theory.

In many cases, we can show that the intersection is not only dense, but open. Suppose we have a finite collection of submanifolds A_i of $J^k(V, M)$. To say that $j^k f$ is transverse to A_i can be regarded as having $j^{k+1} f$ avoid a certain subset, N_i, say, of $J^{k+1}(V, M)$. If the set $F := \bigcup_i N_i$ of non-transverse jets is closed, then by Lemma 4.5.2, the set of maps transverse to all the A_i is indeed open.

A collection of submanifolds A_i of a manifold B is said to be *A-regular* in the sense of Whitney if for each sequence $x_n \in A_i$ converging to a limit $y \in A_j$ and such that the tangent spaces $T_{x_n} A_i$ converge to a limit τ we have $T_y A_j \subset \tau$.

Lemma 4.5.11 *Suppose $\{A_i\}$ a finite A-regular collection of submanifolds of $J^k(V, M)$ with $\bigcup_i A_i$ closed. Then the set $F := \bigcup_i N_i$ of non-transverse jets is closed. Hence the set of maps in $C^\infty(V, M)$ transverse to all the A_i is open.*

Proof Suppose the condition is satisfied but that there is a sequence ξ_n of jets in F with limit $\eta \notin F$. Passing to a subsequence, we may suppose that all the $x_n = \pi_k^{k+1}(\xi_n)$ belong to the same submanifold A_i and that the sequence $T_{x_n} A_i$ of tangent spaces converges to a limit, τ say. Since $\bigcup_i A_i$ is closed, the limit $y = \pi_k^{k+1}(\eta)$ of the x_n belongs to A_j for some j. Since A-regularity holds, $T_y A_j \subset \tau$.

Now ξ_n induces a 1-jet of maps $V \to J^k(V, M)$ and hence a map $d\xi_n : T_{\pi_s(\xi_n)} V \to T_{x_n} J^k(V, M)$, and since $\xi_n \in N_i$, we have $d\xi_n(T_{\pi_s(\xi_n)} V) + T_{x_n} A_i \neq T_{x_n} J^k(V, M)$. Since the ξ_n converge to η, it follows that $d\eta(T_{\pi_s(\eta)} V) + \tau \neq T_y J^k(V, M)$. Hence a fortiori $d\eta(T_{\pi_s(\eta)} V) + T_y A_j \neq T_y J^k(V, M)$, thus $\eta \in N_j \subset F$, a contradiction. □

In the case $k = 0$, where we are given a collection of submanifolds of M, there is even a converse result. We do not give the statement; the crucial point is that any linear map $T_x V \to T_y M$ occurs as a 1-jet. It is however far from true that any linear map $T_x V \to T_y J^k(V, M)$ is induced by a $(k + 1)$-jet, on account of the symmetry of higher derivatives.

Stratifications give important examples of collections of submanifolds, and A-regularity is often defined in this context.

We now define some submanifolds of jet space: the most important are spaces of 1-jets. Recall that a 1-jet with source $P \in V^v$ and target $Q \in M^m$ is determined by the points P, Q and a linear map $g : T_PV \to T_QM$. We partition these according to the rank of the linear map g: it is traditional to write $\Sigma^i(V, M)$ for the set of 1-jets (P, Q, g) such that the rank of g is $v - i$. Write also $\Sigma^i(f) := \{P \in V \mid j^1f(P) \in \Sigma^i(V, M)\}$. Since the rank takes values from 0 to $\min(v, m)$, Σ^i is empty unless

if $v \geq m$, we have $v \geq i \geq v - m$;
if $v \leq m$, we have $v \geq i \geq 0$.

Lemma 4.5.12 *(i) The set of $(v \times m)$ matrices of rank $(v - i)$ is a smooth submanifold of codimension $i(m - v + i)$ in the space of matrices.*

(ii) $\Sigma^i(V, M)$ is a smooth submanifold of codimension $i(m - v + i)$ in $J^1(V, M)$.

Proof (i) In an open subset of the space of matrices, the first $v - i$ columns are linearly independent. The condition for rank $v - i$ is then that the remaining $m - v + i$ columns each lie in a subspace of $\mathbb{R}^v$ of codimension i. The same argument applies if we use a different set of columns.

(ii) Using local coordinates with $U_1 \subset V$ and $U_2 \subset M$, we see that the result holds in the preimage of any $U_1 \times U_2$. □

Thus the Σ^i form a stratification of matrix space, and the $\Sigma^i(V, M)$ a stratification of $J^1(V, M)$. We may think of the closure of Σ^i as a submanifold with singularities: it is the union of the Σ^j with $j \geq i$, and is a variety in the sense of algebraic geometry. A first step in putting a map f into general position is to make it transverse to the Σ^i. This is facilitated by

Lemma 4.5.13 *The stratification Σ^i is A-regular.*

Proof It suffices to consider the submanifolds of the space of matrices, since $J^1(V, M)$ is locally a product of V, M, and $\mathrm{Hom}(T_xV, T_yM)$.

We first show that the tangent space to Σ^i at a map $\phi \in \Sigma^i$ can be decomposed as a sum $S_1 + S_2$, where S_1 is the set of linear maps ψ with $\psi(\mathrm{Ker}\,\phi) = 0$ and S_2 the set of those with $\mathrm{Im}\,\psi \subset \mathrm{Im}\,\phi$. We can take coordinates such that the matrix of ϕ is in normal form. Then the matrix $\begin{pmatrix} A & B \\ C & D \end{pmatrix}$, with A nonsingular and $r \times r$, has rank r if and only if $D = CA^{-1}B$. If we take $A - I$, B, C and D as infinitesimals, then to the first order this condition becomes $D = 0$. Thus we have the sum of the subspaces S_1 $(B = D = 0)$ and S_2 $(C = D = 0)$.

Consider a sequence $\psi_n \to \phi$ with all ψ_n of the same rank. We may suppose that both $\mathrm{Ker}\,\psi_n$ converges to a limit K and $\mathrm{Im}\,\psi_n$ converges to a limit L.

Then $K \subset$ Ker ϕ and Im $\phi \subset L$. We need to show that the tangent space at ϕ is contained in the limit, which is the sum of the set of maps with kernel containing K and that with image contained in L. But this now follows. □

Corollary 4.5.14 *The set of maps $f : V \to M$ with $j^1 f$ transverse to each Σ^i is open in $C^\infty(V, M)$.*

This follows from Lemmas 4.5.13 and 4.5.11.

4.6 Multitransversality

In general, applying the transversality theorem allows us to control the behaviour of a map $f : V \to M$ near a point of V. However, to describe the image of f we must contemplate pairs of points of V with a common image, and multitransversality is designed to enable us to do this.

An advantage of the above proof of the transversality theorem is that the version of Lemma 4.5.4 for multijets is an immediate consequence, so the same argument now leads to the multitransversality theorem.

Theorem 4.6.1 (Multitransversality Theorem) *Let N be a submanifold of ${}_rJ^k(V, M)$. The set of maps $f : V \to M$ such that ${}_rj^k f$ is transverse to N is residual in $C^\infty(V, M)$.*

Proof We follow the same plan as for Theorem 4.5.6.

Step 1: As for Lemma 4.5.4, given a smooth map $f : V^v \to M^m$ and points P_j $(1 \leq j \leq r)$ in V, write $j^k f(P_j) = Q_j$. By that lemma, we have neighbourhoods $\mathcal{W}_j$ of f in $C^\infty(V, M)$, coordinate neighbourhoods (U_{P_j}, φ_j) of P_j in V, and coordinate neighbourhoods (U_{Q_j}, ψ_j) of Q_j in $J^k(V, M)$ such that for each $g \in \mathcal{W}_j$ there is a family $G_j : V \times K_j \to M$ with $G_{j,0} = g$, each $G_{j,u} \in \mathcal{W}_j$, and such that the restriction to $U_{P_j} \times K_j$ of the partial jet map $j_1^k G_j$ takes values in U_{Q_j} and is a submersion.

Since the P_j are distinct, we may suppose their neighbourhoods disjoint, and since G_j agrees with g outside a neighbourhood of P_j, for $g \in \mathcal{W} := \bigcap_j \mathcal{W}_j$ we may combine these deformations to $G : V \times K_0 \to M$ (with $K_0 := \prod_j K_j$), where the value near P_j is given by G_j. Then the restriction to $\prod_j U_{P_j} \times K_0$ of the partial jet map ${}_rj_1^k G$ takes values in $\prod_j U_{Q_j}$ and is a submersion.

Step 2: follow Corollary 4.5.5. We are now given a submanifold N of ${}_rJ^k(V, M)$. Let ${}_rj^k f(P_1, \dots, P_r) = (Q_1, \dots, Q_r) \in N$. For $g \in \mathcal{W}$ we construct G as above. Now since ${}_rj_1^k G$ gives a submersion to ${}_rJ^k(V, M)$, by Lemma 4.5.3 there exist $k \in K$ arbitrarily close to 0 such that $j^r g_k \mid \prod_j U_j$ is transverse to N.

Step 3: By Lemma 1.1.6(i) (adapted to r-tuples), a compact subset K of $V^{(r)}$ can be covered by a finite number of sets $\prod_j U_j^\alpha$ with the U_j^α compact and disjoint. The intersection of the corresponding sets $\mathcal{W}_\alpha$ is an open neighbourhood $\mathcal{W}$ of f, and the subset of $\mathcal{W}$ of functions g with $g|\prod_j U_j^\alpha$ transverse to N is dense and open in $\mathcal{W}$. It follows using the Baire property that the subset $\mathcal{T}$ of g with $g|K$ transverse to N is also dense in $\mathcal{W}$, and since this holds for some neighbourhood $\mathcal{W}$ of any f, is dense in $C^\infty(V, M)$; in fact, a residual set.

The result follows by another application of the Baire property. □

Unlike Theorem 4.5.6, the set given by an application of Theorem 4.6.1 is almost never open. In applications, we often want to prove we have an *open* subset of mapping space, not just a dense one. It is thus necessary in some way to 'fill in' the diagonal. This is usually accomplished by combining the multitransversality condition with a simple transversality condition.

Lemma 4.6.2 *Let A be a closed submanifold of ${}_2J^k(V, M)$ and U an open neighbourhood of $\Delta(V)$ in $V \times V$. Then the set of $f \in C^\infty(V, M)$ with ${}_2j^k f \mid (V^{(2)} \setminus U)$ transverse to A is open in $C^\infty(V, M)$.*

Proof By Lemma 1.1.6(ii), we can find a countable collection of pairs of disjoint compact sets (K_α, K'_α) in V such that $\{K_\alpha, K'_\alpha\}$ is locally finite in V, and such that the $\bigcup_\alpha (K_\alpha \times K'_\alpha) \supseteq V^{(2)} \setminus U$.

The condition that ${}_2j^k f$ is transverse to A at all points of the closed subset $(K_\alpha \times K'_\alpha) \setminus U$ defines an open set in $C^\infty(K_\alpha \times K'_\alpha, M)$ by Lemma 4.5.2, and hence in $C^\infty(V, M)$, since the restriction map $C^\infty(V, M) \to C^\infty(K_\alpha \times K'_\alpha, M)$ is continuous (for as $K_\alpha \times K'_\alpha$ is compact, its inclusion in V is proper).

Now we have a countable family of open conditions on the restrictions of f to members of a locally finite cover of V, so by the definition of the fine topology, the intersection again gives an open set. □

We now have

Proposition 4.6.3 *Suppose $\mathcal{W}$ an open subset of $C^\infty(V, M)$ and A a closed submanifold of ${}_2J^k(V, M)$; write $\mathcal{W}^*$ for the set of $f \in \mathcal{W}$ with ${}_2j^k f$ transverse to A.*

Suppose that, for each $f \in \mathcal{W}^*$, each $x \in V$ has a neighbourhood U_x such that $\{g \in \mathcal{W} \mid {}_2j^k g \mid U_x^{(2)}$ is transverse to $A\}$ is a neighbourhood of f.

Then $\mathcal{W}^*$ is open in $C^\infty(V, M)$.

Proof We first show that, for each $f \in \mathcal{W}^*$, there exist a neighbourhood U_f of $\Delta(V)$ in $(V \times V)$ and an open neighbourhood $\mathcal{W}_f$ of f in $\mathcal{W}$ such that, for all $g \in \mathcal{W}_f$, ${}_2j^k g(U_f \setminus \Delta(V)) \cap A = \emptyset$. By hypothesis we have a neighbourhood

U_x for each $x \in V$; we may suppose these open. Since they cover V, we can pick a locally finite refinement $\{U_\alpha\}$. We set $U_f := \bigcup_\alpha (U_\alpha \times U_\alpha)$. By hypothesis, the set of maps $g \in \mathcal{W}$ satisfying the condition on $U_x^{(2)}$ contains an open neighbourhood of f; the same follows for $U_\alpha^{(2)}$. But by the properties of the fine topology, the intersection $\mathcal{W}_f$ of a family of open sets defined by conditions on members of a locally finite family of subsets U_α of V is open in the W^∞ topology.

By Lemma 4.6.2, the set X_G of maps with ${}_2j^k f \mid (V \times V \setminus U_f)$ transverse to A is open in $C^\infty(V, M)$, so $\mathcal{W}_f \cap X_G$ is open. But this is a neighbourhood of f in $\mathcal{W}^*$. □

Corollary 4.6.4 *The set of injective immersions is open in $C^\infty(V, M)$.*

Proof We can take $\mathcal{W}$ as the set of immersions $V \to M$ and $\mathcal{W}^*$ as the set of injective immersions: then it suffices to show that, for each $f \in \mathcal{W}^*$, each $x \in V$ has a neighbourhood U_x such that $\{g \in \mathcal{W} \mid g\,|U_x \text{ is injective}\}$ is a neighbourhood of f.

But this is clear: we can take coordinates at x and $f(x)$ in which $f|\,U_x$ is the inclusion of the unit disc U in $\mathbb{R}^v$ into $\mathbb{R}^m$; then the maps whose restriction to a closed disc of smaller radius project immersively to $\mathbb{R}^v$ form an open set (we have a compact subset of V and an open subset of $J^1(V, M)$). □

Given two subspaces P_1, P_2 of a vector space Q, we say that they are transversal if $P_1 + P_2 = Q$: this condition is stable under perturbations. The corresponding condition for a set of several subspaces P_i of Q is less familiar. We require each P_i to be transverse to the intersection of the others. The neat formulation is that the set $\{P_i\}$ of linear subspaces of Q is *mutually transversal* if the diagonal map from Q to $\bigoplus_i (Q/P_i)$ is surjective; equivalently, if the map from $Q \bigoplus_i P_i$ to $\bigoplus_i Q$, where the first summand is mapped by the diagonal, is surjective.

All our explicit applications of the multitransversality Theorem 4.6.1 follow a common pattern. Suppose we have submanifolds A_i $(1 \leq i \leq r)$ of jet space $J^k(V, M)$: then define $(A_1, \ldots, A_r)_\Delta$ to be the submanifold of ${}_rJ^k(V, M)$ of multijets $(j_1, \ldots, j_r)$ with each $j_i \in A_i$, and all $\pi_t(j_i)$ equal. (For convenience, we take all submanifolds in the same jet space, but if $k < l$, the preimage of a submanifold $A \subset J^k(V, M)$ in $J^l(V, M)$ is a submanifold A^* of the same codimension, and f is transverse to A^* if and only if it is transverse to A.) Observe that

$$\operatorname{codim}(A_1, \ldots, A_r)_\Delta = \sum_i \operatorname{codim}(A_i) + (r - 1)m.$$

For $f : V \to M$, write $A_i(f) := \{x \in V \mid j^k f(x) \in A_i\}$.

Lemma 4.6.5 *Suppose $P_i \in V$ with $j^k f(P_i) \in A_i$ and $j^k f$ transverse to A_i at P_i for each i, and each $f(P_i) = Q$. Then ${}_rj^k f$ is transverse at $(P_1, \ldots, P_r)$ to*

$(A_1, \ldots, A_r)_\Delta$ *if and only if the subspaces* $df(T_{P_i}A_i f)$ *of* $T_Q M$ *are mutually transversal.*

Proof Write $j_i := j^k f(P_i)$. The tangent space at $(j_1, \ldots, j_r)$ to $(A_1, \ldots, A_r)_\Delta$ is the pullback of the diagonal under the projection $\bigoplus_i T_{j_i} A_i \to \bigoplus_i T_Q M$. Thus transversality holds, i.e. $T(A_1, \ldots, A_r)_\Delta \bigoplus_i T_{P_i} V$ maps onto $\bigoplus_i (T_{j_i} J^k)$ if and only if the map $T_Q M \bigoplus T_{P_i} V \bigoplus T_{j_i} A_i \longrightarrow \bigoplus (T_Q M \oplus T_{j_i} J^k)$ is surjective.

Since transversality holds at each P_i, $T_{P_i} V \oplus T_{j_i} A_i$ surjects to $T_{j_i} J^k$, and we have $T_{P_i}(A_i(f)) = \mathrm{Ker}(T_{P_i} V \to T_{j_i} J^k / T_{j_i} A_i)$. Thus the condition holds if and only if $T_Q M \bigoplus T_{P_i} A_i(f)$ maps onto $\bigoplus_i T_Q M$, which is equivalent to the stated condition. □

Our first application is a simple general result.

Proposition 4.6.6 *The set of self-transverse immersions* $f : V \to M$ *is open and dense in* $\mathrm{Imm}(V, M)$.

Proof First consider the submanifold $(J^0, J^0)_\Delta$ of ${}_2J^0(V, M)$ consisting of pairs of 0-jets with a common target. By Theorem 4.6.1, the set of maps $f : V \to M$ with ${}_2 j^0 f$ transverse to $(J^0, J^0)_\Delta$ is dense in $C^\infty(V, M)$. By Lemma 4.6.5, ${}_2 j^0 f$ is transverse to $(J^0, J^0)_\Delta$ at a point (P_1, P_2) with $f(P_1) = f(P_2) = Q$ if and only if $df(T_{P_1} V) + df(T_{P_2} V) = T_Q M$, i.e. the branches of $f(V)$ at P_1 and P_2 meet transversely at Q.

Since $\mathrm{Imm}(V, M)$ is open in $C^\infty(V, M)$, it follows that the set of immersions f with this property is dense in $\mathrm{Imm}(V, M)$. Higher intersections are dealt with in the same way using $(J^0, \ldots, J^0)_\Delta \subset_r J^0(V, M)$.

For openness we use Proposition 4.6.3. Again we give the details only for the case $r = 2$. The result will follow if for each self-transverse immersion f, each $x \in V$ has a neighbourhood U_x such that $\{g \in \mathcal{W} \mid {}_2 j^k g \mid U_x^{(2)}$ is transverse to $(J^0, J^0)_\Delta\}$ is a neighbourhood of f.

But since f is an immersion, each $x \in V$ has a neighbourhood U_x embedded by f. Since the set of embeddings is open, the set of maps of V restricting to an embedding of U_x is also open. □

4.7 Generic singularities of maps

In this section we apply the general theorems to reduce singularities of maps to general form. We first give applications of jet transversality, then deal with multijets. As well as showing that maps with a certain form are dense in the space of all maps, we also show they form an open set, so that the simplifications do not disappear under small perturbations. We first consider the case $m = 1$.

Theorem 4.7.1 *Non-degenerate functions are dense and open in $C^\infty(V, \mathbb{R})$.*

Proof As $m = 1$, Σ^i is empty unless $i = v$ or $i = v - 1$, and Σ^{v-1} is smooth of codimension v. By Theorem 4.5.6, the set of functions f which are transverse to $\Sigma^{v-1}(V, \mathbb{R})$ is dense and open.

Now $j^1 f(P) \in \Sigma^{v-1}$ if and only if $df_P = 0$: P is a critical point of f. We claim that $j^1 f$ is transverse to Σ^{v-1} if and only if f is non-degenerate: this will imply the result.

Take local coordinates $\{x_i\}$ at P and y on $\mathbb{R}$, and write u_i for the coordinate on $J^1(V, \mathbb{R})$ corresponding to $\partial y/\partial x_i$. Now apply the calculation (4.4.2), which reduces here to $dj^1 f\left(\frac{\partial}{\partial x_i}\right) = \frac{\partial}{\partial x_i} + u^i \frac{\partial}{\partial y} + \sum_k u^{ik} \frac{\partial}{\partial u^k}$. Since Σ^{v-1} is defined by the equations $u_i = 0$, its tangent space is spanned by $\partial/\partial y$ and the $\partial/\partial x_i$. These together with the $dj^1 f\left(\frac{\partial}{\partial x_i}\right)$ span $T_{j^1 f(P)} J^1(V, \mathbb{R})$ if and only if the matrix $u^{ik} = (\partial^2 f/\partial x_i \partial x_k)_P$ is nonsingular, i.e. P is a non-degenerate critical point of f. □

For the case $m = 2$, we have

Theorem 4.7.2 *Maps f with the following properties form a dense open subset of $C^\infty(V^v, M^2)$: $\Sigma^{v-2}(f)$ is empty, $\Sigma^{v-1}(f)$ is a smooth curve, and at each point of $\Sigma^{v-1}(f)$, there are local coordinates in which $j^2 f$ is given by either*
$(x_1, \sum_{i,j=2}^{v} b_{ij} x_i x_j)$ with $(b_{ij})_{i,j=2}^{v}$ nonsingular or
$(x_1, x_1 x_2 + \sum_{i,j=3}^{v} b_{ij} x_i x_j)$ with $(b_{ij})_{i,j=3}^{v}$ nonsingular;
in the latter case, the coefficient of x_2^3 in y_2 is non-zero.

Proof By Lemma 4.5.12, Σ^{v-2} has codimension $2v$ and Σ^{v-1} has codimension $(v - 1)$. It thus follows from Theorem 4.5.6 that the set of maps $f : V^v \to M^2$ such that $\Sigma^{v-2}(f)$ is empty and f is transverse to Σ^{v-1} is dense, and from Corollary 4.5.14 that this set is open.

Since f is transverse to Σ^{v-1}, $\Sigma^{v-1}(f)$ is a smooth curve in V. We now need to calculate. We choose local coordinates at a point of $\Sigma^{v-1}(f)$ such that the 1-jet of f is $(x_1, 0)$. The 2-jet is then of the form

$$\left(x_1 + \sum a_{ij} x_i x_j, \sum b_{ij} x_i x_j\right).$$

Essentially the same calculation as in the preceding proof using (4.4.2) shows that this 2-jet is transverse to $\Sigma^{v-1}(f)$ if and only if the vectors

$$\sum_{j=1}^{v} b_{ij} x_j (2 \leq i \leq v)$$

are independent.

There are now two cases. In general, the matrix $B := (b_{ij})_{i,j=2}^{\upsilon}$ is nonsingular. We may then make a linear substitution $x_j' = x_j + \lambda_j x_1$ to reduce the $b_{1,j}$ $(j > 1)$ to zero, and the further change of coordinates $x_1' = x_1 + \sum a_{ij}x_ix_j$, $y_2' = y_2 - b_{1,1}y_1^2$ reduces the 2-jet to $(x_1, \sum_{i,j=2}^{\upsilon} b_{ij}x_ix_j)$. We label this case $\Sigma^{\upsilon-1,0}$. For f in this form, the tangent space to $\Sigma^{\upsilon-1}(f)$ is the x_1-axis, and the restriction of f to $\Sigma^{\upsilon-1}(f)$ is an immersion.

Otherwise the matrix B has rank $\upsilon - 2$, so by a change of coordinates $x_2, \ldots, x_\upsilon$ we can reduce to the case when $b_{2,i} = 0$ for $2 \le i \le \upsilon$. The transversality condition now implies that $b_{1,2} \neq 0$. Coordinate changes as before allow us to reduce the 2-jet to the form $(x_1, x_1x_2 + \sum_{i,j=3}^{\upsilon} b_{ij}x_ix_j)$. We label this case $\Sigma^{\upsilon-1,1}$. For f in this form, the tangent space to $\Sigma^{\upsilon-1}(f)$ is the x_2-axis, and the restriction of f to $\Sigma^{\upsilon-1}(f)$ is not an immersion.

We have effectively defined $\Sigma^{\upsilon-1,1}$ as a subspace of $J^2(V, M)$: it has codimension 1 in the space of 2-jets defining maps transverse to $\Sigma^{\upsilon-1}$. A further application of the transversality theorem tells us that for a dense set of maps, $j^2 f$ is also transverse to this.

Since $\Sigma^{\upsilon-1,1}$ was defined as a subset of $\Sigma^{\upsilon-1}$ by the vanishing of $\det(B)$, f is transversal to it if and only if $dj^2 f(\partial/\partial x_1)$ maps onto the normal space to this. In the neighbourhood of a matrix B of rank $\upsilon - 2$ and with $b_{2,i} = 0$ for $2 \le i \le \upsilon$, the normal space is spanned by $b_{2,2}$. Since the tangent space to $\Sigma^{\upsilon-1}(f)$ is the x_2-axis, we need to evaluate $dj^2 f(\partial/\partial x_2)$. Again using (4.4.2), we see that the desired condition holds if and only if the coefficient of x_2^3 in y_2 is non-zero.

For openness it suffices by Lemma 4.5.11 to prove that the set of submanifolds of J^2 defined by $\Sigma^{\upsilon-2}$, $\Sigma^{\upsilon-1,0}$ and $\Sigma^{\upsilon-1,1}$ is A-regular; the only non-trivial case is a sequence in $\Sigma^{\upsilon-1,1}$ with limit in $\Sigma^{\upsilon-2}$. But as $\Sigma^{\upsilon-2}$ is the set of jets with zero 1-jet, the inclusion of tangent spaces follows. □

In the final case, the condition that the coefficient of x_2^3 in y_2 is non-zero implies that $f(\Sigma^{\upsilon-1}(f))$ has a simple cusp.

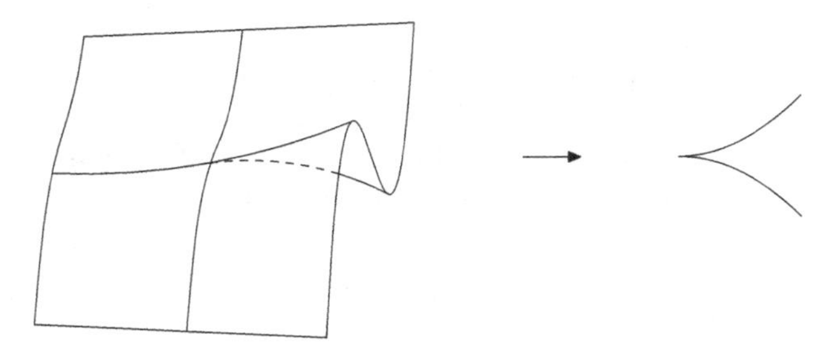

Figure 4.2 A cusp singularity

In Figure 4.2, we illustrate a cusp singularity of a map $M^2 \to \mathbb{R}^2$ as the projection of a surface M embedded in $\mathbb{R}^3$, together with the discriminant set $f(\Sigma^1(f)) \subset \mathbb{R}^2$.

Finally, we consider cases with m large compared to υ.

Theorem 4.7.3 *Maps f with the following properties form a dense open subset of $C^\infty(V^\upsilon, M^m)$:*

if $m \geq 2\upsilon$, f is an immersion,

if $2m \geq 3\upsilon - 3$, $\Sigma^2(f)$ is empty and f is transverse to Σ^1, so $\Sigma^1(f)$ is a smooth submanifold of V of dimension $2\upsilon - m - 1$,

if $2m \geq 3\upsilon - 1$, the 2-jet of f at any point of $\Sigma^1(f)$ can be reduced to the form

$$y_1 = \frac{1}{2}x_1^2,\ y_j = x_j\ (2 \leq j \leq \upsilon),\ y_{i+\upsilon-1} = x_1 x_i\ (2 \leq i \leq m - \upsilon + 1). \tag{4.7.4}$$

Proof By Theorem 4.5.6, the maps transverse to all the Σ^i form a dense set, and by Corollary 4.5.14 it is also open.

Since $m \geq \upsilon$, the codimension of Σ^1 is $m - \upsilon + 1$. Thus if $m \geq 2\upsilon$, the maps avoiding Σ^1, i.e. immersions, are dense and open in $C^\infty(M, V)$. This already sharpens Theorem 4.2.3.

Next, the codimension of Σ^2 is $2(m - \upsilon + 2)$, so provided this exceeds υ, i.e. $2m \geq 3\upsilon - 3$, for a dense open set of maps f, we have $\Sigma^2(f) = \emptyset$ and f is transverse to Σ^1. We choose local coordinates in which the 1-jet of f at P is given by $(0, x_2, \ldots, x_\upsilon, 0, \ldots, 0)$, thus $\partial/\partial x_1$ spans $\ker(df)$. Thus at $j^1 f(P)$, Σ^1 is locally the set of jets such that the first row of (u^i_j) is a linear combination of the rest, and the tangent space of Σ^1 is given by infinitesimal vanishing of u^1_1 and u^1_j for $\upsilon < j \leq m$.

From the calculation (4.4.2) we see that the coefficient of $\partial/\partial u^1_j$ in $dj^1 f(\partial/\partial x_i)$ is u^{1i}_j, i.e. $\partial^2 y_j/\partial x_1 \partial x_i$.

Now f is transverse to Σ^1 if and only if $dj^1 f(T_P V)$ spans the normal space to Σ^1, i.e. the matrix $(\partial^2 f_j/\partial x_1 \partial x_i)_{j=1, \upsilon<j\leq m, 1\leq i\leq \upsilon}$ has rank $m - \upsilon + 1$.

Consider the subvariety $\Sigma^{1,1}$ of $J^2(V, M)$ consisting of jets in Σ^1 at P such that $\ker(df)_P \subset T_P\Sigma^1(f)$. Since $\Sigma^1(f)$ has codimension $m - \upsilon + 1$, and we now impose $m - \upsilon + 1$ further conditions, $\Sigma^{1,1}$ has codimension $2(m - \upsilon + 1)$. By Theorem 4.5.6, provides this exceeds υ, i.e. $2m \geq 3\upsilon - 1$, the condition that $j^2 f(V)$ avoids $\Sigma^{1,1}$, i.e. at each point of $\Sigma^1(f)$, $dj^1 f(\ker df)$ is not tangent to Σ^1, holds for a dense set of maps f.

The condition for this tangency is that $dj^1 f(\partial/\partial x_1)$ lies in the subspace where the coefficients of $\partial/\partial u^1_1$ and $\partial/\partial u^1_j$ for $\upsilon < j \leq m$ vanish, i.e. that

$\partial^2 f_1/\partial x_1^2 = 0$ and $\partial^2 f_j/\partial x_1^2 = 0$ for $\upsilon < j \leq m$. If this condition does not hold, we may suppose, after replacing y_1 and the y_j for $\upsilon < j \leq m$ by linear combinations, that $\partial^2 f_1/\partial x_1^2 = 1$ and $\partial^2 f_j/\partial x_1^2 = 0$ for $\upsilon < j \leq m$.

It now follows that the matrix $(\partial^2 f_j/\partial x_1 \partial x_i)_{\upsilon < j \leq m, 2 \leq i \leq \upsilon}$ has rank $m - \upsilon$. We can thus make a linear transformation of the y_j with $\upsilon < j \leq m$ to arrange that $\partial^2 f_j/\partial x_1 \partial x_i = 1$ for $j = \upsilon - 1 + i$ and vanishes otherwise for $\upsilon < j \leq m$, $2 \leq i \leq \upsilon$. Thus the 2-jets take the form

$$y_1 = \tfrac{1}{2}x_1^2 + Q_1(x_2, \cdots, x_\upsilon),$$
$$y_j = x_j + Q_j(x_1, \cdots, x_\upsilon), \text{ for } 2 \leq j \leq \upsilon,$$
$$y_{i+\upsilon-1} = x_1 x_i + Q_{i+\upsilon-1}(x_2, \cdots, x_\upsilon) \text{ for } 2 \leq i \leq m - \upsilon + 1,$$

where the Q_j are quadratic. Finally, if we make the coordinate changes

$$x'_j = x_j + Q_j(x_1, \cdots, x_\upsilon),$$
$$y'_1 = y_1 - Q_1(y_2, \cdots, y_\upsilon), \text{ and}$$
$$y'_{i+\upsilon-1} = y_{i+\upsilon-1} - Q_{i+\upsilon-1}(y_2, \cdots, y_\upsilon),$$

the quadratic terms drop out too.

For openness we could seek to show that A-regularity continues to hold when we throw in $\Sigma^{1,1}$. It is easier to apply the method of Proposition 4.6.3. Write $\mathcal{W}$ for the set of $f \in C^\infty(V, M)$ transverse to the Σ^i and $\mathcal{W}^*$ for the set of $f \in \mathcal{W}$ with $j^3 f$ transverse to $\Sigma^{1,1}$. Since for any $f \in \mathcal{W}$, $j^2 f(V)$ avoids Σ^2, it avoids a neighbourhood U_f of Σ^2 in $J^3(V, M)$, hence there is an open neighbourhood $\mathcal{W}_f$ of f in $\mathcal{W}$ such that, for all $g \in \mathcal{W}_f$, $j^2 g(V)$ avoids U.

In the complement of U we only need to consider $\Sigma^{2,0}$ and $\Sigma^{2,1}$, and here the A-regularity condition trivially holds (the latter is a smooth submanifold of codimension 1 in Σ^2 and the former is its complement). By Lemma 4.5.11, transversality defines an open subset of U_f. It follows that $\mathcal{W}^* \cap U_f$ is open in U_f, so $\mathcal{W}^*$ contains an open neighbourhood of f. □

We now give applications of multitransversality: we treat the cases in the same order, so begin with functions $f \in C^\infty(V)$. Recall that the critical values of f are the $f(P)$ with P a critical point of f.

Proposition 4.7.5 *Non-degenerate functions with all critical values distinct form a dense open set in $C^\infty(V)$.*

Proof By Theorem 4.7.1, functions with only non-degenerate critical points are dense. As the submanifold $(\Sigma^{\upsilon-1}, \Sigma^{\upsilon-1})_\Delta$ of ${}_2J^1(V, \mathbb{R})$ of pairs of singular jets with the same image has codimension $2\upsilon + 1$, it follows from the Multitransversality Theorem 4.6.1, that functions avoiding it are also dense. As these are both residual sets, so is their intersection.

For openness it suffices, by Proposition 4.6.3, to show that for each non-degenerate function f with distinct critical values, each $x \in V$ has a neighbourhood U_x such that the set of non-degenerate functions g whose restriction to U_x has distinct critical values is a neighbourhood of f. Choose a coordinate neighbourhood U'_x so that f has at most one critical point on U'_x, and let U_x be the neighbourhood defined by a disc of half the radius. Then the set of non-degenerate functions on V with at most one critical point in U_x is open. □

We come to target dimension 2.

Theorem 4.7.6 *For any V^v, M^2, maps with the following properties form a dense and open subset of $C^\infty(V, M)$:*

the singular set of f is a smooth curve $\Sigma(f)$ embedded in V,

$f \mid \Sigma(f)$ is a smooth embedding except that

(a) for a discrete set of points $P \in \Sigma(f)$, the curve $f(\Sigma(f))$ has a cusp at $f(P)$,

(b) for a discrete set of pairs (P, Q) of points in $\Sigma(f)$ (all distinct from the cusps), $f(\Sigma(f))$ has a transverse self-intersection at $f(P) = f(Q)$.

Proof We give the proof of density: openness is more technical and is best established using methods described in the Notes §4.9.

Most of the conclusions were obtained in Theorem 4.7.2, but we have yet to consider double points of $f(\Sigma(f))$.

First apply the multitransversality theorem to the submanifold $(\Sigma^{v-1,1}, \Sigma^{v-1})_\Delta$ of ${}_2J^2(V, M)$. This has codimension $v + (v-1) + 2$, so is avoided by a dense set of maps; thus cusps will not be double points.

Now apply the theorem to $(\Sigma^{v-1}, \Sigma^{v-1})_\Delta$. This has codimension $(v-1) + (v-1) + 2$, so occurs at isolated points. It follows by Lemma 4.6.5 that the self-intersection of $f(\Sigma(f))$ is transverse at such points. □

For the cases of large target dimension, we have

Theorem 4.7.7 *Maps f with the following properties form a dense subset of $C^\infty(V^v, M^m)$ if V is compact, or of $C^\infty_{pr}(V, M)$ in general:*

(i) If $m \geq 2v + 1$, f is an embedding.

(ii) If $m = 2v$, f is an immersion with isolated points of transverse self-intersection.

(iii) If $2m \geq 3v - 3$, $\Sigma^2(f)$ is empty and f is transverse to Σ^1, so $\Sigma^1(f)$ is a smooth submanifold of V of dimension $2v - m - 1$.

(iv) If $2m > 3v$, f is an embedding except as follows. There are double points, forming a submanifold $D(f)$ of dimension $2v - m$, and singular points, forming a submanifold $\Sigma^1(f)$ of dimension $2v - m - 1$. Near $\Sigma^1(f)$, f is given locally by (4.7.4). Hence the closure $\bar{D}(f)$ of $D(f)$ is $D(f) \cup \Sigma^1(f)$ and is smooth, and $f(\bar{D}(f))$ is a submanifold of M with boundary $f(\Sigma^1(f))$.

(v) If $2m = 3v$ the same holds, except that now $D(f)$ is immersed with transverse self-intersection, and $f(D(f))$ can have triple points with transverse self-intersection.

Proof We extend the results of Theorem 4.7.3. For (i), we may suppose f an immersion, and apply multijet transversality to $(J^0, J^0)_\Delta$. Since this has codimension $2v$, it is avoided by a dense set of maps. Thus injective immersions are dense in $C^\infty(V, M)$; now any proper injective immersion is an embedding by Proposition 1.2.10.

Now (ii) follows using Proposition 4.6.6.

We make three further applications of the multitransversality Theorem 4.6.1. First consider the subvariety $(\Sigma^1, J^1)_\Delta$ of ${}_2J^1(V, M)$ consisting of pairs of jets with the same image, one of which (say the first) is singular. As this has codimension $m + (m - v + 1)$, if $2m \geq 3v$, the set of maps avoiding it is dense.

Next consider $(J^0, J^0)_\Delta$: by Lemma 4.6.5, ${}_2j^0 f$ is transverse to this at (P_1, P_2) if and only if $df(V_{P_1}) + df(V_{P_2}) = M_Q$. By the previous paragraph, neither P_1 nor P_2 is a singular point, so we have a transverse intersection of smooth pieces of the image, giving the set $D(f)$ of double points of f.

Finally consider the subvariety $(J^0, J^0, J^0)_\Delta$ of ${}_3J^0$ of triples of jets with the same image. Since this has codimension $2m$, if $2m > 3v$ it follows by multitransversality that the set of maps avoiding it is dense. If $2m = 3v$, this will appear at isolated points, and by Lemma 4.6.5, the three branches at such points are mutually transverse.

We have seen that $D(f)$ is an immersed submanifold; when there are no triple points it is imbedded. That $D(f)$ remains a manifold near $\Sigma^1(f)$, with $\Sigma^1(f)$ as its frontier, follows from the equations (4.7.4). Now $D(f)$ is simply given by $x_i = 0$ $(2 \leq i \leq m - v + 1)$ (modulo higher terms). Moreover $f(D(f))$ is also a submanifold, except perhaps near $f(\Sigma^1(f))$; but there it is locally given by $y_1 \geq 0$, $y_i = 0$ $(2 \leq j \leq m - v + 1)$ and $(v + 1 \leq j \leq m)$.

To prove openness it again suffices by Proposition 4.6.3 to show that, for each f satisfying the conditions, each $x \in V$ has a neighbourhood U_x such that the set of maps g whose restriction to $\Sigma^1(g) \cap U_x$ is injective is a neighbourhood of f.

By Theorem 4.7.3, we may suppose that at the point x, either f is an immersion (in which case the immersions give a neighbourhood of the desired type), or the 2-jet of f has the form (4.7.4): $y_1 = \frac{1}{2}x_1^2$, $y_j = x_j$ for $2 \leq j \leq v$, and

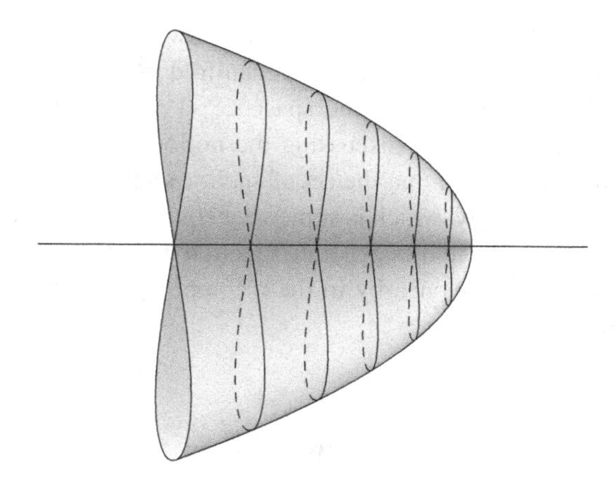

Figure 4.3 A Whitney umbrella

$y_{i+v-1} = x_1 x_i$ for $2 \leq i \leq m - v + 1$; so $\Sigma^1(f)$ is given (to the first order) by $x_i = 0$ for $1 \leq i \leq m - v + 1$. Restricting to a small neighbourhood U we see that for any nearby g, the coordinates x_i for $i > m - v + 1$ are independent on $\Sigma^1(g)$ and define an injective map of it. $\square$

We can now give a fuller statement of Whitney's Embedding Theorem.

Corollary 4.7.8 *For any smooth manifold V^v there exist proper smooth embeddings $V^v \to \mathbb{R}^m$ whenever $m > 2v$. The image of such an embedding is a closed submanifold of $\mathbb{R}^m$.*

The existence of proper maps $V \to \mathbb{R}^m$ is given by Corollary 2.2.10 and of proper smooth maps follows from Proposition 1.1.7; it follows by the theorem that there exist proper smooth embeddings. The final statement follows from Proposition 1.2.10.

It also follows that for a dense open set of maps of a compact smooth surface to 3-dimensional space, the possible types of singularity of the image are the curves D of (transverse) self-intersections of the surface, triple points where three sheets meet transversely, and a set S of isolated singular points, where the map is locally of the form (modulo higher terms, but we will see in Theorem 4.8.5 that these are unnecessary)

$$f(x_1, x_2) = (x_1^2, x_2, x_1 x_2),$$

so here the image is defined by $y_3^2 = y_1 y_2^2$ and D is the curve $y_2 = y_3 = 0,\ y_1 > 0$. Points of this type are known as *Whitney umbrella* points: an example is pictured in Figure 4.3.

Although the results using multitransversality always give a partial description of the picture of the map in the target manifold M, this should be treated with caution unless we restrict to the space $C^\infty_{pr}(V, M)$ of proper maps. We already saw this in §1.2 when discussing the notion of submanifold. Moreover, though we have proved that the set of such 'excellent' maps is open, we have used the W^∞ topology, which is somewhat counterintuitive. For example, it is possible to construct a non-degenerate function with distinct critical values which are dense in $\mathbb{R}$: maps nearby in the C^∞ topology need no longer have distinct critical values.

4.8 Normal forms

We show in this section that in each of the cases studied in the preceding section, we can choose local coordinates to reduce the map f to a precise normal form.

We begin by showing that a mutually transverse set of submanifolds has as local normal form a set of linear subspaces of a vector space.

Lemma 4.8.1 *Suppose the submanifolds V_i of M each contain a point P, and suppose that the subspaces T_PV_i of T_PM are mutually transverse. Then there exists a chart $\varphi : (U, P) \to (\mathbb{R}^m, 0)$, with U a neighbourhood of P in M, such that each $\varphi(V_i \cap U)$ is an open subset of a coordinate subspace of $\mathbb{R}^m$.*

Proof For each i, if V_i has codimension r_i, we know that there is a set of r_i smooth functions on M, each vanishing on V_i, whose differentials at P are linearly independent.

It follows from the definition of mutual transversality that the differentials of all these functions at P are linearly independent, so we can extend them to a basis of $T_P^\vee M$ by adjoining the differentials of a further $m - \sum_i r_i$ smooth functions. It follows from the Inverse Function Theorem that the set of all these functions defines a chart at P, and by construction, this has the desired property on some neighbourhood of P. □

A normal form theorem for non-degenerate functions is proved as follows. First take local coordinates with source $O \in \mathbb{R}^m$ and target $0 \in \mathbb{R}$; then by linear algebra reduce the 2-jet of f to the form $\sum_1^m \varepsilon_i x_i^2$, with each $\varepsilon_i = \pm 1$.

Proposition 4.8.2 (Morse Lemma) *Let f be a smooth function on a neighbourhood of 0 in $\mathbb{R}^n$ with 2-jet $\sum_1^n \varepsilon_i x_i^2$, where each $\varepsilon_i = \pm 1$. Then there is a smooth coordinate change $y = y(x)$ such that $y(0) = 0$, $\left.\frac{\partial y}{\partial x}\right|_0 = I_n$, and near 0, $f(x) = \sum_1^n \varepsilon_i x_i^2$.*

Proof We have $f(0) = 0$, so by Lemma 1.2.3 there exist near 0 smooth functions f_i with $f(x) = \sum x_i f_i(x)$. Also, $f_i(0) = \left.\frac{\partial f}{\partial x_i}\right|_0 = 0$, so we can apply the result again to obtain h_{ij} with $f_i(x) = \sum x_j h_{ij}(x)$. Write $g_{i,j}(x) = \frac{1}{2}(h_{ij}(x) + h_{ji}(x))$. We think of $f(x) = \sum_{ij} g_{i,j}(x) x_i x_j$ as a quadratic form, and diagonalise. Note that

$$g_{i,j}(0) = \frac{1}{2}\left.\frac{\partial^2 f}{\partial x_i \partial x_j}\right|_0 = \begin{cases} 0 & i \neq j \\ \varepsilon_i & i = j. \end{cases}$$

Set $y_1 = (\varepsilon_1 g_{11}(x))^{-1/2}(\sum_{j=1}^n g_{1j} x_j)$. Then

$$\frac{\partial y_1}{\partial x_1} = \pm 1, \ \frac{\partial y_1}{\partial x_i} = 0 \quad \text{if } i > 1, \quad \text{and} \quad f(x) = \pm y_1^2 + \sum_{i,j=2}^{n} g'_{i,j}(x) x_i x_j.$$

We now repeat the reduction, observing only that although $g'_{i,j}(x)$ depends on x_1 we can express x_1 by y_1, and the dependence is smooth. Eventually we obtain the required result. □

For the remaining cases we require further machinery, which is provided by the Malgrange Preparation Theorem. To formulate this, we need some notation. Denote by $\mathcal{E}_n$ the ring of germs at 0 of smooth functions on $\mathbb{R}^n$ under pointwise addition and multiplication. This is a local ring with maximal ideal $\mathfrak{m}_n$ consisting of germs of functions vanishing at 0. This is closely related to our introduction of jets: it follows from Lemma 1.2.3 by a simple induction that a function-germ f on $(\mathbb{R}^n, 0)$ has zero r-jet: $f \sim_r 0$: if and only if $f \in \mathfrak{m}_n^{r+1}$.

Theorem 4.8.3 (Malgrange Preparation Theorem) *For $u : \mathbb{R}^m \to \mathbb{R}^n$ a map-germ and $f_1, \ldots f_p \in \mathcal{E}_m$, the following are equivalent:*
the f_i generate $\mathcal{E}_m$ as module over $\mathcal{E}_n$,
the images of the f_i generate $\mathcal{E}_m / u^* \mathfrak{m}_y . \mathcal{E}_m$ as real vector space.

We omit the proof: see Notes §4.9 for references.

By Theorem 4.7.2, for a dense open set of maps $f : V^v \to M^2$, local coordinates can be taken at any point $P \in V$ such that we have either a submersion, a map with 2-jet $(x_1, \sum_{i,j=2}^{v} b_{ij} x_i x_j)$ with (b_{ij}) nonsingular, or a map with 2-jet $(x_1, x_1 x_2 + \sum_{i,j=3}^{v} b_{ij} x_i x_j)$ with (b_{ij}) nonsingular, and a non-zero coefficient of x_1^3 in y_2.

Theorem 4.8.4 *For a dense open set of maps $f : V^v \to M^2$, local coordinates can be taken at any point $P \in V$ such that f takes one of the forms*

(x_1, x_2),
$(x_1, \sum_{i=2}^{v} \varepsilon_i x_i^2)$, *or*
$(x_1, x_1x_2 + x_2^3 + \sum_{i=3}^{v} \varepsilon_i x_i^2)$.

We give the proof only for $v = 2$.

Proof In each case, y_1 has 1-jet x_1. First simplify by taking $x_1' = y_1(x_1, x_2)$, $x_2' = x_2$. By the Inverse Function Theorem 1.2.5, this is an allowed coordinate change, and it reduces us to the case $y_1 = x_1$.

We recall that by Lemma 1.2.3, if g is a smooth function and $g(0) = 0$, there exist near 0 smooth functions g_i with $g(x) = \sum x_i g_i(x)$. We can thus write $y_2 = x_1A_1 + x_2A_2$. As y_2 has 2-jet x_2^2, each of A_1 and A_2 vanishes at 0, so applying the lemma again gives $y_2 = x_1^2A_{11} + x_1x_2A_{12} + x_2^2A_{22}$.

Thus the ideal $f^*\mathfrak{m}_2.\mathcal{E}_v = \langle y_1, y_2\rangle = \langle x_1, x_1^2A_{11} + x_1x_2A_{12} + x_2^2A_{22}\rangle = \langle x_1, x_2^2A_{22}\rangle$. But $A_{22}(0) \neq 0$, so A_{22} is invertible, hence the ideal coincides with $\langle x_1, x_2^2\rangle$, and the quotient $\mathcal{E}_v/u^*\mathfrak{m}_2.\mathcal{E}_v$ is generated by $\{1, x_2\}$. In case (iii) a similar argument shows that the ideal is equal to $\langle x_1, x_2^3\rangle$, and the quotient is generated by $\{1, x_2, x_2^2\}$.

In case (ii), it follows by Theorem 4.8.3 that $\mathcal{E}_v$ is generated over $\mathcal{E}_2$ by $\{1, x_2\}$. Thus we can write $y_2(x_1, x_2) - \varepsilon x_2^2 = A(y_1 \circ u, y_2 \circ u) + x_2B(y_1 \circ u, y_2 \circ u)$ for some C^∞ functions A, B. Now change coordinates first by $x_2' = x_2 + \frac{\varepsilon_2}{2}B(y_1 \circ u, y_2 \circ u)$ to eliminate B; then by $y_2' = y_2 - A(y_1, y_2)$ to achieve the desired normal form.

In case (iii), $\mathcal{E}_v$ is generated over $\mathcal{E}_2$ by $\{1, x_2, x_2^2\}$. We can thus write

$$x_2^3 = (A \circ f) + x_2(B \circ f) + 3x_2^2(C \circ f),$$

where we omit the explicit dependence of A, B and C on y_1 and y_2. So

$$(x_2 - C)^3 = (A + BC + 2C^3) + (x_2 - C)(B + 3C^2).$$

If we can substitute

$$x_1' = (B + 3C^2) \circ f,\ x_2' = x_2 - C \circ f,\ y_1' = B + 3C^2,\ y_2' = A + BC + 2C^3,$$

we indeed obtain

$$y_1' = x_1, \quad y_2' = x_2^3 - x_1x_2.$$

Equating successively coefficients of x_1, x_1x_2 and x_2^3 shows that the 1-jet of A is y_2 and the 1-jet of B has the form $-y_1 + \alpha y_2$. Hence by the Inverse Function Theorem, the change of y coordinates is legitimate. Now the 1-jet of $B \circ f$ is $-x_1$ and the 1-jet of $C \circ f$ has the form βx_1, so also the change of x coordinates is legitimate. □

In Theorem 4.7.3 we saw that if $2m \geq 3v - 1$, maps f with the following properties form a dense open subset of $C^\infty(V^v, M^2)$: $\Sigma^2(f)$ is empty, f is transverse to Σ^1, and the 2-jet of f at any point of $\Sigma^1(f)$ can be reduced to the form

$$y_1 = \frac{1}{2}x_1^2, \quad y_j = x_j \text{ for } 2 \leq j \leq v, \quad y_{i+v-1} = x_1 x_i \text{ for } 2 \leq i \leq m - v + 1.$$

Theorem 4.8.5 *There exist local coordinates in which f takes precisely this form.*

Proof Here Theorem 4.8.3 gives generators $\{1, x_1\}$. So we can write

$y_1 = \frac{1}{2}x_1^2 + x_1 A_1(y) + B_1(y)$,

$y_j = x_j + x_1 A_j(y) + B_j(y)$ for $2 \leq j \leq v$, and

$y_{i+v-1} = x_1 x_i + x_1 A_{i+v-1}(y) + B_{i+v-1}(y)$ for $2 \leq i \leq m - v + 1$;

moreover equating terms of order 2 shows that the B_* have zero 1-jet, and the 1-jet of each A_i is a linear combination of the y_i with $i = 1$ or $i > v$.

First substitute $x_1' = x_1 + (A_1 \circ f)$; this reduces the map to a map of the same form, but with A_1 absent. We continue to write A_j, B_j, etc. for the new terms.

Next substitute $x_j' = x_j + x_1(A_j \circ f) + (B_j \circ f)$ for $2 \leq j \leq v$; this eliminates A_j and B_j but gives $y_{i+v-1} = x_1(x_i' - x_1 A_i(y) - B_i(y)) + x_1 A_{i+v-1}(y) + B_{i+v-1}(y)$ for $2 \leq i \leq m - v + 1$. Now set $y_{i+v-1}' = y_{i+v-1} + 2y_1 A_i(y)$ to eliminate the term in x_1^2, and renotate as before.

Thirdly write $x_i' = x_i + (A_{i+v-1} \circ f)$ for $2 \leq i \leq m - v + 1$. We now have

$y_1 = \frac{1}{2}x_1^2 + B_1(y)$,

$y_j = x_j - A_{i+v-1}(y)$ for $2 \leq j \leq v$, and

$y_{i+v-1} = x_1 x_i + B_{i+v-1}(y)$ for $2 \leq i \leq m - v + 1$.

By the Inverse Function Theorem 1.2.5, the equations $y_1' = y_1 - B_1(y)$, $y_j' = y_j + A_{i+v-1}(y)$, $y_{i+v-1}' = y_{i+v-1} - B_{i+v-1}(y)$ can now be solved to give a coordinate transformation; making these substitutions reduces f to the stated form. □

4.9 Notes on Chapter 4

§4.1 Sard's work followed that of Brown [32] which obtained a weaker result sufficient for most applications. There is a neat account of the proof in Milnor's little book [100].

§4.2 Whitney's great paper [175], although written in terms of explicit inequalities, effectively also introduced the W^∞ topology on the space of maps.

§4.3 The elementary argument here essentially goes back to Monge. A similar account is given by Milnor [98, I]. The importance of non-degenerate functions will appear in §5.1.

§4.4 Jets were first introduced by Ehresmann [48]. Their application to singularities was pioneered by Whitney, and systematically promoted by Thom [153].

A general discussion of these function space topologies, with references for proofs, is given, for example, in [121, §3.4]. A number of proofs are given in [73, §2.1] (but with a number of errors); another useful reference is [57]. The account in §A.4 includes proofs for the C^0 cases, which can be adapted to the C^∞ case.

There is no general agreement on terminology for these topologies. Some authors refer to the Thom topology for C^∞ and to the Whitney topology for W^∞, though neither of these authors formally introduced these topologies. Indeed, the first formal use of W^∞ seems to be in [88]. A discussion of their origins is given on [47, p. 59].

§4.5 The idea and use of transversality was introduced by Thom in [150]. The original proof was somewhat clumsy, but soon evolved to essentially the one presented here. An abstract form of the argument was given by Abraham [1].

Direct construction of families allowing use of Lemma 4.5.3 was given in many cases in [168].

The submanifolds Σ^i were first introduced by Thom, as were extensions to higher orders. A precise account, with the notations $\Sigma^{i,j}$ etc., was given by Boardman [19]. Whitney's regularity conditions were first formulated in [180].

The transversality Theorem 4.5.6 can be adapted to obtain results about 1-parameter families of mappings. We consider such a family as a map $F : V \times \mathbb{R} \to M \times \mathbb{R}$ of the form $F(x, t) = (f(x, t), t)$: F is compatible with projection on $\mathbb{R}$; we say that it is level-preserving. If N is a submanifold of $J^r(V \times \mathbb{R}, M \times \mathbb{R})$, we wish to make $j^r F$ transverse to N allowing only perturbation of F through level-preserving maps.

The idea of the proof of Theorem 4.5.6 is to embed g in a family $G : V \times U \to M$ such that the partial jet map $j^r_1 G : V \times U \to J^r(V, M)$ is a submersion, and then apply Lemma 4.5.3; moreover we constructed G by piecing together maps locally constructed as $G' : X \times Y \to \mathbb{R}^m$ defined by $G'(x, y) := g(x) + B(x)y(x)$, where X is a coordinate chart for V and Y is the set of polynomial maps $y : \mathbb{R}^v \to \mathbb{R}^m$ of degree $\leq r$.

To adapt this to 1-parameter families, we must replace Y by the set Y^{lp} of level-preserving polynomial maps $\mathbb{R}^{v+1} \to \mathbb{R}^{m+1}$ of degree $\leq r$. We then need to require that N is a submanifold of $J^r(V \times \mathbb{R}, M \times \mathbb{R})$ transverse to the set of level-preserving jets. In practice, it is more efficient to use the methods of Mather mentioned below.

Consider in particular $M = \mathbb{R}$ and $N = \Sigma^{v-1}$. A generic map f meets this only at its (isolated) critical points, all non-degenerate. It can be shown that a generic homotopy F can be locally put in one of the forms of Theorem 4.7.2: (x_1, x_2), $(x_1, \sum_{i=2}^{v} \pm x_i^2)$ or $(x_1, x_1x_2 + x_2^3 + \sum_{i=3}^{v} \pm x_i^2)$, with $x_1 = t$ the parameter in $\mathbb{R}$. In the first case, f_t has no critical point; in the second there is a critical point at the origin; in the third, there are no critical points if $t > 0$, but if $t = -3u^2$ there are two critical points, at $(\pm u, 0, \ldots, 0)$. This gives the model for the deformation of a function corresponding to the handle cancellations considered in §5.4.

In Lemma 4.5.13 I offered a direct proof of A-regularity: however, it follows from the fact that the strata are the orbits of the natural action of $GL(V) \times GL(M)$ that the stratification is locally trivial, which is stronger than A-regularity.

§4.6 Versions of transversality involving several source points were current in the early 1960s (and indeed examples were given in the original version of these notes) but the formulation in terms of multitransversality is due to Mather [88] III in 1969. Some of the openness lemmas are new.

§4.7 We have just focussed on the examples needed later. Proving openness, as well as density, is harder than is often given credit for. A useful general criterion was given by Looijenga (see [56, p. 146], [47, Theorem 3.4.11]).

We have presented the results in three stages, following the natural progression. Thom used the term 'source genericity' for the results obtained from transversality (for example, Theorem 4.7.3) and 'target genericity' for those using multitransversality (for example, Theorem 4.7.7); we go on to normal forms (for example, Theorem 4.8.5). These cases ($2m > 3v$) are due to Haefliger, who used them in his original proof [60] of Theorem 6.4.11.

§4.8 In general, given v and m, we can think of a generic map of V^v to M^m as one which satisfies all the transversality conditions which can be stated in terms of v, m alone (using no special facts about V, M).

This vague idea is made precise in §4.7 and §4.8 above in the cases $m = 1$, $m = 2$, and $2m \geq 3v$. In each of these cases we have a class of maps with the four properties of characterisation, local normal form, density, and stability. It can be shown [88] that (at least if V is compact) maps with these properties are C^∞ stable in the sense that nearby maps are equivalent up to diffeomorphisms of the source and target, and that in these dimensions, C^∞ stable maps are dense in $C^\infty(V, M)$. The case $v = m = 2$, motivating the search for results in higher dimensions was obtained by Whitney [179], and the case $m = 2v - 1$ by Whitney [176].

A general survey and discussion was given by Thom in [153]. However, he found that to describe general map-germs $\mathbb{R}^{16} \to \mathbb{R}^{16}$ a finite list of normal forms does not suffice: one needs to allow a parameter. The simplest example is for $\mathbb{R}^8 \to \mathbb{R}^6$: for a general map in these dimensions, $\Sigma^4 f$ consists of isolated points, and to describe the 2-jet of f at such a point involves a homogeneous quadratic map $\mathbb{R}^4 \to \mathbb{R}^2$; and the classification of such maps involves a parameter.

The above method of direct reduction to normal form is somewhat clumsy. A more general approach was introduced by Mather [88] III. Here one uses the Malgrange preparation theorem to construct vector fields, and then integrates these to find changes of coordinates. I have written an expository account of this approach in [169]. Malgrange gave a proof of his preparation theorem in Cartan seminars in 1962–63; full details appear in his book [86]. There have been many further proofs: four appear in the volume [166].

Mather's work created a full theory of C^∞ stability: see [88], also [121]. The final conclusion is that stable maps are dense in $C^\infty(V^v, M^m)$ (V compact), and a finite explicit list of normal forms analogous to the above can be given, if and only if the pair (v, m) belongs to the so-called nice dimensions, which are given if $m - v \geq 4$ by $7v < 6m + 8$; otherwise by

$m - v$	3	2	1	0	-1	-2	≤ -3
m	< 30	< 23	< 16	< 9	< 8	< 6	< 7.

5

Theory of handle decompositions

A handle decomposition is perhaps the simplest way to build a manifold from elementary pieces. The existence of such decompositions is obtained by analysing the geometry associated to a non-degenerate function on the manifold.

In the first section we prove the existence of handle decompositions for compact manifolds: in the next few sections we will show how to operate on such decompositions. In §5.2 we normalise the decomposition; then, after a section on the homology of handles, we manipulate the decompositions: there are results on adding handles, and on removing or introducing complementary pairs of handles. The technical details use the results treated in Chapter 2.

The definition of a handle decomposition is analogous to that of a CW complex. Also the results we establish run in parallel with operations on finite CW complexes that can be performed in homotopy theory. We will see below that up to a point the theory of handle presentations parallels that of cell decompositions and even to an important extent to that of algebraic operations on chain complexes.

The high point of this development is the h-cobordism theorem, which gives an effective criterion for diffeomorphism of compact manifolds. We prove this result in §5.5. Then we give a number of applications, discuss what is known in low dimensions, and outline what modifications need to be made to the theory when the fundamental group is non-trivial. In some places we anticipate Theorem 6.4.11, but Chapter 6 is independent of this chapter.

In this chapter, all manifolds will be compact unless otherwise stated.

5.1 Existence

Let W be a manifold, and suppose $\partial_- W$ and $\partial_+ W$ disjoint manifolds with union ∂W. Then we call the pair $(W, \partial_- W)$ a *cobordism* and the pair $(W, \partial_+ W)$ the

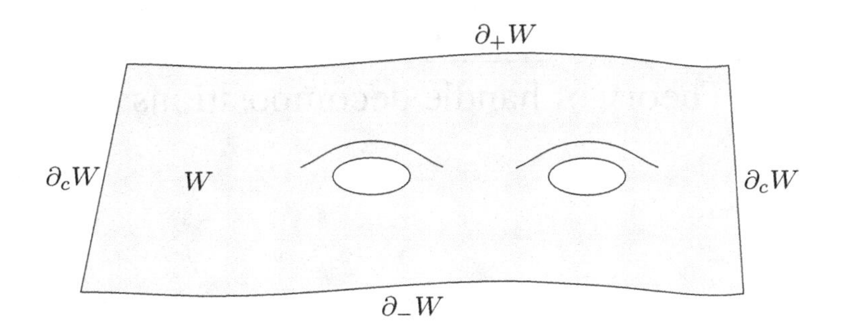

Figure 5.1 A cobordism

dual cobordism; we also call W a cobordism of ∂_-W to ∂_+W, and say that ∂_-W and ∂_+W are *cobordant*. If W is a manifold with corner, and ∂_-W, ∂_cW, ∂_+W are parts of the boundary such that ∂_-W and ∂_+W are disjoint and

$$\partial\partial_cW = \angle W = \partial(\partial_-W \cup \partial_+W),$$

we still call W a cobordism of ∂_-W to ∂_+W. We shall usually denote a cobordism by a single letter and often just call it a manifold. A picture of a cobordism is offered in Figure 5.1. For example, we usually regard a product $M \times I$ as a cobordism, with $\partial_-(M \times I) = M \times 0$, $\partial_+(M \times I) = M \times 1$; if M has boundary, write $\partial_c(M \times I) = \partial M \times I$.

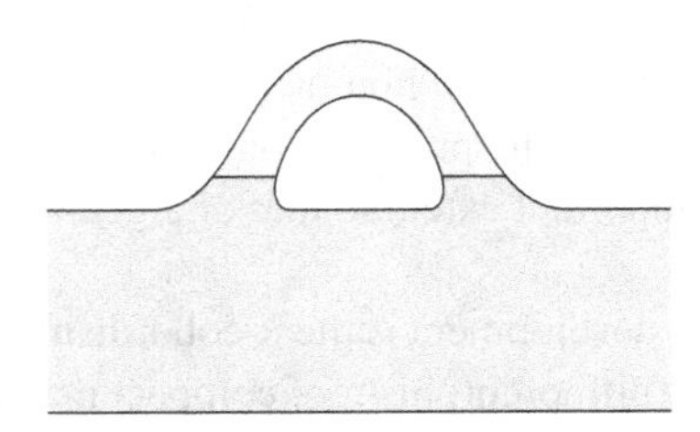

Figure 5.2 A handle

Suppose W^n a cobordism, $f : S^{r-1} \times D^{n-r} \to \partial_+W$ an embedding. Introduce a corner (as in Lemma 2.6.3) along $f(S^{r-1} \times S^{n-r-1})$. Now glue $D^r \times D^{n-r}$ to W by f. The result is unique up to diffeomorphism, and is denoted $W \cup_f h^r$; it has the same corners as W. We describe it as W *with an r-handle attached by* f. We call f the *attaching map* of the handle, and r the *dimension* of the handle. Figure 5.2 offers a picture of a handle. We define $\partial_+(W \cup h^r) = (\partial_+W \setminus \operatorname{Im} f) \cup (D^r \times S^{n-r-1})$.

If we have a sequence of attached handles:

$$W^* = W \cup_{f_1} h^{r_1} \cup \cdots \cup_{f_k} h^{r_k},$$

we describe this as a *handle presentation* of W^* on W; if the maps f_i are not specified, as a *handle decomposition*. In particular, if $W = M \times I$, we speak of a handle decomposition of W^* with *base* M (here, M may be empty).

To prove existence, we use non-degenerate functions.

Lemma 5.1.1 *Any cobordism W with $\partial_c W = \emptyset$ admits a non-degenerate function f, with all critical values distinct, attaining an absolute minimum on $\partial_- W$ only, and an absolute maximum on $\partial_+ W$ only. The same holds if $\partial_c W$ is a product $M \times I$.*

Proof Let $\partial_- W \times I$, $\partial_+ W \times I$ be disjoint collar neighbourhoods of $\partial_- W$ and $\partial_+ W$. Define $g : W \to [0, 1]$ by:

$$g(x, t) = \begin{cases} \frac{1}{3}t & \text{for } x \in \partial_- W, \\ 1 - \frac{1}{3}t & \text{for } x \in \partial_+ W, \end{cases} \tag{5.1.2}$$

and extend to a continuous function taking only values between $\frac{1}{3}$ and $\frac{2}{3}$ elsewhere: this is possible since W is normal. By Proposition 1.1.7, we can approximate g by a smooth function h, agreeing with g near ∂W. Now approximate h by a non-degenerate function f with distinct critical values, agreeing with h, and so g, near ∂W. This is possible by Proposition 4.5.10 since g and h have no critical points in a neighbourhood of ∂W.

For the case $\partial_c W = M \times I$ we use the same argument: here we use Proposition 1.5.6 to find the collars, and to ensure that along $\angle W$ they agree with the product structure on $\partial_c W$, and Proposition 1.1.7 allows us to suppose that h and hence f agree on $M \times I$ by the projection on I; the proof now concludes as above. □

The proof shows that we may suppose that for x close to ∂W, f is defined by the formula (5.1.2).

Now we give W a Riemannian structure adapted to the boundary; for convenience we suppose, as in Proposition 2.3.7 that it is a product metric in some neighbourhood of ∂W. Then the differential 1-form df induces at each $P \in W$ an element df_P of $T_P W^\vee$; using the Riemannian structure, this is identified with an element of $T_P W$, a tangent vector. Thus df gives a vector field, which we call ∇f.

In $\mathring{W}$, we can use Theorem 1.4.2 to integrate f and obtain a flow $\varphi_t(P)$, defined for certain values of (t, P). Near a point of $\partial_- W$, we can take coordinates $x_1, \ldots, x_n$ such that W is defined by $x_1 \geq 0$, x_1 is the t-coordinate in the tubular neighbourhood, so that $f(x) = x_1 - 1$ and the Riemannian structure is of the form $ds^2 = dx_1^2 + \sum_{i,j=2}^n g_{i,j} dx_i dx_j$. Hence ∇f agrees with $\partial/\partial x_1$ in such a neighbourhood, and orbits are of the form

$$\varphi_t(x_1, \ldots, x_n) = (x_1 + t, x_2, \ldots, x_n) \qquad x_1 \geq 0, x_1 + t \geq 0.$$

Thus for $P \in \partial_- W$, $\varphi_t(P)$ is defined for small positive values of t, and φ is defined on a neighbourhood of $\partial_- W \times 0$ in $\partial_- W \times \mathbb{R}^+$ and gives a collar neighbourhood of $\partial_- W$. Similarly for $\partial_+ W$.

If we regard $\varphi_t(P)$ as a point parametrised by t, it is smooth, and we have a metric, so can speak of speed.

Lemma 5.1.3 *We have (a)* $\frac{d}{dt} f(\varphi_t(P))|_{t=0} = \|df_P\|^2$,
(b) The speed of $\varphi_t(P)$ at $t = 0$ *is* $\|df_P\|$.

Proof (a)
$$\left.\frac{d}{dt} f(\varphi_t(P))\right|_{t=0} = \nabla f(f)|_P \quad \text{by definition of } \varphi$$
$$= df(\nabla f)|_P$$
$$= \langle df_P, df_P \rangle = \|df_P\|^2$$

in the Riemannian inner product on $T_P W$, since this defined ∇f.

(b) Take coordinates $(x_1, \ldots, x_m)$ at P, so that P has coordinates $(0, \ldots, 0)$ and at P the Riemannian metric agrees with the standard metric in $\mathbb{R}^n$. Let $df = \sum a_i dx_i$: then $\nabla f = \sum a_i \partial/\partial x_i$ (at P). Thus, at P, $\frac{\partial \varphi_t(P)}{\partial x_i} = a_i$, so the speed of $\varphi_t(P)$ is just $(\sum a_i^2)^{1/2} = \|df_P\|$. □

Now suppose $P \in \mathring{W}$, and that the maximum range of t in which $\varphi_t(P)$ is defined is (a, b).

Lemma 5.1.4 *Suppose W is compact. Then either*
a is finite and as $t \to a$, $\varphi_t(P)$ *tends to a point on* $\partial_- W$, *or*
$a = -\infty$ *and, for any K, the closure of* $\varphi_t^{-1}(-\infty, -K)$ *contains a critical point of f.*
Similarly for b.

Proof If a is finite, by Lemma 5.1.3 (b), the points $\varphi_t(P)$ form a Cauchy sequence as $t \to a$ (since W is compact, $\|df_P\|$ is bounded); since W is complete, they tend to a limit point Q. If Q was interior to W, it would follow that Q was on the orbit, which could then be extended: thus Q is on ∂W. Since by Lemma 5.1.3 (a), f increases along each orbit, $f(Q) < f(P)$, so Q is on $\partial_- W$.

Now let $a = -\infty$. Then by Lemma 5.1.3 (a),

$$\int_{-\infty}^{0} \|df_{\varphi_t(P)}\|^2 dt$$

converges. So $\|df_{\varphi_t(P)}\|$ has infimum zero as $t \to -\infty$. Outside any open neighbourhood of the set of critical points, $\|df\|$ is non-zero, and attains its lower bound (by compactness), so $\varphi_t(P)$ meets any such neighbourhood. But the set of critical points is compact, and so meets the closure of the orbit. □

We are now ready to analyse the function f of Lemma 5.1.1. For $a \in \mathbb{R}$, write

$$W_a = \{P \in W : f(P) \leq a\}$$
$$M_a = \{P \in W : f(P) = a\}$$

thus for

$a = 0$	$W_a = \partial_- W$	$M_a = \partial_- W$
$a = \varepsilon$	$W_a = \partial_- W \times [0, \varepsilon]$	$M_a = \partial_- W \times \varepsilon$
$a = 1 - \varepsilon$	$W_a = W \setminus (\partial_+ W \times [0, \varepsilon))$	$M_a = \partial_+ W \times \varepsilon$
$a = 1$	$W_a = W$	$M_a = \partial_+ W$

provided that ε is so small that $\partial_\eta W \times [0, \varepsilon]$ $(\eta = +, -)$ are contained in the collar neighbourhoods described earlier. Clearly, for $a < b$, $W_a \subset W_b$; we next investigate how W_b is formed from W_a.

Lemma 5.1.5 *Suppose that for $a \leq c \leq b$, c is not a critical value of f. Then $f^{-1}[a, b]$ is diffeomorphic to $M_a \times [a, b]$ and W_b is diffeomorphic to W_a.*

Observe that since a, b are not critical values, it follows from Lemma 4.5.1 that M_a, M_b and $f^{-1}[a, b]$ are submanifolds.

Proof The first assertion follows at once by applying Theorem 1.5.4 to the vector field ∇f.

Thus W_b is obtained from W_a by glueing on $M_a \times I$ along M_a. The result now follows from Lemma 2.7.2. □

This shows that 'as long as a does not pass through a critical value, the diffeomorphism type of W_a remains constant'. We now have to investigate the critical value.

Theorem 5.1.6 *Suppose that for $a \leq f(P) \leq b$ there is just one critical point Q, which is non-degenerate and with $f(Q) = c$ $(a < c < b)$. Then W_b is diffeomorphic to W_a with a handle attached.*

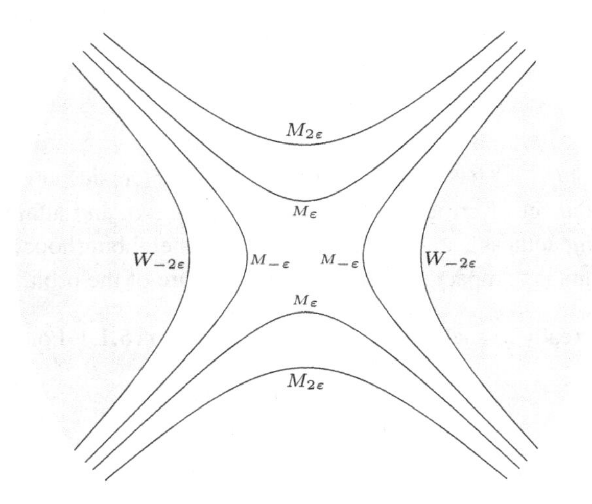

Figure 5.3 Level sets

Proof Our discussion of orbits in Theorem 1.5.4 remains valid except for those orbits with Q as a limit point. We must therefore investigate a neighbourhood of Q. By the Morse Lemma 4.8.2 there exist local coordinates $x_1, \ldots, x_n$ such that in a neighbourhood of Q f is given by

$$f(x) = c - x_1^2 - \cdots - x_\lambda^2 + x_{\lambda+1}^2 + \cdots + x_n^2.$$

The integer λ is called the *index* of f of the critical point 0. Using a partition of unity, we choose a Riemannian structure which agrees with the Euclidean structure in this coordinate system. With respect to this, the gradient vector field ∇f is given by

$$\nabla f = \sum_1^\lambda -x_i \frac{\partial}{\partial x_i} + \sum_{\lambda+1}^n x_i \frac{\partial}{\partial x_i}.$$

For example, if $f(x) = -x_1^2 + x_2^2$, the curves M_a are hyperbolae with asymptotes $x_1^2 = x_2^2$, except for M_0 which is this line-pair, and as a increases up to zero, W_a increases without essential change, but it engulfs the origin when $a = 0$. Figure 5.3 shows the evolution of M_a for $a = -2\epsilon,\ -\epsilon,\ \epsilon$, and 2ϵ.

Choose ε so small so that for $\|x\| \leq 5\varepsilon$, the above formulae are valid. We now modify $W_{-\varepsilon}$ by first introducing a corner, then attaching a handle, to obtain something close to W_ε: the procedure is illustrated in Figure 5.4.

More precisely, write $x = (\xi, \eta)$, where $\xi = (x_1, \ldots, x_\lambda)$, $\eta = (x_{\lambda+1}, \ldots, x_n)$, $f(x) = c - \|\xi\|^2 + \|\eta\|^2$. Define the handle H to be the set $\|\eta\| \leq \varepsilon$, $\|\xi\| \leq \varepsilon$. Let V be a smooth manifold with corner which

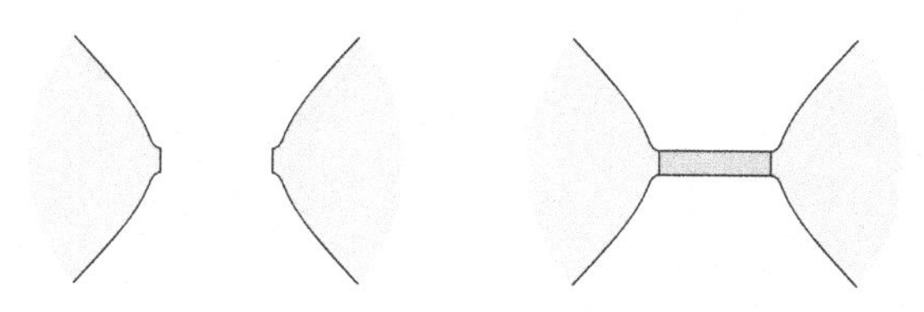

Figure 5.4 Attaching a handle

(i) coincides with $\|\xi\| \leq \varepsilon$, $\|\eta\| \geq \varepsilon$ near $\|\xi\| = \varepsilon$ (this includes the corner $\|\xi\| = \|\eta\| = \varepsilon$);

(ii) coincides with $W_{-\varepsilon}$ when $\|x\| \geq 5\varepsilon$, and contains $W_{-\varepsilon}$;

(iii) has ∂V everywhere transverse to the orbits of ∇f.

Such a V may be constructed using a bump function. Then by Proposition 2.6.4, $M_{-\varepsilon}$ is obtained from V by straightening the corner – or equivalently, by Lemma 2.6.3, V from $M_{-\varepsilon}$ by introducing one. Now H is diffeomorphic to the product $D^\lambda \times D^{n-\lambda}$, and $\partial_- H := H \cap V$ is given by $\|\xi\| = \varepsilon$, $\|\eta\| \leq \varepsilon$, hence is a copy of $S^{\lambda-1} \times D^{n-\lambda}$ in $\partial H \cap \partial V$. Since the union $H \cup V$ is smooth, and H and V are defined by cutting it along $H \cap V$, it follows by Proposition 2.7.3 that $H \cup V$ is obtained by glueing these.

Now $H \cup V$ is a smooth manifold, transverse to the orbits, with no critical points between it and M_b; thus it follows from Theorem 1.5.4 that W_b is diffeomorphic to $H \cup V$. But $H \cup V$ consists of W_a with a λ-handle attached. □

The following are immediate consequences.

Corollary 5.1.7 *(i) If the Hessian of f at c has index λ, we attach a λ-handle.*

(ii) If there are several non-degenerate critical points at level c, we attach several handles.

(iii) W has a handle decomposition based on $\partial_- W$,

for we can apply the above argument in a neighbourhood of each critical point.

With a little care, the arguments may also be applied to non-compact manifolds: we give one sample result.

Lemma 5.1.8 *Any manifold W can be expressed as the union of an infinite sequence of handles attached one at a time.*

Proof By Corollary 2.2.10 there is a proper map $f : W \to \mathbb{R}$; as before, we may suppose f smooth and non-degenerate. Then each set W_a is compact, and we may apply the same arguments as above. □

It is also possible to proceed in the opposite direction.

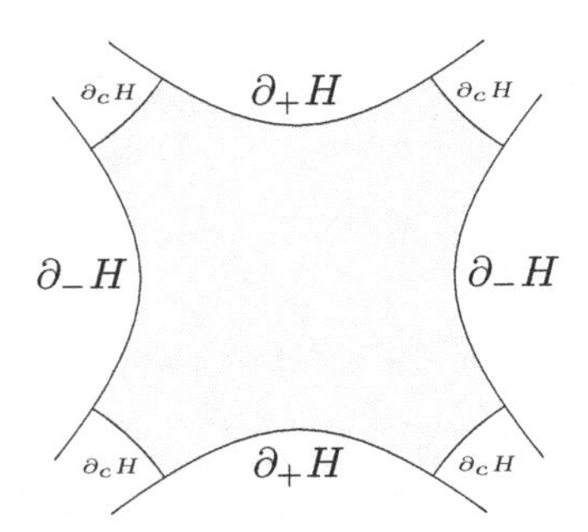

Figure 5.5 Alternative picture of a handle

Theorem 5.1.9 *Given a handle decomposition of W on $\partial_- W$, there is a non-degenerate function f on W (as in Lemma 5.1.1) with just one critical point of index λ for each λ-handle.*

Proof The result is proved by induction on the number of handles: if there are none, $W \cong \partial_- W \times I$, and we take f as the projection on I. Now let V be defined by attaching all but the last handle: by the induction hypothesis, f can be defined on V, constant on $\partial_+ V$. So if we can define f on $(\partial_+ V \times I) \cup h^\lambda$ we can glue back (using collar neighbourhoods of $\partial_+ V$ on which f reduces to a projection) to make f smooth. Hence it suffices to consider the case when W is formed from $\partial_- W \times I$ by attaching one handle.

Now let $g : S^{\lambda-1} \times D^{n-\lambda} \to \partial_- W$ be the attaching map of a λ-handle. Write K for the closure of the complement of the image. Write H for the subset of $\mathbb{R}^\lambda \times \mathbb{R}^{n-\lambda}$ defined by

$$-1 \leq -\|x\|^2 + \|y\|^2 \leq 1, \quad \|x\|^2 \|y\|^2 \leq 2.$$

Then the function defined on H by $F(x, y) = -\|x\|^2 + \|y\|^2$ attains its minimum value -1 on $\partial_- H$, say, and its maximum $+1$ on $\partial_+ H$. Write $\partial_c H$ for the subset given by $\|x\|^2 \|y\|^2 = 2$.

We have a diffeomorphism $G_- : \partial_- H \to S^{\lambda-1} \times D^{n-\lambda}$ given by $G_-(x, y) = \left(\frac{x}{\|x\|}, y\right)$: its inverse is given by $G_-^{-1}(u, v) = \left((1 + \|v\|^2)^{1/2} u, v\right)$. We also have a diffeomorphism $G_c : \partial_c H \to S^{\lambda-1} \times S^{n-\lambda-1} \times [-1, 1]$ given by $G_c(x, y) = \left(\frac{x}{\|x\|}, \frac{y}{\|y\|}, F(x, y)\right)$: its inverse is $G_c^{-1}(u, v, t) = (au, bv)$, where $b^2 - a^2 = t$ and $b^2 + a^2 = \sqrt{t^2 + 8}$. This description goes with the picture of a handle offered in Figure 5.5.

Now attach H to $K \times [-1, 1]$ by the map G_c to form a manifold W'. The function $f : W' \to [-1, 1]$ defined by $f|H = F$, $f|K =$ projection is a smooth function, whose only critical point is the non-degenerate one in H. We have a diffeomorphism h of $\partial_- W'$ to $\partial_- W$ given by the identity on $K \times -1$, and by $g \circ G_-$ on $\partial_- H$.

Finally, we have a diffeomorphism of W' on W. For each is obtained by attaching a λ-handle to the lower boundary: W by hypothesis and W' by Theorem 5.1.6. By construction, the attaching maps of the handles correspond under h, so the identity map of $\partial_- W$ extends to a diffeomorphism. □

5.2 Normalisation

We could proceed immediately to make various deductions about smooth manifolds from the existence of a handle decomposition. First, however, it is convenient to normalise a presentation. We have defined $W \cup_f h^r$ by attaching $D^r \times D^{n-r}$ to W using an embedding $f : S^{r-1} \times D^{n-r} \to M := \partial_+ W$; however it will usually be more convenient to regard the handle H as consisting of a collar $M \times I$ to which $D^r \times D^{n-r}$ is attached. The *attaching sphere* (or *a*-sphere) of H is the sphere $f(S^{r-1} \times 0)$ in $\partial_- H$. The *belt sphere* (or *b*-sphere) is the sphere $0 \times S^{n-r-1}$ in $\partial_+ H$. The *core* is the disc $D^r \times 0$.

It follows at once from Theorem 2.4.2 (diffeotopy extension) that $W \cup_f h^r$ is determined up to diffeomorphism by the diffeotopy class of f, for if g is a diffeomorphism of W, g induces a diffeomorphism of $W \cup_f h^r$ with $W \cup_{gf} h^r$. By Theorem 2.5.5 (tubular neighbourhood), it is even determined by the diffeotopy class of $\bar{f} = f|S^{r-1} \times 0$ together with a homotopy class of normal framing of $f(S^{r-1} \times 0)$ in $\partial_+ W$.

Lemma 5.2.1 *Let $r \leq s$. Then $(W \cup_f h^s) \cup_g h^r$ is diffeomorphic to manifolds obtained from W by attaching the handles simultaneously, or in the reverse order.*

Proof Let $n = \dim W$, $Q = \partial_+(W \cup_f h^s)$. Then we have in Q the a-sphere S^{r-1} of h^r and the b-sphere S^{n-s-1} of h^s. Since

$$(r-1) + (n-s-1) = n - 1 - (s + 1 - r) < n - 1 = \dim Q,$$

by Theorem 4.5.6, S^{r-1} may be approximated by a sphere not meeting S^{n-s-1}: by Proposition 4.4.4 if the approximation is close enough, we still have an imbedded sphere, diffeomorphic to the old one. By further diffeotopies, we may make S^{r-1} avoid the tubular neighbourhood $D^s \times S^{n-s-1}$ (using the diffeotopy extension theorem, and the fact that the tubular neighbourhood may be shrunk to avoid S^{r-1}) and shrink the tubular neighbourhood $S^{r-1} \times D^{n-r}$ so that, this, too, avoids $D^s \times S^{n-s-1}$. But now the attaching map of the r-handle is disjoint from the s-handle: its image lies in ∂W, and the handles may be added in either order. □

Corollary 5.2.2 *Any W has a handle decomposition on $\partial_- W$ with the handles arranged in increasing order of dimension.*

This follows at once by induction.

From now on we shall generally assume that handles have been arranged in order of increasing dimension: this is in some sense the usual case. Indeed, since we can always reduce to this case, a handle decomposition without this property carries extra information.

We now introduce the notation

$$W_{r+\frac{1}{2}} = (\partial_- W \times I) \cup \text{ all } s - \text{handles for } s \leq r,$$

where we use Lemma 5.2.1 and attach all r-handles simultaneously. Also set $M_{r+\frac{1}{2}} = \partial_+ W_{r+\frac{1}{2}}$.

This is related to our previous notation as follows. It follows from Lemma 5.2.1, in conjunction with the relation between handles and non-degenerate functions on W, that there exist non-degenerate functions f with the property that, for each critical point P of f of index λ, we have $f(P) = \lambda$. Such functions are called *self-indexing*. If we use a self-indexing function f on W, the two definitions of $W_{r+\frac{1}{2}}$ coincide.

In $M_{r+\frac{1}{2}}$ we have the a-spheres S^r of the $(r+1)$-handles and the b-spheres S^{n-r-1} of the r-handles, which have complementary dimensions. By Theorem 4.5.6, the embedding of a sphere S^r may be approximated by a map transverse to S^{n-r-1}, and if the approximation is close enough, we have merely altered the embedding by a diffeotopy. Since the dimensions are complementary, and the map transverse, intersections are isolated points; since S^r is compact, there are only finitely many. We can thus modify the presentation by a diffeotopy so that all these a-spheres are transverse to all these b-spheres. We will say that a presentation with this property is in normal position. We have shown

Lemma 5.2.3 *Any handle presentation of $(W, \partial_- W)$ may be modified by diffeotopies so that the handles are arranged in increasing order of dimension, and any two handles of consecutive dimensions are in normal position.*

In this situation, for each such transverse intersection P of S^r with S^{n-r-1}, it follows from Lemma 4.8.1 that there is a chart for $M_{r+\frac{1}{2}}$ meeting the a-sphere in $D^r \times \{0\}$ and the b-sphere in $\{0\} \times D^{n-r-1}$. Since the tubular neighbourhoods are unique up to diffeotopy, we may suppose that they both meet this chart in $D^r \times D^{n-r-1}$, with the projections being those on the factors.

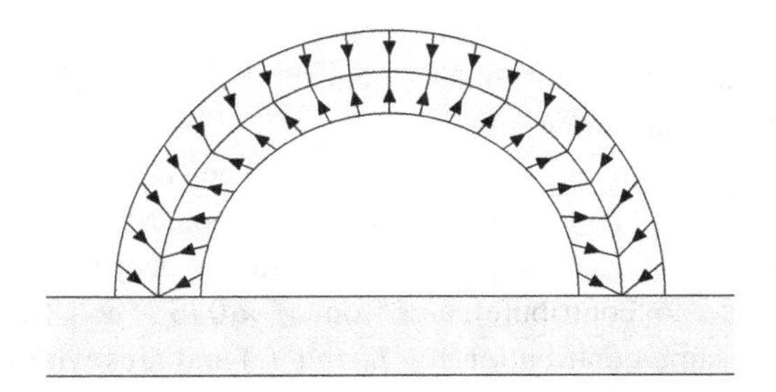

Figure 5.6 Retracting a handle on its core

5.3 Homology of handles and manifolds

For each r—handle attached to W, using a deformation retraction of $D^r \times D^{n-r}$ on $(S^{r-1} \times D^{n-r}) \cup (D^r \times \{0\})$ (which may be obtained from a deformation retraction of $I \times I$ on $(\{1\} \times I) \cup (I \times \{0\})$ by rotating about both axes), we have a deformation retraction of $W \cup_f h^r = W \cup_f (D^r \times D^{n-r})$ on $W \cup_f (D^r \times \{0\})$. Thus, up to homotopy, attaching a handle is the same as attaching a cell (its core). The deformation retraction is pictured in Figure 5.6.

This gives a very close connection between handle decompositions and cell complexes. In particular, we deduce the following from Corollary 5.2.2.

Proposition 5.3.1 *If W is closed, it has the homotopy type of a finite CW complex. In general, $(W, \partial_- W)$ has the homotopy type of a finite CW pair.*

Proof The first statement follows by taking a normalised handle decomposition of W and replacing each handle by an equivalent cell. In fact it is not difficult to show that W is homeomorphic to an appropriate finite CW complex.

For the second statement, note that by the first, we can regard $\partial_- W$ as a finite cell complex, and again apply Corollary 5.2.2. □

Before continuing, it is convenient to recall some basic results about the homology of manifolds: we focus on the simplest case when M^m is closed, connected, and oriented. Then $H_m(M; \mathbb{Z})$ is infinite cyclic (this is a special case of the Poincaré duality Theorem 5.3.5 below). If M is triangulated and we use simplicial homology, a generator is represented by the sum of the m-simplices, where each must be given the orientation induced by that of M. We denote this generator by $[M]$.

A map $f : M \to N$, where M and N are both closed, oriented, and connected has degree d, where d is the integer such that $f_*[M] = d[N]$. The degree may be determined as follows. Suppose f smooth and transverse to a point $Q \in N$. Then $f^{-1}(Q)$ is a finite set of points P_i and (by Lemma 4.5.1) for each i, the tangent map induces an isomorphism $T_{P_i}M \to T_Q N$. Set $\varepsilon_i = \pm 1$ according as this map preserves the given orientations or not. Then $d = \sum_i \varepsilon_i$.

We can see directly that this is independent of choices: a homotopy F of f_0 to f_1 may be made transverse to Q, and the preimage of Q is then a collection of loops (which do not contribute), arcs from $M \times 0$ to $M \times 1$ (whose two end points make the same contribution for f_0 and f_1) and arcs with both ends on $M \times 0$ (or on $M \times 1$) (whose two end points contribute opposite signs ε, so cancel each other).

To see that $d = \sum_i \varepsilon_i$, choose a disc neighbourhood D of Q such that its preimage consists of discs D_i round the P_i each mapped by a diffeomorphism. Inclusion induces isomorphisms $H_m(D, \partial D) \to H_m(N, N \setminus \mathring{D} \to H_m(N)$. In M there are similar isomorphisms, but now $H_m(M, M \setminus f^{-1}(\mathring{D}))$ is the direct sum of the $H_m(M, M \setminus \mathring{D}_i))$, and the result follows by adding up the local contributions.

The same considerations apply to intersection numbers: again we describe only the simplest case when we have compact oriented submanifolds V_1, V_2 of the oriented manifold M, of dimensions v_1 and v_2 with $m = v_1 + v_2$. At a point P where V_1 and V_2 intersect transversely we define a local intersection number $\varepsilon(P)$ to be ± 1 according as a base tor $T_P V_1$ giving the chosen orientation of V_1, followed by a corresponding base for $T_P V_2$ defines the given orientation of M or not. If V_1 and V_2 meet transversely everywhere, $\sum_{P \in V_1 \cap V_2} \epsilon(P)$ gives the intersection number $V_1.V_2$. Again, arguing by making a homotopy transverse, we see that this depends at most on the diffeotopy classes of V_1 and V_2.

Each compact oriented submanifold V^v of M^m defines a homology class $i_*[V] \in H_v(M : \mathbb{Z})$ and hence by duality 5.3.5 a cohomology class in $H^{m-v}(M; \mathbb{Z})$, which we temporarily denote by $\{V\}$. In the situation of the preceding paragraph, the cup product $\{V_1\}\{V_2\}$ is equal to $V_1.V_2$ times the class $\{P\}$ of a point. More generally we can see that if V_1 and V_2 (still closed and oriented, with $v_1 + v_2 \geq m$) intersect transversely along a submanifold W, then $\{V_1\}\{V_2\} = \{W\}$. This principle extends in a natural way (subject to appropriate technical conditions) when we allow boundaries and cease to require orientations.

It follows from the remark preceding Proposition 5.3.1 that, up to homotopy, we may replace handles by cells, and may calculate homology using the chain groups

$$C_r(W, \partial_- W) = \oplus\, \mathbb{Z},$$

where the summands are indexed by the r-handles. We will denote by α_r the number of r-handles, equal to the rank of $C_r(W, \partial_- W)$. We need to calculate the boundary homomorphism

$$\partial : C_{r+1}(W, \partial_- W) \to C_r(W, \partial_- W).$$

This is determined by incidence numbers, one for each r- and each $(r+1)$-handle.

Lemma 5.3.2 *The incidence number of handles h^{r+1} and h^r equals the intersection number in $M_{r+\frac{1}{2}}$ of the a-sphere of h^{r+1} and the b-sphere of h^r.*

Proof We need some care with signs: a choice of orientation of the cell $(D^{r+1} \times 0)$ in the cell complex induces orientations of the bounding a-sphere S^r and of the *normal bundle* of the corresponding b-sphere. If an a-sphere S^r and a b-sphere S^{n-r-1} meet transversely at a point, we take the sign $+$ or $-$ according as the orientation of S^r does or does not agree with that in the normal bundle of S^{n-r-1}. If W (and hence M) is oriented, orienting the normal bundle of a b-sphere is equivalent to orienting the sphere, and we can count multiplicities in the usual way.

We may suppose that S^r meets S^{n-r-1} transversely: then the intersection number agrees with the (local) degree of the projection of S^r on the normal disc D^r. But this degree coincides with the incidence number in the cell complex. □

If F is a field of coefficients (for example, $\mathbb{Q}$ or $\mathbb{Z}_2$), we define the Betti numbers β_i (strictly, $\beta_i(W, \partial_- W; F)$) as the ranks of the F-vector spaces $H_i(W, \partial_- W; F)$. Since these may be calculated from the chain groups

$$C_i(W, \partial_- W; F) := C_i(W, \partial_- W) \otimes F,$$

which have ranks α_i, we have

Lemma 5.3.3 (Morse inequalities) *We have*

$$\sum_0^n (-1)^i \alpha_i = \sum_0^n (-1)^i \beta_i$$

and, for each $0 \le j \le n$,

$$\sum_0^j (-1)^{j-i} \alpha_i \ge \sum_0^j (-1)^{j-i} \beta_i.$$

Proof Write r_i for the rank of the boundary map

$$C_i(W, \partial_- W; F) \to C_{i-1}(W, \partial_- W; F).$$

The definition of homology gives $\alpha_i = r_{i+1} + \beta_i + r_i$. Hence

$$\sum_0^j (-1)^{j-i}\alpha_i = r_{j+1} + \sum_0^j (-1)^{j-i}\beta_i. \qquad \square$$

We now discuss duality. Observe that with f, $-f$ is also non-degenerate. Its critical points coincide with those of f, but if f has index λ at 0, it has locally the form

$$f(x) = c - x_1^2 - \cdots - x_\lambda^2 + x_{\lambda+1}^2 + \cdots + x_n^2$$

and $-f$ has index $n - \lambda$. Using the correspondence (Theorems 5.1.6 and 5.1.9) between non-degenerate functions and handle decompositions, we find the following.

Proposition 5.3.4 *Suppose W has a handle decomposition on $\partial_- W$ with α_r r-handles for $0 \le r \le n$. Then it also has one on $\partial_+ W$, with α_r $(n-r)$-handles.*

If we ignore corners, we may identify the handles in the two cases, and observe that in the reversal, a- and b-spheres are interchanged.

Theorem 5.3.5 (Lefschetz Duality Theorem)
Suppose either that W is orientable or we use $\mathbb{Z}_2$ for coefficients: then we have isomorphisms $H_r(W, \partial_- W) \cong H^{n-r}(W, \partial_+ W)$.
In particular, $H_r(W) \cong H^{n-r}(W, \partial W)$ and $H^r(W) \cong H_{n-r}(W, \partial W)$.
If $\partial W = \varnothing$, then (Poincaré Duality) $H_r(W) \cong H^{n-r}(W)$.

Proof By Proposition 5.3.4 we can identify the chain groups of $(W, \partial_- W)$ with the chain or cochain groups of $(W, \partial_+ W)$. By Lemma 5.3.2 the incidence numbers are the same up to sign (only a-spheres and b-spheres are interchanged) and the isomorphism identifies the one boundary with the other coboundary. $\square$

The proof above is reminiscent of the earliest proofs of the result (see, for example, the account in [84]), but of course is only valid for compact smooth manifolds.

As a special case of homology groups, we mention connectivity. We retain the notation of Lemma 5.3.2. The a-sphere S^{-1} of a 0-handle is the empty set; in fact a 0-handle consists precisely of an n-disc, disjoint from $\partial_- W \times I$. The a-sphere S^0 of a 1-handle is a pair of points: these may or may not be in the same component of $W_{1/2}$. If not, the 1-handle connects the two components; but if they are, the corresponding handle does not affect connectivity.

If $\partial_- W$ is non-orientable then so, of course, is W. If, however, $\partial_- W$ is orientable, so is $W_{\frac{1}{2}}$, since adding a disjoint set of discs has no effect. Nor does adding a set of 1-handles which connect different components of $W_{\frac{1}{2}}$ (here we think of 1-handles as being added in turn, not simultaneously). However, the attaching map for a 1-handle is a map of $S^0 \times D^{n-1}$ – i.e. of a pair of discs. If these are mapped into the same component of $W_{\frac{1}{2}}$ with opposite orientations, then the orientation of $W_{\frac{1}{2}}$ can be extended over the handle; but if with the same orientation, $W_{1\frac{1}{2}}$ is non-orientable. Thus if, say, $W_{\frac{1}{2}}$ is connected and orientable, we may speak of orientable and of non-orientable 1-handles. Now r-handles for $r \neq 1$ do not affect orientability; for they introduce no new (potentially orientation-reversing) elements of the fundamental group.

For a 1-handle with both ends in the same component of $W_{\frac{1}{2}}$, we can deform both components of $S^0 \times D^{n-1}$ into a disc in $M_{\frac{1}{2}}$: as for the Disc Theorem, the diffeotopy class is determined by the orientations. Attaching an orientable 1-handle to D^n gives $S^1 \times D^{n-1}$, so we have $W_{1\frac{1}{2}} = W_{\frac{1}{2}} + (S^1 \times D^{n-1})$. In the non-orientable case, we have the sum with a non-orientable bundle over S^1 with fibre D^{n-1}.

5.4 Modifying decompositions

In this section we discuss several modifications that can be made to handle decompositions. We will see that (under suitable hypotheses) any elementary change of the chain complex $C_*(W, \partial_- W)$ can be effected by a change in the handle decomposition. The basic moves are introduction or cancellation of a complementary pair of handles, and addition of handles. We suppose throughout that W is a compact manifold, perhaps with boundary.

The results are simplest for 0-handles. If W has α_i i-handles, then $W_{\frac{1}{2}} \cong (\partial_- W \times I) \cup_{\alpha_0} D^n$. Attaching a 1-handle affects connectivity only if its a-sphere S^0 has the two points in different components of $W_{1/2}$.

Suppose that W is connected: since r-handles for $r \geq 2$ do not affect connectedness, $W_{\frac{1}{2}}$ is connected. Rearrange the 1-handles (Lemma 5.2.1) such that the first few each connect different components of $W_{\frac{1}{2}}$. For each of these, we have two manifolds with boundary, and a disc imbedded in the boundary of each. Attaching $D^{n-1} \times I$ is the same (§2.7) as glueing along the $(n-1)$-discs, i.e. forming the boundary sum. Moreover, by Proposition 2.7.6, for any manifold N^n, $N^n + D^n \cong N^n$. So the 0-handles are just cancelled out, and the remaining components of $\partial_- W \times I$ added together. We observe that each use of $N^n + D^n \cong N^n$ to simplify the decomposition removes just one 0-handle and one 1-handle.

We have shown

Proposition 5.4.1 *A connected manifold W admits a handle presentation of the following kind.*

If $\partial_- W = \varnothing$, there is just one 0-handle D^n.

If $\partial_- W$ has components $M_{(i)}$, $1 \leq i \leq k$, there are no 0-handles, then $(k-1)$ 1-handles connecting the components to give $(M_{(1)} \times I) + \cdots + (M_{(k)} \times I)$, then a further number of 1-handles.

We turn to cancellation of handles in general, and first describe a model.

Lemma 5.4.2 *Let $\varphi : D^{n-r-1} \to D^{n-r}$ be the embedding, by stereographic projection from $(0, \ldots, 0, -1)$ on the boundary of the upper hemisphere. Then $(S^r \times D^{n-r}) \cup_{1\times\varphi} h^{r+1} \cong D^n$.*

Proof If we attach D^{r+1} along the boundary to $S^r \times I$, we clearly have another $(r+1)$-disc. Multiplying by D^{n-r-1} shows that there exists a homeomorphism of the desired type. However to obtain a result up to diffeomorphism requires care with rounding corners systematically.

We first give the proof for $r = 0$, $n = 2$. Let E be the ellipse $\frac{1}{2}x^2 + y^2 = 1$ and H the confocal hyperbola $2x^2 - 2y^2 = 1$. Write Int and Ext for the (closed) interior and exterior regions of E. We shall show that $\operatorname{Int} E \cap \operatorname{Ext} H$ is obtained from $S^0 \times D^2$ by introducing a corner along $S^0 \times D^1$; that $\operatorname{Int} E \cap \operatorname{Ext} H$ is diffeomorphic to $D^1 \times D^1$, and that the attaching map $1 \times \varphi$ becomes the identity. It follows that the required manifold is diffeomorphic to $\operatorname{Int} E$, which is diffeomorphic to D^2 by $(x, y) \mapsto (2^{-1/2}x, y)$. Now E meets H at $(\pm 1, \pm 1/\sqrt{2})$. Consider the component of $\operatorname{Int} E \cap \operatorname{Ext} H$ in $x > 0$; it has the focus $(1, 0)$ as interior point.

Rays through the focus define a vector field everywhere transverse to the boundary, which may therefore be used for straightening the corner. A smooth cross-section is given by $(x-1)^2 + y^2 = 1/4$, which meets the rays through the corner in $(1, \pm 1/2)$. Thus the disc component is obtained from a disc by introducing corners at opposite ends of a diameter, as stated. It may be helpful to imagine these constructions using Figure 5.7.

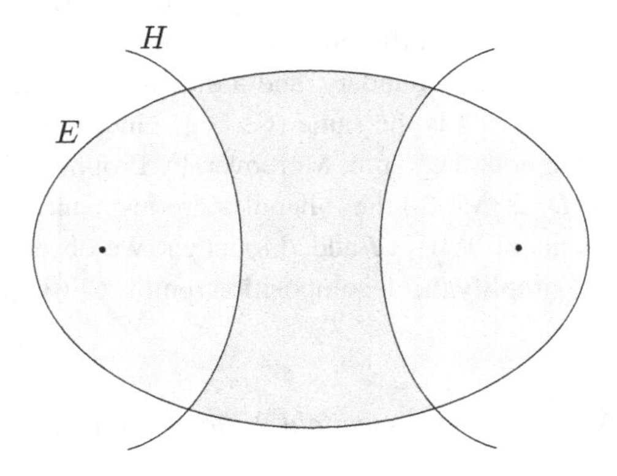

Figure 5.7 A confocal ellipse and hyperbola

In Int $E \cap$ Ext H we use confocal coordinates. Each point (x, y) of the plane with $xy \neq 0$ lies on just two of the conics

$$x^2/(\lambda+1) + y^2/\lambda = 1 :$$

one hyperbola, given by $-1 < \lambda_1 < 0$, and one ellipse, given by $0 < \lambda_2$. However, these meet in 4 points. So we write $\mu^2 = a + \lambda_1$, $\nu^2 = \lambda_2$, and obtain

$$x = \mu\sqrt{1+\nu^2} \quad y = \nu\sqrt{1-\mu^2},$$

where the positive square roots are to be taken, and $-1 < \mu < 1$. It is easy to verify that this transformation is smooth, with non-zero Jacobian, injective, and onto the whole plane except for $y = 0$, $x^2 \geq 1$. Hence, in particular, it induces a diffeomorphism of the rectangle $|\mu| \leq 1/\sqrt{2}$, $|\nu| \leq 1$ onto Int $E \cap$ Ext H, as required.

Now return to the case of general r and n, which is obtained by rotating the figures about x- and y-axes. Write

$$x = (x_1, \ldots, x_{r+1}) \qquad y = (y_1, \ldots, y_{n-r-1})$$
$$\mu = (\mu_1, \ldots, \mu_{r+1}) \qquad \nu = (\nu_1, \ldots, \nu_{n-r-1})$$

and $\|x\|^2 = \sum_1^{r+1} x_i^2$, etc. Then the transformation given by

$$x_i = \mu_i\sqrt{1+\|\nu\|^2}, \qquad y_i = \nu_i\sqrt{1-\|\mu\|^2}$$

induces a diffeomorphism of the $D^{r+1} \times D^{n-r-1}$ given by $\|\mu\|^2 \leq 1/2$, $\|\nu\|^2 \leq 1$ onto the intersection $\frac{1}{2}\|x\|^2 + \|y\|^2 \leq 1$, $2\|x\|^2 - 2\|y\|^2 \leq 1$.

Likewise in the intersection $\frac{1}{2}\|x\|^2 + \|y\|^2 \leq 1$, $2\|x\|^2 - 2\|y\|^2 \leq 1$, consider the field formed by rays through the r-sphere $y = 0$, $\|x\| = 1$ and perpendicular to it (and not produced beyond their intersection with $x = 0$). This certainly is a vector field (except on the sphere and on $x = 0$), and is transverse to the boundary, so can be used for rounding the corner. Rounding it, we obtain the manifold

$$(\|x\| - 1)^2 + \|y\|^2 \leq 1/4,$$

where the corner is to be introduced along $\|x\| = 1$, $\|y\| = 1/2$ (in fact $S^r \times S^{n-r-2}$).

Consider $S^r \times D^{n-r} \subset \mathbb{R}^{r+1} \times \mathbb{R}^{n-r-1} \times \mathbb{R}^1$ with coordinates (u, w, t), so $\|u\| = 1$, $\|w\|^2 + |t|^2 \leq 1$. We define inverse diffeomorphisms between this and the manifold above by

$$u = x/\|x\| \qquad w = 2y \qquad t = 2(\|x\| - 1)$$
$$x = u(1 + t/2) \qquad y = w/2.$$

Since $\|x\|$ is nowhere zero, both it and its inverse are smooth. The corner $\|x\| = 1$, $\|y\| = 1/2$ becomes the locus $\|w\| = 1, t = 0$.

Finally we must identify the attaching map. The sphere $S^r \times 0$ given by $\|\mu\|^2 = 1/2, \nu = 0$ maps (via $x_i = \mu_i$) to $\|x\|^2 = 1/2, y = 0$, then rounding the corner multiplies x_i by $2^{-1/2}$ and leaves y at 0. Finally we obtain $u = x/\|x\| = \mu/\|\mu\|$ and $v = (w, t) = (0, -1)$; modulo the obvious identifications, we have the identity map. The attaching map is a tubular neighbourhood of this, and a normal direction $\partial/\partial\nu_i$ maps to some positive multiple of $\partial/\partial v_i$; using the tubular neighbourhood theorem, it follows that the attaching map is, up to a diffeotopy, as stated. □

Theorem 5.4.3 (Handle Cancellation Theorem) *Suppose that for $W^n \cup_f h^r \cup_g h^{r+1}$, the a-sphere of h^{r+1} meets the b-sphere of h^r transversely in one point. Then we can suppose $\partial_+ W$ contains a disc D^{n-1} to which both handles are added. Thus we can write $W^n \cong W^n + D^n$, with the handles added to D^n, and so $W^n \cup h^r \cup h^{r+1} \cong W^n + (D^n \cup h^r \cup h^{r+1}) \cong W^n + D^n \cong W^n$.*

Proof It clearly suffices to consider the case $W = M \times I$. By hypothesis, in $M_{r+\frac{1}{2}}$ the a-sphere and b-sphere of the handles meet transversely at a single point P. It follows from Lemma 4.8.1 that there is a chart for $M_{r+\frac{1}{2}}$ at P meeting the a-sphere in $D^r \times \{0\}$ and the b-sphere in $\{0\} \times D^{n-r-1}$. Since the tubular neighbourhoods are unique up to diffeotopy, we may suppose that they both meet this chart in $D^r \times D^{n-r-1}$, with the projections being those on the factors.

The r-handle is attached to M by an embedding $f : S^{r-1} \times D^{n-r} \to M$; W is formed from $M \times I$ by attaching $D^r \times D^{n-r}$ to $M \times \{1\}$ by f and rounding the corner. Here Figure 5.8 represents an r-handle with an $(r + 1)$-handle being sewn on as a patch.

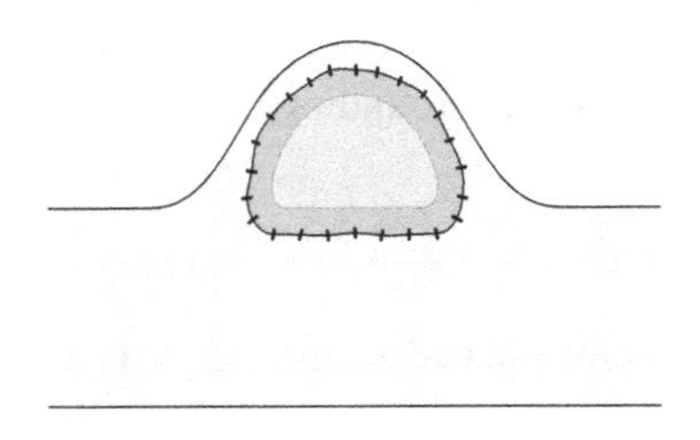

Figure 5.8 Cancelling a handle

The $(r + 1)$ handle is attached by an embedding $g : S^r \times D^{n-r-1} \to M_{r+\frac{1}{2}}$, so there is an embedding of D^r as a hemisphere of S^r, which we may take as S^r_+: thus $S^r_+ \times D^{n-r-1}$ maps onto the outer edge of the r-handle. The closed complementary region $S^r_- \times D^{n-r-1}$ is mapped to the closed complement in M

of the image of f. As before, we may choose the maps to identify D^{n-r-1} with a hemisphere S_+^{n-r-1}.

Thus the subset of M affected by the handles is the union of the embedded images $f(S^{r-1} \times D^{n-r})$ and $g(S_-^r \times D^{n-r-1})$, modulo rounding the corner. The latter image is a disc; since we can isotope D^{n-r} inside a neighbourhood of a point $X \in S_+^{n-r-1}$ and hence $S^{r-1} \times D^{n-r}$ inside a neighbourhood of $S^{r-1} \times X$, there is a disc in M containing both the embedded images. The result now follows from Lemma 5.4.2. □

A pair of handles of consecutive dimensions, with the a-sphere of the second meeting the b-sphere of the first transversely in one point, is called a *complementary pair*.

We can thus paraphrase Theorem 5.4.3 briefly by saying that a complementary pair of handles may always be cancelled. The converse result is now trivial.

Theorem 5.4.4 *At any point of a handle decomposition of a manifold, a complementary pair of handles can be introduced.*

Proof 'At any point' means when we have constructed some manifold W, say. Now $W \cong W + D$ by Proposition 2.7.5 and by Lemma 5.4.2, we can add a complementary pair of handles to D, hence also to W. □

We will see that adding two complementary handles in succession to W has the effect on $V = \partial_+ W$ of performing consecutively spherical modifications of types $(r, n-r)$, leading to V', say, and $(r+1, n-r-1)$: returning to V. 'Reversing' the second of these shows that we can also go from V to V' by a modification of type $(n-r-1, r+1)$. The condition on the first modification necessary for this replacement to be possible was the existence of a complementary handle; arguing as above shows that this is equivalent to requiring the a-sphere to span a disc in V, such that the inward normal vector to the sphere in the disc agrees with the first vector of the chosen normal framing of the a-sphere.

Since $\partial_- W$ need not be simply-connected, an $(r-1)$-sphere in it does not necessarily have a well-defined homotopy class. Here we will ignore this point and focus on homology. This allows us to give a much simpler account, and still obtain full results in the simply-connected case. We reserve comments on the general case until §5.7.

We next discuss 'addition' of handles in a homology sense.

Theorem 5.4.5 (Handle Addition Theorem) *Suppose $\partial_+ W = M$ connected, $2 \le r \le m-2$. Let $f, g : \partial D^r \times D^{m-r} \to M$ be disjoint embeddings, determining homology classes x, $y \in H_{r-1}(M; \mathbb{Z})$. Then for $\varepsilon = \pm 1$ there is an* *embedding* $h_\varepsilon : \partial D^r \times D^{m-r} \to M$, *disjoint from* f, *and determining* $y + \varepsilon x \in H_{r-1}(M; \mathbb{Z})$ *such that* $W \cup_f h^r \cup_g h^r \cong W \cup_f h^r \cup_{h_\varepsilon} h^r$.

Moreover, if the classes of the handles in $H_r(W \cup_f h^r \cup_g h^r, W; \mathbb{Z})$ *are* ξ, η *for the first decomposition, those for the second are* ξ, $\eta + \varepsilon\xi$.

Proof We observe that x maps to zero in $H_{r-1}(W \cup_f h^r; \mathbb{Z})$; the idea of the proof is to deform the second handle 'across' the first, by a diffeotopy of the attaching map in $\partial_+(M \cup_f h^r)$; we know that this will not affect the diffeomorphism class of the result.

Since M is connected, there is a path λ joining $f(1 \times 1)$ and $g(1 \times 1)$ (in the non-simply-connected case, it is important to note that this path may be taken in any homotopy class). By the general position arguments of §4.5, we can make the path an embedding, disjoint from the images of $\bar{f}$ and $\bar{g}$; we can choose it to start along the outward normals to $\operatorname{Im} f$ and $\operatorname{Im} g$, and we can deform it off tubular neighbourhoods of $\operatorname{Im} \bar{f}$ and $\operatorname{Im} \bar{g}$, so that it meets $\operatorname{Im} f$ and $\operatorname{Im} g$ only at its ends.

Choose a normal framing $e_1, \ldots, e_{m-2}$ for λ so that $e_1, \ldots, e_{r-1}$ gives the standard orientation of $g(S^{r-1} \times 1)$ at $g(1 \times 1)$. Since $r \leq m - 2$, we can also change this framing so that $e_1, \ldots, e_{r-1}$ agrees with the opposite orientation of the $(r-1)$-sphere. By Proposition 1.5.6 (ii) we can choose a Riemannian metric in which $f(S^{r-1} \times 1)$ and $g(S^{r-1} \times 1)$ are totally geodesic. Then exponentiating normal vectors to λ gives an embedding $\varphi' : I \times D^{r-1} \to M$ with

$$\varphi'(0 \times D^{r-1}) \subset g(S^{r-1} \times 1), \qquad \varphi'(1 \times D^{r-1}) \subset f(S^{r-1} \times 1).$$

Extend λ by a diameter of $D^{r-1} \times 1$ in $\partial_+(M \cup_f h^r)$, and φ' correspondingly to an embedding $\varphi : [0, 2] \times D^{r-1} \to \partial_+(M \cup_f h^r)$.

The properties of the bump function ensure that the formulae

$$\begin{aligned} \bar{g}_t(x) &= x \quad \text{if} \quad x \notin \varphi(0 \times D^{r-1}), \\ \bar{g}_t\varphi(0, y) &= \varphi(2tBp(1 - \|y\|), y) \end{aligned}$$

fit to give a smooth diffeotopy of $\bar{g}$. This 'pulls' the cell $\varphi(0 \times D^{r-1}) \subset g(S^{r-1} \times 1)$ across part of the disc $D^r \times 1$, covering the central point.

This procedure (with $r = 1$) is illustrated in Figure 5.9: here, to add the handle, deform the attaching sphere of the left handle along the dotted path to give a new attaching sphere.

Since $g(S^{r-1} \times 0)$ is diffeotopic to $g(S^{r-1} \times 1)$, we also obtain a diffeotopy of $\bar{g}$, which we can extend to one of g such that the final embedding h is disjoint from $0 \times S^{n-r-1}$. But we can think of the (f-) handle as shrunk to a small neighbourhood of this b-sphere (c.f. proof of 5.2.3), so $h(S^{r-1} \times D^{n-r})$ lies in M again.

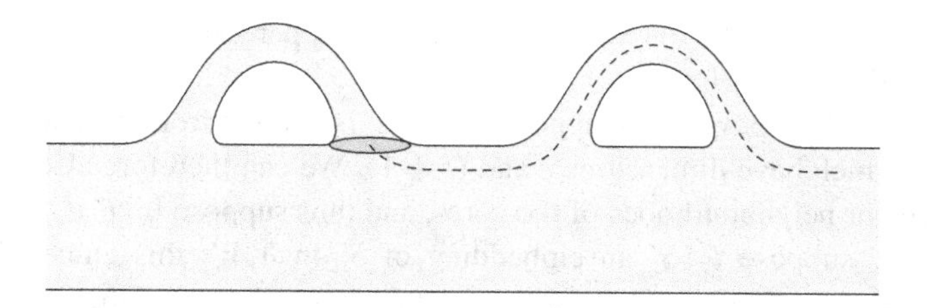

Figure 5.9 Adding one handle to another

Since our diffeotopy has degree 1 on the attached cell, the homology class of h is that of g plus or minus that of f, the sign depending on an orientation chosen earlier. □

5.5 Geometric connectivity and the h-cobordism theorem

In the last section we gave methods of changing handle decompositions under geometric assumptions. We now obtain corresponding results under algebraic hypotheses: this will enable us to operate with handles using only homotopy data. We recall that a CW-pair (Y, X) is called r-connected if any map $f : (B, A) \to (Y, X)$ with $\dim(B) \leq r$ is homotopic relative to A to a map into X; equivalently, if the relative homotopy set $\pi_i(Y, X)$ is trivial for $0 \leq i \leq r$. Moreover this holds if and only if the pair (Y, X) is homotopy equivalent to a pair with Y' obtained from X by attaching cells of dimension $> r$.

We focus first on results showing the existence of handle decompositions without i-handles for $i \leq r$: if W admits such a decomposition, we say that $(W, \partial_- W)$ is *geometrically r-connected.*

We start with a technique of handle replacement. It is interesting to note that this closely resembles a technique of Whitehead, with CW complexes. Although it may seem that it would be more efficient to simply cancel handles, handle replacement bypasses arguments involving fundamental groups, which otherwise would confuse the issue for low-dimensional cases.

Proposition 5.5.1 *Suppose $n \geq 2r + 3$, $W^n = (M \times I) \cup h^r \cup lh^{r+1}$, and $\pi_r(W, M) = 0$. Then $W \cong (M \times I) \cup lh^{r+1} \cup h^{r+2}$.*

Proof The case $r = 0$ follows from Proposition 5.4.1; otherwise we may suppose M connected.

We identify h^r with $D^r \times D^{m-r}$. Since $n \geq 2r + 2$, we can perform a diffeotopy to ensure that the attaching maps of the h^{r+1} avoid $D^r \times 1$. The disc $D^r \times 1$ determines an element of $\pi_r(W, M)$, which is zero by the hypothesis. Hence this disc is homotopic in W (relative to its boundary) to one in M; i.e.

there is a map $F : D^{r+1} \to W$, which takes the upper hemisphere of S^r onto $D^r \times 1$ and the lower into M.

Since $n \geq 2r+3$, we may suppose that $\operatorname{Im} F$ is disjoint from the cores of the handles, which have dimensions r and $(r+1)$. We can therefore also deform F off tubular neighbourhoods of the cores, and thus suppose $\operatorname{Im} F \subset \partial_+ W$.

We may suppose $F \mid S^r$ an embedding of S^r in $\partial_+ W$: this embedding is homotopic to zero, hence also diffeotopic (since $n \geq 2r+3$, a map $S^r \times I \to \partial_+ W \times I$ may be supposed an embedding). So by Theorem 5.4.4, we can use $F(S^r)$ for the a-sphere of the first of a complementary pair of handles h_A^{r+1}, h_B^{r+2}, where h_A^{r+1} is disjoint from the other h^{r+1}. But h_A^{r+1} is also complementary to h^r, so

$$\begin{aligned} W &\cong (M \times I) \cup h^r \cup lh^{r+1} \cup (h_A^{r+1} \cup h_B^{r+2}) && \text{(Theorem 5.4.4)} \\ &\cong (M \times I) \cup (h^r \cup h_A^{r+1}) \cup lh^{r+1} \cup h_B^{r+2} \\ &\cong (M \times I) \cup lh^{r+1} \cup h_B^{r+2} && \text{(Theorem 5.4.3).} \end{aligned}$$ □

It is possible, with some difficulty, to sharpen the proof of Theorem 5.5.1 to cover also the case $n = 2r+2$, $r \neq 1$: the points to be addressed are the deformation of F off the cores of the h^{r+1} and obtaining a diffeotopy.

Theorem 5.5.2 *If* $W = V \cup kh^r \cup lh^{r+1}$, $\pi_r(W, V) = 0$, $\pi_1(\partial_+ V) \cong \pi_1(V)$, $n \geq 2r+3$, *then* $W \cong V \cup lh^{r+1} \cup kh^{r+2}$.

Proof Write $V' := V \cup (k-1)h^r$, $M' := \partial_+(V')$, so W is the union of V' and $W' := (M' \times I) \cup h^r \cup lh^{r+1}$. Since $\pi_r(W, V)$ and $\pi_{r-1}(V', V)$ vanish, so does $\pi_r(W, V')$. If we show that $\pi_r(W', M') = 0$, we can apply Proposition 5.5.1 to replace the r-handle in W' by an $(r+2)$-handle, so $W \cong V \cup (k-1)h^r \cup lh^{r+1} \cup h^{r+2}$. Since the $(r+2)$-handle does not affect the calculation of π_r, the result will follow by induction.

Now $W = V' \cup W'$ and $M' = V' \cap W'$. Since $n \geq r+4$, the fundamental groups of $\partial_+ W$, W', W, M' and hence of V' are isomorphic. Thus the universal covers of W', V', M' (which we denote by affixing a tilde) are induced from that of W.

Thus by the Hurewicz isomorphism theorem (see B.3 (i)),

$$\pi_r(W', M') \cong H_r(\tilde{W}', \tilde{M}') \cong H_r(\tilde{W}, \tilde{V}) \cong \pi_r(W, V) = 0,$$

where the middle isomorphism holds by excision. □

We next extend the result by a more direct application of the Handle Cancellation Theorem. To avoid technicalities, we restrict to the simply-connected case: in §5.7 we indicate what is needed to remove this restriction.

Proposition 5.5.3 *Proposition 5.5.1 continues to hold if the hypothesis '$n \geq 2r+3$' is replaced by '$n \geq r+4$ and M is simply-connected.'*

Proof Since $H_r(W, M) = 0$, the class y of the r-handle in the chain complex $C_*(W, M)$ is a boundary, so if the $(r+1)$-handles have classes x_i, there are coefficients $c_i \in \mathbb{Z}$ with $\partial(\sum_i c_i x_i) = y$.

Use Theorem 5.4.4 to add a complementary pair of handles h_A^{r+1}, h_B^{r+2} away from the existing handles. Now use Theorem 5.4.5 to add c_i copies of the ith $(r+1)$-handle to h_A^{r+1} for each i. The a-sphere of the resulting handle has intersection number 1 with the b-sphere S_b of h^r.

Hence by Theorem 6.3.2 (i), provided $r \geq 3$ and $n \geq r+4$, we can perform a diffeotopy to reduce the number of intersections to one. But then h^r and $h_A^{\prime r+1}$ are complementary, so can be cancelled by Theorem 5.4.3.

The cases $r \leq 1$ follow from Proposition 5.5.1, also the case $r = 2$ except if $n = 6$. But if $r = 2$, we can use (ii) of the theorem, provided we show that the complement of S_b is simply-connected. But we have a diffeomorphism of $\partial_+((M \times I) \cup h^r) \setminus S_b$ with $M \setminus S^1$, where S^1 is the a-sphere of h^2. By hypothesis M is simply-connected, and deleting an embedded circle does not affect this property. □

We can go a little further.

Proposition 5.5.4 *The result also holds if $n = r+3$, provided $r \geq 3$ and $\partial_+ W$ is simply connected.*

Proof The above argument remains valid, except in the use of Theorem 6.3.2 (i). Again, we can use (ii) of the theorem, provided we show that the complement of S_a is simply-connected.

In the dual decomposition, we attach to $\partial_+ W$ first h_B^1, then a complementary h_A^2 and other 2-handles, then a 3-handle. Thus the fundamental group remains trivial at each stage after the second, hence the boundaries are simply-connected. But the complement of S_a in $\partial_+((M \times I) \cup h^r)$ is diffeomorphic to the complement of the belt sphere (a circle) in $\partial_+((M \times I) \cup h^r \cup h_A^{r+1})$, so is simply-connected. □

Theorem 5.5.5 *Suppose (W, V) r-connected, $\partial_+ V$, V and W simply-connected, and either $n \geq r+4$ or $n = r+3$, $r \geq 3$ and $\partial_+ W$ is simply-connected.*

Then W has a handle decomposition on V with no i-handles for $i \leq r$.

Proof By induction on r, we may suppose there are no i-handles for $i < r$. A second induction shows that it will suffice to remove or replace a single r-handle.

Write W' for the union of V and all but one of the r-handles, W'' for the union of the remaining r-handle and all $(r+1)$-handles, and W''' for the rest. The conclusion will follow if we show that Proposition 5.5.3 can be applied to W''.

We thus need to show that $\partial_- W''$ is simply-connected and $H_r(W'', \partial_- W'') = 0$. As V is simply-connected and W' is obtained from V by attaching r-handles with $r \neq 1$, W' is simply-connected. Since W' has no handle of index $n-2$, $\partial_+ W' = \partial_- W''$ also is simply-connected.

From the exact homology sequence

$$0 = H_r(W, V) \to H_r(W, W') \to H_{r-1}(W', V) = 0$$

we see that $H_r(W, W') = 0$; since handles of index $> r+1$ do not change H_r, we have $0 = H_r(W' \cup W'', W') = H_r(W'', \partial_- W'')$. □

The culmination of the theory developed in this chapter is the so-called h-cobordism theorem. Let W be a cobordism. If the inclusions of $\partial_- W$, $\partial_+ W$ in W are homotopy equivalences, W is called an *h-cobordism*. Provided all of $\partial_- W$, $\partial_+ W$, and W are simply-connected, it suffices if the relative homology groups $H_i(W, \partial_- W; \mathbb{Z})$ vanish, since by duality the $H_i(W, \partial_+ W; \mathbb{Z})$ also vanish, so both inclusions are homotopy equivalences. We have

Theorem 5.5.6 (h-cobordism Theorem) *If W^n is a simply-connected h-cobordism with $n \geq 6$, then $W \cong \partial_- W \times I$, so $\partial_+ W \cong \partial_- W$.*

Proof Take a handle decomposition, and choose r with $2 \leq r \leq n-3$. By Theorem 5.5.5, we can inductively replace each i-handle for $i < r$ by an $(i+2)$-handle. Now apply the same argument to the dual handle decomposition to eliminate all j-handles with $j > r+1$. Observe that if $r = 2$ or $r = n-3$ we need to use Theorem 6.3.2 (ii), so cannot allow both equalities together.

We now only have r- and $(r+1)$-handles, so the chain complex $C_*(W, \partial_- W)$ reduces to a single map $\partial : C_{r+1} \to C_r$ which, since we have an h-cobordism, is an isomorphism. Performing handle additions has the effect of row operations on the matrix of ∂. Consider the first column: by the Euclidean algorithm, we can repeatedly subtract smaller from larger entries to reduce until there is a single non-zero entry, which must be ± 1.

This shows that one of the a-spheres and one of the b-spheres have intersection number ± 1. Now if $2 < r < n-3$ we can use Theorem 6.3.2 (i) to reduce the number of intersections to one, and then cancel the corresponding handles using Theorem 5.4.3. As above if one (but not both) equality holds, we can instead use (ii). Repeating the argument, we can remove all the handles. □

5.6 Applications of h-cobordism

We can formulate the h-cobordism theorem a little more generally as follows.

Theorem 5.6.1 *If $V^n \subset W^n$ is a homotopy equivalence with $\partial_- V = \partial_- W$, V, $\partial_+ V$ and $\partial_+ W$ are all simply-connected and $n \geq 6$, then $V^n \cong W^n$.*

Proof We may write W as the union of two cobordisms V and V' with a common boundary $\partial_+ V$; since this and W are simply-connected, so is V'. It now follows as each $H_i(V', \partial_- V') \cong H_i(W, V) = 0$ that V' is an h-cobordism, hence by Theorem 5.5.6 is diffeomorphic to $\partial_+ V \times I$. By Lemma 2.7.2, W is diffeomorphic to V.

The argument applies even allowing $\partial_- V$ to have a boundary X. Here we need first to adjust corners so that $\partial_c V \cong X \times I$ and also $\partial_c V' \cong X \times I$. □

We have as simple application,

Theorem 5.6.2 (Disc Bundle Theorem) [139] *Suppose M^{n-c} a submanifold of W^n, $\partial M = \varnothing$, $c \geq 3$, $n \geq 6$, $M \subset W$ a homotopy equivalence, and M, ∂W simply-connected. Then W has the structure of a disc bundle with M as zero cross-section.*

Proof Take V as a tubular neighbourhood of M. Since $c \geq 3$, ∂V is simply connected. The result thus follows from the preceding theorem. □

Taking M to be a point gives

Corollary 5.6.3 *If W^n is contractible, $n \geq 6$, $\pi_1(\partial W) = 0$, then $W^n \cong D^n$.*

We call a closed manifold a *homotopy sphere* if it is homotopy equivalent to a sphere.

Corollary 5.6.4 *If Σ^n is a homotopy sphere, $n \geq 6$, then Σ^n may be obtained by glueing two discs together along the boundary. Thus Σ^n is homeomorphic to S^n.*

Proof Let W^n be the closure of the complement of a disc D^n in Σ^n. Then W is homotopic to $\Sigma^n \setminus \{point\}$, so is simply-connected, and its reduced homology groups vanish, so W is contractible. By Corollary 5.6.3, $W^n \cong D^n$.

Since D^n is homeomorphic to the cone over S^{n-1}, any homeomorphism of S^{n-1} extends, by taking the cone, to a homeomorphism of D^n. Since Σ^n is homeomorphic to the union of two copies of D^n glued by a homeomorphism of S^{n-1}, it follows that we have a homeomorphism on S^n. □

The Generalised Poincaré Conjecture states that any homotopy sphere Σ^n is homeomorphic to the sphere S^n: the original conjecture referred to the case $n = 3$. We have just proved this if $n \geq 6$. The cases $n \leq 5$ are discussed in the next section §5.7. We will return to the question of diffeomorphism in the final section §8.8.

Proposition 5.6.5 *(i) Suppose M, M' compact, simply-connected and without boundary, $f : M \to M'$ a homotopy equivalence and $2c \geq m$. Then $M' \times D^c$ is a disc bundle over M.*

(ii) Suppose in addition that $c \geq m + 1$ and $f^*(\mathbb{T}(M') + 1) \cong \mathbb{T}(M) + 1$. Then $M \times D^c \cong M' \times D^c$.

Proof If $c < 3$ then $m \leq 1$, M and M' are homotopy equivalent to a circle or a point, and the result is trivial. Now let $c \geq 3$. Then by Theorem 6.4.11, we can approximate f by an embedding of M in $M' \times D^c$. The result now follows from Theorem 5.6.2.

(ii) In this case, the normal bundle of $g(M)$ in $M' \times D^c$ is stably trivial and of fibre dimension $\geq m + 1$, hence (by §B.3(xi)) is trivial. □

Proposition 5.6.6 *Let Σ^{n-c} be a homotopy sphere embedded in S^n ($n \geq 6$, $c \geq 3$), N a tubular neighbourhood of Σ, V the closure of its complement. Then V is diffeomorphic to $S^{c-1} \times D^{n-c+1}$.*

Proof Let N' be a tubular neighbourhood of Σ with N in its interior, D^c a fibre, S^{c-1} its boundary. Since S^{c-1} bounds the contractible D^c, its normal bundle is trivial. We assert that the inclusion of S^{c-1} in V is a homotopy equivalence; indeed, both are simply-connected (V since S^n is, and $S^n \setminus \Sigma^{n-c}$ since $c \geq 3$) and the complement of $V \cup D^c$ is the interior of $N \setminus D^c$, a cell bundle over a cell and so contractible. By duality, $V \cup D^c$ is contractible, and $0 = H_r(V \cup D^c, D^c) = H_r(V, V \cap D^c)$. But $V \cap D^c$ is an annulus with S^{c-1} as deformation retract, hence $H_r(V, S^{c-1}) = 0$.

If $c \neq n - 1$, $\partial V = \partial N$ is simply-connected, and $n - c + 1 \geq 3$, so the result follows by applying Theorem 5.6.2 to $S^{c-1} \subset V$. If $c = n - 1$, Σ is a circle, and unknots, so the result is trivial. □

We can adapt some of the above arguments to give a relative result.

Theorem 5.6.7 *(i) Suppose W^n a simply-connected h-cobordism, $n \geq 6$, V^{n-c} a submanifold, $c \geq 3$, such that $V^{n-c} \cong \partial_- V \times I$. Then $(W, V) \cong (\partial_- W, \partial_- V) \times I$.*

(ii) Two h-cobordant pairs of homotopy spheres $(\Sigma_i^{n+c}, \Sigma_i^n)(i = 0, 1)$ with $n \geq 5$, $c \geq 3$ are diffeomorphic.

Proof As in Lemma 5.1.1, we can find a non-degenerate function on W whose restriction to V has no critical points; the proof of Lemma 5.1.1 is only changed by using the given product structure to define g near V. We can now carry out all the handle decomposition and cancellation arguments in $W \setminus V$.

More precisely, write N for a tubular neighbourhood of V in W, $\mathring{N}$ for its interior, $X = W \setminus \mathring{N}$ and $Y = N \cap X = \partial_c N = \partial_c X$.

Since $c \geq 3$ is the codimension of V in W (and of $\partial_- V$ in $\partial_- W$, $\partial_+ V$ in $\partial_+ W$), removing V does not alter the fundamental groups.

So it is enough to check that $\partial_- X \subset X$ is a homotopy equivalence, and so enough to show that $H_*(X, \partial_- X) = 0$. Since $\partial_- V$ is a deformation retract of V, and N is a disc bundle, $\partial_- N$ is a deformation retract of N, also of $\partial_- N \cup Y$. Hence $0 = H_*(N, \partial_- N \cup Y) \cong H_*(W, X \cup \partial_- W)$ by excision. But $H_*(W, \partial_- W)$ is trivial, so using the homology exact sequence of the triple $\partial_- W \subset X \cup \partial_- W \subset W$, we deduce that $H_*(X \cup \partial_- W, \partial_- W)$ is trivial. It follows by excision that $H_*(X, \partial_- X) = 0$. The result follows.

(ii) By the h-cobordism theorem, the h-cobordism of the Σ_i^n is a product, so the result follows from (i). □

A different relative form of these results can also be obtained, giving a topological unknotting theorem for pairs of spheres.

Proposition 5.6.8 *(i) Let $M^m \subset W^{m+c}$ be a proper embedding of contractible manifolds with $c \geq 3$, $m + c \geq 6$. Assume that either $M^m \cong D^m$ or $m \geq 6$. Then the pair (W^{m+c}, M^m) is diffeomorphic to (D^{m+c}, D^m).*

(ii) Let $T^m \subset \Sigma^{m+c}$ be an embedding of homotopy spheres with $c \geq 3$, $m + c \geq 6$: assume either that $T^m \cong S^m$ or that $m \geq 6$. Then the pair (Σ^{m+c}, T^m) is homeomorphic *to (S^{m+c}, S^m).*

Proof (i) Take a tubular neighbourhood V of M in W: then V is contractible, so we can apply Theorem 5.6.1 (where we set $\partial_- V = \partial_- W = V \cap \partial W$) to the inclusion $V \subset W$ to infer that W is obtained from V by adding a collar.

(ii) Choose an embedding $(D^{m+c}, D^m) \to (\Sigma^{m+c}, T^m)$ (it is essentially unique by Lemma 2.5.11), and delete the interior to give a pair as in (i): by that result, we have another copy of (D^{m+c}, D^m). These copies are attached by a diffeomorphism of the boundary (S^{m+c-1}, S^{m-1}). But as in Corollary 5.6.4, any such diffeomorphism extends, taking the cone, to a *homeomorphism* of (D^{m+c}, D^m). □

We now proceed to obtain minimal handle decompositions in general.

Theorem 5.6.9 *Suppose W^n ($n \geq 6$) such that $\partial_- W$, $\partial_+ W$ and W are simply-connected. Let $H_i(W, \partial_- W) \cong F + T$, where F is a free abelian group of rank*

β_i and T is a finite group with $\tau_{i+\frac{1}{2}}$ generators. Then W has a handle decomposition on $\partial_- W$ with $\tau_{i-\frac{1}{2}} + \beta_i + \tau_{i+\frac{1}{2}}$ i-handles for each i.

Proof By Corollary 5.4.1, there is a handle decomposition with no 0- or 1-handles. Similarly, we can dispense with $(n-1)$- and n-handles. This gives a chain-complex of free abelian groups whose homology is that of $H_*(W, \partial_- W)$. By making changes of basis of the chain groups, we can put this chain-complex into normal form, i.e. a direct sum of elementary subcomplexes, each with rank 1 or 2, and differential either

$$0 \to \mathbb{Z} \to 0 \qquad \text{or} \qquad 0 \to \mathbb{Z} \xrightarrow{\theta} \mathbb{Z} \to 0.$$

Now the required changes of base can be induced by a sequence of elementary automorphisms of the chain groups, and by Theorem 5.4.5, each of these can be induced by a change in handle decomposition. It remains only to remove the elementary subcomplexes with $\theta = 1$. But it follows as above from Theorem 5.4.3 that such pairs of handles may be cancelled. □

This allows us in favourable cases to obtain classifications up to diffeomorphism. It follows at once from Theorem 5.6.9 that

Lemma 5.6.10 *Suppose M^m, with $m \geq 6$, such that M and ∂M are simply-connected, $H_r(M)$ is free abelian of rank k, and $\tilde{H}_i(M) = 0$ for $i \neq r$. Then M admits a handle decomposition with one 0-handle, k r-handles, and no others.*

Such a manifold is called a *handlebody* . By Lemma 5.2.1, it can be obtained from D^m by simultaneous attachment of all k r-handles, so is determined by an embedding

$$F : \bigcup_{i=1}^{k} (S^{r-1} \times D^{m-r})_i \to S^{m-1}.$$

We can take this in two stages: first study the restriction $\overline{F}$ of F to the union of the spheres $S^{r-1} \times \{0\}$, and then thicken the spheres up to their tubular neighbourhoods.

The classification in the case $m > 2r$ is straightforward.

Theorem 5.6.11 *A handlebody M with $m > 2r$ is a boundary sum of k $(m-r)$–disc bundles over S^r. M is determined up to diffeomorphism by the values of k, r, m, and the subgroup of $\pi_{r-1}(SO)$ generated by the classes of the bundles.*

Proof Since $m > 2r$, it follows by general position that any two embeddings $\overline{F}$ are diffeotopic. In particular, the components of the image are contained in disjoint $(m-1)$–discs in S^{m-1}. It follows that the handlebody is a boundary sum. Each summand is obtained by attaching a single handle, so (for example, by Theorem 5.6.2) is a disc bundle over S^r.

Such disc bundles are classified by $\pi_{r-1}(SO_{m-r})$. Since $m > 2r$, this is isomorphic to the stable homotopy group $\pi_{r-1}(SO)$. Hence the bundle is determined by the restriction to the central sphere of the (stable) tangent bundle of M, which in turn is determined by the classifying map $M \to B(SO)$. Since M is homotopy equivalent to a bouquet of r-spheres, this comes to the same as a collection of maps $S^r \to B(SO)$. If we change the handle decomposition using the Handle Addition Theorem 5.4.5, the elements of $\pi_r(B(SO))$ add correspondingly.

We now recall the result of Bott [21] (see §B.3(xii)) that the group $\pi_{r-1}(SO)$ is cyclic, infinite if $r \equiv 0$ $(mod\ 4)$, of order 2 if $r \equiv 1, 2$ $(mod\ 8)$, and zero otherwise. Thus we can change the basis of $H_r(M)$ to ensure that all but the first basis element map to zero, and the image of the first generates the subgroup of $\pi_{r-1}(B(SO))$. □

In the case $m = 2r$, there are two extra points: the embedding $\overline{F}$ is no longer unique up to diffeotopy, and the group $\pi_{r-1}(SO_{m-r}) = \pi_{r-1}(SO_r)$ lies in exact sequences (see B.3.2):

$$\mathbb{Z} = \pi_r(S^r) \xrightarrow{\partial} \pi_{r-1}(SO_r) \xrightarrow{i_*} \pi_{r-1}(SO_{r+1}) = \pi_{r-1}(SO).$$

$$\xrightarrow{\partial} \pi_{r-1}(SO_{r-1}) \xrightarrow{i_*} \pi_{r-1}(SO_r) \xrightarrow{\pi_*} \pi_{r-1}(S^{r-1}) = \mathbb{Z}.$$

If r is even, the first map in the first sequence is injective, and both points are accommodated by taking into account the intersection pairing on $H_r(M)$. We have

Theorem 5.6.12 *Let M^{2n} be a manifold with M and ∂M simply-connected, $n \geq 3$, with $\tilde{H}_r(M)$ vanishing for $r \neq n$ and free abelian for $r = n$. The diffeomorphism type of M is determined by the following invariants:*

a free abelian group $H := H_n(M; \mathbb{Z})$,

a $(-1)^n$-symmetric bilinear map $H \times H \to \mathbb{Z}$ given by intersection numbers,

a map $\alpha : H \to \pi_{n-1}(SO_n)$.

These satisfy

(i) $x.x = \pi(\alpha(x))$ for $x \in H$, and

(ii) $\alpha(x+y) = \alpha(x) + \alpha(y) + xy(\partial\iota_n)$ for $x, y \in H$.

Proof We first define α. It follows from Theorem 6.4.11 that each $x \in H_n(M) \cong \pi_n(M)$ is represented by an embedding $f_x : S^n \to M$, and that for

$n \geq 4$ such an embedding is unique up to diffeotopy. We may thus define $\alpha(x)$ as the characteristic class of the normal bundle of $f_x(S^n)$. In the case $n = 3$, the group $\pi_{n-1}(SO_n) = \pi_2(SO_3)$ is trivial, so α is unique.

To see (i), note that $\pi : \pi_{n-1}(SO_n) \to \mathbb{Z}$ coincides with the natural map to $\pi_{n-1}(S^{n-1})$. Now $x.x$ is the intersection number of $f_x(S^n)$ with a nearby perturbation. Since f_x is an embedding, this is the primary obstruction to finding a cross-section of the bundle with fibre S^{n-1} associated to the normal bundle, hence with the image of $\alpha(x)$ under π. As to (ii), we may join the embedded spheres f_x and f_y by a tube to obtain an immersed sphere representing $x + y$. This has normal bundle given by $\alpha(x) + \alpha(y)$ and self-intersection $x.y$. Now as in §6.3 performing a homotopy to remove a single self-intersection will add $\partial\iota_n$ to the normal bundle.

We must now show that these invariants determine M up to diffeomorphism. Choose a handle presentation as above: it will suffice to show that F is determined up to diffeotopy. First consider $\overline{F}$, and note that classifying embeddings into S^{2n-1} is equivalent to classifying embeddings into $\mathbb{R}^{2n-1}$. It follows from Theorem 6.4.11 that an embedding $S^{n-1} \to \mathbb{R}^{2n-1}$ is unique up to diffeotopy.

According to Corollary 6.4.10, if $2m > 3(v + 1)$, diffeotopy classes of smooth embeddings $f : V^v \to \mathbb{R}^m$ correspond bijectively to equivariant homotopy classes of equivariant maps $V \times V \setminus \Delta(V) \to S^{m-1}$, where an embedding f determines the equivariant map f_δ defined by $f_\delta(x, y) = (f(x) - f(y))/\|f(x) - f(y)\|$. Taking $V = \bigcup_{i=1}^k S_i^{n-1}$ and $m = 2n - 1$, we see that the dimension condition is $n > 2$; the result for $k = 1$ shows that we can ignore the components $(S_i^{n-1} \times S_i^{n-1})$; and if $i \neq j$, an equivariant homotopy of a map of $(S_i^{n-1} \times S_j^{n-1}) \cup (S_j^{n-1} \times S_i^{n-1})$ is equivalent to a homotopy of $(S_i^{n-1} \times S_j^{n-1})$.

Since homotopy classes of maps $S^{n-1} \times S^{n-1} \to S^{2n-2}$ are determined by their degree, an integer, for each pair $i \neq j$ we have an integer $c_{i,j}$, which can be interpreted as the linking number of S_i^{n-1} and S_j^{n-1} in S^{2n-1}, so is equal to the intersection number of the corresponding n-spheres in M, and is $(-1)^n$-symmetric.

For each component, the choice of extension of the map $\overline{f}$ on S^{n-1} to an embedding f of $S^{n-1} \times D^n$ is equivalent to choosing an element of $\pi_{n-1}(SO_n)$, and making an appropriate normalisation, this element coincides with the characteristic class $\alpha(x_i)$ of the normal bundle of the corresponding sphere S^n in M. Hence indeed the invariants determine F up to diffeotopy, hence M up to diffeomorphism. □

It follows by a short calculation that if (and only if) the intersection form is nonsingular, the boundary of M is homotopy equivalent – and hence homeomorphic – to S^{2n-1}. This was the case of prime interest in [159], where I also considered the question of when ∂M is diffeomorphic to S^{2n-1}.

There is a corresponding classification for handlebodies in the metastable range. The proof is essentially the same, but the arguments for (i) and (ii) are somewhat more delicate, and we omit the details.

Theorem 5.6.13 *Let M^m be a handlebody with handles of dimension $s \geq 2$, $m \geq 6$, $2m \geq 3s + 3$. Then the diffeomorphism type of M is determined by invariants $H := H_s(M; \mathbb{Z})$, a $(-1)^s$-symmetric bilinear map $\lambda : H \times H \to \pi_s(S^{m-s})$, and a map $\alpha : H \to \pi_{s-1}(SO_{m-s})$, satisfying*

(i) $\lambda(x, x) = S\pi_*\alpha(x)$ for $x \in H$, and

(ii) $\alpha(x + y) = \alpha(x) + \alpha(y) + \partial_*\lambda(x, y)$ for $x, y \in H$.

Here ∂_* is the boundary map in the homotopy exact sequence (B.3.2) of the fibre bundle $SO_{m-s} \to SO_{m-s+1} \to S^{m-s}$; $\pi : SO_{m-s} \to S^{m-s-1}$ is the projection, and $S : \pi_{s-1}(S^{m-s-1}) \to \pi_s(S^{m-s})$ the suspension map.

5.7 Complements

In this section we first summarise (without proofs) what is known in dimensions $n \leq 5$. Then we indicate what changes need to be made if we drop the simply-connected hypothesis.

The cases $n \leq 1$ are trivial. For $n = 2$ it follows from Proposition 5.4.1 that a connected closed 2-manifold M has a handle decomposition with just one 0- and one 2-handle. Write α_1 for the number of 1-handles, so that $\chi(M) = 2 - \alpha_1$. In particular, if M is a homotopy sphere, $\alpha_1 = 0$. Since any diffeomorphism of S^1 is diffeotopic to the identity (or a reflection) and hence extends to one of D^2, it follows that $M \cong S^2$.

Otherwise we analyse M by induction on α_1. If $\alpha_1 \geq 1$, choose a 1-handle $D^1 \times D^1$, join the ends $\partial D^1 \times \{0\}$ of the arc $D^1 \times \{0\}$ (the a-sphere) by a smooth arc in D^2 to form an embedded circle C, and cut M along this circle to give N'. There are three possibilities which are illustrated in Figure 5.10.

(a) C is 1-sided, so $\partial N'$ is a circle, the double cover of C. Adding a disc along this boundary gives a closed surface N with $\chi(N) = 1 + \chi(M)$, so $\alpha(N) = \alpha(M) - 1$. Moreover the procedure to recover M from N shows that we have a connected sum $M = N\#P^2(\mathbb{R})$.

(b) C is 2-sided and separates M into two pieces, N_1' and N_2', each with boundary C. Adding a disc to the boundary yields closed surfaces N_1, N_2 with $M \cong N_1\#N_2$. Since $\chi(M) = \chi(N_1) + \chi(N_2) - 2$, we have $\alpha(M) = \alpha(N_1) + \alpha(N_2)$. It follows from our construction of C that neither N_i can be S^2, so each $\alpha(N_i) < \alpha(M)$.

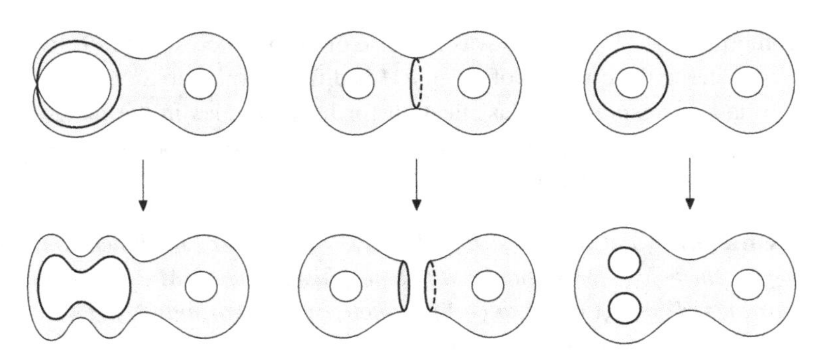

Figure 5.10 The effect of cutting a surface

(c) C is 2-sided, but does not separate M. Filling each component of $\partial N'$ by a disc gives a closed surface N; now we can choose a disc in N containing each of these in its interior, and so express N as connected sum of N^*, say, with a manifold obtained from S^2 by removing two discs, and identifying the two boundaries together. This yields a torus $S^1 \times S^1$ if orientable or a Klein bottle K^2 if not. Calculating as above gives $\chi(M) = \chi(N^*) + 2$.

It follows by induction that M is the connected sum of a collection (possibly empty) of copies of $P^2(\mathbb{R})$, $S^1 \times S^1$ and K^2. The classification is completed by the easy proof that $K^2 \cong P^2(\mathbb{R})\#P^2(\mathbb{R})$ and $P^2(\mathbb{R})\#K^2 \cong P^2(\mathbb{R})\#(S^1 \times S^1)$. The conclusion can also be formulated by saying that Theorem 5.6.12 applies in this case.

For $n = 3$, a decomposition with just one 0- and one 3-handle is essentially equivalent to a Heegard decomposition, i.e. expressing M as the union of two handlebodies, which by itself does not tell us much. However the theory of (compact) 3-manifolds is highly developed, and the principal structural result is Thurston's Geometrisation Principle, which was established by Grigori Perelman [120] in 2003, and which includes the original Poincaré Conjecture. An account of the proof in book form was given by Morgan and Tian [107]. Thurston's own work [154] gives a more leisurely and very geometric account giving some insight into how he was led to the Principle.

The case $n = 4$ is the one where our methods yield the least. To avoid repetition below, let us write $\mathcal{C}_4$ for the class of closed, simply-connected 4-manifolds. The obvious invariant of any $X \in \mathcal{C}_4$ is the symmetric bilinear form λ given by intersection numbers on $H_2(X; \mathbb{Z})$; this has rank $\beta_2(X)$ and signature $\sigma(\lambda) = \sigma(X)$; it follows from duality that λ is nonsingular. The type of λ is even if each $\lambda(x, x)$ is even (equivalently if $w_2(X) = 0$, thus iff X has a spinor structure) and odd otherwise.

It is known from the general theory of quadratic forms that two indefinite nonsingular forms of the same rank, signature, and type are isomorphic (this fails badly for definite forms), so the matrix of an indefinite odd form can be diagonalised. For even forms, $\sigma(\lambda)$ is divisible by 8 (see Proposition 7.3.3) and an example with $\sigma = 8$ is given by the E_8 matrix (7.3.4). A result going back to Rokhlin, and corresponding to the 4-dimensional case of Proposition 8.8.6, tells us that for a closed spinor 4-manifold, σ is divisible by 16. A well-known example with $\sigma = 16$ (and with $\beta_2 = 22$) is a so-called K3 surface, for example, the one given in $P^3(\mathbb{C})$ by $z_0^4 + z_1^4 + z_2^4 + z_3^4 = 0$.

The author proved in [162] that if $X_1, X_2 \in \mathcal{C}_4$ have isomorphic intersection forms they are h-cobordant, and I deduced that they become diffeomorphic after taking connected sums with a number of copies of $S^2 \times S^2$. Up to 1980 it still seemed plausible that this implied diffeomorphism, and that any nonsingular symmetric bilinear form could occur. Indeed the *topological* triviality of an h-cobordism of manifolds $X_1, X_2 \in \mathcal{C}_4$ (and hence the $n = 4$ case of the Generalised Poincaré Conjecture) was proved by Michael Freedman [53] in 1982, and he also proved that indeed any nonsingular symmetric bilinear form is the intersection form of some $X \in \mathcal{C}_4$ (but not in general smooth): see also [54].

The picture changed dramatically with the work of Donaldson. His first paper [42] proved that if $X \in \mathcal{C}_4$ is smooth and its intersection form λ is positive definite, then λ can be diagonalised (and so agrees with the intersection form of a connected sum of copies of $P^2(\mathbb{C})$); in particular, unless $\beta_2 = 0$, λ cannot be even. Next in [43] Donaldson proved non-existence of diffeomorphisms for certain pairs $X_1, X_2 \in \mathcal{C}_4$ of smooth manifolds with isomorphic intersection forms, and hence h-cobordant: thus the h-cobordism theorem fails for such manifolds. Donaldson's techniques are well outside the scope of this book (and beyond the competence of this author), but here is a brief indication of what is involved.

Let X be a closed oriented 4-manifold; write $\beta_2^+(X)$ for the dimension of a maximal subspace of $H_2(X; \mathbb{R})$ which is positive definite for the intersection form. The details require $\beta_2^+(X)$ to be odd, and extra complications arise if $\beta_2^+(X) = 1$. Principal SU_2-bundles P over X are classified by $k = \langle c_2(P), [X] \rangle$. Choose a Riemannian metric g on X; and consider the space of connections A on P. The so-called Yang–Mills equations require that the self-dual part of the curvature tensor of A vanishes. The quotient of the set of solutions of these equations by the group of bundle automorphisms of P ('gauge equivalence') is the moduli space $M_k(g)$. It is shown (with some effort) that for a generic metric g this moduli space is a smooth manifold of dimension $2d_k = 8k - 3(1 + \beta_2^+(X))$, An orientation of this maximal subspace induces one of

$M_k(g)$, and the homology class of $M_k(g)$ in the space of gauge equivalence classes of connections determines a symmetric multilinear function $q(d_k)$ of degree d_k on $H^2(X; \mathbb{R})$ which is independent of the metric. These functions give new diffeomorphism invariants for X.

The paper [81] by Kronheimer and Mrowka assembled these invariants into a generating function $q = \sum_k q(d_k)/(d_k)!$ which is regarded as a power series over $H_2(X; \mathbb{R})$; their first main theorem states that if X is 1-connected and of simple type, there exists a finite list $K_1, \ldots, K_p \in H^2(X; \mathbb{Z})$ and non-zero $a_1, \ldots, a_p \in \mathbb{Q}$ such that

$$q = \exp\left(\tfrac{1}{2}\lambda \sum_{s=1}^{p} a_s e^{K_s}\right),$$

where λ denotes the intersection form. The 'simple type' condition was somewhat ad hoc, but at least allowed large families of examples. The classes K_s are called *basic classes*. They all satisfy $K.K = 2\chi(X) + 3\sigma(\lambda)$. If X is a minimal complex algebraic surface of general type, then the only basic classes are $\pm K$, where K is the canonical class: thus K (up to sign) is a diffeomorphism invariant, and the basic classes in general can be regarded as a diffeomorphism invariant version of the canonical class.

A formula relating the invariants of X to those of the blow-up $X\#P^2(\mathbb{C})$ was obtained in general by Fintushel and Stern [51]. It involves elliptic functions which, when X is of simple type, specialise to the trigonometric functions in the above formula.

Shortly afterwards, a new theory was introduced by Witten [181], based on the so-called Seiberg–Witten equations. Here there is an additional element of structure. We start with a $Spin^c$-structure on X: this induces a pair of vector bundles $W^{\pm}$, a complex line bundle L over X, and isomorphisms $\Lambda^2 W^+ \cong \Lambda^2 W^- \cong L$. The Seiberg–Witten equations define a subset of the space of pairs (A, ψ) with A a unitary connection on L and ψ a section of W^+. Again we form the space M of equivalence classes of solutions under gauge equivalence into a moduli space and need to show that for a generic metric on X, M is smooth of the expected dimension $2s(L)$, where $s(L) = \frac{1}{8}(c_1(L)^2 - (2\chi(X) + 3\sigma(\lambda)))$ (there are in fact possible isolated singularities corresponding to 'reducible solutions'), compact, oriented, and deforms well under change of metric; in fact it seems these points are somewhat easier to deal with here than in the preceding case. There is a canonical class $h \in H^2(M)$ and we obtain an invariant $n_L = \langle h^s, [M]\rangle$. There are only finitely many line bundles L with $n_L \neq 0$. This description assumes $\beta_2^+ > 1$; otherwise the invariant depends on a choice of a chamber in the cohomology of X and there is a wall-crossing formula for moving to a neighbouring chamber.

This method led to a flurry of papers which are surveyed in [45]. The review of this paper mentions conjectures of Witten that the list of basic classes coincides with the set of first Chern classes of $Spin^c$-structures on X having invariant $n_L \neq 0$, that the corresponding coefficients a_s agree (up to a normalising factor) with the Seiberg–Witten invariants, and also hints at a formula for q valid without restriction. Although nearly twenty years have now elapsed, and there is by now a large literature in this area, these questions still seem to be open.

These developments led to refinements of Donaldson's original theorems restricting the intersection form, with the simple-connectivity hypothesis weakened. The best result known in 2015 seems to be that of Furuta [55]: that if the intersection form λ of a spinor 4-manifold X is not definite, then $\beta_2(X) \geq \frac{5}{4}|\sigma(X)| + 2$ (if λ *is* definite, a theorem of Donaldson implies $\beta_2(X) = \sigma(X) = 0$). (The conjecture that $\beta_2(X) \geq \frac{11}{8}|\sigma(X)|$ remains open.)

The second major result in [81], again for X simply-connected and of simple type, asserts that if Σ is a connected surface of genus g, smoothly embedded in X, and with $\Sigma.\Sigma > 0$, then $2g - 2 \geq \Sigma.\Sigma + max_s(K_s.\Sigma)$. This gives a clear indication that there is no simple substitute for the Whitney trick of §6.3 for obtaining embeddings of surfaces in 4-folds. In particular it establishes (as was conjectured by Thom) that no surface smoothly embedded in $P^2(\mathbb{C})$ has lower genus than a smooth projective curve of the same degree.

In contrast to all these results, NO effective general technique is known (in 2015) for proving that two given closed smooth 4-manifolds are diffeomorphic.

For $n = 5$, in the presence of simple connectivity, we can cancel 1- and 4-handles, but the Whitney trick does not apply to allow us to cancel 2- and 3-handles. However any closed oriented 5-manifold with $w_2 = 0$ is the boundary of a 6-manifold, and (see Chapter 7) we can simplify the 6-manifold by surgery. In particular, one can show that any homotopy sphere Σ^5 bounds a contractible W^6, and hence is diffeomorphic to S^5. Similar arguments lead to a complete classification of closed simply-connected 5-manifolds up to diffeomorphism: see §7.9.

We next give brief indications of the changes needed to be made in the main results of this chapter to accommodate the fundamental group.

First, where we use the Whitney trick to remove intersections of spheres of complementary dimensions, it does not suffice to measure intersections in M by a single number: we must take account of the paths joining intersection points. Each intersection is then associated to a sign ± 1 and an element of $\pi_1(M)$, and we add to obtain an element of the integer group ring $\Lambda := \mathbb{Z}[\pi_1(M)]$.

In the discussion of the homology of handles, we must now consider chains in the universal cover of W, giving chains with coefficients in $\mathbb{Z}[\pi_1(W)]$.

We can partly compensate for this by improving the handle addition Theorem 5.4.5 to incorporate a path following an arbitrary element of $\pi_1(M)$.

Now Proposition 5.5.2 goes through without the hypothesis of simple connectivity, and Theorem 5.5.5 with the additional requirement $\pi_1(\partial_+ V) \cong \pi_1(V)$, and if $n = r + 3$ also $\pi_1(\partial_+ W) \cong \pi_1(W)$.

The Euclidean algorithm used in Theorem 5.5.6 fails. Here we have the matrix of ∂ over Λ and moves of the following kinds can be realised geometrically:

(a) Add some multiple of a row to another row (use handle addition, Theorem 5.4.5).

(b) Multiply some row by an element of π, or by -1 (change the path from $*$ to an a-sphere, or the orientation of a cell).

(c) Take the direct sum of the matrix with (1) (insert a complementary pair of handles, Theorem 5.4.4).

The operations (a) and (b) generate a normal subgroup $E_N(\Lambda)$ of the general linear group $GL_N(\Lambda)$; we stabilise using (c) to obtain $E_\infty(\Lambda) \lhd GL_\infty(\Lambda)$, and the quotient defines the Whitehead group $Wh(\pi_1(M))$. There is an obstruction in this group to completing the proof of Theorem 5.5.6. An h-cobordism is called an *s-cobordism* (and the map $\partial_- W \to W$ a simple homotopy equivalence) if this obstruction vanishes.

It is known that $Wh(\pi)$ vanishes if π is free or free abelian, or an elementary 2-group or if $\pi \cong \mathbb{Z}_3, \mathbb{Z}_4$, and many other calculations are known: a survey of results for π finite is given by Oliver [116].

The results in §5.6 remain valid if the simple connectivity hypotheses are replaced as follows:

Theorem 5.6.1 (i) $\partial_+ V \subset V$ and $\partial_+ W \subset W$ induce isomorphisms of π_1; (ii) $M \subset W$ a simple homotopy equivalence, and $\partial W \subset W$ induces an isomorphism of π_1.

For Theorem 5.6.7 it suffices to require that W is an s-cobordism.

For Theorem 5.6.9 there is no direct analogue: the same argument shows that any chain complex chain homotopy equivalent to $C_*(W, \partial_- W)$ can be realised by a handle presentation, subject to compatibility with presentations of $\pi_1(W)$. To formulate this precisely comes to saying that we can imitate construction of a CW-complex of the desired (simple) homotopy type by a handle presentation. The most satisfactory results in this direction are the following, due to Mazur [90], which can be regarded as generalisations of Theorem 5.6.1 (ii).

Let M^m be a compact manifold, K^k a finite complex. We call an embedding $f : K \subset M$ *tame* if M is covered by coordinate neighbourhoods $\varphi_\alpha : U_\alpha \to \mathbb{R}^m$ such that each $\varphi_\alpha | f^{-1}(U_\alpha) : f^{-1}(U_\alpha) \to \mathbb{R}^m$ is linear on each simplex.

A submanifold U^m of $\mathring{M}^m$ is a *simple neighbourhood* of $f(K)$ if $K \subset \mathring{U}$, the inclusion $K \subset U$ is a simple homotopy equivalence, and $\pi_1(\partial U) \cong \pi_1(U \setminus K)$.

Theorem 5.7.1 (Simple Neighbourhood Theorem) *(i) A smooth regular neighbourhood is a simple neighbourhood.*

(ii) A smooth regular neighbourhood has a finite handle decomposition with one h^i corresponding to each simplex σ^i of K.

(iii) Let $m \geq 6$, codim $K \geq 3$. Then if U_1, U_2 are simple neighbourhoods of K, there is a diffeotopy of M, constant near K and away from $U_1 \cup U_2$, which moves U_1 to U_2.

Theorem 5.7.2 (Non-stable Neighbourhood Theorem) *Suppose W^n has a handle decomposition with no i-handles for $i > n-2$. Assume $\pi_1(W) \cong \pi_1(\partial_+ W)$, $n \geq 6$. Let X be a CW complex with no i-cells for $i > n-2$ and $f : X \to W$ a simple homotopy equivalence. Then W has a handle decomposition with cells corresponding to those of X.*

There is also a relative version.

5.8 Notes on Chapter 5

§5.1 Although this decomposition has its roots in the nineteenth century, and a version was used by Poincaré, the modern version is essentially due to Morse [108]; however the accurate formulation first appeared in work of Smale [138] and Wallace [171].

§5.3 The Poincaré duality theorem has its origins in work of Poincaré, though in his time homology groups had yet to be invented, so the result obtained was an equality of Betti numbers $\beta_r = \beta_{n-r}$. The Morse inequalities 5.3.3 are due to Morse, who in [109] applied the existence theorem to obtain results on the homology. See [98, I] for a similar account. The extension of duality to manifolds with boundary is due to Lefschetz.

§5.2, §5.4, §5.5 Apparently h-cobordism was first defined by Thom.

This development in these sections is due to Smale [138], [139]: the first paper proved the Generalised Poincaré Conjecture, the second went on to the h-cobordism theorem. Smale had been working on dynamical systems, and was seeking to simplify them.

The preprint version had an error (which was soon corrected but annoying) in the treatment of the fundamental group; in the above account we have bypassed the difficulty by using the handle replacement technique (Proposition 5.5.1).

Another account of the proof of the h-cobordism theorem (in terms of functions rather than handles) is given in the little book [101] by Milnor.

§5.6 We have included examples to illustrate that the h-cobordism theorem is an effective tool for obtaining classification results up to diffeomorphism. These are taken from the author's papers [159] and [160].

§5.7 In the lecture notes from which this book originated, I was at pains to obtain results in maximum generality, and in particular, to remove all restrictions on the fundamental group. Here I have tried to supply enough to give the interested reader a taste of what is involved.

6

Immersions and embeddings

We saw in Chapter 4 that a map $V^v \to M^m$ in general position is already an embedding if $m > 2v$. If this condition fails, we still have effective techniques for constructing embeddings, and will describe some of the main results in this chapter.

For immersions, the results give a complete reduction of the problem to a problem in homotopy theory. The proof of this major result is somewhat technical, and the details will not be required elsewhere.

We will now need to assume rather more familiarity with homotopy theory than in earlier chapters, and refer to Appendix B for a summary of the relevant definitions and results.

The theory of embeddings begins with a technique introduced by Whitney for removing pairs of self-intersections of a smooth n-manifolds in a $2n$-manifold (if $n \geq 3$). We describe this in some detail in §6.3: it was used as a key tool in §5.5. We then apply it to discuss embeddings of S^n in $2n$-manifolds.

The essential idea of this technique was generalised by Haefliger to maps $V^v \to M^m$ whenever $2m \geq 3(v + 1)$ – a condition we call the *metastable* range. There are several related results giving homotopy theoretic criteria for deforming a map to an immersion, or to an embedding, or for finding a regular homotopy of an immersion to an embedding; each one also has a simplified form when the target is Euclidean space, and also a companion criterion for finding a diffeotopy of the constructed embeddings. We describe these results, but confine ourselves to an outline of the rather involved proof.

6.1 Fibration theorems

A map $f : E \to B$ is said to be a *fibration* if given a space K, a map $a : K \to E$ and a homotopy $b : K \times I \to B$ such that $b|(K \times 0) = p \circ a$, there exists a homotopy $c : K \times I \to X$ such that $a = c \mid (K \times 0)$ and $b = p \circ c$. We also say that f has the covering homotopy property (CHP). If this holds for K a finite CW-complex, it follows for any CW-complex; it also follows if (K, L) is a CW-pair that c can be chosen to extend a lift already given on $L \times I$. It suffices to require this condition for pairs $(K, L) = (D^n, S^{n-1})$.

If $f : E \to B$ is a fibration, $*$ a point of B, and $F = f^{-1}(*)$ (called the *fibre*), there is an exact homotopy sequence $\ldots \pi_n(F) \to \pi_n(E) \to \pi_n(B) \to \pi_{n-1}(F) \ldots$.

The term 'fibration' recalls the fact that (see Lemma B.1.1) the projection map of a fibre bundle has the CHP.

If $p : E \to B$ and $p' : E' \to B'$ are fibrations, a map $f : E \to E'$ is called a *fibre map* if $p(e_1) = p(e_2)$ implies $p'(f(e_1)) = p'(f(e_2))$, so that there is a map $g : B \to B'$ with $g \circ p = p' \circ f$.

In this section I give fibration theorems for spaces of cross-sections and of (smooth) embeddings to prepare the way for the next section.

Theorem 6.1.1 *Let M be a smooth manifold, $V \subset M$ a compact submanifold. Then the map Diff(M) $\to$ Emb(V, M) is a fibration.*

This is an upgrading of the Diffeotopy Extension Theorem 2.4.2, and the same proof goes through with minor changes.

Proof We may suppose given a space P, a map $f : P \to \mathrm{Diff}(M)$. and a homotopy $g : P \times I \to \mathrm{Emb}(V, M)$ of the restriction of f, and need to lift g to a homotopy of f. Denote by $i : V \to M$ the inclusion, $f' : P \times M \to M$ and $g' : P \times I \times V \to M$ the maps associated to f and g (thus $f'(p, x) = f(p)(x)$). Thus for $p \in P, x \in V$ we have $g'(p, 0, x) = f'(p, x)$.

For each $(p, x) \in P \times V$ we have a path $g'(p, t, x)$ in M; denote the tangent vector to this path at $g'(p, t, x)$ by $\xi(p, t, x)$. We need to construct a tangent vector field $\eta(p, t, y)$ to M for each $(p, t) \in P \times I$, depending smoothly on $y \in M$ and continuously on p and t, and extending ξ.

The argument of Theorem 2.4.2 now goes through, but (i) allowing the additional parameter $p \in P$ and (ii) not insisting on smoothness as a function of the variables p and t: these do not significantly affect the argument. □

As for Theorem 2.4.2, compactness of V is essential to the argument. In Cerf [36] and Palais [118] we find a more precise result: the fibration is locally trivial, where the spaces of sections have the C^∞ topology.

Lemma 6.1.2 *Let $f : E \to B$ be a fibration; let $K \subset L \subset B$ be CW-complexes. Write $\Gamma(K)$ for the space of cross-sections of f over K. Then restriction defines a fibration $\Gamma(L) \to \Gamma(K)$.*

Proof We may suppose given a map $P \to \Gamma(L)$ and a homotopy of the composed map to $\Gamma(K)$; i.e. $f : P \times L \to E$ and $g : P \times K \times I \to E$ with $g(p, a, 0) = f(p, a)$ if $a \in K$. We seek to construct a homotopy $P \times L \times I \to E$ covering the projection on B and also extending g. But this exists since f has the CHP for the pair $(P \times L, P \times K)$. □

In case $K \subset L$ are smooth manifolds, there is a corresponding result for spaces of smooth sections.

A map $f : X \to Y$ is said to be a *weak homotopy equivalence* if, for any CW-pair (K, L) and maps $a : L \to X$ and $b : K \to Y$ with $b \mid L = f \circ a$ there exists $c : K \to X$ with $c \mid L = a$ and $f \circ c$ homotopic to b keeping L fixed. For this it suffices to consider pairs $S^{k-1} \subset D^k$ instead of $L \subset K$; thus for X connected it suffices if f induces isomorphisms $f_* : \pi_r(X) \to \pi_r(Y)$ of homotopy groups.

Lemma 6.1.3 *Suppose given a commutative diagram*

$$\begin{array}{ccc} E & \xrightarrow{h} & E' \\ {\scriptstyle p}\downarrow & & \downarrow{\scriptstyle p'} \\ B & \xrightarrow{g} & B' \end{array}$$

with p and p′ fibrations and g a weak homotopy equivalence. Then if h is a weak homotopy equivalence, so its restriction to each fibre of p.

Conversely, if the fibre map h induces a weak homotopy equivalence on each fibre, g is a weak homotopy equivalence.

This result is an easy deduction from the homotopy exact sequences of the fibrations and the five lemma.

6.2 Geometry of immersions

If $f : V \to M$ is an immersion, at each $P \in V$ the map $df_P : T_PV \to T_{f(P)}M$ is injective. When we were discussing submanifolds, we remarked that the restriction of $\mathbb{T}(M)$ to V had $\mathbb{T}(V)$ as a sub-bundle, and described the quotient as $\mathbb{N}(M/V)$. If f is an immersion, instead of the restriction of $\mathbb{T}(M)$ we have its pullback $f^*\mathbb{T}(M)$ by f, and an embedding of $\mathbb{T}(V)$ as a sub-bundle of $f^*\mathbb{T}(M)$. The main result about immersions is a converse to this statement.

A homotopy $g_t : V \to M$ is called a *regular homotopy* if g_t is an immersion for each t. We also seek to classify immersions up to regular homotopy. In fact, not only is the main result stated in more precise terms, but the result is a special case of a principle, formulated by Gromov [59] and called by him the h-principle, which holds for a variety of other geometric structures as well as immersions.

Again let $f : V \to M$ be an immersion, then $j^1 f : V \to J^1(V, M)$ avoids the set $\Sigma^*(V, M) := \bigcup_{i>0} \Sigma^i(V, M)$ of singular jets, and so defines a section g to the projection map $\pi_1^* : J^1(V, M) \setminus \Sigma^*(V, M) \to V$, and this section carries the information about bundles. Not every section of $\pi_1 : J^1(V, M) \to V$ has the form $j^1 f$ for a map $f : V \to M$: apart from the requirement of differentiability, these sections satisfy the additional equations given in local coordinates as $u^i_j = \partial y_j / \partial x_i$. Nevertheless, we will see that in many situations any section of π_1^* can be deformed to one of the form $j^1 f$, hence arising from an immersion f.

In fact the proof gives a stronger result, and this strengthening is key to the proof. Instead of considering a single map, we consider spaces of maps. Taking 1-jets defines a map J from the space $\mathrm{Imm}(V, M)$ of immersions to the space $\Gamma(V, M)$ of sections of π_1^*. We now state the main theorem.

Theorem 6.2.1 *Provided that either $v < m$ or V^v is open, the map*

$$J : Imm(V, M) \to \Gamma(V, M)$$

is a weak homotopy equivalence.

This will usually only be applied in the following form.

Corollary 6.2.2 *Any section of $\Gamma(V, M)$ is homotopic to one induced by an immersion, and two immersions are regularly homotopic if and only if the corresponding sections of $\Gamma(V, M)$ are homotopic.*

The necessity of the condition in the theorem is clear: for example, there is certainly no immersion $S^1 \to \mathbb{R}^1$, since any map has bounded image, while the image of an immersion would be open.

We can state the result in a more concrete way. Recall that $\Gamma(V, M)$ is the space of sections of $\pi_1 : J^1(V, M) \to V$, and that a 1-jet with source P and target Q is determined by these points and a linear map $T_P V \to T_Q M$. Thus a section σ of $\Gamma(V, M)$ assigns to each point $P \in V$ a point $f(P) = Q \in M$ and a linear map $g(P) : T_P V \to T_Q M$. The component f is a smooth map $V \to M$, and g gives a map $\mathbb{T}(V) \to f^*\mathbb{T}(M)$, which we require to be injective on each fibre.

We pause to introduce the Stiefel manifold $V_{m,v}$, defined as the set of isometric embeddings $\mathbb{R}^v \to \mathbb{R}^m$, and hence diffeomorphic to O_m / O_v. This is a deformation retract of the space of linear embeddings $\mathbb{R}^v \to \mathbb{R}^m$, which we denote $V'(m, v)$ and call the weak Stiefel manifold.

The above bundle map g is injective on each fibre if and only if it is a section of a bundle ξ'_f over V with fibre $V'_{m,v}$. Thus the existence part of the criterion can be formulated as follows.

Proposition 6.2.3 *Provided that either $v < m$ or V^v is open, a map $f : V^v \to M^m$ is homotopic to an immersion if and only if the above bundle ξ'_f over V admits a section.*

A corresponding formulation for the uniqueness part concerns a homotopy between maps f_0, f_1 and a homotopy of sections g_i lying over this.

Note also that homotopy classes of sections of ξ'_f correspond to classes of sections of a bundle ξ_f over V with fibre $V_{m,v}$.

In the case $M = \mathbb{R}^m$, the tangent bundle $\mathbb{T}(M)$ is trivial and ξ_f is associated to the normal bundle $\mathbb{N}(M/V)$ of V. An easy application is to the case when $\mathbb{T}(V)$ is trivial.

Corollary 6.2.4 *If V^v has trivial tangent bundle, there is an immersion of V in $\mathbb{R}^{v+1}$; if V is open, it immerses in $\mathbb{R}^v$.*

The idea of the proof of the theorem is to build V up as a union of stages V^i and to show, by induction on i, that the result holds at each stage. At each stage we attach a k-handle for some k, and need to show that the property remains true. This step is established by induction on k.

We recall the notation $D^k(a)$ for the disc $\{x \in \mathbb{R}^k \mid \|x\| \le a\}$; we now also write $D^k(a, b) := \{x \in \mathbb{R}^k \mid a \le \|x\| \le b\}$.

The theorem will be deduced from three lemmas.

Lemma 6.2.5 *The theorem holds if $V = D^v$ is a disc.*

Lemma 6.2.6 *Suppose V^+ obtained from V by introducing a corner or attaching a collar. Then the restriction maps $Imm(V^+, M) \to Imm(V, M)$ and $\Gamma(V^+, M) \to \Gamma(V, M)$ are weak homotopy equivalences.*

Lemma 6.2.7 *The restriction map $Imm(D^k(2) \times D^{v-k}, M) \to Imm(D^k(1, 2) \times D^{v-k}, M)$ is a fibration.*

Proof of Theorem 6.2.1 By Corollary 5.1.7 (if V is compact) and Lemma 5.1.8 (if not), V has a handle decomposition. If $v < m$, this can have no m-handles; if $v = m$ and V is open we may suppose by Proposition 5.4.1 that there are none. First suppose V compact.

We will prove by induction on k that the result holds for any V' which has only j-handles for $j < k$. Lemma 6.2.5 provides the start of the induction. We also induct on the number of handles of V: write V_i for the manifold with i handles, and suppose V_{i+1} obtained by attaching a k-handle. By the description in §5.1, V_{i+1} is obtained from V_i by first introducing a corner to obtain V_i^+, say, then attaching a copy of $D^k \times D^{v-k}$, which we take as $D^k(2) \times D^{v-k}$. Consider the diagram

$$\begin{array}{ccc} \mathrm{Imm}(D^k(2) \times D^{v-k}, M) & \xrightarrow{J} & \Gamma(D^k(2) \times D^{v-k}, M) \\ \downarrow & & \downarrow \\ \mathrm{Imm}(D^k(1,2) \times D^{v-k}, M) & \xrightarrow{J} & \Gamma(D^k(1,2) \times D^{v-k}, M) \end{array} . \quad (6.2.8)$$

By Lemma 6.2.7, the left-hand downward map is a fibration; the right-hand one is by Lemma 6.1.2. By Lemma 6.2.5 together with Lemma 6.2.6, the upper map J is a weak homotopy equivalence. By inductive hypothesis, so is the lower map J. Hence by Lemma 6.1.3 J induces a weak homotopy equivalence on each fibre. Now the diagram

$$\begin{array}{ccc} \mathrm{Imm}(V_{i+1}, M) & \xrightarrow{J} & \Gamma(V_{i+1}, M) \\ \downarrow & & \downarrow \\ \mathrm{Imm}(V_i^+, M) & \xrightarrow{J} & \Gamma(V_i^+, M) \end{array} \quad (6.2.9)$$

maps by restriction to diagram (6.2.8). The vertical maps in (6.2.9) are the pull-backs of the vertical maps in (6.2.8) which are fibrations; hence they too are fibrations. The restriction of J to each fibre in (6.2.9) is a weak homotopy equivalence. Since the lower map J is a weak homotopy equivalence by Lemma 6.2.6, it follows that the upper also is.

For the case when V is not compact, so we have an infinite number of handles, we note that $\mathrm{Imm}(V, M)$ is the inverse limit of the $\mathrm{Imm}(V_i, M)$, $\Gamma(V, M)$ is the inverse limit of the $\Gamma(V_i, M)$, and apply Lemma B.1.3. □

Proof of Lemma 6.2.5 Since the disc is contractible, the space $\Gamma(D^v, M)$ of sections of $\pi_1^* : J^1(D^v, M) \setminus \Sigma^*(D^v, M) \to D^v$ is homotopy equivalent to the space of sections over the origin, which is the space W of injective linear maps from $\mathbb{R}^v$ to the tangent space $T_Q M$.

We need to look at a map from D^k to $\Gamma(D^v, M)$ and a lift to $\mathrm{Imm}(D^v, M)$ of its restriction to S^{k-1}. So for each $x \in D^k$ we have an injective linear map from $\mathbb{R}^v$ to some $T_Q M$, i.e. a 1-jet j_x^1 at 0 of a map $f_x : D^v \to M$. To see that we can choose the f_x to depend continuously on x, take a closed embedding $h : M \to \mathbb{R}^K$ (which exists by Corollary 4.7.8), a tubular neighbourhood of its image with image N, and hence a smooth retraction $\rho : N \to M$, the projection of the tube. Now j_x^1, composed with the inclusion h gives a 1-jet of map $\mathbb{R}^v \to \mathbb{R}^K$ which has polynomial (in fact, linear) representative g_x. Composing with ρ gives $f_x = \rho \circ g_x$, defined on a neighbourhood of 0, and depending continuously on x. Moreover since D^k is compact we can choose the same neighbourhood for all $x \in D^k$.

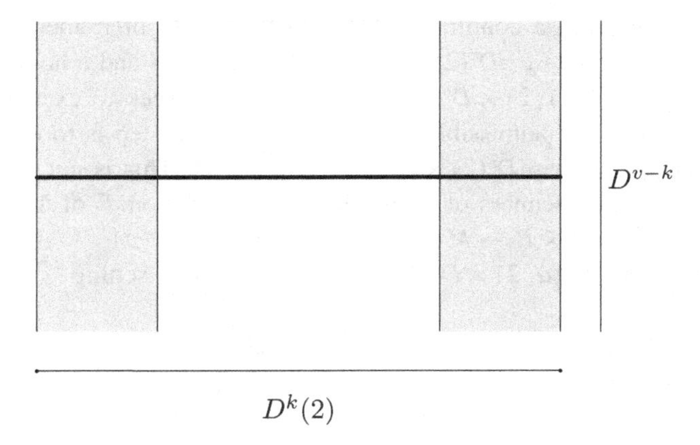

Figure 6.1 The core of $D^k(2) \times D^{v-k}$

We have constructed a lift, but on a smaller disc $D^v(\varepsilon)$. Composing with a diffeotopy J_t of D^v, equal to the identity near 0, and compressing D^v inside $D^v(\varepsilon)$ by a diffeotopy, we obtain a lift of maps of the larger disc.

This does not yet agree on the boundary S^{k-1} with the given lift. For $x \in \partial D^k$ we have the given immersion $g_x : D^v \to M$ and the map f_x just constructed, both with the same 1-jet at 0. Working again in $\mathbb{R}^N$, we consider the linear interpolation $\lambda g_x(y) + (1-\lambda) f_x(y)$ and compose with ρ to obtain a homotopy in M. Since the 1-jets at $0 \in D^v$ are constant, this restricts to a regular homotopy on a smaller disc $D^v(\varepsilon')$. Using J_t again, we obtain a regular homotopy on D^v. □

The proof of Lemma 6.2.6 is simple: the second result holds since $V \subset V^+$ is a homotopy equivalence; as to the first, we have embeddings $V \to V^+ \to V$ with composite diffeotopic to the identity.

Proof of Lemma 6.2.7 The proof that

$$\mathrm{Imm}(D^k(2) \times D^{v-k}, M) \to \mathrm{Imm}(D^k(1,2) \times D^{v-k}, M)$$

is a fibration is the key to the whole result. Define the *core* (of $D^k(2) \times D^{v-k}$) to be $C := (D^k(2) \times \{0\}) \cup (D^k(1,2) \times D^{v-k})$: this is pictured in Figure 6.1.

The parameter space P plays very little part below (we just use the fact that P is compact). Nor does M: we have to make sure that each map into M is an immersion, but can use the fact that immersions form an open set. Let us call a map $\phi : A \times B \to M$, with A a submanifold of $D^k(2) \times D^{v-k}$, *admissible* if, for each $b \in B$, the induced map $a \to \phi(a, b)$ is an immersion.

We are thus given a continuous P-parameter family of immersions giving an admissible map $g : D^k(2) \times D^{v-k} \times \{0\} \times P \to M$ and a homotopy of its restriction $f : D^k(1,2) \times D^{v-k} \times I \times P \to M$, and seek to extend this to a homotopy (through admissible maps) of g. The first step is to extend the maps f and g to a map $D^k(2) \times D^{v-k} \times I \times P \to M$. This is not in general admissible, but by openness of immersions, the restriction f' of this map to $D^k(\alpha, 2) \times D^{v-k} \times I \times P \to M$ is admissible for some $\alpha < 1$.

Next define $K : D^k(\alpha, 2) \times D^{v-k} \times I \times P \times I \to M$ by setting

$$\begin{aligned} K(x, y, t, z, T) &= f'(x, y, t, z) && \text{if } T = t \text{ or if } \|x\| > (\alpha + 2)/3 \\ &= f'(x, y, T, z) && \text{if } |\|x\| - \alpha| < (1 - \alpha)/3 \end{aligned}$$

and extending smoothly to other values. This defines an admissible map on some neighbourhood of $T = t$, hence for some $\varepsilon > 0$ whenever $|t - T| < \varepsilon$. We can thus find $0 = t_0 < t_1 < \ldots < t_s = 1$ such that $k_i(x, y, t, z) := K(x, y, t, z, t_i)$ is admissible for $t_i \leq t \leq t_{i+1}$.

We will now inductively construct an admissible extension g_n for $0 \leq t \leq t_n$ which is equal to f' on $\|x\| \geq a_n$, where $\alpha = a_0 < \cdots a_{n-1} < a_n < 1$. We start the induction by setting

$$\begin{aligned} g_0(x, y, p, t) &= g(x, y, p, t) && \text{if } \|x\| \leq \alpha, \\ &= k_0(x, y, p, t) && \text{if } \|x\| \geq \alpha. \end{aligned}$$

The key step is now a diffeotopy $h_n : D^k(2) \times D^{v-k} \times [0, t_n] \to D^k(2) \times D^{v-k}$ such that

(i) $h_n(x, y, t) = (x, y)$ on a neighbourhood of $\|x\| = a_n,\ y = 0$, and outside a neighbourhood of $a_{n-1} \leq \|x\| \leq a_{n+1}$,

(ii) $h_n(x, y, t) = (x, y)$ if $t \leq t_{n-1}$,

(iii) $h_n(*, *, t_n)$ maps $S^{k-1}(a_{n+1}) \times \{0\}$ onto $S^{k-1}(a) \times \{0\}$.

To construct this it is essential that $k < v$, so that dim $D^{v-k} > 0$ and there is enough space within D^{v-k} to move one point past another. The effect of using this diffeotopy is to introduce folds in the immersion, thus giving extra space to move.

We define g_{n+1} by

$$\begin{aligned} g_{n+1}(x, y, t, z) &= g_n(x, y, t, z) && \text{if } 0 \leq t \leq t_n,\ \|x\| \leq a_n \\ &= f'(h_n(x, y, t), t, z) && \text{if } 0 \leq t \leq t_n,\ a_n \leq \|x\| \leq 2 \\ &= k_n(h_n(x, y, t_n), t, z) && \text{if } t_n \leq t \leq t_{n+1},\ a_{n+1} \leq \|x\| \leq 2 \\ &= g_{n+1}(x, y, t_n, z) && \text{if } t_n \leq t \leq t_{n+1},\ \|x\| \leq a_{n+1} \end{aligned}$$

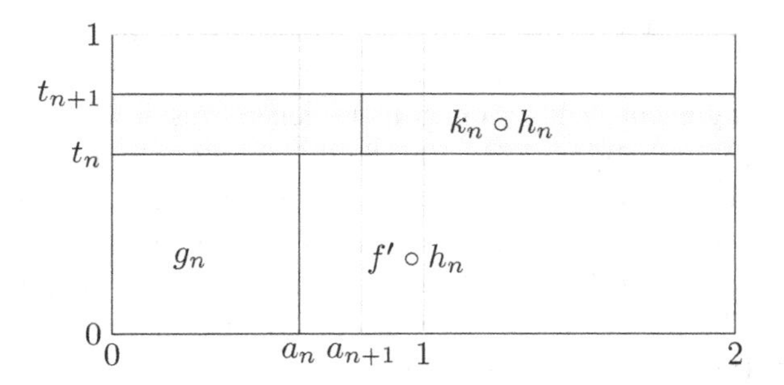

Figure 6.2 Piecing together the construction of g_{n+1}

We offer Figure 6.2 to help follow this 'piecing together' construction of g_{n+1}: note also that g_{n+1} has already been defined for $t = t_n$ by the lines above, and in the upper left rectangle, the value is independent of t.

We now check that, at least for (x, y) in some neighbourhood C^* of C, these formulae agree on the intersections of the different domains of definition.

First consider $0 \leq t \leq t_n$, $\|x\| = a_n$. Since $h_t(x, y)$ is constant near $\|x\| = a_n$, $y = 0$, the second formula here reduces to f', as does the first.

Along $t = t_n$, we have
(i) $g_n(x, y, t_n, z)$ $(\|x\| \leq a_n)$,
(ii) $f'(h_{t_n}(x, y), t_n, z)$ $(a_n \leq \|x\| \leq 2)$,
(iii) $k_n(h_{t_n}(x, y), t_n, z) = g'(h_{t_n}(x, y), t_n, z, t_n) = f'(h_{t_n}(x, y), t_n, z)$ $(a_{n+1} \leq \|x\| \leq 2)$, while the final formula agrees by definition, so indeed all match up.

Finally consider $t_n \leq t \leq t_{n+1}$, $\|x\| = a_{n+1}$. Again we need only check in a neighbourhood of $y = 0$, and use (iii) above. The fourth formula is independent of t in this range, and we have already checked agreement at $t = t_n$. Now $k_n(h_{t_n}(x, y), t, z) = g'(h_{t_n}(x, y), t, z, t_n)$, and indeed this is independent of t if t_n is near to a.

Choose a diffeotopy H_t of $D^k(2) \times D^{\upsilon-k}$ into itself which is the identity on a neighbourhood of C and has $H_1(D^k(2) \times D^{\upsilon-k}) \subset C^*$. We use this to re-parametrise our maps to obtain the desired extension: the map $g_{n+1}(H_1(x, y), p, t)$ is admissible, and allows us to continue the induction. □

Since the proof proceeded by attaching successively to V handles of dimension less than υ, it follows without further argument that if we have already constructed an immersion on a closed submanifold W of V of the same dimension, this can be extended over the rest of V, provided dim $V <$ dim M or no component of $W \setminus V$ has compact closure. Applying this to the case when W is a collar neighbourhood of ∂V, we deduce that the theorem extends to the case of immersions $(V, \partial V) \to (M, \partial M)$: given a map covered by an injective map of tangent spaces, then if $v < m$ we can construct an immersion.

A similar argument yields a much more general result. Let $E(M)$ be any bundle over M naturally associated to the differential structure, in the sense that any diffeomorphism $M \to N$ induces an isomorphism $E(M) \to E(N)$ and for any open set $U \subset M$, the restriction of $E(M)$ to U is naturally isomorphic to $E(U)$: for example, for any manifold P we can take $E(M) = J^k(M, P)$. Write $E^r(M)$ for the bundle of r-jets of sections of $E(M)$. Let $E_0^r(M)$ be an open subbundle of $E^r(M)$ with the same invariance property.

Define $\Gamma E_0^r(M)$ to be the space of sections of $E_0^r(M)$ and $\Gamma_0 E(M)$ to be the space of sections σ of $E(M)$ whose r-jet is a section of $E_0^r(M)$. We assign these spaces the C^∞ topologies. The following result is called by Gromov the *h-principle.*

Theorem 6.2.10 *[59] If M is open, the map $j^r : \Gamma_0(M) \to \Gamma E_0^r(M)$ is a weak homotopy equivalence.*

For example, if we have an immersion $V^v \to \mathbb{R}^k$ with a continuous section to the bundle of unit normal vectors, we have an immersion $V^v \to \mathbb{R}^{k-1}$. We will use this to show in Theorem 6.3.6 that any manifold immerses in $\mathbb{R}^{2v-1}$.

A smooth map $f : V \to M$ is called a k-mersion if, at each point $P \in V$, the map $df_P : T_P V \to T_P M$ has rank $\geq k$. The h-principle applies to k-mersions: given a map $f : V \to M$ covered by a map $\mathbb{T}(V) \to f^*\mathbb{T}(M)$ having rank $\geq k$ at each point then f is homotopic to a k-mersion.

6.3 The Whitney trick

We have seen from general position arguments that any manifold M^m embeds in $\mathbb{R}^{2m+1}$ and immerses in $\mathbb{R}^{2m}$. It was shown by Whitney [177] that in fact M^m embeds in $\mathbb{R}^{2m}$ and, again by Whitney, in [178] that M^m immerses in $\mathbb{R}^{2m-1}$.

In this section we explain the construction used by Whitney to establish these results. It has further applications, which will be frequently used in Chapter 7.

By Corollary 4.5.8 to the transversality theorem, if we have two embeddings of compact manifolds $f : V^v \to M^m$ and $f' : V'^{v'} \to M^m$ with $m = v + v'$ we may suppose, up to a (small) diffeotopy of f, that the images are distinct except that there are finitely many pairs $P_i \in V$ and $P_i' \in V'$ with $f(P_i) = f'(P_i') = R_i$ and the two intersections are transverse.

Similarly, by Theorem 4.7.7, if $m = 2v$, for a dense set of maps $g : V^v \to M^m$, g is an immersion, and fails to be injective only insofar as there are finitely many pairs of distinct points (P_i, Q_i) in V with $g(P_i) = g(Q_i) = R_i$, but the two branches of V are transverse.

The so-called Whitney trick is a procedure which, under some conditions, perturbs the embeddings f and f' to be disjoint, or perturbs the map g to obtain an embedding of V in M. Given orientations, each intersection is assigned (at least locally) a sign. The construction will enable us to cancel a pair of intersections of opposite signs. Moreover this is achieved by a diffeotopy of f or a regular homotopy of g.

A second construction, also due to Whitney, allows us to introduce a single self-intersection (of either sign) of g by taking connected sum with a standard map inside a coordinate neighbourhood. Combining the two constructions gives a further chance to modify g to an embedding.

Suppose given orientations of V, V' and M. At a point R where V and V' intersect transversely, we have $T_RM = T_RV \oplus T_RV'$. Choose bases $(e_1, \ldots, e_v)$ of T_RV and $(e'_1, \ldots, e'_{v'})$ of T_RV' defining the given orientations. Then the local intersection number of V and V' at R is defined to be $+1$ if the basis $(e_1, \ldots, e_v, e'_1, \ldots, e'_{v'})$ of T_RM defines the given orientation of M and -1 if it does not.

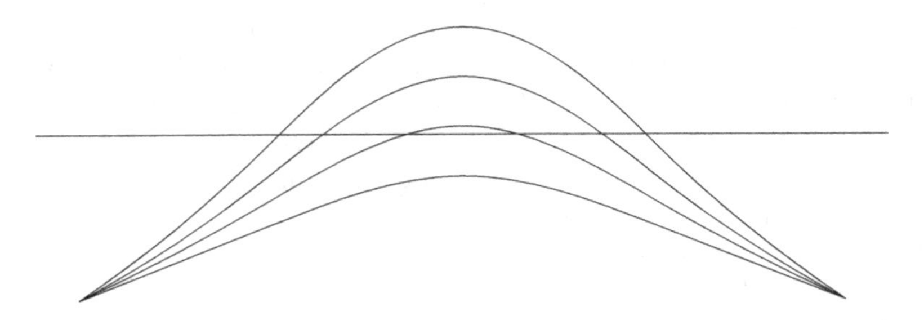

Figure 6.3 Model of the deformation

The model picture, which is illustrated in Figure 6.3, is to take the line $C : y = 0$ in the plane and the curve $C' : y = x^2 - 1$ intersecting it at the two points $A_1 = (1, 0)$, $A_2 = (-1, 0)$ and deform the curve C' vertically to C_t given by $y = x^2 - 1 + t$: for $t > 1$ the intersections have disappeared. More precisely, we choose a deformation $y(x, t)$ for $|x| \leq 1 + 2\varepsilon$ and $t \in I$ such that for $|x| \leq 1$ we have $y(x, t) = x^2 - 1 + t$ and for $|x| \geq 1 + \varepsilon$ we have $y(x, t) = x^2 - 1$. Write D^* for the region spanned by the two arcs, and D^+ for a neighbourhood of D^* in $\mathbb{R}^2$.

Suppose given connected manifolds V, V' of dimensions v, v', and embeddings $\phi : V \to M^m$, $\phi' : V' \to M^m$ with $m = v + v'$ which intersect transversely. The key idea is to embed the above model in the manifold.

Take two points of intersection, say $\phi(P_i) = \phi'(P_i') = R_i$ $(i = 1, 2)$. Choose paths $f : (I, 0, 1) \to (V, P_1, P_2)$ and $f' : (I, 0, 1) \to (V', P_1', P_2')$. Since v and v' are each ≥ 3, general position shows that we may take each of f, f' to be a smooth embedding. Taking x as parameter on C and C' allows us to regard the paths as maps $f : (C, A_1, A_2) \to (V, P_1, P_2)$ and $f' : (C', A_1, A_2) \to (V', P_1', P_2')$. Then $\phi \circ f$ and $\phi' \circ f'$ together define a loop $F : C \cup C' \to M$.

Proposition 6.3.1 *Suppose also that V, V' and M are all orientable, that the intersections at R_1 and R_2 have opposite signs, and that either*

(i) v, $v' \geq 3$ and F defines a nullhomotopic loop in M or

(ii) $v > 2 = v'$ and F defines a nullhomotopic loop in $M \setminus V$.

Then there is an embedding $\phi : D^+ \times \mathbb{R}^{v-1} \times \mathbb{R}^{v'-1} \to M$ such that $\phi^{-1}(V) = (D^+ \cap C) \times \mathbb{R}^{v-1} \times \{0\}$ and $\phi^{-1}(V') = (D^+ \cap C') \times \{0\} \times \mathbb{R}^{v'-1}$.

Proof Since F is nullhomotopic in M, the map of the two arcs extends to a map of the disc D^*, and hence to a map g of a neighbourhood D^+, which we can take as smooth. We next put the map g in general position. Since $m \geq 5$, we may suppose that g is an embedding.

In case (i) as v, $v' > 2$ we may suppose using general position that the only intersections of $g(D^+)$ with V and V' occur along the images of C and C'. In case (ii) we can avoid V' by general position, and the fact that the extension g, outside a neighbourhood of C, can be taken to avoid V holds by our hypothesis.

For short, write C for $g(C \cap D^+)$ and C' for $g(C' \cap D^+)$. Write ζ for the tangent vector $\zeta = d(\phi \circ f)(\partial/\partial t)$ along C. Let η_1 be the vector field along C, normal to C, and inwards tangent to $g(D^*)$. Similarly write ζ' for the tangent along C' and ξ_1 for the normal pointing inwards along $g(D^*)$. Observe that we have $\xi_1 = \zeta$ and $\eta_1 = \zeta'$ at R_1, and $\xi_1 = -\zeta$, $\eta_1 = -\zeta'$ at R_2. These steps are illustrated in Figure 6.4.

We next construct smooth vector fields ξ_i $(2 \leq i \leq v)$ and η_j $(2 \leq j \leq v')$ on $g(D^+)$ such that

(i) at each point, they form a base for the normal space to $g(D^+)$,

(ii) along C the ξ_i $(2 \leq i \leq v)$ are tangent to V, and

(iii) along C' the η_j $(2 \leq j \leq v')$ are tangent to V'.

First choose vectors $\xi_2, \ldots, \xi_v$ tangent to V at R_1 such that $(\zeta, \xi_2, \ldots, \xi_v)$ is a base defining the orientation of V. Since C is contractible, we can extend this to give a base of $T_P V$ at all $P \in C$. We can also extend $\xi_1, \xi_2, \ldots, \xi_v$ to give a base of the normal space $N_P(M/V')$ at all $P \in C'$, but now need compatibility at R_2. Since the intersections at R_1 and R_2 have opposite signs, the two orientations

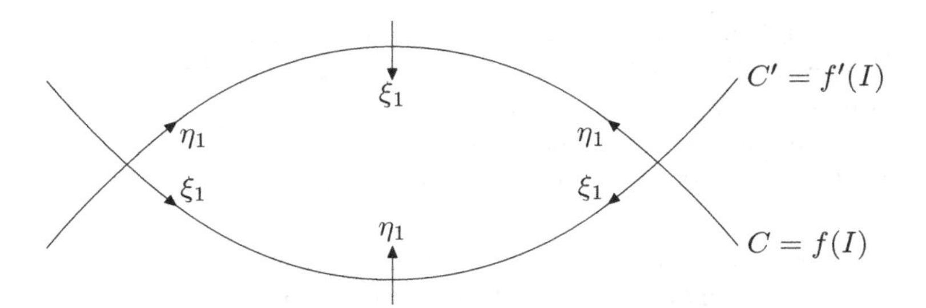

Figure 6.4 Using the model to construct a deformation

of $T_{R_2}V$ differ. But as $\zeta = -\xi_1$ at R_2, the two bases $\xi_2, \ldots, \xi_v$ at R_2 have the same orientation, so by choosing a different extension above, we may suppose that they coincide, so these vectors are defined over $C \cup C'$.

The bundle over D^+ of orthonormal $(v-1)$-frames orthogonal to D^+ is a trivial bundle whose fibre is the Stiefel manifold $V_{v+v'-2,v-1}$. We have constructed a section over a circle; since $v' \geq 3$, $V_{v+v'-2,v-1}$ is simply connected (see §B.3(xv)), so we can extend the section over D^+.

Since D^+ is contractible, all bundles over D^+ are trivial. We may thus extend $\eta_2, \ldots, \eta_{v'}$ to a base for the bundle of vectors normal to D^+ and to the ξ_i. It follows from our choice of these that these satisfy (iii) above. We have thus constructed normal vector fields $(\xi_2, \ldots, \xi_v, \eta_2, \ldots, \eta_{v'})$ along D^+ such that $(\xi_2, \ldots, \xi_v)$ are tangent to V at all points of V $(\eta_2, \ldots, \eta_{v'})$ are tangent to V' at all points of V'.

By Theorem 2.5.5, a neighbourhood of D^+ is diffeomorphic to a disc bundle over D^+, which must be trivial, hence diffeomorphic to $D^+ \times \mathbb{R}^{v+v'-2}$; corresponding statements hold for C and C'.

By Proposition 2.5.10 there exists a tubular neighbourhood of C in M whose restriction gives a tubular neighbourhood of C in V; by the remark following that result, we may suppose this neighbourhood compatible with D^+. We use this to define ϕ on a neighbourhood of C.

We argue similarly for C'; moreover, these neighbourhoods are constructed by glueing together pieces, so if we begin with charts at P and P', we can ensure that these maps agree, thus defining ϕ on a neighbourhood of $C \cup C'$.

As above, we can use general position to extend ϕ over D^+. Finally, the above maps may be regarded as defining a tubular neighbourhood for D^+ on a neighbourhood of $C \cup C'$, and the proof of Proposition 2.5.10 shows how to extend this over D^+: note that our construction of bases for the normal spaces shows that these partial tubular neighbourhoods do indeed define an embedding of our model. □

Theorem 6.3.2 *Suppose we have connected orientable manifolds V^v, $V'^{v'}$ and embeddings $\phi : V \to M^{v+v'}$, $\phi' : V' \to M^{v+v'}$ which intersect transversely, with two intersection points $\phi(P_i) = \phi'(P'_i) = R_i$ ($i = 1, 2$) of opposite signs. Choose paths giving smooth embeddings $f : (C, A_1, A_2) \to (V, P_1, P_2)$ and $f' : (C', A_1, A_2) \to (V', P'_1, P'_2)$ defining a loop $F : C \cup C' \to M$. Suppose also that either*

(i) $v, v' \geq 3$ and F defines a nullhomotopic loop in M or

(ii) $v > 2 = v'$ and F defines a nullhomotopic loop in $M \setminus V$.

Then there is a diffeotopy of $\phi : V \to M$, supported on $\phi(I)$, such that $h_1(V) \cap V'$ agrees with $V \cap V'$ less the points R_1, R_2.

Proof It suffices to construct a diffeotopy in the model $D^+ \times \mathbb{R}^{v-1} \times \mathbb{R}^{v'-1}$ which is constant near the boundary, then transport it into M by the embedding ϕ.

Begin with the diffeotopy $\phi_t : C \times I \to D^+$, modified using a bump function to be the identity outside a neighbourhood of D^*. Now define

$$\Phi : C \times \mathbb{R}^{v-1} \times I \to D^+ \times \mathbb{R}^{v-1} \times \mathbb{R}^{v'-1}$$

by $\Phi_t(x, y) = (\phi_{t\alpha(y)}(x), y, 0)$, where $\alpha(y)$ is equal to 1 when $y = 0$ and to 0 for $\|y\| \geq \varepsilon$. □

Although the details involve local orientations, the hypothesis of global orientability is not needed for this argument. If V, for example, is non-orientable, by replacing f by its composite with an orientation reversing loop we can change the local orientation, so in this case we do not need to assume the two intersections of opposite signs. However the condition that (i) or (ii) holds is essential.

The same construction is used, taking $V = V'$, to eliminate self-intersections of an n-manifold in a $2n$-manifold. Here we can do somewhat more. First suppose n odd. Then the sign of the intersection number is changed if we reverse the order of the two branches V, V' at R. Thus (provided $n > 2$ and M is simply connected) we can eliminate any pair of transverse self-intersections by a regular homotopy, as we can start by joining P_1 to P'_2 and P'_1 to P_2 instead.

We also have

Proposition 6.3.3 *There is a self-transverse immersion $\Xi : S^n \to S^{2n}$ with a single transverse self-intersection, and with normal bundle isomorphic to the tangent bundle of S^n.*

Proof We begin with the immersion $f : S^1 \to \mathbb{R}^2$ given by $f(\theta) = (\sin\theta, \frac{1}{2}\sin 2\theta)$, where θ denotes the angular coordinate on $S^1 =$

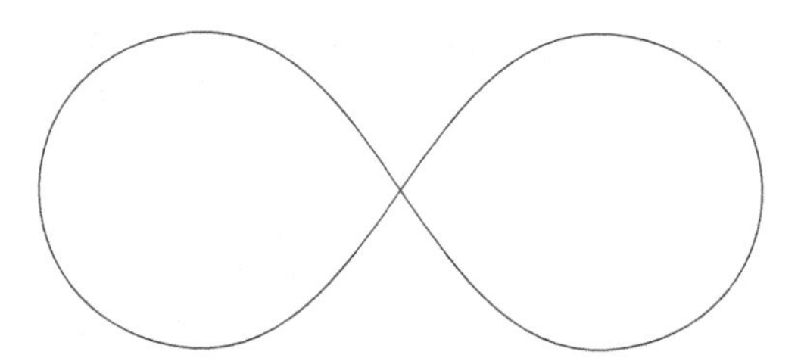

Figure 6.5 An immersed sphere

$\{(\sin\theta, \cos\theta)\}$. (This is essentially the same as an example already used in §1.2.) The curve passes through the origin when $\theta = 0$ and when $\theta = \pi$, and the two branches are transverse.

Rotating this, define $F : S^{n-1} \times [0, \pi] \to \mathbb{R}^{2n}$ by $F(\xi, \theta) = (\xi \sin\theta, \frac{1}{2}\xi \sin 2\theta)$, where we regard S^{n-1} as a subset of $\mathbb{R}^n$ and identify $\mathbb{R}^{2n}$ with $\mathbb{R}^n \times \mathbb{R}^n$. We observe that $F(\xi, \theta) = F(-\xi, -\theta)$. All the points with $\theta = 0$ or π are again mapped to the origin. Hence F factors as $F = G \circ p$, where $p : S^{n-1} \times [0, \pi] \to S^n$ is defined by $p(\xi, \theta) = (\xi \sin\theta, \cos\theta)$. I claim that $G : S^n \to \mathbb{R}^{2n}$ is a smooth immersion.

Near the point $(0, \dots, 0, 1)$ on S^n we can write $x = \xi sin\,\theta$: then $\|x\| = sin\,\theta$, so $y = cos\,\theta = \sqrt{1 - \|x\|^2}$ and $G(x, y) = (x, x\sqrt{1 - \|x\|^2})$. Here we can take x as giving local coordinates, and $\sqrt{1 - \|x\|^2}$ is a smooth function of x near $x = 0$. Thus the map is smooth at this point, and the image has tangent the diagonal $\{(x, x)\}$. A similar calculation deals with the point $(0, \dots, 0, -1)$ (here $\theta = \pi$). The image (for $n = 1$) is pictured in Figure 6.5.

For the immersion f, both tangent and normal bundle are trivial. We can define an isomorphism between them by rotating each tangent vector through an angle $+\frac{\pi}{2}$ in the plane. A corresponding rotation can be made in $\mathbb{R}^{2n}$, using the matrix (in block form) $\begin{pmatrix} 0 & -I \\ I & 0 \end{pmatrix}$, and again this gives an isomorphism of the tangent space to $G(S^n)$ on its normal space. That this also works for each branch at the origin follows from the above calculation of the tangent space there.

Composing G with an embedding $\mathbb{R}^{2n} \subset S^{2n}$ given the desired map Ξ. □

Given any immersion $V^n \to M^{2n}$ we can take a connected sum with J, at a smooth point of each submanifold, and this produces (up to diffeomorphism) another immersion $V^n \to M^{2n}$, but now with an additional point of self-intersection. Moreover, changing the orientation, this point can be supposed to have intersection number of either sign.

Theorem 6.3.4 *Given compact smooth manifolds with V^n connected, M^{2n} simply-connected, and $n \geq 3$, any map $f : V^n \to M^{2n}$ is homotopic to a smooth embedding.*

Proof We may suppose by general position that f is an immersion, and is self-transverse, so that there are finitely many points P_i of self-crossing.

First suppose V orientable. Choose orientations of V and M, and hence a sign ± 1 attached to each P_i. As just observed, we may introduce a further self-intersection point of either sign. Introduce such points until there are equal numbers of both signs.

Given two points of opposite sign, we apply Theorem 6.3.2 to construct a diffeotopy of the neighbourhood of an arc in V joining the points, hence inducing a regular homotopy of V, which removes these intersection points and introduces no new ones.

If V is non-orientable, we first introduce a self-intersection point, if necessary, to make the number of such points even. Now we can cancel any pair of P_i by the same construction: we just need to choose the arc in V of the desired parity. □

The hypotheses are necessary. If V has two components, they may have non-zero intersection number in M: for example, consider $(S^n \times *) \cup (* \times S^n)$ in $S^n \times S^n$. If M is not simply-connected, counting self-intersections more carefully gives an obstruction lying in the group ring $\mathbb{Z}[\pi_1(M)]$: see [167, §5]. For a counterexample, we can take the above map Ξ with M a neighbourhood of its image. If $n = 2$, the whole Whitney trick fails.

For any manifold M^{2n}, we know by general position that any map $S^n \to M^{2n}$ is homotopic to an immersion, and from Theorem 6.2.1 that immersions in a given homotopy class are classified up to regular homotopy by $\pi_n(E)$, where $E \to M$ is the bundle associated to $\mathbb{T}M$ and with fibre the Stiefel manifold $V_{2n,n}$. We have an exact sequence

$$\pi_n(V_{2n,n}) \to \pi_n(E) \to \pi_n(M) \to \{1\};$$

by §B.3(xvi), the first term is cyclic of order ∞ or 2 according as n is even or odd.

If n is even, the immersion ϕ determines a homology class $[\phi]$ with self-intersection number $[\phi].[\phi]$. It also has a normal bundle, with Euler class giving a number $e(\phi)$. We may suppose ϕ has transverse self-intersections; summing the intersection numbers at these points gives a further integer $I(\phi)$.

Lemma 6.3.5 *We have $[\phi].[\phi] = e(\phi) + 2I(\phi)$.*

Proof The homological self-intersection is the intersection number of the image of ϕ with a small perturbation of it; this is the sum of the contribution $e(\phi)$ of the self-intersection in the normal bundle and a contribution of 2 from each point of self-intersection of ϕ. □

Taking the connected sum with the example of Proposition 6.3.3 has the effect of changing the regular homotopy class by the action of a generator of $\pi_n(V_{2n,n})$, and of adding 2 to $e(\phi)$ and subtracting 1 from $I(\phi)$.

If n is odd, there are two regular homotopy classes of immersions in each homotopy class of maps $S^n \to M^{2n}$, one in which the number of self-intersections is even, and one with this number odd: since the parity is invariant under regular homotopy, the map $\pi_n(V_{n,n}) \to \pi_n(E)$ is injective. If also $n \geq 3$ and M is simply-connected, just the former regular homotopy class contains embeddings. The two classes have different normal bundles if $\mathbb{T}(S^n)$ is non-trivial, i.e. if $n \neq 3, 7$.

Similar conclusions to these apply with any V^n in place of S^n.

Now consider immersions of N^n in Euclidean spaces. By Proposition 6.2.3, a map $f : N^n \to M^m$ with $n < m$ is homotopic to an immersion if and only if a certain bundle η over N with fibre $V_{m,n}$ admits a section. Obstruction theory tells us that obstructions to the existence of sections lie in $H^i(N; \pi_{i-1}(V_{m,n}))$, and by §B.3(xv) $V_{m,n}$ is $(m-n-1)$-connected. So the primary obstruction is in $H^{m-n+1}(N; \pi_{m-n}(V_{m,n}))$; and by (xvi), $\pi_{m-n}(V_{m,n})$ is isomorphic to $\mathbb{Z}$ if $(m-n)$ is even and to $\mathbb{Z}_2$ if $(m-n)$ is odd. This obstruction is denoted $W_{m-n+1}(\eta)$; its image in $H^{m-n+1}(N; \mathbb{Z}_2)$ is the Stiefel–Whitney class $w_{m-n+1}(\eta)$ (see §8.6).

First take $m = 2n$: since $V_{2n,n}$ is $(n-1)$-connected, there is (as expected) no obstruction.

Theorem 6.3.6 *For $n \geq 2$, any smooth n-manifold immerses in $\mathbb{R}^{2n-1}$.*

Proof The result is due to Whitney [178]; we follow the account of Hirsch [70].

We set $m = 2n-1$ in the above. Since $V_{2n-1,n}$ is $(n-2)$-connected, the only obstruction is the primary obstruction, which lies in $H^n(N : \pi_{n-1}(V_{2n-1,n}))$.

If N is non-compact, or has boundary, then the obstruction lies in a zero group, so vanishes, and immersions exist. Otherwise the obstruction lies in the group $H^n(N; \mathbb{Z})$, where the coefficients are twisted if N is non-orientable, hence the group is isomorphic to $\mathbb{Z}$ in both cases.

Now proceed indirectly, and start with an immersion in $\phi : N^n \to \mathbb{R}^{2n}$. If we find a non-zero normal vector field to this immersion, this implies the existence of a section to the bundle with fibre $V_{2n-1,n}$ and hence of an immersion in $\mathbb{R}^{2n-1}$. Such a normal vector field exists if and only if the normal Euler class $e(\phi)$

vanishes. (We refer to [103, VIII] for background on this class.) The class $e(\phi)$ also lies in $H^n(N; \mathbb{Z})$ and can be identified with the above obstruction.

If n is even, we recall the equation $[\phi].[\phi] = e(\phi) + 2I(\phi)$. As intersection numbers in $\mathbb{R}^{2n}$ vanish, it follows that $e(\phi)$ is even. Taking the connected sum with a suitable numbers of copies of the example of Proposition 6.3.3 gives an immersion ψ with $e(\psi) = 0$. We can thus find a non-vanishing normal vector field to $\psi(N)$, and hence obtain an immersion in $\mathbb{R}^{2n-1}$.

If n is odd, the Euler class satisfies $2e(\phi) = 0$. Since it lies in a group isomorphic to $\mathbb{Z}$, it vanishes, so a normal vector field exists. □

Some results can be obtained for the problem of immersibility of N^n in $\mathbb{R}^{2n-2}$. It was shown in [70] that if $n \equiv 1 \pmod 4$, an immersion exists if and only if $W_{n-1}(\eta) = 0$.

Among the more interesting problems is to determine the lowest dimensions into which one can immerse or embed the real projective spaces $P^n(\mathbb{R})$. By studying the conditions on bundles, Atiyah [13] proved that, if we define $\sigma(n)$ to be the greatest integer s such that $2^{s-1}\binom{n+s}{s}$ is *not* divisible by $2^{\phi(n)}$ (where ϕ denotes Euler's phi function) then $P^n(\mathbb{R})$ cannot be immersed in $\mathbb{R}^{n+\sigma(n)-1}$ or embedded in $\mathbb{R}^{n+\sigma(n)}$.

6.4 Embeddings and immersions in the metastable range

Given an embedding $f : V^v \to \mathbb{R}^m$, as in §4.2 we associate to any pair of distinct points P, Q of V the non-zero vector $\Delta_f(P, Q) := f(Q) - f(P) \in \mathbb{R}^m$, and hence the unit vector $\delta_f(P, Q) := \mathbf{u}(\Delta_f(P, Q)) \in S^{m-1}$. Recalling the notation $V^{(2)}$ for the set of pairs of distinct points of V, we have a map $\delta_f : V^{(2)} \to S^{m-1}$ with the property $\delta_f(Q, P) = -\delta_f(P, Q)$. If f is an immersion, the same formula defines a map on $U \setminus \Delta(V)$ for some neighbourhood U of $\Delta(V)$ in $V \times V$.

Since we will have a number of similar conditions in this section, let us agree that we have standard actions of the group $\mathbb{Z}_2$ of order 2, given on $V \times V$ or on $V^{(2)}$ by interchange of the factors, and on a vector bundle by the map which is minus the identity on each fibre. Thus when we say a map is 'equivariant' we mean with respect to this action. We also say that an equivariant map $\alpha : A \to B$ is *isovariant* if the preimage of the fixed set (of the action) in B is the fixed set in A.

For any map $f : V \to M$ the product $f \times f : V \times V \to M \times M$ is equivariant. It is isovariant if and only if $f(x) = f(y)$ implies $x = y$, i.e. if f is injective. In this case, $f \times f$ restricts to an isovariant map $f^{(2)} : V^{(2)} \to M^{(2)}$.

If f is an immersion rather than an embedding, we know at least that it is locally injective, so there is a neighbourhood U of the diagonal $\Delta(V)$ in $V \times V$ such that the restriction of $f \times f$ to U is isovariant and the map $f^{(2)}$ is defined on $U \setminus \Delta(V)$. We will thus consider isovariant maps $g : U \to M \times M$ defined on an unspecified neighbourhood U of $\Delta(V)$ in $V \times V$; for such g denote by $g_\Delta : V \to M$ the map such that $g(x, x) = (g_\Delta(x), g_\Delta(x))$ for $x \in V$. If g, g' are isovariant maps defined on U, U' we say that they are *isovariant germ homotopic* if there is an isovariant homotopy of their restrictions to some unspecified neighbourhood of $\Delta(V)$ in $V \times V$. We will also refer to equivariant germ homotopy classes for their restrictions to sets $U \setminus \Delta(V)$.

In this section we will describe results showing that conversely, the existence of a suitable isovariant map implies existence and uniqueness up to diffeotopy or regular homotopy of an embedding or immersion giving rise to a map homotopic to the given map. These results hold under the condition that $2m > 3v + \epsilon$, where ϵ is a small number (0, 1 or 2) whose exact value depends on precisely which result is in question. We refer to a dimensional condition of this type as the *metastable range*, in contrast to the stable range $m > 2v + \epsilon$ in which any map is homotopic to an embedding (or immersion), unique up to diffeotopy (or regular homotopy).

Such results will imply simplified statements for the special case when the target is Euclidean space as follows.

Lemma 6.4.1 *There is a natural bijection between isovariant homotopy classes of isovariant maps $g : V \times V \to \mathbb{R}^m \times \mathbb{R}^m$ and equivariant homotopy classes of equivariant maps $F : V^{(2)} \to S^{m-1}$.*

There is a natural bijection between isovariant germ homotopy classes of isovariant maps defined on some neighbourhood U of $\Delta(V)$ in $V \times V$ and equivariant germ homotopy classes of equivariant maps defined on $U \setminus \Delta(V)$ for some neighbourhood U of $\Delta(V)$ in $V \times V$.

Proof Given an isovariant map g, we define $r(g) : V^{(2)} \to S^{m-1}$ by $r(g)(P, Q) := \mathbf{u}(s(g(P, Q)))$, where $s : \mathbb{R}^m \times \mathbb{R}^m \to \mathbb{R}^m$ denotes the subtraction map $s(x, y) = x - y$. Conversely, given an equivariant map $F : V^{(2)} \to S^{m-1}$, define an isovariant map by

$$l(F)(P, Q) := \rho(P, Q)(F(P, Q), -F(P, Q)),$$

where ρ denotes the distance in some Riemannian metric. For any F, we have $r(l(F)) = F$. For any g, $s(l(r(g)))$ is a non-zero multiple of $s(g)$, hence $l(r(g))$ is isovariantly homotopic to g.

The second assertion follows from the same argument, by restricting the maps to appropriate neighbourhoods of $\Delta(V)$. □

We first treat immersions, and start from the above Corollary 6.2.2, which we can restate as:

Any injective bundle map $\mathbb{T}(V) \to \mathbb{T}(M)$ is homotopic to one induced by an immersion, and two immersions are regularly homotopic if and only if the corresponding bundle maps are homotopic through injective maps.

The first step is to choose metrics on V and M, and observe that (by an easy homotopy argument) the space of injective bundle maps is homotopy equivalent to the space of bundle maps which preserve distances in the fibres. Thus in each fibre we have an isometric embedding $\mathbb{R}^v \to \mathbb{R}^m$ and hence a map $f : \mathbb{R}^v \to \mathbb{R}^m$ with $f^{-1}(0) = 0$ and having the (equivariance) property that $f(-x) = -f(x)$. We call fibre maps with this property *skew maps*, and homotopies preserving this condition *skew homotopies*.

The next step, which will be accomplished in Proposition 6.4.3, is to replace the space of isometric bundle maps by the space of skew maps.

Now $V'_{m,v}$ is the space of injective linear maps $\mathbb{R}^v \to \mathbb{R}^m$; denote by $W_{m,v}$ the space of skew maps $\mathbb{R}^v \to \mathbb{R}^m$, and by $\rho_{m,v} : V'_{m,v} \to W_{m,v}$ the inclusion.

Lemma 6.4.2 *The map $\rho_{m,v}$ is $(2m - 2v - 1)$-connected.*

Proof The Stiefel manifold $V_{m,v}$ is a deformation retract of $V'_{m,v}$; similarly the subspace $Y_{m,v} \subset W_{m,v}$ of radial skew maps that preserve length along each radius is a deformation retract: the retraction is given by taking the skew map $f : \mathbb{R}^v \to \mathbb{R}^m$ to g, where for $t > 0$, $||x|| = 1$, we have $g(tx) = tf(x)/||f(x)||$. A deformation is given by

$$h(u, tx) = \left(\frac{t}{||f(x)||}\right)^{1-u} f(t^u x).$$

We can identify $Y_{m,v}$ with the space of maps $S^{v-1} \to S^{m-1}$ that commute with the antipodal map; $V_{m,v}$ is the subspace of isometric embeddings.

We prove the result by induction on v; for $v = 1$, we have $X_{1,m} = Y_{1,m} = S^{m-1}$. We have the diagram

$$\begin{array}{ccc} V_{m,v} & \longrightarrow & Y_{m,v} \\ p_X \downarrow & & p_Y \downarrow \\ V_{v-1,m} & \longrightarrow & Y_{v-1,m} \end{array},$$

where the vertical arrows are induced by restriction to S^{v-2}. The map p_X is the projection of a fibre bundle; p_Y is a fibration (compare Lemma 6.1.2: the covering homotopy property for p_Y follows from the homotopy extension property for $S^{v-2} \subset S^{v-1}$). Let $x \in V_{v-1,m}$ have image $y \in Y_{v-1,m}$: then the result will follow from the homotopy exact sequences of the fibrations if we can show that $p_X^{-1}(x) \to p_Y^{-1}(y)$ is $(2m - 2v - 1)$-connected.

Now $p_X^{-1}(x)$ is homeomorphic to S^{m-v}. The space $p_Y^{-1}(y)$ consists of equivariant maps $S^{v-1} \to S^{m-1}$ agreeing with y on S^{v-2}, hence determined by the extension of y to one of the hemispheres bounded by S^{v-2}. The space of such extensions has the same homotopy type as when the map $S^{v-2} \to S^{m-1}$ is constant, and hence is homotopy equivalent to the iterated loop space $\Omega^{v-1}(S^{m-1})$. Thus we need to show that $S^{m-v} \to \Omega^{v-1}(S^{m-1})$ is $(2m-2v-1)$-connected, or equivalently that $\pi_r(S^{m-v}) \to \pi_{r+v-1}(S^{m-1})$ is surjective for $r \leq 2m-2v-1$ and bijective for $r \leq 2m-2v-2$. But this is the standard stability property of the suspension map (see §B.3(vi)). □

Now let $\pi : E \to B$ and $\pi' : E' \to B'$ be vector bundles over CW-complexes B, B' with respective fibres $\mathbb{R}^v$ and $\mathbb{R}^m$.

Proposition 6.4.3 *(i) If* $\dim(B) \leq 2m-2v-1$ *and* $\phi : E \to E'$ *is a skew map, there is a bundle map* $\psi : E \to E'$*, with* $\overline{\phi} = \overline{\psi}$*, skew homotopic to* ϕ*.*

(ii) If $\dim(B) < 2m-2v-1$ *and* $\phi_0, \phi_1 : E \to E'$ *are skew homotopic bundle maps, there is a bundle homotopy of* ϕ_0 *to* ϕ_1*, covering the given homotopy of maps* $B \to B'$*.*

Proof The skew maps $\phi : E \to E'$ that cover $\overline{\phi}$ are in bijective correspondence with the cross-sections of the bundle $\mathcal{W}$ over B whose fibre over $x \in B$ is the space of skew maps of E_x to E'_x, which can be identified with $W_{m,v}$; correspondingly for the bundle $\mathcal{L}$ of fibrewise injective bundle maps, with $V'_{m,v}$. By Lemma 6.4.2, we have $\pi_r(W_{m,v}, V'_{m,v}) = 0$ for $r < 2m-2v-1$. Since the obstructions to deforming a cross-section of $\mathcal{W}$ into $\mathcal{L}$ lie in these groups, the results follow. □

Now choose a complete metric on V. By Proposition 2.2.6, the map $e_V : \mathbb{T}(V) \to V \times V$ given by $e_V(\xi) = (\exp(\xi), \exp(-\xi))$ is a local diffeomorphism along $\Delta(V)$ and there exist neighbourhoods A_V of $\mathbb{T}^0(V)$ in $\mathbb{T}(V)$ and O_V of $\Delta(V)$ in $V \times V$ such that e_V gives a diffeomorphism of A_V on O_V; make corresponding choices for M.

Proposition 6.4.4 *There is a natural bijection between isovariant germ homotopy classes of isovariant maps* $F : U \to M \times M$ *defined on some neighbourhood* U *of* $\Delta(V)$ *in* $V \times V$ *and skew homotopy classes of skew bundle maps* $\mathbb{T}(V) \to \mathbb{T}(M)$*.*

Proof Let $F : U \to M \times M$ be an isovariant map, with U a neighbourhood of $\Delta(V)$ in $V \times V$. The composite $F_1 := e_M^{-1} \circ F \circ e_V$ is an isovariant map $\mathbb{T}(V) \to \mathbb{T}(M)$ defined on a neighbourhood of $\mathbb{T}^0(V)$. The restriction of F to the diagonal defines a map $F_0 : V \to M$ agreeing with the restriction of F_1 to zero vectors.

We next deform F_1 to a fibre map over F_0. If we had a trivialisation $\mathbb{T}(M) \cong M \times \mathbb{R}^m$ we could separate base and fibre components, write $F_1(X) = (F_b(X), F_f(X))$ and simply define $F_2 := (F_0, F_f)$. In general we define a natural metric on $\mathbb{T}(M)$: at a point given by a tangent vector v at $x \in M$ we can identify $T_v(\mathbb{T}(M)$ with $T_x(M) \oplus T_x(M)$ and use the Riemann metric on M twice. For $X \in \mathbb{T}(M)$ and $y \in M$ consider the point $\sigma(X, y) \in T_y(M)$ closest in $\mathbb{T}(M)$ to X. It follows from Theorem 2.3.2 that we have a smooth map $\sigma : \mathbb{T}(M) \times M \rightsquigarrow \mathbb{T}(M)$ defined on a neighbourhood of the diagonal. Moreover σ is a submersion along (and hence near) the set of points (X, x) with $X \in T_x(M)$, so the preimage of the zero cross-section is smooth, hence coincides (near the diagonal) with $T^0(M) \times M$.

Now define $F_2(X) := \sigma(F_1(X), F_0(\pi(X)))$. It follows that if X is a non-zero vector, so is $F_2(X)$. Thus F_2 is isovariant and, in some neighbourhood of $\mathbb{T}^0(V)$, is homotopic to F_1 through isovariant maps.

Using a partition of unity, we construct a positive continuous function $\varepsilon_V(X)$ on $\mathbb{T}(V)$ such that $\varepsilon_V(X) = 1$ for X in a neighbourhood of $\mathbb{T}^0(V)$, $\varepsilon_V(X)X \in U \cap A_V$ for all $X \in \mathbb{T}(V)$, and $\varepsilon_V(-X) = \varepsilon_V(X)$ for all X.

Now define $F_3 : \mathbb{T}(V) \to \mathbb{T}(M)$ by $F_3(X) = \varepsilon_V(X)^{-1}(F_2(\varepsilon_V(X)X))$. This is still isovariant, and is defined on all of $\mathbb{T}(V)$.

The converse procedure is more straightforward: if $G : \mathbb{T}(V) \to \mathbb{T}(M)$ is a skew map, $e_M \circ G \circ e_V^{-1}$ already gives an isovariant map on some neighbourhood of the diagonal. □

Putting these results together, we have

Theorem 6.4.5 *(i) If $2m > 3v$, U_V is a neighbourhood of $\Delta(V)$ in $V \times V$ and $F : U_V \to M \times M$ is isovariant, F_Δ can be approximated by immersions $f : V \to M$ such that F and $f^{(2)}$ are isovariantly germ homotopic.*

(ii) If $2m > 3v + 1$, two immersions $f, g : V \to M$ are regularly homotopic if and only if $f^{(2)}$ and $g^{(2)}$ are isovariantly germ homotopic.

Proof By Theorem 6.2.1, regularly homotopy classes of immersions $V \to M$ correspond bijectively to homotopy classes of fibrewise injective linear maps $\mathbb{T}(V) \to \mathbb{T}(M)$. It follows from Proposition 6.4.3 that if $\dim(V) < 2m - 2v - 1$ these correspond bijectively to skew homotopy classes of skew bundle maps $\mathbb{T}(V) \to \mathbb{T}(M)$. Finally by Proposition 6.4.4 there is a natural bijection between these and isovariant germ homotopy classes of isovariant maps $F : U \to M \times M$. A corresponding argument yields (i). □

Corollary 6.4.6 *If $2m > 3v + 1$ the classification of immersions $V \to M$ depends only on the topology of V and M (not on the differential structure).*

In the case when M is Euclidean space, the statement can be simplified using Lemma 6.4.1. Recall that an immersion $f : V \to \mathbb{R}^m$ induces a map $f^{(2)} : V^{(2)} \to (\mathbb{R}^m)^{(2)}$ whose restriction to some neighbourhood U of $\Delta(V)$ is isovariant, hence induces an equivariant map, $f_\delta : U \setminus \Delta(V) \to S^{m-1}$.

Corollary 6.4.7 *(i) If $2m > 3v$, U_V is a neighbourhood of $\Delta(V)$ in $V \times V$ and $F : U_V \setminus \Delta(V) \to S^{m-1}$ is equivariant, there is an immersion $f : V \to \mathbb{R}^m$ such that F and f_δ are equivariantly germ homotopic.*

(ii) If $2m > 3v + 1$, two immersions $f, g : V \to \mathbb{R}^m$ are regularly homotopic if and only if f_δ and g_δ are equivariantly germ homotopic.

We come to embeddings. As promised above, the main result is

Theorem 6.4.8 *Let V^v, M^m be manifolds with the former compact. Then*

(i) If $2m \geq 3(v + 1)$, a continuous map $f : V \to M$ is homotopic to a smooth embedding if and only if $f \times f$ is equivariantly homotopic to an isovariant map.

(ii) If $2m > 3(v + 1)$, two smooth embeddings $f_0, f_1 : V \to M$ are diffeotopic if and only if $f_0 \times f_0$ and $f_1 \times f_1$ are isovariantly homotopic.

In view of Theorem 6.4.5, this will be an immediate consequence of

Theorem 6.4.9 *Let V^v, M^m be manifolds with the former compact. Then*

(i) If $2m \geq 3(v + 1)$, an immersion $f : V \to M$ is regularly homotopic to a smooth embedding if and only if there is an equivariant homotopy H of $f \times f$ to an isovariant map such that $\Delta(V)$ is open in $H_t^{-1}(\Delta(M))$ for each t.

(ii) If $2m > 3(v + 1)$, a regular homotopy f_t between two smooth embeddings $f_0, f_1 : V \to M$ is regularly homotopic to a diffeotopy if and only if there is a map $H : V \times V \times I \times I \to M \times M$ {write $H_{t,u}(v, w)$ for $H(v, w, t, u)$} such that $H_{t,0} = f_t \times f_t$, $H_{0,u} = f_0 \times f_0$, $H_{1,u} = f_1 \times f_1$, $H_{t,1}$ is isovariant and $\Delta(V)$ is open in $H_{t,u}^{-1}(\Delta(M))$ for each (t, u).

Proof The proof of this result follows the same lines as that of the Whitney trick. We need to construct a model for the deformation, then show how to embed the model in M. As the details are somewhat involved, we confine ourselves here to an outline of the key points of the proof, and refer to the original paper [63] for a careful account. We deal only with (i), in the case when V has no boundary, and try to keep our notation close to that of §6.3.

The core of the model is a smooth manifold C together with an involution σ of C and a σ-invariant function $\lambda : C \to \mathring{D}^1$. The double point set will be $C_0 := \lambda^{-1}(0)$. The core C is smoothly embedded in the source manifold V, so a tubular neighbourhood is determined by a vector bundle L over C, which we may take to be an orthogonal bundle, and consists of the set L_ε of vectors in L of length $\leq \varepsilon$.

We will slide the image of C across the space D defined as the quotient of $C \times \mathring{D}^1$ by identifying, for $x \in C$ and $-1 < t < 1$, $(x, t) \sim (\sigma(x), -t)$: write $[x, t]$ for the image of (x, t). The map $\phi : C \to D$ is given by $\phi(x) = [x, \lambda(x)]$ (thus C_0 is indeed the double point set) and the deformation by $\phi(x, t) = [x, \lambda(x) - t\mu(x)]$ for a suitable μ. Since the first component remains at x, this is a regular homotopy; provided μ is σ-invariant, the double point set is given by $\lambda(x) - t\mu(x) = 0$, so ϕ_1 is an embedding provided $\mu(x) > \lambda(x)$ for all x; we also need $\mu(x) < \lambda(x) + 1$ for the map to be defined.

As before, we expect the normal bundle of D in M to be locally the direct sum of two bundles, one restricting on C to the isomorphic image of L and the other to the normal bundle along C of the image of V in M; but the roles of the summands at C_0 interchange. Explicitly, define $L \oplus_\sigma L$ to be the pull-back of the external direct sum $L \times L$ over $C \times C$ by the antidiagonal map $x \mapsto (x, \sigma(x))$, then let W be the bundle over D given as the quotient of the bundle $(L \oplus_\sigma L) \times \mathring{D}^1$ over $C \times \mathring{D}^1$ by the identification $(e_x, e_{\sigma(x)}, t) \sim (-e_{\sigma(x)}, -e_x, -t)$; again use square brackets to denote a point in the quotient.

The regular homotopy ϕ now extends to the map $\Phi : L_\varepsilon \times I \to W_\varepsilon$ given by $\Phi_t(e_x) = [e_x, 0, \lambda(x) - t\alpha(\|e_x\|)\mu(x)]$: here we require α to be a bump function with $\alpha(0) = 1$ and $\alpha(y) = 0$ for $|y| \geq \varepsilon$: for example, we can take $\alpha(y) = Bp(1 - (|y|/\varepsilon)^2)$.

This concludes the construction of the model; now we need to embed it in V and M. We extend the given homotopy H to a map $V \times V \times [-1, 1] \to M \times M$; away from $\Delta(V) \times [-1, 1]$, we may suppose by transversality that H is transverse to $\Delta(M)$, so that $X := H^{-1}(\Delta(M)) \setminus \Delta(V)$ is a closed sub-manifold of $V \times V \times [-1, 1]$, of dimension $2v + 1 - m$. The first projection defines a map $p_1 : X \subset V \times V \times [-1, 1] \to V$. Since $v > 2(2v + 1 - m)$, we may suppose by general position that p_1 is an embedding. Since H is equivariant, the second projection p_2 is also an embedding, with the same image.

Define C to be $p_1(X)$, $\sigma : C \to C$ to be $p_2 \circ p_1^{-1}$ and $\lambda : C \to \mathring{D}^1$ to be $p_3 \circ p_1^{-1}$. Since $H_0 = f \times f$, the double point set C_0 of the model is indeed the double point set of f. Taking L to be the normal bundle of C in V, the choice of a tubular neighbourhood of C gives an embedding $L \to V$. This completes the constructions in V.

We begin the construction of a map $\psi : D \to M$ by defining

$$\psi[x, t] := p_1(H(x, \sigma(x), \lambda(x) - t)) \quad \text{whenever } 0 \leq t \leq \lambda(x).$$

Since the subset $0 \le t \le \lambda(x)$ is a deformation retract of D, this can be extended to a continuous map ψ of D. Recalling that $2m \ge 3(v+1)$ and that $\dim D = 2v - m + 1$, we may suppose in turn

(i) that ψ is smooth,

(ii) that along the image of C, ψ is an embedding not tangent to V (it is 'transverse' in the sense that the intersection of the tangent spaces to the images of D and of V is that of C) this we can ensure using general position since $\dim D < m - v$,

(iii) that ψ is an embedding, by general position, since $m > 2\dim D$,

(iv) that the image of ψ meets V only along C, again by general position, since $m > \dim D + \dim V$.

Now define $\xi : C \to M$ by $\xi(x) = \psi[x, 0]$.

We now need to identify W with the normal bundle of $\psi(D)$ in M. Since p_1 and p_2 are embeddings, the normal bundle $\mathbb{N}(V \times V \times [-1, 1]/X)$ splits into components, leading to an isomorphism of the normal bundle $\mathbb{N}(V \times V \times [-1, 1]/X)$ onto $\mathbb{N}(V/C) \oplus_\sigma \mathbb{N}(V/C) \oplus \mathbb{T}(C) \oplus E$, where E is a trivial line bundle. On the other hand, since X is the transverse preimage of $\Delta(M)$ it follows by Lemma 4.5.1 that this normal bundle is the pullback of $\mathbb{N}((M \times M)/M) \cong \mathbb{T}(M)$. We thus have an isomorphism

$$\Xi : \mathbb{N}(V/C) \oplus_\sigma \mathbb{N}(V/C) \oplus \mathbb{T}(C) \oplus E \to \mathbb{T}(M).$$

We can identify L with $\mathbb{N}(V/C)$ and D as the quotient of $C \times \mathring{D}^1$ by $\mathbb{Z}_2$. We would now like to identify the summands $\mathbb{T}(C) \oplus E$ of $\mathbb{T}(M)$ with $\mathbb{T}(D)$ and (hence) the normal bundle $\mathbb{N}(M/D)$ with $\mathbb{N}(V/C) \oplus_\sigma \mathbb{N}(V/C) \cong L \oplus_\sigma L$ and hence with W.

A number of details need attention. The restriction of Ξ to C_0 is equal to $df \oplus_\sigma (-df) \oplus 0$ on $\mathbb{N}(V/C) \oplus_\sigma \mathbb{N}(V/C) \oplus 0$. It is now not difficult to identify the bundle maps over C_0.

If σ denotes the involution of $\mathbb{N}(V/C) \oplus_\sigma \mathbb{N}(V/C) \oplus \mathbb{T}(C) \oplus E$ given by $\sigma(e_x, e_{\sigma(x)}, f_x, r) = (e_{\sigma(x)}, e_x, f_{\sigma(x)}, r)$, it follows from equivariance of H that $\Xi \circ \sigma = -\Xi$. Using again the dimension condition, we can extend to an embedding η of $\mathbb{N}(V/C) \oplus_\sigma \mathbb{N}(V/C)$ in $\mathbb{T}(M)$, covering ξ, with $\eta(e_x, e_{\sigma(x)}) = -\eta(e_{\sigma(x)}, e_x)$ and such that, for $x \in C_0$, $\eta(e_x, e_{\sigma(x)}) = df(e_x) - df(e_{\sigma(x)})$.

To construct the desired isomorphism $\chi : L \oplus_\sigma L \to W$ over ψ and agreeing with η on $(L \oplus_\sigma L) \times [0, 1]$, we first restrict to $(L \oplus 0) \times [0, 1]$, and define $\chi(e_x, 0, \lambda(x)) = df(e_x)$ if $\lambda(x) \ge 0$, and check that the obstructions to extending over $C \times I$ lie in zero groups. Then extend using $\chi(0, -e_x, -\lambda(x)) = df(e_x)$ for $\lambda(x) \le 0$.

As in the proof of Proposition 6.3.1, we can now use Proposition 2.5.10 to construct embeddings of a neighbourhood of C in L into V and of a neighbourhood of D in W into M compatible with the given maps. We have already shown how to construct a regular homotopy to an embedding. A final calculation is necessary to check compatibility of this embedding with the given isovariant map. □

In view of Lemma 6.4.1, we now have

Corollary 6.4.10 *Let V^v be a compact manifold.*

(i) If $2m \geq 3(v+1)$, there is a smooth embedding $f : V \to \mathbb{R}^m$ if and only if there is an equivariant map $(V \times V) \setminus \Delta(V) \to S^{m-1}$.

(ii) If $2m > 3(v+1)$, two smooth embeddings $f_0, f_1 : V \to \mathbb{R}^m$ are diffeotopic if and only if $(f_0)_\delta$ and $(f_1)_\delta$ are equivariantly homotopic.

One can also formulate a Euclidean version of Theorem 6.4.9.

The above are not the only important results about embedding in the metastable range. The following result is also due to Haefliger, and was originally proved using the normal forms for singularities obtained in Theorem 4.8.5.

Theorem 6.4.11 *Let V^v be a compact connected manifold (without boundary), M^m a manifold and $f : V \to M$ a $(k+1)$-connected map.*

(a) If $m \geq 2v-k$ and $2m \geq 3(v+1)$, f is homotopic to an embedding.

(b) If $m > 2v-k$ and $2m > 3(v+1)$, any two embeddings homotopic to f are diffeotopic.

This is deduced in [62] from Theorem 6.4.8 and the following

Proposition 6.4.12 *If $f : V \to M$ is $(2v-m+1)$-connected, V is closed and $m > v$, then $f \times f$ is equivariantly homotopic to an isovariant map.*

This Proposition is proved by an obstruction theory argument. Some applications of Theorem 6.4.11 were given in §5.6: we now give others following [62]. Taking M to be Euclidean space, we deduce

Corollary 6.4.13 *Let V^v be a compact k-connected manifold (without boundary).*

(a) If $m \geq 2v-k$ and $2m \geq 3(v+1)$, V embeds in $\mathbb{R}^m$.

(b) If $m > 2v-k$ and $2m > 3(v+1)$, any two embeddings of V in $\mathbb{R}^m$ are diffeotopic.

Corollary 6.4.14 *If $2m > 3(v+1)$, any two embeddings of S^v in $\mathbb{R}^m$ are diffeotopic.*

For the equivariant homotopy class of $(S^v \times S^v) \setminus \Delta(S^v) \to S^{m-1}$ is unique. We will see in §8.8 that this result is best possible.

Corollary 6.4.15 *Provided $2m > 3\max(p, q) + 1$, the isotopy classes of embeddings of $S^p \cup S^q$ in $\mathbb{R}^m$ correspond bijectively to $\pi_{p+q}(S^{m-1})$.*

For, using the preceding result, we just need isovariant homotopy classes of $(S^p \times S^q) \cup (S^q \times S^p) \to S^{m-1}$, i.e. homotopy classes $[S^p \times S^q : S^{m-1}] = \pi_{p+q}(S^{m-1})$.

In some situations, the results of Corollary 6.4.13 can be sharpened: we refer the interested reader to [72]. In particular ([72, Theorem 8]).

Proposition 6.4.16 *Let V^v be a compact k-connected manifold and $N \geq \max(2v - 2k - 1, \frac{1}{2}(3v - k), v + 2)$. Then V embeds in $\mathbb{R}^N$ if and only if $W_{N-v+1} = 0$.*

6.5 Notes on Chapter 6

§6.1 The main result in the following section is best stated at the level of function spaces. We have collected here some fundamental definitions and results, so as not to interrupt the exposition in the next section.

§6.2 The breakthrough in obtaining a general theory of immersions was made by Steve Smale – his lecture at the International Congress in 1958 was one I found particularly exciting. His work appeared in [137], and was quickly generalised by Moe Hirsch [70]. This theory is often referred to as Smale–Hirsch theory.

The next major step was taken by Misha Gromov [58], who created a general theory. The account given above follows closely the version in lectures by André Haefliger [65]. Another account is given in [2].

§6.3 Whitney had used general position arguments to show in [175] that any m-manifold embeds in $\mathbb{R}^{2m+1}$ and immerses in $\mathbb{R}^{2m}$. He introduced the 'Whitney trick' in [177] to show that any m-manifold embeds in $\mathbb{R}^{2m}$. In the same paper he gave a construction of an m-sphere immersed in $\mathbb{R}^{2m}$ with a single self-intersection. He went on in [178] to show that any m-manifold immerses in $\mathbb{R}^{2m-1}$.

The Whitney trick fails if $m = 2$: here finding the embedding of the 2-disc required for Proposition 6.3.1, which is given by general position if $m \geq 3$, is a problem of the same type as the theorem it seeks to establish. Not only the proof but the result fails: see Section 5.7 for more details. As a result of this failure, the study of 4-manifold topology has a completely different nature to that in higher dimensions. In studying the Whitney trick, Casson was led to introduce an infinite sequence of such problems, leading to the notion of 'grope', and to new results in topological topology. Smooth manifolds behave differently and require new techniques, for which we refer to [46].

Theorem 6.3.4 becomes trivial if $n = 1$: here V must be I or S^1 and $M = S^2$. I do not know what happens if $n = 2$.

§6.4 The first major result on embeddings in the metastable range was obtained by Haefliger [60]. In this impressive paper Haefliger, following the idea of the proof of the Whitney trick, uses the description in Theorem 4.7.3 of singularities of maps in the metastable range to construct a model for a deformation of a map to an embedding.

All the results in this section are due to Haefliger, some in collaboration with Hirsch. For this account we have followed [66] to construct immersions, and [63] for embeddings. A different approach is used in [67]. These theorems are so powerful that much of the subsequent literature is devoted to calculations required for applications. We will return to embeddings of spheres in Euclidean space in the final section §8.8.

7
Surgery

In this chapter we discuss a method of constructing manifolds, or more precisely, of adapting a given manifold to satisfy certain conditions. This method is due to Milnor and Kervaire. In the paper [102] where they introduced the method, the objective was to simplify the homotopy type of the manifold, so the procedure was called 'killing homotopy groups'. However since the procedure can be seen as removing a piece of a manifold and replacing it by something else, it has come to be known as 'surgery'.

It was observed by Novikov that the method could be applied to the more general situation, given a manifold M and a map $f : M \to X$, to change both M and f to make f more like a homotopy equivalence, by killing the homotopy groups of f. The method was then codified and further extended by Browder and by the author.

In more detail, the manifold M will be changed by a cobordism. As we saw in §5.1, we may choose a handle decomposition of this cobordism, so the procedure is broken into a sequence of operations, each corresponding to a single handle. Although we may think of M as a closed manifold, the discussion will apply to any compact manifold M.

In the first section we analyse a single step in the procedure: both the conditions for performing the step and its effect. In §7.2, we show how to modify a map $f : M \to X$ to kill all homotopy groups of f in dimensions below the middle.

In view of duality, any change to the homology of M is reflected by a corresponding change in the dual dimension. We next discuss the algebraic results we need on bilinear and quadratic forms, then in §7.4 formulate duality in the setting of CW-complexes.

In order to perform surgery to make f a homotopy equivalence, we must also require X to satisfy duality and it is convenient to suppose f a 'normal map'. As in Chapter 5, we discuss in detail in this book only the case when X is simply-connected. We treat in turn the cases when the dimension of M is even (when there is an obstruction in $\mathbb{Z}$ or $\mathbb{Z}_2$ to performing surgery) or odd (when there is none).

The finer details of the results depend on deeper results in homotopy theory, which we give in §7.7. Here we show how Spivak's fibration theorem permits a reformulation of the classification of normal maps. We proceed to Brown's interpretation of the Kervaire invariant.

In §7.8 we apply the results to discuss the homotopy types of smooth manifolds: the aim (not quite fully accomplished) is to reduce the problem of classification of smooth manifolds entirely to homotopy theory.

The author has already written a monograph [167] on surgery, in which no restriction is placed on the fundamental group. The account here is intended to be introductory rather than complete but is, of course, informed by the same view of the topic.

7.1 The surgery procedure: a single surgery

Let M^m be a compact manifold M (perhaps with boundary), $\phi : S^{r-1} \times D^{m-r+1} \to M \setminus \partial M$ an embedding. The operation of removing the interior of the image of ϕ, and attaching $D^r \times S^{m-r}$ to the result by $\phi|(S^{r-1} \times S^{m-r})$ is called a simple surgery, or *spherical modification* of M, of type $(r, m-r+1)$. The aim of surgery is to perform a series of spherical modifications on M to simplify M in a way to be made explicit.

The effect of a spherical modification is determined by ϕ, and even by the diffeotopy class of ϕ (by Theorem 2.4.2). The modification gives a manifold M' with the same boundary as M: in particular, if M is closed so is M'.

The manifold $W = (M \times I) \cup_f h^r$ (with corner, if M has a boundary) thus has $\partial_- W = M$, $\partial_+ W = M'$: it is a cobordism between M and M', called the *supporting manifold* of the modification. Also, $\partial_c W = \partial M \times I$. If M' is obtained from M by a spherical modification of type $(r, m-r+1)$, we can obtain M from M' by one of type $(m-r+1, r)$. We have the same supporting manifold for both modifications. It follows from the existence (see §5.1) of handle decompositions that M and N are cobordant if and only if one may be obtained from the other by a series of spherical modifications.

The procedure begins with a manifold M and a continuous map $f : M \to X$. Let W be obtained by attaching an r-handle to $M \times I$, with attaching map $\phi : S^{r-1} \times D^{m-r+1} \to M \setminus \partial M$. If we can extend f to a map $F : W \to X$, we say that the manifold $M' := \partial_+ W$ and the map $f' := F|M'$ are obtained from (M, f) by surgery.

We write ϕ_0 for the restriction of ϕ to $S^{r-1} \times \{0\}$. Up to homotopy, W is obtained from M by attaching an r-cell, and extending f over the handle amounts to giving a map $g : D^r \to X$ whose restriction to the boundary is $f \circ \phi_0$. The pair (ϕ, f) defines an element of the relative homotopy group $\pi_r(f)$.

Conversely, suppose given an element $\xi \in \pi_r(f)$. For this to arise as above, the class $\partial\xi \in \pi_{r-1}(M)$ must be represented by an embedding of S^{r-1} in M. The existence of such embeddings is guaranteed by general position if $m > 2(r-1)$; otherwise more work is required. Moreover, we need to extend the embedding of S^{r-1} to an embedding of $S^{r-1} \times D^{m-r+1}$, so need the normal bundle of the embedded sphere to be trivial. Provided $m \geq 2r - 1$ this follows if the bundle is stably trivial, hence if the restriction to the sphere of the tangent bundle $\mathbb{T}(M)$ is trivial. Since the sphere is nullhomotopic in X, a neat way to ensure this is to require that $\mathbb{T}(M)$ is itself induced from a bundle over X. It is convenient to weaken this slightly, giving the following definition.

A *normal map* consists of a map $f : M \to X$, a vector bundle ν over X and a trivialisation T of the bundle $\mathbb{T}(M) \oplus f^*\nu$. A *normal cobordism* is a normal map $(g : W \to X, \nu, T)$ with the manifold W a cobordism. We can extend this definition in a natural way to the case of a manifold with boundary.

Theorem 7.1.1 *Let $(f : M \to X, \nu, T)$ be a normal map. Then any $\xi \in \pi_r(f)$ determines a regular homotopy class of immersions $\phi : S^{r-1} \times D^{m-r+1} \to M$, and given any embedding in this class we can do surgery to obtain another normal map.*

Proof Suppose ϕ an embedding whose restriction to $S^{r-1} \times \{0\}$ represents $\partial\xi$: then we can use ϕ to attach an r-handle to $M \times I$ and use ξ to extend f to a map $g : (M \times I) \cup h^r \to X$ (more precisely, we first use ξ to extend f to the union of $M \times I$ and the disc $D^r \times \{0\}$, and then use a retraction of $D^r \times D^{m-r+1}$ on $(S^{r-1} \times D^{m-r+1}) \cup (D^r \times \{0\})$ – see Figure 5.6 – to extend to the rest of the handle).

Since the handle $D^r \times D^{m-r+1}$ is contractible, the restriction to it of $g^*\nu$ is trivial, so extending $T \oplus 1$ to a trivialisation of $\mathbb{T}(W) \oplus g^*\nu$ is equivalent to trivialising the sum of a trivial bundle with the restriction to $S^{r-1} \times D^{m-r+1}$ of $\mathbb{T}(M)$. Using stability, such trivialisations correspond to those of this restriction, and hence to isomorphisms of it to $\mathbb{T}(S^{r-1} \times D^{m-r+1})$. But by Corollary 6.2.2, such isomorphisms correspond bijectively to regular homotopy classes of immersions $\phi : S^{r-1} \times D^{m-r+1} \to M$. □

There are two approaches to analysing the effect on homology of a spherical $(r, m-r+1)$-modification: we can use the supporting manifold $W = (M \times I) \cup_f h^r$ or the intersection $X = M \cap M'$ (obtained from M by removing the interior of the image of ϕ). Up to homotopy W is obtained from M by attaching an r-cell, and from M' by attaching an $(m-r+1)$-cell. On the other hand, M is obtained from X by attaching an $(m-r+1)$- and an m-cell, and M' from X by attaching an r- and an m-cell.

The inclusions $(M', X) \subset (W, M \times I) \supset (W, M)$ induce isomorphisms of relative homology groups in dimensions $\neq m$. For the inclusions

$$(D^r \times S^{m-r}, S^{r-1} \times S^{m-r}) \subset (M', X);$$

$$(D^r \times D^{m+1-r}, S^{r-1} \times D^{m+1-r}) \subset (W, M \times I)$$

induce homology isomorphisms by excision. Thus it suffices to consider

$$(D^r \times S^{m-r}, S^{r-1} \times S^{m-r}) \subset (D^r \times D^{m+1-r}, S^{r-1} \times D^{m+1-r}),$$

and here both relative groups vanish except in dimensions r, m; in dimension r we have an isomorphism. It follows that

Lemma 7.1.2 *Let $r \leq m-r$. Then M and M' have the same $(r-2)$-type (in particular, if $r \geq 3$, the same fundamental group). If $r < m-r$, and x is the homology class of the a-sphere $f(S^{r-1} \times 0)$ in M, then $H_{r-1}(M')$ is the quotient of $H_{r-1}(M)$ by the subgroup generated by x.*

We can now express the homology relations by a single diagram.

Proposition 7.1.3 *We have the following exact sequences for $i < m-1$:*

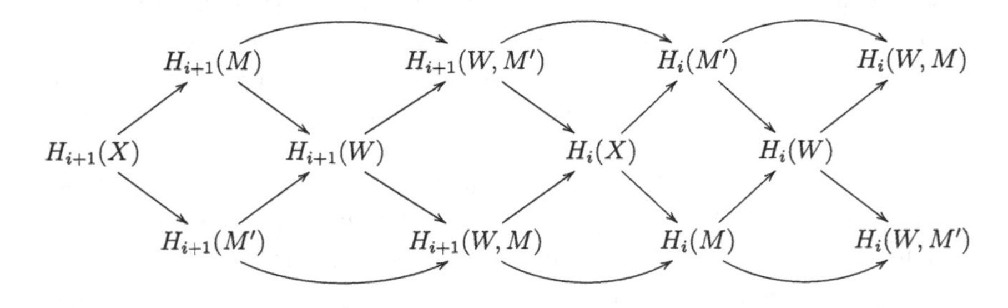

Proof Since we can identify $H_j(M', X) = H_j(W, M)$ and dually $H_j(M, X) = H_j(W, M')$ for $j \leq m-1$, it suffices to write out the homology exact sequences of the four pairs (M, X), (M', X), (W, M), and (W, M'). □

7.2 Surgery below the middle dimension

We now show how we can perform surgery on a normal map $f : M \to X$ to make the homotopy type of M closer to that of X.

Theorem 7.2.1 *If X is a finite CW-complex and $m \geq 2k$, any normal map $(f : M \to X, \nu, T)$ is normally cobordant to a normal map $(f' : M' \to X, \nu, T')$ such that f' is k-connected.*

Proof Let X' be the mapping cylinder of f, obtained from the disjoint union $(M \times I) \cup X$ by identifying each point $(x, 1) \in M \times \{1\}$ with $f(x) \in X$. The inclusion $X \subset X'$ is a homotopy equivalence, so ν extends to a bundle ν' over X'. The inclusion of M as $M \times \{0\}$ is homotopic to f. Thus replacing X by X', ν by ν' and T by the induced trivialisation, we have made no essential change, but may now take f to be an inclusion.

Set $X_0 := M$, and let X_i be a sequence of subcomplexes of X formed by attaching one at a time to X_0 the cells of X of dimension $\leq k$ not already in X_0. As X is finite, this process terminates, in X_K, say.

We now show, by induction on i, that we can add to $M \times I$ a sequence of handles yielding manifolds N_i and extend the inclusion of M in X to homotopy equivalences $f_i : N_i \to X_i$ and normal cobordisms $(f_i' : N_i \to X, \nu, T_i)$, where f_i' is the composite of f_i and the inclusion. We have $\partial_- N_i = M$ and set $M_i := \partial_+ N_i$. We start the induction with $N_0 = M \times I$; f_0' is f composed with the projection; similarly for T_0.

Suppose inductively (N_i, f_i, T_i) already constructed; let X_{i+1} be obtained from X_i by attaching an r-cell. This cell defines an element of $\pi_r(X, X_i)$ hence, since f_i is a homotopy equivalence, of $\pi_r(f_i')$. Denote by $f_i'' : M_i \to X$ the restriction of f_i': we claim that the map $\pi_r(f_i'') \to \pi_r(f_i')$ is an isomorphism.

Since N_i is obtained from $M \times I$ by attaching handles of dimension $\leq k$, it is obtained from M_i by attaching handles of dimensions $\geq m + 1 - k$; hence (N_i, M_i) is $(m - k)$-connected and hence, since $r \leq k < m - k$, it is r-connected. The claim thus follows from the exact sequence

$$\pi_r(N_i, M_i) \to \pi_r(f_i'') \to \pi_r(f_i') \to \pi_{r-1}(N_i, M_i).$$

The r-cell thus defines an element of $\pi_r(f_i'')$ and hence, by Theorem 7.1.1, a regular homotopy class of immersions $S^{r-1} \times D^{m-r+1} \to M_i$. Since $m > 2(r - 1)$, it follows from Theorem 4.7.7 that this class contains embeddings. We may thus perform surgery to obtain a normal cobordism. Since the r-cell of the cobordism maps to the homotopy class of the cell in X_{i+1}, the homotopy equivalence $N_i \to X_i$ extends to a homotopy equivalence $N_{i+1} \to X_{i+1}$. The induction is complete.

At the final step, since X is obtained from X_K by attaching cells of dimensions $\geq k+1$, the map $X_K \to X$, hence also $f'_K : N_K \to X$, is k-connected. Since, as we have just seen, (N_K, M_K) is $(m-k)$-connected, the map f''_K is also k-connected. □

An important special case is when X is a point.

Corollary 7.2.2 *If $\mathbb{T}(M)$ is stably trivial, and $m \geq 2k$, we can perform surgery on M to make it $(k-1)$-connected.*

For we apply the theorem, taking X to be a point; then since $M' \to X$ is k-connected, M' is $(k-1)$-connected.

In general the tangent bundle of M is induced by a map $M \to B(O_m)$. Suppose given a bundle ν^s and a framing T of $\mathbb{T}(M) \oplus \nu$ (so ν is a normal bundle for M). Choose a classifying map $f : M \to B(O_s)$, so that if υ is the universal bundle over $B(O_s)$ we have $\nu \cong f^*\upsilon$. Then $(f : M \to B(O_s), \upsilon, T)$ is a normal map, so we can perform surgery on M to obtain a k-connected map $f' : M' \to B(O_s)$. The mod 2 Betti numbers of M' below the middle dimension thus coincide with those of $B(O_s)$, hence with those of $B(O)$; those above the middle are determined by duality. It follows, for example, that if $w \in H^j(B(O); \mathbb{Z}_2)$ (with $2j > m+1$) is such that for any $w' \in H^{m-j}(B(O); \mathbb{Z}_2)$ the Stiefel–Whitney number $f'^*(ww')[M']$ vanishes, then also $f'^*w = 0$. Corresponding remarks hold for oriented manifolds with $B(O)$ replaced by $B(SO)$ and the coefficient field $\mathbb{Z}_2$ by $\mathbb{Q}$.

A different type of application arises by fixing k. If our object is to make f' induce an isomorphism $\pi_1(M') \to \pi_1(X)$, it suffices to have f' 2-connected, and we can achieve this provided $m \geq 4$.

With a little more care, we can construct embeddings. The following result includes a characterisation of possible fundamental groups of complements in S^m of embedded copies of S^{m-2} (provided $m \geq 5$).

Theorem 7.2.3 *Let (K, L) be a CW pair of dimension $k \geq 3$ with K contractible and K obtained from L by adding a 2-cell. Then provided $m \geq 2k-1$ there exist a smooth embedding of S^{m-2} in S^m with complement C and an $(m-k)$-connected map $C \to L$.*

Proof Set $M := S^1 \times D^{m-1}$, $X = L$, and define $f : M \to X$ to be projection on S^1 composed with the attaching map of the 2-cell. We define a normal map by taking ν to be trivial and using a trivialisation of $\mathbb{T}(M)$.

Now apply the result proved inductively in Theorem 7.2.1. We obtain a manifold N formed by attaching handles of dimensions $\leq k$ to $M \times I$ and a normal cobordism $(g : N \to L, \nu, T)$ which is a homotopy equivalence. Set

$M' := \partial_+ N$. Since N is formed from $M' \times I$ by attaching handles of dimensions $\geq m+1-k \geq k \geq 3$, $\pi_1(M') \to \pi_1(N)$ is an isomorphism.

Define W by attaching $D^2 \times D^{m-1}$ along $M \times \{0\}$. Up to homotopy we have attached a 2-cell, so the homotopy equivalence $g : N \to L$ extends to a homotopy equivalence $W \to L \cup e^2 = K$; thus W is contractible. Now $\partial W = (D^2 \times S^{m-2}) \cup \partial_c N \cup M'$, so $\pi_1(\partial W) \cong \pi_1(M' \cup e^2) \cong \pi_1(N \cup e^2)$, so is trivial. Since $m \geq 5$, it follows from Corollary 5.6.3 that $W^{m+1} \cong D^{m+1}$.

We now have $S^{m-2} \times \{0\} \subset S^{m-2} \times D^2 \subset \partial W \cong S^m$, and its closed complement may be taken as $\partial_c N \cup M'$ or as M'; the inclusion $M' \subset N$ is $(m-k)$-connected and $g : N \to L$ is a homotopy equivalence. □

Corollary 7.2.4 *Given $m \geq 5$ and a group G, there exist a smooth embedding $f : S^{m-2} \to S^m$ and an isomorphism $G \to \pi_1(S^m \setminus f(S^{m-2}))$ if and only if G is finitely presented, $H_1(G) \cong \mathbb{Z}$, $H_2(G) = 0$, and there is an element $x \in G$ whose conjugates generate the whole group.*

Here, and in the proof, all homology has coefficient group $\mathbb{Z}$.

Proof If $f : S^{m-2} \to S^m$ is a smooth embedding and $C := S^m \setminus f(S^{m-2})$, then S^m is obtained from C by attaching a 2-cell and an m-cell. The 2-cell is attached by a map $S^1 \to C$ with homotopy class x, say: the fundamental group is changed by factoring out the normal closure of x (the m-cell has no effect), and becomes trivial. If $G := \pi_1(C)$, then $H_1(G) \cong H_1(C) \cong \mathbb{Z}$ and $H_2(G)$ is a quotient of $H_2(C)$, which is zero.

Conversely, given G and $x \in G$, choose a finite presentation of G and construct a CW-complex with L' with $\pi_1(L) \cong G$ by taking 1-cells given by generators and attaching 2-cells corresponding to relators. Adding a further 2-cell e^2 along x gives a simply-connected space K'; since this is 2-dimensional, it is homotopy equivalent to a bouquet of 2-spheres.

In the sequence $0 \to H_2(L') \to H_2(K') \to H_2(K', L') \to H_1(L')$ the group $H_2(K', L')$ is infinite cyclic, generated by the class of e, hence maps isomorphically to $H_1(L') = H_1(G)$. Hence $H_2(L') \cong H_2(K')$ is free abelian, and we can pick a free basis $\{y_i\}$. It now follows from the exact sequence $\pi_2(L') \to H_2(L') \to H_2(G) = 0$ that we can represent the y_i by maps $f_i : S^2 \to L'$. Define L by attaching 3-cells to L' by the f_i. Then $H_2(L)$, $H_3(L)$ and all higher homology groups vanish. The space $K = L \cup e^2$ is now simply-connected with vanishing homology, hence is contractible.

We can now apply Theorem 7.2.3, taking $k = 3$. This yields a smooth embedding of S^{m-2} in S^m with complement C and an $(m-k)$-connected map $C \to L$. Since $m \geq 2k-1$, $m-k \geq k-1 \geq 2$, so $\pi_1(C) \cong \pi_1(L) \cong G$ as required. □

7.3 Bilinear and quadratic forms

In order to proceed further with surgery, we need to take account of duality. In this section we will introduce the purely algebraic notions, and thus make a digression to discuss results about symmetric and skew-symmetric bilinear forms which will play a role below. I aim to give enough details for the discussion to make sense, but will not give full details of all proofs.

We consider abelian groups $G, G', \ldots$, a value group V, and bilinear maps $\lambda : G \times G' \to V$, which we sometimes call pairings. Denote by $G^\vee$ the dual group $\mathrm{Hom}(G, V)$, by $\lambda^t : G' \times G \to V$ the transpose, given by $\lambda^t(g', g) = \lambda(g, g')$ and by $A\lambda : G \to G'^\vee$ the associated homomorphism given by $A\lambda(g)(g') = \lambda(g, g')$. The map λ is called *nonsingular* if $A\lambda$ is an isomorphism.

If $G' = G$ and $\epsilon = \pm 1$, we call λ ϵ-symmetric if $\lambda^t = \epsilon\lambda$. From now on we consider only pairings which are either symmetric ($\epsilon = 1$) or skew-symmetric ($\epsilon = -1$).

We also suppose that the natural map $G \to (G^\vee)^\vee$ is an isomorphism: this holds, for example, in the following situations:

V a field, G a finite dimensional vector space,

$V = \mathbb{Z}$, G a finitely generated free abelian group,

$V = \mathbb{Q}/\mathbb{Z}$ or the circle group $\mathbb{R}/\mathbb{Z}$, G a finite abelian group.

We call $g, g' \in G$ *orthogonal* if $\lambda(g, g') = 0$; for any subgroup $H \subset G$, its *annihilator* is defined by $H^o := \{g \in G \mid \forall g' \in G, \lambda(g, g') = 0\}$. Thus $H \subseteq H^o$ if and only if $\lambda(H \times H) = 0$. If $H^o = H$, H is called *Lagrangian*. We have

Lemma 7.3.1 *If $\lambda : G \times G \to V$ is nonsingular and ϵ-symmetric, and $H \subset G$ such that $\lambda \mid H \times H$ is nonsingular, then G splits as $H \oplus H^o$.*

We say that the form λ is *even* if, for each $g \in G$, there exists $v \in V$ with $\lambda(g, g) = v + \epsilon v$.

Lemma 7.3.2 *If $\lambda : G \times G \to V$ is nonsingular, ϵ-symmetric and even, and $H \subset G$ is Lagrangian, then there is a Lagrangian subgroup H^* such that $G = H \oplus H^*$. We can identify H^* with $H^\vee$, so that λ is given by*

$$\lambda((g, h), (g', h')) = h(g) + \epsilon h'(g).$$

Thus in this case, the form is determined up to isomorphism by H.

For symmetric bilinear forms over $\mathbb{R}$, it is well known that one can choose a basis $\{e_i \mid 1 \leq i \leq r\}$ for G such that $\lambda(e_i, e_j) = 0$ for $i \neq j$; if the form is nonsingular then each $a_i = \lambda(e_i, e_i) \neq 0$, and the form is classified up to isomorphism by the signature, which is given by $\sigma(\lambda) = \#\{i \mid a_i > 0\} - \#\{i \mid a_i < 0\}$. There is a Lagrangian subspace if and only if $\sigma(\lambda) = 0$.

For symmetric bilinear forms over $\mathbb{Z}$, we can tensor with $\mathbb{R}$ to obtain a form over $\mathbb{R}$ and thus define a signature. The following are known.

Proposition 7.3.3 *A nonsingular symmetric bilinear form λ over $\mathbb{Z}$ has a Lagrangian subspace if and only if $\sigma(\lambda) = 0$.*

If λ is a nonsingular even symmetric bilinear form over $\mathbb{Z}$, then $\sigma(\lambda)$ is divisible by 8.

Necessity of the condition $\sigma(\lambda) = 0$ is trivial. An example of an even form with signature 8 is given by the form with the matrix (the 'E_8 matrix')

$$\begin{pmatrix} 2 & -1 & 0 & 0 & 0 & 0 & 0 & 0 \\ -1 & 2 & -1 & 0 & 0 & 0 & 0 & 0 \\ 0 & -1 & 2 & -1 & 0 & 0 & 0 & 0 \\ 0 & 0 & -1 & 2 & -1 & 0 & 0 & 0 \\ 0 & 0 & 0 & -1 & 2 & -1 & 0 & -1 \\ 0 & 0 & 0 & 0 & -1 & 2 & -1 & 0 \\ 0 & 0 & 0 & 0 & 0 & -1 & 2 & 0 \\ 0 & 0 & 0 & 0 & -1 & 0 & 0 & 2 \end{pmatrix}. \tag{7.3.4}$$

The skew-symmetric case is easily handled.

Proposition 7.3.5 *Let λ be a skew-symmetric bilinear form over $\mathbb{Z}$. Then H has a basis $\{e_i, f_i \mid 1 \le i \le r\}$ $\{g_j\}$ such that $\lambda(x, y) = 0$ for all pairs of basis elements* except *that $\lambda(e_i, f_i) = a_i$ for each i.*

Proof Write $H^o := \{x \mid \forall y \in H, \lambda(x, y) = 0\}$ for the radical of λ. Since, for $0 \neq k \in \mathbb{Z}$, $nx \in H^o$ implies $x \in H^o$, H^o is a direct summand of H. We may thus take a basis $\{g_j\}$ of H^o and extend to a basis for H.

We have reduced to the case when H^o is trivial, so $A\lambda$ is injective, with finite cokernel. Choose $e_1 \in H$ with coset modulo $A\lambda(H)$ of maximal order a_1. Then a_1 is the highest common factor of the $\lambda(e_1, x)$ for $x \in H$. Choose $f_1 \in H$ with $\lambda(e_1, f_1) = a_1$. Now for any $x \in H$ we can write $\lambda(e_1, x) = pa_1$ and $\lambda(f_1, x) = qa_1$: then $x + qe_1 - pf_1$ is orthogonal to both e_1 and f_1. Thus H is the orthogonal direct sum of $\mathbb{Z}\langle e_1, f_1\rangle$ and its orthogonal complement and we can proceed by induction. □

In particular,

Lemma 7.3.6 *Any nonsingular skew-symmetric bilinear form over $\mathbb{R}$ or $\mathbb{Z}$ has a Lagrangian subspace.*

Thus we may take a basis $\{e_i, f_i \mid 1 \le i \le r\}$ of H such that $\lambda(e_i, e_j) = \lambda(e_i, f_j) = \lambda(f_i, f_j) = 0$ for all i, j *except* that $\lambda(e_i, f_i) = 1$ for each i: such a basis is called a *symplectic basis*.

Now suppose given a nonsingular skew-symmetric bilinear form λ on a free abelian group H together with a map $\mu : H \to \mathbb{Z}_2$ with $\mu(0) = 0$ and satisfying the identity

$$\mu(x + x') = \mu(x) + \mu(x') + \lambda(x, x') \quad (\lambda(x, x') \text{ taken mod } 2). \tag{7.3.7}$$

The classification is given by

Lemma 7.3.8 *Given (H, λ, μ) as above, choose a symplectic basis $\{e_i, f_i \mid 1 \le i \le r\}$ of (H, λ). Then the number $Arf(\mu) := \sum_i \mu(e_i)\mu(f_i) \in \mathbb{Z}_2$ is an invariant of (H, λ, μ), and two such triples (of the same rank) are isomorphic if and only if the invariants $Arf(\mu)$ agree.*

Moreover, $Arf(\mu) = 0$ if and only if H has a Lagrangian subgroup on which μ vanishes.

Proof If $\mu(e_i) = 1$ and $\mu(f_i) = 0$, replacing e_i by $e_i' := e_i + f_i$ changes $\mu(e_i)$ to 0 without affecting the other values; similarly with e_i and f_i interchanged; thus we may reduce to the case $\mu(e_i) = \mu(f_i)$ for each i.

If $\mu(e_1) = \mu(f_1) = \mu(e_2) = \mu(f_2) = 1$ we substitute $e_1' := e_1 + e_2$, $f_2' := -f_1 + f_2$, preserving λ, with $\mu(e_1') = \mu(f_2') = 0$, and then deal with $\mu(f_1)$ and $\mu(e_2)$ as above. We may thus reduce to a normal form where μ vanishes on all basis elements except perhaps e_1 and f_1.

To prove $\mathrm{Arf}(\mu)$ an invariant, we note that $\mu : H \to \mathbb{Z}_2$ factors through $H/2H = H \otimes \mathbb{Z}_2$, and can check using the normal form that the number of elements of $H/2H$ on which μ takes the value 1 is $2^{2r-1} + 2^{r-1}$ if $\mathrm{Arf}(\mu) = 1$ and $2^{2r-1} - 2^{r-1}$ if $\mathrm{Arf}(\mu) = 0$.

The final result follows by inspection. □

We now consider the case when G is a finite group and λ takes values in $\mathbb{Q}/\mathbb{Z}$. Here instead of working with a free basis, we write G as the direct sum of subgroups of prime power order; each of these is a direct sum of cyclic subgroups.

Proposition 7.3.9 *Given a nonsingular skew-symmetric form λ on a finite group G, we may express G as a direct sum of mutually orthogonal subgroups, each of which is either of order 2 with $\lambda(x, x) = \frac{1}{2}$ or is isomorphic to $\mathbb{Z}_{p^k} \oplus \mathbb{Z}_{p^k}$ (for some prime p and integer k), with generators x, x' satisfying $\lambda(x, x) = 0$ or $\frac{1}{2}$, $\lambda(x', x') = 0$, $\lambda(x, x') = p^{-k}$.*

Further, we may suppose there is at most one summand of order 2.

Proof For each p, choose $x \in G$ of order p^k with k maximal. Since the form is nonsingular, we can find $x' \in G$ with $\lambda(x, x') = p^{-k}$. If p is odd, since $\lambda(x, x) = \lambda(x', x') = 0$, the form λ is nonsingular on $\langle x, x' \rangle$, so by Lemma 7.3.1 (G, λ) is the orthogonal direct sum of this and another subgroup, so we can proceed by induction.

If $p = 2$ and $k > 1$, each of $\lambda(x, x)$ and $\lambda(x', x')$ may be either 0 or $\frac{1}{2}$, but the same argument applies; if each is $\frac{1}{2}$ we can substitute $x + x'$ for x' to reduce $\lambda(x', x')$ to 0.

If $p^k = 2$ and $\lambda(x, x) = 0$, we may proceed as above, but if $\lambda(x, x) = \frac{1}{2}$, λ is already nonsingular on $\langle x \rangle$, so we can split this off as an orthogonal direct summand.

Finally observe that if we have two such summands $\lambda(x, x) = \lambda(y, y) = \frac{1}{2}$ and $\lambda(x, y) = 0$, we can start with $z = x + y$ to reduce to the preceding case. □

Proposition 7.3.10 *Given a nonsingular symmetric form λ on a finite group G, we may express G as a direct sum of mutually orthogonal subgroups, each of which is either cyclic of prime power order or is isomorphic to $\mathbb{Z}_{2^k} \oplus \mathbb{Z}_{2^k}$ (for some k), with generators x, x' satisfying $2^{k-1}\lambda(x, x) = 2^{k-1}\lambda(x', x') = 0$, $\lambda(x, x') = 2^{-k}$.*

Proof Again it suffices to consider the case when G is a p-group. Let k be the greatest integer such that G has an element of order p^k: choose such an element x. If $\lambda(x, x)$ has order p^k, the restriction of λ to the subgroup H generated by x is nonsingular, and we may apply Lemma 7.3.1. Otherwise, choose an element y with $\lambda(x, y) = p^{-k}$: then y has order p^k. If p is odd, either $z = y$ or $z = x + y$ is such that $\lambda(z, z)$ has order p^k and we may proceed as above.

If $p = 2$, it may be that $\lambda(y, y)$ and $\lambda(x + y, x + y)$ both have order $< 2^k$. In this case, the restriction of λ to the subgroup H generated by x and y is nonsingular, and we may again apply Lemma 7.3.1. □

One can now proceed to further analysis of each of these types of summand.

We define a nonsingular quadratic form on the finite group G to be a pair (λ, μ) with $\lambda : G \times G \to \mathbb{Q}/\mathbb{Z}$ and $\mu : G \to \mathbb{Q}/2\mathbb{Z}$ satisfying

- λ is nonsingular symmetric bilinear,
- $\mu(0) = 0$, $\mu(-x) = \mu(x)$ and
- $2\lambda(x, y) = \mu(x + y) - \mu(x) - \mu(y)$.

It follows that $\mu(x) \equiv \lambda(x, x) \pmod 1$. Note that $\lambda(x, y) \in \mathbb{Q}/\mathbb{Z}$ determines $2\lambda(x, y)$ modulo 2. The classification of quadratic forms is close to that of symmetric bilinear forms (if $|G|$ is odd, there is no essential difference). In analogy with the above, call a subgroup $H \subset G$ Lagrangian if μ vanishes on H (and hence λ vanishes on $H \times H$) and $|G| = |H|^2$. As before, H then coincides with its annihilator under λ and $A\lambda$ induces an isomorphism of G/H on $H^\vee$.

Here there is a new feature. We define the Gauss sum $\mathfrak{G}(\mu) := \sum_{x\in G} e^{i\pi\mu(x)}$.

Theorem 7.3.11 *Suppose (λ, μ) a nonsingular quadratic form on the finite group G. Then $\mathfrak{G}(\mu)$ has the form $A(\mu)\sqrt{|G|}$, with $A(\mu)^8 = 1$. If there is a Lagrangian subgroup H, $\mathfrak{G}(\mu) = |H|$ and $A(\mu) = 1$.*

Proof We prove the second statement first. We split the sum over G into a sum over cosets of H. There are two cases. For the trivial coset, $\sum_{h\in H} e^{i\pi\mu(h)} = \sum_{h\in H} e^0 = |H|$. For any other coset, we have $\sum_{h\in H} e^{i\pi\mu(y+h)} = e^{i\pi\mu(y)} \sum_{h\in H} e^{i\pi\lambda(y,h)}$. Now if y has order k in G/H, as λ is nonsingular, there exists $z \in H$ with $\lambda(y, z) = \frac{1}{k}$, and as z varies, each value $\frac{j}{k}$ is taken $|H|/k$ times. But the sum $\sum_{j \bmod k} e^{2\pi ij/k}$ vanishes unless $k = 1$. Thus the sum over H vanishes. Summing over all cosets, we just have $|H|$.

Given two triples (G, λ, μ) and (G', λ', μ') we can form the direct sum $G'' := G \oplus G'$ and define $\lambda''((x, x'), (y, y')) := \lambda(x, y) + \lambda'(x', y')$ and $\mu''(x, x') := \mu(x) + \mu'(x')$. This has the above properties, and we see that $\mathfrak{G}(\mu'') = \mathfrak{G}(\mu)\mathfrak{G}(\mu')$.

Now take $G' = G$, $\lambda' = -\lambda$ and $\mu' = -\mu$: then $\mathfrak{G}(\mu') = \overline{\mathfrak{G}(\mu)}$. In the direct sum $G'' := G \oplus G'$, the diagonal is a Lagrangian subgroup. Hence $|\mathfrak{G}(\mu)|^2 = \mathfrak{G}(\mu)\overline{\mathfrak{G}(\mu)} = \mathfrak{G}(\mu'') = |G|$.

The calculation of the argument is more sophisticated. For p an odd prime, it was shown by Gauss that the sum $\sum_{j \bmod p} e^{2\pi ij^2/p}$ is equal to $i\sqrt{p}$: other cases when G is cyclic of order a power of p follow easily. The calculations for the case $p = 2$ can be done ad hoc: for example, if G has order 2, and $\lambda(x, x) = \frac{1}{2}$, $\mu(x) = \frac{1}{2}$, we have $\mathfrak{G}(\mu) = 1 + i = e^{i\pi/4}\sqrt{2}$. □

Lemma 7.3.12 *For (λ, μ) a nonsingular quadratic form on G, the function $\mu_z(x) := \mu(x) + 2\lambda(x, z)$ is a quadratic form if and only if $2z = 0$. We have $\mathfrak{G}(\mu_z) = e^{-i\pi\mu(z)}\mathfrak{G}(\mu)$.*

Proof For necessity note that $\mu_z(x) - \mu_z(-x) = \mu(x) - \mu(-x) + 2\lambda(x, 2z)$, and since λ is nonsingular, $2\lambda(x, 2z) \in 2\mathbb{Z}$ for all x if and only if $2z = 0$.

Now since $\mu_z(x) := \mu(x) + 2\lambda(x, z) = \mu(x + z) - \mu(z)$, we can write $\mathfrak{G}(\mu_z) = \sum_{x\in G} e^{i\pi\mu_z(x)} = \sum_{x\in G} e^{i\pi(\mu(x+z)-\mu(z))} = e^{-i\pi\mu(z)} \sum_{x\in G} e^{i\pi\mu(x+z)}$, which equals $e^{-i\pi\mu(z)}\mathfrak{G}(\mu)$. □

In Lemma 7.3.8 we considered pairs (λ, μ) with $\lambda : H \times H \to \mathbb{Z}$ a nonsingular skew-symmetric form and $\mu : H \to \mathbb{Z}_2$ satisfying (7.3.7). If we set

$G := H/2H$ and define $\lambda_S : G \times G \to \mathbb{Q}/\mathbb{Z}$ by $\lambda_S(g, g') = \frac{1}{2}\lambda(x, x')$ and $\mu_S : G \to \mathbb{Q}/2\mathbb{Z}$ by $\mu_S(g) = \mu(x)$ (where x, x' are lifts of g, g'), then (λ_S, μ_S) is a nonsingular quadratic form on the finite group G. Using a symplectic basis, we see at once that $A(\mu) = (-1)^{\mathrm{Arf}(\mu_S)}$.

Now consider an ϵ-symmetric pairing $\lambda : H \times H \to \mathbb{Z}$, but now suppose only that $A\lambda : H \to H^\vee$ is injective; denote by G its cokernel. We can tensor with $\mathbb{Q}$ to embed λ in a nonsingular pairing λ_Q on H_Q to $\mathbb{Q}$: denote by H' the subgroup of H_Q corresponding to $H^\vee$: thus G is identified with H'/H.

We can lift elements $g, g' \in G$ to elements $x, x' \in H'$ and form $\lambda_Q(x, x')$: denote its image in $\mathbb{Q}/\mathbb{Z}$ by $\overline{\lambda}(g, g')$.

Lemma 7.3.13 *The class of $\overline{\lambda}(g, g')$ in $\mathbb{Q}/\mathbb{Z}$ depends only on g, g'. This defines a nonsingular ϵ-symmetric pairing $\overline{\lambda} : G \times G \to \mathbb{Q}/\mathbb{Z}$.*

Proof It is immediate that $\overline{\lambda}$ is well defined, bilinear, and ϵ-symmetric. If $A\overline{\lambda}(g) = 0$, so for each $g' \in G$, $\overline{\lambda}(g, g') = 0$, then the lift x of g is such that for each $x' \in H'$ we have $\lambda(x, x') \in \mathbb{Z}$, so $A\lambda(x) \in H'^\vee$ so $x \in H$ and $g = 0$. Since $A\overline{\lambda} : G \to G^\vee$ is injective, and both these finite groups have the same order, it is an isomorphism. □

In the case $\epsilon = -1$, we note the additional property $\overline{\lambda}(g, g) = 0$ for all $g \in G$.

If the above form λ is even as well as symmetric, we can enhance $\overline{\lambda}$ by defining $\mu : G \to \mathbb{Q}/2\mathbb{Z}$ by $\mu(g) = \lambda(x, x) \pmod{2\mathbb{Z}}$, where x is a lift of g. It is immediate that μ is a quadratic form in the sense defined above. The following is a deeper result.

Theorem 7.3.14 *Let the even symmetric bilinear form λ on H induce the quadratic map $(\overline{\lambda}, \mu)$ on $G = H^\vee/H$ as above. Then $\mathfrak{G}(\mu) = e^{i\pi\sigma(\lambda)/4}\sqrt{|G|}$.*

This result is given by van der Blij [155], together with a short proof that depends on manipulation of divergent integrals. We observe that it ties in with the result in Proposition 7.3.3 that if λ is nonsingular (so G is trivial), then $\sigma(\lambda)$ is divisible by 8.

7.4 Poincaré complexes and pairs

As noted above, we cannot expect to improve the result of Theorem 7.2.1 without imposing some conditions on X. If we can construct a homotopy equivalence $M \to X$ with M a closed manifold, then X also must satisfy Poincaré duality as in Theorem 5.3.5. A corresponding conclusion applies for manifolds with boundary. We begin with a formal definition of the duality property, then explore some consequences in the form of pairings on its homology groups. We turn to the study of homological properties of maps of degree 1.

We make the following definitions. A *Poincaré complex* of formal dimension m consists of a finite CW-complex X and a homology class $[X] \in H_m(X; \mathbb{Z})$ (which we call the *fundamental class*) such that cap product with $[X]$ induces isomorphisms $H^r(X; \mathbb{Z}) \to H_{m-r}(X; \mathbb{Z})$ for all $r \in \mathbb{Z}$.

Thus a first necessary condition for a finite CW complex X to have the homotopy type of a closed manifold is that X be a Poincaré complex. We will see that, if X is simply-connected, there is a natural way to enhance this to obtain a sufficient condition.

We have defined Poincaré complexes in terms of homology. We will see in §7.8 how they can be defined from a homotopy theoretic viewpoint.

The above definition is not adequate if X is not simply-connected, and does not even include non-orientable closed manifolds. For the definition in the general case, see [164].

A *Poincaré pair* consists of a finite CW-pair (Y, X) and a homology class $[Y] \in H_{m+1}(Y, X)$ such that cap product with $[Y]$ induces isomorphisms

$$H^r(Y; \mathbb{Z}) \to H_{m+1-r}(Y, X; \mathbb{Z}), \quad H^r(Y, X; \mathbb{Z}) \to H_{m+1-r}(Y; \mathbb{Z})$$

for all r. It follows that if $[X] := \partial[Y] \in H_m(X; \mathbb{Z})$, then $(X, [X])$ is a Poincaré complex. Indeed, in view of the five lemma, the commutative diagram of exact sequences

$$\begin{array}{ccccccccc}
H^{r-1}(X) & \to & H^r(Y,X) & \to & H^r(Y) & \to & H^r(X) & \to & H^{r+1}(Y,X) \\
[X]\downarrow & & [Y]\downarrow & & [Y]\downarrow & & [X]\downarrow & & [Y]\downarrow \\
H_{m+1-r}(X) & \to & H_{m+1-r}(Y) & \to & H_{m+1-r}(Y,X) & \to & H_{m-r}(X) & \to & H_{m-r}(Y)
\end{array}$$

shows conversely that if we assume X a Poincaré complex, the two conditions defining Poincaré pairs are equivalent.

We regard Poincaré complexes and pairs as the homotopy-theoretic analogue to compact manifolds. Many theorems valid for manifolds have analogues in this context. Corresponding to the Disc Theorem 2.5.6, we have

Lemma 7.4.1 *Let Z be a Poincaré complex of formal dimension $n \geq 3$. Then there exist a Poincaré pair (Y, X) and a homotopy equivalence $f : S^{n-1} \to X$ such that the space $Y \cup_f e^n$ obtained by glueing (Y, X) to (D^n, S^{n-1}) is homotopy equivalent to Z.*

Proof ([164, Theorem 2.4]) We give details here only in the simply connected case. Then Z is homotopy equivalent to a finite CW-complex, and we may suppose that this has no cell of dimension greater than n, and only one n-cell. Now pick an embedding of D^n in the interior of this cell, and define

Y by deleting its interior. Thus inclusions induce isomorphisms $H_n(Z) \to H_n(Z, Y)$, $H_n(D^n, S^{n-1}) \to H_n(Z, Y)$ preserving the fundamental classes $[Z]$ and $[D^n, S^{n-1}]$. Cap products with the fundamental class give maps of the Mayer–Vietoris cohomology sequence of $(Z; Y, D^n; S^{n-1})$ to the homology sequence; these are isomorphisms for Z and for (D^n, S^{n-1}), hence also for (Y, S^{n-1}). □

For any X, cup product gives a $(-1)^k$-symmetric bilinear pairing

$$H^k(X; \mathbb{Z}) \times H^k(X; \mathbb{Z}) \to H^{2k}(X, \mathbb{Z}).$$

If X is a Poincaré complex of formal dimension $2k$, we have $H^{2k}(X, \mathbb{Z}) \cong \mathbb{Z}$, and so a bilinear pairing of $H^k(X; \mathbb{Z})$, which is $(-1)^k$-symmetric. Since the map $[X]\cap : H^k(X; \mathbb{Z}) \to H_k(X; \mathbb{Z})$ is an isomorphism, we also obtain a pairing on $H_k(X; \mathbb{Z})$. The pairing is obtained by composing $[X]\cap$ with the natural map $H^k(X; \mathbb{Z}) \to \mathrm{Hom}(H_k(X; \mathbb{Z}), \mathbb{Z})$. If we extend coefficients from $\mathbb{Z}$ to $\mathbb{Q}$ this becomes an isomorphism, so the pairings become nonsingular. When X is a $2k$-manifold, the self-pairing of $H_k(X; \mathbb{Z})$ can be geometrically interpreted as intersection numbers.

When k is even, the question arises whether the form on $H^k(X; \mathbb{Z})$ is even, in the sense that for each $x \in H^k(X; \mathbb{Z})$, $x.x[X]$ is even. If we reduce mod 2, we obtain the cup product pairing on $H^k(X; \mathbb{Z}_2)$. We have the Wu relations (see §B.4) $x^2[X] = xv_k[X]$ for $x \in H^k(X; \mathbb{Z}_2)$. Thus the vanishing of the characteristic class v_k is necessary and sufficient for the form on $H^k(X; \mathbb{Z})$ to be even.

If (Y, X) is a Poincaré pair of formal dimension $2k + 1$, in the exact sequence

$$H^k(Y; \mathbb{Q}) \to H^k(X; \mathbb{Q}) \to H^{k+1}(Y, X; \mathbb{Q})$$

the two maps are dual to each other, so have the same rank, and the pairing vanishes on the image of $H^k(Y; \mathbb{Q})$ since this factors through the zero map $H^{2k}(Y; \mathbb{Q}) \to H^{2k}(X; \mathbb{Q})$: thus this image is a Lagrangian subspace. In the case when k is even, it follows from Lemma 7.3.2 that the pairings have signature $\sigma = 0$.

Now let X be a Poincaré complex of odd formal dimension $2k + 1$; first suppose $H_k(X; \mathbb{Q}) = 0$, so that $H_k(X; \mathbb{Z})$ is a finite group. Since by duality $H_{k+1}(X; \mathbb{Q}) = 0$, the map $H_{k+1}(X; \mathbb{Q}/\mathbb{Z}) \to H_k(X; \mathbb{Z})$ is an isomorphism, while by duality

$$H_{k+1}(X; \mathbb{Q}/\mathbb{Z}) \cong H^k(X; \mathbb{Q}/\mathbb{Z}) \cong \mathrm{Hom}(H_k(X; \mathbb{Z}), \mathbb{Q}/\mathbb{Z}).$$

Composing these maps gives a nonsingular pairing of $H_k(X; \mathbb{Z})$ with itself to $\mathbb{Q}/\mathbb{Z}$.

When X is a $(2k+1)$-manifold, this too can be interpreted geometrically. If $x \in H_k(X;\mathbb{Z})$ has order θ, we represent x by a k-cycle ξ; then $\theta\xi$ is a boundary, say $\theta\xi = \partial\zeta$. Given another class $y \in H_k(X;\mathbb{Z})$ represented by a cycle η disjoint from ξ, we may suppose η transverse to ζ and count the intersections. Then $\lambda(x, y) = \frac{1}{\theta}(\zeta.\eta) \pmod{\mathbb{Z}}$.

Either from the algebraic or geometric approach we can see that when the hypothesis $H_k(X;\mathbb{Q}) = 0$ is dropped we obtain a nonsingular pairing of the torsion subgroup Tors $H_k(X;\mathbb{Z})$ with itself to $\mathbb{Q}/\mathbb{Z}$. It follows from the symmetry property of cup products over $\mathbb{Q}$ that this form is $(-1)^{k+1}$-symmetric. Thus if k is even, $x \mapsto b(x, x)$ defines a homomorphism c : Tors $H_k(X;\mathbb{Z}) \to \frac{1}{2}\mathbb{Z}/\mathbb{Z}$. It can be shown that we have $c(x) = \langle v_k, x\rangle$.

We saw in §7.1 that to facilitate surgery it is natural to consider normal maps $(f : M \to X, \nu, T)$. We now suppose X a Poincaré complex and impose the further condition $f_*[M] = [X]$ or, as we will say, that f has degree 1. Observe that if $(g : N \to X, \nu, U)$ is a normal cobordism of f to f' and f has degree 1, then so has f'. We thus study the homology of maps of degree 1.

Proposition 7.4.2 *Let $\phi : M \to X$ be a map of degree 1 of Poincaré complexes. Then the diagram*

$$\begin{array}{ccc} H^r(M;\mathbb{Z}) & \xleftarrow{\phi^*} & H^r(X;\mathbb{Z}) \\ \downarrow{\scriptstyle [M]\cap} & & \downarrow{\scriptstyle [X]\cap} \\ H_{m-r}(M;\mathbb{Z}) & \xrightarrow{\phi_*} & H_{m-r}(X;\mathbb{Z}) \end{array}$$

is commutative, so $.[M]\cap$ induces an isomorphism of the cokernel $K^r(M;\mathbb{Z})$ of ϕ^* on the kernel $K_{m-r}(M;\mathbb{Z})$ of ϕ_*. In particular, if ϕ is k-connected, ϕ_* and ϕ^* are isomorphisms for $r < k$ and for $r > m-k$.

A map $\phi : (N, M) \to (Y, X)$ of degree 1 of Poincaré pairs induces split surjections of homology groups $M \to X$, $N \to Y$ and $(N, M) \to (Y, X)$ with kernels K_* and split injections of cohomology groups with cokernels K^*. The duality map $.[N]\cap$ induces isomorphisms

$$K^*(N) \to K_*(N, M), \qquad K^*(N, M) \to K_*(N).$$

The homology (cohomology) exact sequence of (N, M) is isomorphic to the direct sum of the sequence for (Y, X) and a sequence of groups $K_*(K^*)$.

Commutativity of the diagram follows from naturality of cap products. Since the vertical maps are isomorphisms, ϕ^* is a split injection and ϕ_* a split surjection. The other assertions are immediate consequences. The same holds if $\mathbb{Z}$ is replaced by any coefficient group.

If $\phi : (M, \partial M) \to (X, \partial X)$ is a map of degree 1 of Poincaré pairs inducing a homotopy equivalence $\partial M \to \partial X$, it follows that, as in the case of closed manifolds, $.[M]\cap$ induces an isomorphism of the cokernel $K^r(M; \mathbb{Z})$ of ϕ^* on the kernel $K_{m-r}(M; \mathbb{Z})$ of ϕ_*.

It follows that if M has dimension $2k$, we have a $(-1)^k$-symmetric bilinear form on $K_k(M; \mathbb{Z})$ to $\mathbb{Z}$. If moreover the map $f : M \to X$ is k-connected, so the groups K_* vanish in lower dimensions, this pairing is nonsingular. It follows from the commutative diagrams and the characteristic property of v_k that $v_k(M) = \phi^* v_k(X)$. Hence if k is even, the self-pairing of $K_k(M; \mathbb{Z})$ is even.

If M has odd dimension $2k + 1$, we have a $(-1)^{k+1}$-symmetric bilinear form on Tors $K_k(M; \mathbb{Z})$ to $\mathbb{Q}/\mathbb{Z}$. Again, if the map $f : M \to X$ is k-connected, this pairing is nonsingular.

Now suppose given a Poincaré complex X of formal dimension m, and a normal map $(f : M \to X, \nu, T)$ of degree 1 (or more generally $(X, \partial X)$ a Poincaré pair and $f : (M, \partial M) \to (X, \partial X)$ inducing a homotopy equivalence $\partial M \to \partial X$). By Theorem 7.2.1, if $m \geq 2k$, we may perform surgery to make f k-connected. Then $K_r(M)$ vanishes for $r < k$. It now follows from Proposition 7.4.2, together with duality, that if $m = 2k$, $K_r(M)$ vanishes except if $r = k$, while if $m = 2k + 1$, the exceptions are $r = k,\ k + 1$.

Now let (Y, X) be a Poincaré pair of formal dimension n and $\phi : (N, M) \to (Y, X)$ a normal map of degree 1. We may first apply Theorem 7.2.1 to $\phi \mid M$, extend to a normal cobordism of ϕ, and then apply the Theorem to ϕ. This kills all the K groups except those in the sequence: if $n = 2k$,

$$0 \to K_k(M) \to K_k(N) \to K_k(N, M) \to K_{k-1}(M) \to 0 :$$

and if $n = 2k + 1$,

$$0 \to K_{k+1}(N) \to K_{k+1}(N, M) \to K_k(M) \to K_k(N) \to K_k(N, M) \to 0.$$

The following extension of Theorem 7.2.1 will be useful.

Proposition 7.4.3 *([167, p. 15]) Suppose $n = 2k + 1$, $k \geq 2$, and both X and Y are simply-connected; then ϕ is normally cobordant to a k-connected normal map such that $K_k(N, M) = 0$.*

Proof We may suppose ϕ k-connected. Since $k \geq 2$ and a 2-connected map induces an isomorphism of fundamental groups, both M and N are simply-connected. Thus $K_k(N, M) \cong \pi_{k+1}(\phi)$. Choose a finite set $\{e_i\}$ of generators. As in Theorem 7.1.1 we can represent each e_i by a framed immersion $f_i : (D^k, S^{k-1}) \to (N, M)$. By general position, we may suppose the f_i disjoint embeddings. We extend these to disjoint embeddings $F_i : (D^k, S^{k-1}) \times D^{k+1} \to (N, M)$.

Since these represent elements of $\pi_{k+1}(\phi)$, they are nullhomotopic in (Y, X), so ϕ is homotopic to a map taking the image of each F_i to a point. We obtain (N', M') from (N, M) by deleting the interiors of the images of the F_i and rounding the corners. We inherit a normal map $\phi' : (N', M') \to (Y, X)$, and a normal cobordism of ϕ' to ϕ is obtained from $\phi \times I$ by adjusting the corners as in Figure 8.1. Although this has not been described as a surgery, it can equivalently be obtained by first performing surgery on the boundary using the $F_i : S^{k-1} \times D^{k+1} \to M$ to add handles, then interior surgery on the k-spheres created by this.

Denote by A the free abelian group with basis $\{e_i\}$. We have an exact sequence $A \to K_k(N, M) \to K_k(N', M') \to 0$; since the first map is surjective, we have $K_k(N', M') = 0$. □

Lemma 7.4.4 *In the above situation, if $K_k(N, M) = 0$, $K_{k+1}(N, M)$ is a Lagrangian subspace of $K_k(M)$.*

Proof We again relativise the arguments of Lemma 7.1.1. We have an isomorphism $\pi_{k+2}(\phi) \to K_{k+1}(N, M)$, so each element α of $K_{k+1}(N, M)$ is represented by a map $g_\alpha : (D^{k+1}, S^k) \to (N, M)$ together with a nullhomotopy of the composed map to (Y, X). Now D^{k+1} is contractible, so has trivial tangent bundle, and the nullhomotopy shows that $g_\alpha^* \mathbb{T}(N)$ is trivial. We thus have a stable isomorphism of $\mathbb{T}(D^{k+1})$ with $g_\alpha^* \mathbb{T}(N)$, which restricts to a stable isomorphism of $\mathbb{T}(S^k)$ with $g_\alpha^* \mathbb{T}(M)$. By the remark following the proof of Theorem 6.2.1, such isomorphisms correspond bijectively to regular homotopy classes of framed immersions $i_\alpha : (D^{k+1}, S^k) \to (N, M)$.

We have now shown that α is represented by a framed immersion i_α. By Proposition 4.6.6, we may suppose this immersion self-transverse; the same goes for the immersion $i_\alpha \cup i_\beta$ of the union of two discs. The double point set is then a 1-manifold, so consists of a collection of embedded circles and arcs whose end points are the (self)-intersection points of the boundary spheres in M.

Now if $\alpha \neq \beta$, each intersection arc has two end points, which make contributions of opposite signs to the intersection number of the two spheres in M. It follows that $\partial\alpha.\partial\beta = 0$, so indeed the image of $K_{k+1}(N, M)$ in $K_k(M)$ is self-annihilating. For k even, this proves the result; if k is odd, $\mu(\alpha)$ is the number of self-intersection points of $i_\alpha(S^k)$, and this vanishes by the same argument. □

7.5 The even dimensional case

Suppose X a Poincaré complex of formal dimension $m = 2k$, and $(f : M \to X, \nu, T)$ a k-connected normal map, or more generally that $(X, \partial X)$ is a Poincaré pair and f a normal map inducing a homotopy equivalence $\partial M \to \partial X$. It follows from Proposition 7.4.2 that all $K_r(M)$ and $K^r(M)$ vanish except for $r = k$. Duality implies further that these groups are free abelian, and the isomorphism $[M]\cap$, together with the dual pairing, gives a nonsingular bilinear form

$$\lambda : K_k(M) \times K_k(M) \to \mathbb{Z},$$

which is symmetric if k is even and skew-symmetric if k is odd.

From now on we make the further assumption that X is simply-connected. Then the Hurewicz Theorem gives an isomorphism $h : \pi_{k+1}(f) \cong K_k(M; \mathbb{Z})$. By Theorem 7.1.1, any $\xi \in \pi_{k+1}(f)$ induces a regular homotopy class of immersions $S^k \times D^k \to M$, and given any embedding in this class we can perform surgery. Write $x := h(\xi)$. Then

Lemma 7.5.1 *In this situation, if $k \geq 3$ is even, surgery on ξ is possible if and only if $\lambda(x, x) = 0$.*

If k is odd, there is an invariant $\mu(x) \in \mathbb{Z}_2$, and if $k \geq 3$, surgery on ξ is possible if and only if $\mu(x) = 0$.

Proof We recall the discussion in §6.3.2 of immersions ϕ of S^k in $2k$-manifolds.

For k even, if $e(\phi)$ denotes the number given by the Euler class of the normal bundle and $I(\phi)$ the signed sum of the intersection numbers at points of self-intersection of $\phi(S^k)$ (which we may assume transverse), then by Lemma 6.3.5, we have $[\phi].[\phi] = e(\phi) + 2I(\phi)$. In the present situation, we have an immersion of $S^k \times D^k$, so $e(\phi) = 0$. Since $x := s(\xi)$ is the homology class $[\phi]$, we have $\lambda(x, x) = 2I(\phi)$. Finally by Theorem 6.3.2, provided $k \geq 3$, if $I(\phi) = 0$, ϕ is regularly homotopic to an embedding.

For k odd, there are two regular homotopy classes of immersions in each homotopy class of maps $S^k \to M^{2k}$, which are distinguished by the parity of the number $I(\phi)$ of self-intersection points of ϕ. Define $\mu(x)$ to be $I(\phi)$ mod 2, where ϕ is in the regular homotopy class determined by ξ. The conclusion again follows from the results in §6.3.2. □

We now calculate the effect of a surgery on homology. Let $(g : N \to X, \nu, T'')$ be a normal cobordism of $(f : M \to X, \nu, T)$ to $(f' : M \to X, \nu, T')$. Since we may regard g as a map $(N, M, M') \to (X \times I, X \times \{0\}, X \times \{1\})$, we may use the groups K_* defined above. Observe that $K_*(N, M) \cong H_*(N, M)$. Thus for a single surgery as above, the K_* exact sequence of (N, M) reduces to

$$0 \to K_{k+1}(N) \to K_{k+1}(N, M) \to K_k(M) \to K_k(N) \to 0,$$

in which $K_{k+1}(N, M) \cong \mathbb{Z}$, and a generator maps to $x \in K_k(M)$. Thus for $x \neq 0$, $K_k(N)$ is the quotient of $K_k(M)$ by the class of x.

We have an exact sequence

$$0 \to K_k(M') \to K_k(N) \to K_k(N, M') \to K_{k-1}(M') \to 0,$$

with $K_k(N, M') \cong \mathbb{Z}$. The map $K_k(N) \to K_k(N, M')$ can be identified with $K^{k+1}(N, \partial N) \to K^{k+1}(N, M)$, so is induced by intersection with the $(k+1)$-cell representing a generator of $K_{k+1}(N, M)$ or with x, its boundary. Thus provided for some $x' \in K_k(M)$ we have $\lambda(x, x') = 1$, this map is surjective, and hence $K_{k-1}(M') = 0$.

If k is even, the intersection pairing λ on $H_k(X; \mathbb{R})$ is symmetric, so determines a signature invariant $\sigma(X) \in \mathbb{Z}$. We saw above that if (Y, X) is a Poincaré pair, then $\sigma(X) = 0$. The main result here is

Theorem 7.5.2 *If $(f : M \to X, \nu, T)$ is a normal map of degree 1 with X a simply-connected Poincaré pair of formal dimension $2k$ with $k \geq 4$ even, and $\partial M \to \partial X$ a homotopy equivalence, then surgery to obtain a homotopy equivalence is possible if and only if $\sigma(M) = \sigma(X)$. Moreover, it then suffices to perform surgeries on spheres of dimension $\leq k$.*

Proof First suppose $(g : N \to X, \nu, T'')$ a normal cobordism of f to a homotopy equivalence $(f' : M' \to X, \nu, T')$. Since ∂N is the disjoint union of M' and M with orientation reversed, $0 = \sigma(\partial N) = \sigma(M') - \sigma(M) = \sigma(X) - \sigma(M)$.

Conversely, suppose $\sigma(M) = \sigma(X)$. We may suppose by Theorem 7.2.1 that f is k-connected. It follows from the proof of Theorem 7.4.2 that in the decomposition $H_k(M) \cong K_k(M) \oplus H_k(X)$ the two summands are mutually orthogonal for the intersection form. Hence the induced pairing λ on $K_k(M)$ has signature zero.

It follows from Proposition 7.3.3 that if λ is a nonsingular symmetric bilinear form on H over $\mathbb{Z}$, of signature zero, there exists a basis $\{e_i, f_i\}$ $(1 \leq i \leq r)$ of H such that $\lambda(e_i, e_j) = \lambda(e_i, f_j) = 0$ for all i, j *except* that $\lambda(e_i, f_i) = 1$ for each i.

Since $\lambda(e_1, e_1) = 0$, by Lemma 7.5.1, we can do surgery on e_1. It follows from the above calculations that for the resulting normal cobordism, $K_k(N)$ is the quotient of $K_k(M)$ by the class of e_1, and that $K_k(M')$ is the subgroup of $K_k(N)$ which is the quotient of the subgroup of $K_k(M)$ consisting of classes orthogonal to e_1: viz. with $\lambda(y, e_1) = 0$. Hence $K_k(M')$ looks like $K_k(M)$ but with base corresponding to $\{e_i, f_i\}$ $(2 \leq i \leq r)$. Thus at the end of r simple surgeries we have arrived at a situation with $K_k = 0$ and hence a homotopy equivalence. □

The details for k odd are somewhat subtler. We first need a closer study of μ, which we defined as a map $K_k(M) \to \mathbb{Z}_2$. First we have

Lemma 7.5.3 *For $x, x' \in K_k(M)$, we have $\mu(x+x') = \mu(x) + \mu(x') + \lambda(x, x')$, where $\lambda(x, x')$ has to be reduced modulo 2.*

Proof Set $x = h(\xi)$, $x' = h(\xi')$: then ξ, ξ' determine immersions ϕ, $\phi' : S^k \times D^k \to M$ up to regular homotopy.

We may think of two disjoint $(k+1)$-discs in $\mathbb{R}^{k+1}$ joined by a thickened arc: the boundary of the union is a model for the connected sum of two k-spheres. Now ξ, ξ' give maps of the spheres to M which extend to maps of the discs to X: joining along the arc gives maps representing $\xi + \xi'$. The ingredients (ν, T) of the normal maps also pass to the union. We conclude that an immersion ϕ'' representing $\xi + \xi'$ may be obtained from ϕ and ϕ' by joining the spheres along the neighbourhood of an arc (which may be taken disjoint from the spheres).

The self-intersections of ϕ'' thus consist of those of the two spheres together with their mutual intersections. The result follows by counting up. □

We thus have a nonsingular skew-symmetric bilinear form λ on a free abelian group $H := K_k(M)$ together with a map $\mu : H \to \mathbb{Z}_2$ satisfying the identity $\mu(x+x') = \mu(x) + \mu(x') + \lambda(x, x')$. According to Lemma 7.3.8, the classification of such triples (H, λ, μ) is given by the rank of H and the invariant $\mathrm{Arf}(\mu) \in \mathbb{Z}_2$.

We are now ready to give a first version of the main result for the case k odd.

Theorem 7.5.4 *If $(f : M \to X, \nu, T)$ is a normal map of degree 1 with X a simply-connected Poincaré pair of formal dimension $2k$ with $k \geq 3$ odd, and $\partial M \to \partial X$ a homotopy equivalence, then surgery to obtain a homotopy equivalence is possible provided that $\mathrm{Arf}(\mu) = 0$. If surgery is possible, it suffices to perform surgeries on spheres of dimension $\leq k$.*

Proof We may suppose after preliminary surgery that f is k-connected, so $K_k(M)$ is free abelian and supports a nonsingular skew-symmetric intersection form λ. Choose a symplectic basis $\{e_i, f_i \mid 1 \leq i \leq r\}$ of $(K_k(M), \lambda)$. Since $\mathrm{Arf}(\mu) = 0$, the proof of Lemma 7.3.8 shows that we may adjust this basis so that μ vanishes on each basis element.

Thus by Lemma 7.5.1, we may perform surgery on e_1. As in the proof of Theorem 7.5.2, $K_k(M')$ for the resulting manifold M' is the quotient by $\langle e_1 \rangle$ of the subgroup of $K_k(M)$ orthogonal to e_1, thus has basis $\{e_i, f_i \mid 2 \leq i \leq r\}$. The result now follows by induction on r. □

Theorem 7.5.4 is incomplete: we have neither given an à priori definition of $\mathrm{Arf}(\mu)$ nor proved necessity of its vanishing for completing surgery. We will offer an invariant form of $\mathrm{Arf}(\mu)$ in §7.7 after introducing further concepts. We now take up the other point.

Consider a normal map $(f : M \to X, \nu, T)$ of degree 1 with X a simply-connected Poincaré pair of formal dimension $2k$ with k odd, and $\partial M \to \partial X$ a homotopy equivalence: we can perform surgery to replace f by a k-connected map $f' : M' \to X$, and then define λ and μ on $K_k(M')$ as above.

Proposition 7.5.5 *In the above situation, $\mathrm{Arf}(\mu)$ is an invariant of the normal cobordism class of (f, ν, T).*

Proof It suffices to consider two normally cobordant normal maps and show that the corresponding values of Arf are the same. Denote by $F : W \to X \times I$ the normal cobordism; we may suppose each of $\partial_- W \to X$ and $\partial_+ W \to X$ k-connected.

We next wish to use Proposition 7.4.3 to allow us to do surgery to kill $K_k(W, \partial W)$. There is a minor technical point: if X has no boundary, ∂W is disconnected. To deal with this, use Lemma 7.4.1 to write $X = X_0 \cup_g e^{2k}$, with (X_0, S^{2k-1}) a Poincaré pair and correspondingly delete the interior of an embedded disc from $\partial_\pm W$. This reduces to the case when $\partial X_0 = S^{2k-1}$, and now ∂W is connected.

The result of the surgery is a manifold W' with $K_k(\partial W')$ the orthogonal direct sum of $K_k(\partial_- W)$, $K_k(\partial_+ W)$ and a number of copies of $\mathbb{Z} \oplus \mathbb{Z}$, with one produced by each surgery on the boundary. Its Arf invariant is the sum of those of the summands, which are respectively equal to $\mathrm{Arf}(\partial_- W)$, $\mathrm{Arf}(\partial_+ W)$ and zero. But now $K_{k+1}(W', \partial W')$ provides a Lagrangian subspace, so the Arf invariant is zero. Hence indeed $\mathrm{Arf}(\partial_- W) = \mathrm{Arf}(\partial_+ W)$ as required. □

The invariant $\mathrm{Arf}(\mu)$ of a normal cobordism is known as the *Kervaire invariant*, and denoted $\mathrm{Kerv}(f, \nu, T)$. By Theorem 7.5.4, if $k \geq 3$ it is the only obstruction to completing surgery.

7.6 The odd dimensional case

Let X be a simply-connected Poincaré complex of formal dimension $m = 2k + 1$, and suppose again that $(f : M \to X, \nu, T)$ is a k-connected normal map inducing a homotopy equivalence $\partial M \to \partial X$. Again the Hurewicz Theorem gives an isomorphism $h : \pi_{k+1}(f) \cong K_k(M; \mathbb{Z})$; by Theorem 7.1.1, any $\xi \in \pi_{k+1}(f)$ induces a regular homotopy class of immersions $S^k \times D^{k+1} \to M$, given any embedding in this class we can perform surgery, and we write

$x := h(\xi)$. In this case, for any $x \in K_k(M : \mathbb{Z})$ we can find such embeddings, but these are not unique up to diffeotopy. The object of this section is to prove:

Theorem 7.6.1 *Let $(f : M \to X, \nu, T)$ be a normal map of degree 1, inducing a homotopy equivalence $\partial M \to \partial X$, with X a simply-connected Poincaré complex of formal dimension $2k+1$, with $k \geq 2$. Then surgery to obtain a homotopy equivalence is possible. Moreover, it suffices to perform surgeries on spheres of dimension $\leq k$.*

We first calculate the effect of a single surgery on homology. If N is a normal cobordism of M to M' we have the diagram of Proposition 7.1.3, and by Proposition 7.4.2, we may simplify this by replacing terms H_* by K_*. The only ones which remain non-zero are those in the diagram

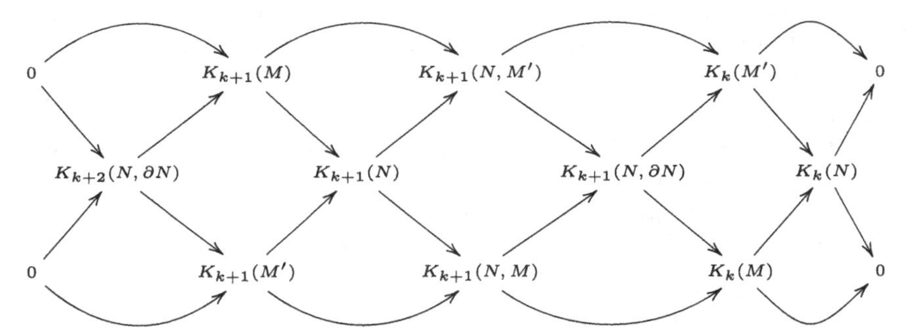

Here the groups $K_{k+1}(N, M; \mathbb{Z})$ and $K_{k+1}(N, M'; \mathbb{Z})$ are isomorphic to $\mathbb{Z}$. If we take coefficients $\mathbb{Q}$, the isomorphism $[N, \partial N]\cap$, together with dual pairings, shows that groups symmetrically placed in the diagram have equal ranks. Writing r for the rank of $K_k(N; \mathbb{Q})$ we find just three possibilities for the ranks of all groups in the diagram: either

(i) $K_k(M)$, $K_k(M')$, $K_k(N)$ and their duals have rank r, $K_{k+1}(N, \partial N)$ and $K_{k+1}(N)$ have rank $r+1$;

(ii) $K_k(M)$, $K_k(N)$ and their duals have rank r, $K_k(M')$, $K_{k+1}(N, \partial N)$ and their duals have rank $r+1$; or

(ii') as (ii) but with M and M' interchanged.

Moreover the map $K_{k+1}(N) \to K_{k+1}(N, \partial N)$ induced by the intersection pairing has rank 1 in case (i) and 0 in cases (ii), (ii'). Since the intersection pairing is skew-symmetric if k is even, it follows that here case (i) cannot arise.

We now consider the torsion in $K_k(M)$ in more detail. In the following we use coefficients $\mathbb{Z}$ for homology throughout.

Lemma 7.6.2 *Let $x \in K_k(M)$ be indivisible, so that there is a homomorphism $\phi : K_k(M) \to \mathbb{Z}$ with $\phi(x) = 1$. Then if we perform surgery on x, the map*

$K_k(M') \to K_k(N)$ is an isomorphism, so $K_k(M')$ is the quotient of $K_k(M)$ by the class of x.

Proof We can identify ϕ with a class in $K^k(M)$, hence by duality with a class y in $K_{k+1}(M)$. We claim that the image of y in $K_{k+1}(N, M')$ is a generator, so that the map $K_{k+1}(N, M') \to K_{k+1}(N, \partial N)$ and hence also the map to $K_k(M')$ vanishes, which implies the result.

To calculate this image, we may consider the intersection of y with a generator z of $K_{k+1}(N, M)$. This equals the intersection in M of y with the boundary of z, namely x. But by hypothesis this is 1. □

We can now give the proof of Theorem 7.6.1 in the case when k is even.

Proof First perform preliminary surgeries to make f k-connected: as $k \geq 2$, all manifolds we encounter from now on are simply-connected. Next perform an induction on the rank r of $K_k(M)$. If $r > 0$, choose $x \in K_k(M)$ of infinite order and not divisible by an integer > 1, and perform surgery on x. By Lemma 7.6.2, $K_k(M')$ is isomorphic to the quotient of $K_k(M)$ by the class of x, so has lower rank. By induction, we may reduce the rank to 0.

We now have $K_k(M)$ finite, and conclude with a second induction on the order $|K_k(M)|$. Choose any non-zero $x \in K_k(M)$ and perform surgery. Here we have case (ii), so $K_k(N)$ is the quotient of $K_k(M)$ by the class of x, so has lower order than $K_k(M)$, and we have a short exact sequence

$$0 \to \mathbb{Z} \cong K_{k+1}(N, M') \to K_k(M') \to K_k(N) \to 0.$$

Thus $K_k(M')$ is isomorphic to the direct sum of $\mathbb{Z}$ and a finite group G, and as the map $G \to K_k(N)$ is injective, we have $|G| \leq |K_k(N)| < |K_k(M)|$.

Take a generator y of the summand $\mathbb{Z}$ of $K_k(M')$ and perform surgery on y. This yields a normal cobordism N', say, of M' to M''; $K_k(N')$ is the quotient of $K_k(M')$ by the class of y, so is isomorphic to G, and by Lemma 7.6.2, the map $K_k(M'') \to K_k(N')$ is an isomorphism. Since $K_k(M'')$ has lower order than $K_k(M)$, the desired result follows by induction. □

The case when k is odd requires further arguments. We consider the effect of a single surgery and recall the commutative diagram (7.6). The handle H is a smooth submanifold H of N with $H \cap M = \partial_- H \cong (S^k \times D^{k+1})$, $\partial_c H \cong (S^k \times S^k \times [-1, 1])$, and $H \cap M' = \partial_+ H \cong (D^{k+1} \times S^k)$ (compare Figure 5.5). Write V for the closure of $N \setminus H$; then $V \cong \partial_- V \times I$. We can extend the K_* notation to V, etc., by expressing X as the union of a complex X^* with a $2k$-cell attached, so that f maps V to X^* and H to the extra cell.

We have an exact sequence

$$K_{i+1}(N, V \cup M \cup M') \to K_i(V, \partial_- V \cup \partial_+ V) \to K_i(N, M \cup M') \\ \to K_i(N, V \cup M \cup M'),$$

and since by excision $H_i(N, V \cup M \cup M') \cong H_i(H, \partial H)$, so vanishes except for $i = 2k+2$, we have an isomorphism $K_{k+1}(V, \partial_- V \cup \partial_+ V) \to K_{k+1}(N, M \cup M')$.

We may thus replace the term $K_{k+1}(N, \partial N)$ in (7.6) by $K_{k+1}(V, \partial_- V \cup \partial_+ V) \cong K_k(\partial_- V)$. We have an a-sphere $S^k \times \{0\} \subset S^k \times D^{k+1} \cong \partial_- H$ diffeotopic in H to $S^k \times * \times \{-1\} \subset \partial_c H$ which bounds a disc in M' and a b-sphere $\{0\} \times S^k \subset D^{k+1} \times S^k \cong \partial_+ H$ diffeotopic in H to $* \times S^k \times \{1\}$ which bounds a disc in M. Denote the corresponding classes in $K_{k+1}(N, \partial N)$ by x_a and x_b.

Further calculations depend on the self-pairing λ of $K_k(M; \mathbb{Z})$ to $\mathbb{Q}/\mathbb{Z}$.

Lemma 7.6.3 *Suppose we perform surgery on a sphere representing $x \in K_k(M; \mathbb{Z})$. Then in the exact sequence*

$$\mathbb{Z}\,(\cong K_{k+1}(N, M')) \to K_k(\partial_- V)\,(\cong K_{k+1}(N, \partial N)) \to K_k(M) \to 0,$$

the generator 1 of $\mathbb{Z}$ maps to x_b, and if $y \in K_k(M)$ has order q, $p/q \in \mathbb{Q}$ projects to $\lambda(x, y) \in \mathbb{Q}/\mathbb{Z}$, and $w \in K_k(\partial_- V)$ maps to y, we have $qw = p'x_b$ for some $p' \equiv p\ (mod\ q)$.

Proof Represent y by a k-cycle η and let $q\eta = \partial\phi$ for some $(k+1)$-chain ϕ in M. The class x is represented by the a-sphere $S^k \times *$, and by definition of λ, this has p' intersections with ϕ for some $p' \equiv p \pmod q$. We may suppose these transverse; then each one is the centre of a disc $* \times D^{k+1}$ (in H). Removing these discs gives a chain ϕ^* in $\partial_- V$ with boundary consisting of $q\eta$ and p' spheres $* \times S^k$ each parallel to the b-sphere, and with reversed orientation. Thus $qw - p'x_b$ vanishes in $H_k(\partial_- V)$. □

We now give the proof of Theorem 7.6.1 in the case when k is odd.

Proof As in the case when k is even, we may suppose preliminary surgeries performed so that f is k-connected and moreover that $K_k(M)$ is finite; again we proceed by induction on $|K_k(M)|$.

We will perform surgery on $x \in K_k(M)$. Since $r = 0$ we cannot have case (ii') above, so the first map in the exact sequence of 7.6.3 is injective. That Lemma calculates the extension, but does not determine the class x_a.

In fact, the class is not determined by the choice of x: the embedding $\phi : S^k \times D^{k+1} \to M$ is determined up to regular homotopy, but not up to diffeotopy. For any map $g : S^k \to SO_{k+1}$ we may form the twist ϕ_g by $\phi_g(x, y) := \phi(x, g(x)(y))$. If g is nullhomotopic in SO_{k+2}, ϕ and ϕ_g define homotopic embeddings of the tangent bundle of $S^k \times D^{k+1}$ in that of M, and hence, by Corollary 6.2.2, regularly homotopic immersions. Since $SO_{k+2}/SO_{k+1} = S^{k+1}$, we have an exact sequence

$$\pi_{k+1}(S^{k+1}) \to \pi_k(SO_{k+1}) \to \pi_k(SO_{k+2}),$$

so may twist by any element in the image of $\pi_{k+1}(S^{k+1}) \cong \mathbb{Z}$. Since k is odd, twisting by the image of $s \in \mathbb{Z}$ induces the self-map of $H_k(S^k \times S^k)$ with $x_a \mapsto x_a + 2sx_b$, $x_b \mapsto x_b$.

First suppose x of order q and that $\lambda(x, x) = p/q$ with p prime to q: then $K_k(M)$ is the direct sum of the group $\mathbb{Z}_q$ generated by x and its orthogonal complement, G, say. By Lemma 7.6.3, $K_k(\partial_- V)$ is the direct sum of an infinite cyclic group with generator z, say, where $qz = x_b$ and a group isomorphic to G. Write x_a in the form $mz + g$ with $m \in \mathbb{Z}$, $g \in G$. Taking x_a for w and x for y in the lemma, we see $m \equiv p \pmod q$.

Twisting as above, we may suppose $|m| < q$. Now $K_k(M')$ is (isomorphic to) the quotient of $K_k(\partial_- V)$ by the class of x_a. It thus has order $m|G|$, so as $|m| < q$ we have reduced the order.

In view of Proposition 7.3.10, we see that it will now suffice to consider the case when $x, y \in K_k(M)$ have order 2^k, $\lambda(x, y) = \frac{1}{2^k}$, and $2^k\lambda(x, x) = 2^k\lambda(y, y) = 0$. Then $K_k(M)$ is the orthogonal direct sum of the group generated by x and y and a finite group G. Applying Lemma 7.6.3, we can write $K_k(\partial_- V)$ in the form $\mathbb{Z} \oplus \mathbb{Z}_{2^k} \oplus G$, where the first summand is generated by a class z which projects to y, the second by v say, we have $2^k z = x_b$, and $x_a = mz + v$ with m even. Twisting, we may suppose $|m| \leq 2^k$.

If $m = 0$, we obtain for $K_k(M')$ the direct sum of $\mathbb{Z}$ and G; by Lemma 7.6.2 a further surgery will kill the $\mathbb{Z}$, so we have reduced the order of K_k. If $m \neq 0$, $K_k(M')$ has order $2^k m|G|$, so if $|m| < 2^k$ we have again reduced the order. Finally if $m = 2^k$, although we have not reduced the order we have the direct sum of G and a cyclic group of order 2^{2k}, so we have reduced to the first case. □

7.7 Homotopy theory of Poincaré complexes

In this section we prepare for the reformulation in the next of the results of surgery in more general terms. We also complete our discussion of the Kervaire obstruction. The proofs of these results involve somewhat technical homotopy theory arguments. We will present an outline of the ideas, but give references for the detailed proofs.

First consider an orthogonal vector bundle ξ, with fibre dimension k and base B. Write A_ξ for the associated disc bundle: the set of vectors in the total space of ξ of length ≤ 1, and S_ξ for its boundary sphere bundle. The space obtained from A_ξ by identifying the subspace S_ξ to a point is called the *Thom space* of ξ and denoted $T(\xi)$. There is a natural isomorphism, called the *Gysin isomorphism*

$$H^r(X) \to H^{k+r}(A_\xi, S_\xi) \cong \tilde{H}^{k+r}(T(\xi)).$$

The class in $H^k(T(\xi); \mathbb{Z})$ corresponding to the unit in $H^0(X; \mathbb{Z})$ is called the *Thom class*, and traditionally denoted by U. Geometrically, if B is a CW complex, $T(\xi)$ has a natural decomposition with just one $(k+r)$-cell for each r-cell of B, as well as the base point.

If V is a submanifold of M, we can choose a tubular neighbourhood of V, which consists of an orthogonal vector bundle ξ over V together with an embedding h of A_ξ in M. Composing h^{-1} with the map $A_\xi \to T(\xi)$ gives a map of $h(A_\xi)$ which sends its boundary $h(S_\xi)$ to the base point, and hence extends to a map $M \to T(\xi)$ which sends the rest of M to this point. This is known as the Thom construction. As it is the foundation of the study of cobordism, we will treat it more fully in §8.1.

In particular, if V^v is a compact submanifold of Euclidean space $\mathbb{R}^{v+k}$ with normal bundle ν, since we can regard $\mathbb{R}^{v+k}$ as obtained from S^{v+k} by deleting a point, we obtain a map $F : S^{v+k} \to T(\nu)$; moreover, this map has degree 1 in the sense that it induces an isomorphism on the top non-vanishing homology group H_{v+k}.

If we start with a Poincaré complex, rather than a manifold, there are no immediately visible bundles. We generalise, replacing sphere bundles $S_\xi \to B$ by fibrations $\pi : X \to B$ with fibres homotopy equivalent to the sphere S^{k-1}. In general we use the term *spherical fibration* for a fibration with fibres homotopy equivalent to a sphere. The role of the disc bundle A_ξ is now played by the mapping cylinder of π, and we define $T(\pi)$ to be the mapping cone of π.

A decisive step is given by the following result of Spivak [142].

Theorem 7.7.1 *If X is a Poincaré complex of formal dimension m, and $k > m+1$, there exist a fibration π^k over X with fibre homotopy equivalent to S^{k-1} and a map $F : S^{m+k} \to T(\pi)$ of degree 1. Moreover, the pair (π^k, F) is unique up to suspension and homotopy equivalence.*

We describe the construction in the simplest case when X is a finite simply connected CW complex. Choose an embedding $i : X \to \mathbb{R}^{m+k}$ for some k. Take a regular neighbourhood N of $i(X)$, and form the space EN of paths (continuous maps) $\alpha : I \to N$. As discussed in §B.1, the projection $p_0 : EN \to N$ given by $p_0(\alpha) = \alpha(0)$ is a homotopy equivalence, and its fibres are contractible. Thus if $P := p_0^{-1}(\partial N)$, p_0 gives a homotopy equivalence $P \to \partial N$.

Now define $p_1 : P \to N$ by $p_1(\alpha) = \alpha(1)$: this too is a fibration. The key lemma states that the fibres of p_1 are homotopy equivalent to S^{k-1} if and only if X satisfies Poincaré duality with formal dimension n. We omit the proof, which depends on an examination of the spectral sequence of the fibration. We can now either use the fact that N is homotopy equivalent to X or restrict to $p_1^{-1}(X)$ to obtain the desired fibration over X.

Now $T(\pi)$ is the mapping cone of π, hence is homotopy equivalent to the union of X, or equivalently, N, and the cone on P, which is homotopy equivalent to ∂N. Hence $T(\pi)$ is homotopy equivalent to the space formed from N by identifying ∂N to a point. But this is obtained from S^{m+k} by identifying ∂N and everything outside N to a point, so indeed we have a map $S^{m+k} \to T(\pi)$ of degree 1.

The existence proof in the general case depends on the same idea, but there are more details to check. The result also extends to Poincaré pairs. We will discuss the uniqueness shortly.

In general the Thom space of an external direct sum of two bundles is

$$\begin{aligned} T(\xi \oplus \eta) = \frac{A_{\xi\oplus\eta}}{S_{\xi\oplus\eta}} &= \frac{A_\xi \times A_\eta}{(S_\xi \times A_\eta) \cup (A_\xi \times S_\eta)} \\ &= \frac{T(\xi) \times T(\eta)}{(\{\infty\} \times T(\eta)) \cup (T(\xi) \times \{\infty\})}, \end{aligned}$$

and this is a space called the *smash product* of $T(\xi)$ and $T(\eta)$ and denoted $T(\xi) \wedge T(\eta)$. The same goes for spherical fibrations if we interpret $\oplus$ as the fibrewise join. Since the bundle ε^1 over a point has Thom space S^1, we have $T(\xi \oplus \varepsilon^1) = T(\xi) \wedge S^1$, which is the suspension of $T(\xi)$, which we denote $ST(\xi)$. In particular, a pair (π^k, F) as in Theorem 7.7.1 defines a suspended pair $(\pi^k \oplus \varepsilon^1, SF)$.

We introduce a related notation: for any space X, write X^+ for the disjoint union of X and a point $*$, which we take as base point. Then if Z has a base point ∞, we have

$$X^+ \wedge Z = \{(X \cup \{*\}) \times Z\}/\{(\{*\} \times Z) \cup (X \times \{\infty\})\} = (X \times Z)/(X \times \{\infty\});$$

in particular, $X^+ \wedge Y^+ = (X \times Y)^+$.

Now let X be a Poincaré complex of formal dimension m, ν a spherical fibration over X, and $F : S^{m+k} \to T(\nu)$ a map of degree 1. If $\nu = \alpha \oplus \beta$, we have a map $S^N \to T(\nu) = T(\alpha) \wedge T(\beta)$.

Theorem 7.7.2 *The Thom spaces $T(\alpha)$ and $T(\beta)$ are (S,N)-dual in the sense of Spanier and Whitehead [141].*

A proof appears in [14]. Taking slant product with the homology class of the image of the fundamental homology class $[S^N]$ induces a map $\tilde{H}^q(T(\alpha)) \to \tilde{H}_{n-q}(T(\beta))$. Somewhat as in the proof of Proposition 7.4.2, using the fact that we have maps in both directions, it can be deduced that these maps are isomorphisms. But this is the condition defining (S,N)-duality.

It is more usual to speak of Spanier–Whitehead duality. Here we adhere to the earlier terminology since it is useful to make the dualising dimension N explicit. A textbook account of this duality appears in Adams' book [7].

Essentially the same argument yields the relative case: if (Y, X) is a Poincaré pair, ν a bundle or spherical fibration over Y such that there is a map $S^N \to T(\nu)/T(\nu \mid X)$ of degree 1, and $\nu = \alpha \oplus \beta$, then $T(\alpha)$ is (S,N)-dual to $T(\beta)/T(\beta \mid X)$.

In the simply-connected case, the converse follows: if a map $S^N \to T(\nu)$ induces an S-duality, then X satisfies Poincaré duality.

S-duality is a duality in stable homotopy theory. If X and Y are spaces (finite CW complexes will suffice here) the set of (based) homotopy classes of maps $X \to Y$ is denoted $[X : Y]$. The set of morphisms in stable homotopy theory is the limit

$$\{X : Y\} := lim_n[S^n X : S^n Y],$$

which is an abelian group. If X and X^*, and Y and Y^* are (S,N)-dual, there are isomorphisms

$\{X^* : Y^*\} \cong \{Y : X\}$;

$\{X : Y\} \cong \{S^N : X^* \wedge Y\}$.

If $\phi : (N, M) \to (Y, X)$ is a normal map of degree 1 of Poincaré pairs, the S-dual gives a map $\psi : T_{Y/X}(\nu) \to T_{N/M}(\nu)$. Composing with the Gysin isomorphism gives the map $\psi^* : H^*(N, M) \to H^*(Y, X)$ dual to the homology map ϕ_* which we used in Proposition 7.4.2.

We can now deal with the question of uniqueness in Theorem 7.7.1. Suppose we have spherical fibrations ν and ν' over X and maps $S^{m+k} \to T(\nu)$ and $S^{m+k'} \to T(\nu')$, both of degree 1. Since X is a finite CW-complex, if r is large enough, the suspension $\nu \oplus \varepsilon^r$ is fibre homotopy equivalent to $\alpha \oplus \beta$ with α fibre homotopy equivalent to ν'. Hence $T(\alpha)$ is (S,$(m + k + r)$)-dual to $T(\beta)$.

The map $S^{m+k'} \to T(\nu') \simeq T(\alpha)$ of degree 1 is (S,$(m + k + r)$)-dual to a map $T(\beta) \to S^{r+k-k'}$ inducing an isomorphism of the homology group $H^{r+k-k'}$. Now we can obtain $T(\beta)$ from the total space of the spherical fibration $\beta \oplus \varepsilon^1$ by identifying the cross-section in the summand ε^1 to a point. We thus have a map of this total space to a sphere which induces an isomorphism of $H^{r+k-k'}$, and hence a homotopy equivalence on each fibre. But this shows that $\beta \oplus \varepsilon^1$ is a fibre homotopically trivial bundle. As $\nu \oplus \varepsilon^{r+1}$ is fibre homotopy equivalent to $\nu' \oplus \beta \oplus \varepsilon^1$, the desired equivalence of ν and ν' (together with the maps of degree 1) is established.

We now turn to the Kervaire invariant. The following account is a simplified version of [34], with most proofs omitted. Define $W_k(n)$ to be the mapping fibre of the map $K(\mathbb{Z}_2, k) \to K(\mathbb{Z}_2, k+n+1)$ given by the cohomology operation $\chi(Sq^{n+1})$, and define a Wu orientation of a bundle ξ^n to be a map $T(\xi) \to W_k(n)$ such that the class in dimension k pulls back to the Thom class U.

If X^{2n} is a Poincaré complex, and ν its normal Spivak fibration, the Wu class $v_{n+1}(X) = 0$ since Sq^{n+1} vanishes on n-dimensional classes. It follows that $\chi(Sq^{n+1})U = 0$. Thus ν admits a Wu orientation.

It follows from the (S,$2n+k$)-duality between X^+ and $T(\nu)$ that there are bijections

$\{X^+ : S^n\} \to \{S^{2n+k} : S^n \wedge T(\nu)\}$ and

$\{X^+ : K(\mathbb{Z}_2, n)\} \to \{S^{2n+k} : T(\nu) \wedge K(\mathbb{Z}_2, n)\}$.

Homotopy calculations yield isomorphisms

$\{S^{2n} : K(\mathbb{Z}_2, n)\} \cong \mathbb{Z}_2$:

we denote by χ the image in $\{X^+ : K(\mathbb{Z}_2, n)\}$ of the non-zero element;

$z : \{S^{2n+k} : W_k(n) \wedge K(\mathbb{Z}_2, n)\} \cong \mathbb{Z}_4$.

Composing the map z with a Wu orientation α for ν gives a homomorphism $\{S^{2n+k} : T(\nu) \wedge K(\mathbb{Z}_2, n)\} \to \mathbb{Z}_4$ and hence, by (S,($2n+k$))-duality, a map $h : \{X^+ : K(\mathbb{Z}_2, n)\} \to \mathbb{Z}_4$. We now define $\phi : H^n(X) \to \mathbb{Z}_4$ by $\phi(u) := h(\{u\})$.

Further calculations show that, if $j : \mathbb{Z}_2 \subset \mathbb{Z}_4$ denotes the (unique) injective map, $h(\chi) = j(1) \neq 0$ (this key point depends on the Wu orientation); next that (cf. (7.3.7))

$$\phi(u+v) = \phi(u) + \phi(v) + j\{u.v[X]\}.$$

Thus ϕ is a quadratic form on $H^k(X; \mathbb{Z}_2)$ in the sense of Theorem 7.3.11. By that result, the Gauss sum $\mathfrak{G}(\phi) := \sum_{x \in H^k(X;\mathbb{Z}_2)} i^{\phi(x)}$ has the form $R(\phi)\eta_X$, where $R(\phi) = \sqrt{|H^k(X; \mathbb{Z}_2)|}$ and $\eta_X^8 = 1$. Writing η_X as $e^{2\pi i a(X)/8}$ defines an invariant $a(X) \in \mathbb{Z}_8$ of (X, α).

It is also shown that if (Y, X) is a Poincaré pair with a Wu orientation, ϕ vanishes on the image of $H^k(Y)$. Hence this image is a Lagrangian subgroup of $H^k(X)$. It follows from Theorem 7.3.11 that in this case $a(X) = 0$.

Now suppose given a normal map $(f : M \to X, \nu, T)$. A Wu orientation α for ν pulls back to one for M, and the above map h factors as $h : \{X^+ : K(\mathbb{Z}_2, n)\} \to \{M^+ : K(\mathbb{Z}_2, n)\} \to \mathbb{Z}_4$. Now $H^k(M; \mathbb{Z}_2) = K^k(M; \mathbb{Z}_2) \oplus H^k(X : \mathbb{Z}_2)$, and the restriction to $H^k(X : \mathbb{Z}_2)$ of the map ϕ_X coincides with the map ϕ_M defined in the same way for M. A further calculation shows that the restriction to $K^k(M; \mathbb{Z}_2)$ corresponds by duality to the map $\mu : K_k(M; \mathbb{Z}_2) \to \mathbb{Z}_2$ of Lemma 7.5.3. From this we deduce the equality

$$\eta_M = (-1)^{A(\mu)}\eta_X. \tag{7.7.3}$$

This gives a calculation of the Kervaire invariant which depends only on the Wu orientation of X.

It is also shown in [34] that choices of Wu orientation for X correspond bijectively to forms ϕ satisfying (7.3.7). Since the intersection form is nonsingular, it follows that a second such form can be written as $\phi_z(x) = \phi(x) + j(x.z)[X]$ for some z. It follows from Lemma 7.3.12 that $\mathfrak{G}(\phi_z) = e^{-i\pi\phi(z)}\mathfrak{G}(\phi)$. Now if M is given the induced Wu orientation, then changing ϕ to ϕ_z will multiply each of η_M and η_X by the same factor $e^{-i\pi\phi(z)}$. Thus the value of $A(\mu)$ is independent of this choice.

There exist Wu orientations for the universal $Spin_{4n+2}$ bundles, and (uniquely) for the universal SU_{4n+2} bundle. Choosing these give maps of the corresponding cobordism groups (studied in the next chapter) to $\mathbb{Z}_8$.

There is also a choice of a Wu orientation for the universal SO_{4n} bundle such that the corresponding invariant is just the signature mod 8.

An interesting special case is $n = 2$. In the case when ν has fibre dimension 1, there is a canonical Wu orientation. Geometrically, we have a framing of $\mathbb{T}(M) \oplus \nu^1$, hence an immersion $M^2 \to \mathbb{R}^3$. In this case, the map ϕ can be geometrically interpreted by representing $u \in H_1(M; \mathbb{Z}_2)$ by an immersion of S^1, and counting the number of half-twists of M in $\mathbb{R}^3$ as you go round the circle.

7.8 Homotopy types of smooth manifolds

A natural problem is to seek to characterise the set of homotopy types of closed smooth manifolds. The first necessary condition on X for being homotopy equivalent to a closed manifold is that it be a Poincaré complex. If X is itself a manifold, the identity map, together with the normal bundle ν of some embedding in Euclidean space and the induced trivialisation of $\mathbb{T}(X) \oplus \nu$ is a normal map of degree 1. Thus a second necessary condition is the existence of a normal map $(f : M \to X, \nu, T)$ of degree 1, with M a smooth manifold. We next reformulate this condition. Denote by $\Omega_m(X, \nu^k)$ the set of normal cobordism classes of normal maps $(f : M^m \to X, \nu^k, T)$. Surgery replaces a normal map by another normally cobordant to it, and thus in the same class in $\Omega_m(X, \nu^k)$.

Given a normal map $(f : M^m \to X, \nu^k, T)$, we can add a trivial bundle ϵ^1 to ν^k; then T together with the induced trivialisation of $f^*\epsilon^1$ induces a trivialisation T' of $\mathbb{T}(M) \oplus f^*(\nu^k \oplus \epsilon^1)$, defining a suspended normal map $(f : M^m \to X, \nu^k \oplus \epsilon^r, T')$. The same goes for normal cobordisms, so we have a suspension map $\Omega_m(X, \nu^k) \to \Omega_m(X, (\nu^k \oplus \epsilon^1))$. It is easily shown that this is an isomorphism for $k > m$.

Proposition 7.8.1 *For any finite CW-complex X and vector bundle ν^k over X, if $k > m + 1$ there is a natural bijection $\Phi_m(X, \nu^k)$ from $\Omega_m(X, \nu^k)$ to the homotopy group $\pi_{m+k}(T(\nu^k))$. If X is a Poincaré complex, $\Phi_m(X, \nu^k)$ preserves degree.*

Proof Given an element $\alpha \in \Omega_m(X, \nu^k)$, choose a representative $(f : M^m \to X, \nu^k, T)$, where T is a trivialisation of $\mathbb{T}(M) \oplus f^*\nu^k$. Let E^{m+k} be the total space of a disc bundle associated to $f^*\nu$: then T defines a trivialisation of $\mathbb{T}(E)$. By immersion theory (see Corollary 6.2.2), this corresponds to an immersion $E^{m+k} \to \mathbb{R}^{m+k} \subset S^{m+k}$. Since $k > m + 1$ we may suppose by general position (see, for example, Proposition 4.6.6) that this gives an embedding of M and hence of a neighbourhood of M in E, which we may choose to be given by a disc sub-bundle of $f^*\nu$. As above, identifying the boundary sphere bundle and all outside to a point gives a map $t : S^{m+k} \to T(f^*\nu^k)$. The map f induces a map $T(f) : T(f^*\nu^k) \to T(\nu^k)$; composing gives a map $T(f) \circ t : S^{m+k} \to T(\nu^k)$ defining $\beta \in \pi_{m+k}(T(\nu))$.

If we begin with a normal cobordism $(g : N^{m+1} \to X, \nu, T^*)$ between two normal maps, we can follow through the same construction leading to an immersion, then an embedding $F^{m+k+1} \to S^{m+k} \times I$ inducing the chosen embeddings in $S^{m+k} \times \{0\}$ and $S^{m+k} \times \{1\}$, and a map $S^{m+k} \times I \to T(g^*\nu^k) \to T(\nu^k)$, and conclude that the two maps $S^{m+k} \to T(\nu^k)$ are homotopic. We may thus define $\Phi_m(X, \nu^k)$ by setting $\Phi_m(X, \nu^k)(\alpha) = \beta$.

To prove $\Phi_m(X, \nu^k)$ surjective, we first choose a smooth manifold Y with a homotopy equivalence $h : Y \to X$: this is possible by Lemma 1.2.9. Under h a bundle ν over X induces a bundle ν' over Y.

A class $\beta \in \pi_{m+k}(T(\nu))$ is represented by a map $S^{m+k} \to T(\nu)$ and so by a map $\phi : S^{m+k} \to T(\nu')$. The space $T(\nu')$ is not a smooth manifold, but contains a point $*$ whose complement is an open disc bundle over Y and so is a smooth manifold. Using Proposition 2.3.4, we modify the map ϕ keeping it fixed on $\phi^{-1}(*)$ so as to make it smooth on the complement. Next by Proposition 4.5.10, modify ϕ further to make it transverse to Y, embedded as the zero cross-section.

Now set $M := \phi^{-1}(Y)$ and write $f = h \circ (\phi|M) : M \to X$. It follows from the basic property Lemma 4.5.1 of transversality that M is a smooth manifold of dimension m, and its normal bundle in S^{m+k} is the pullback of ν' and hence of ν: thus a framing of the tangent bundle of $\mathbb{R}^{m+k}$ induces one of $\mathbb{T}(M) \oplus f^*\nu$. We have thus constructed a normal map $(f : M \to X, \nu, T)$, defining $\alpha \in \Omega_m(X, \nu)$. Since we have effectively reversed the above construction, it follows that $\Phi_m(X, \nu^k)(\alpha) = \beta$.

We argue similarly to prove $\Phi_m(X, \nu^k)$ injective. Given two normal maps leading to homotopic maps $f^\nu \circ t$ and $g^\nu \circ t$, we take a homotopy $S^{m+k} \times I \to T(\nu')$ and, keeping it fixed at the ends, make it smooth away from the preimage of $*$ and then transverse to Y. The preimage of Y then gives a normal cobordism between the given normal maps.

The fact that corresponding maps have the same degree follows since $T(f) : T(f^*\nu^k) \to T(\nu^k)$ has the same degree as $f : M \to X$ and $t : S^{m+k} \to T(f^*\nu^k)$ has degree 1. □

This result reduces us from the somewhat mysterious set of normal cobordism classes of normal maps to an explicit homotopy group. We now make a further reduction. There is a classifying space for vector bundles (see §B.2): isomorphism classes of vector bundles ν^k over X correspond to homotopy classes of maps $X \to B(O_k)$. There is a corresponding result for (fibre homotopy classes of) spherical fibrations, with a classifying space $B(G_k)$. By Spivak's Theorem 7.7.1, if X is a Poincaré complex, there is a well-defined spherical fibration π over X, which determines a map $\tau_X : X \to B(G_k)$ up to homotopy.

We write (following standard notation) G_k for the monoid of self-homotopy equivalences of S^{k-1}. For homotopy purposes, we can treat this as a topological group, and $B(G_k)$ as its classifying space.

A normal map $(f : M \to X, \nu^k, T)$ of degree 1 determines a class of maps $S^{m+k} \to T(\nu^k)$ of degree 1, and hence by the uniqueness in Theorem 7.7.1, a fibre homotopy equivalence $\nu \to \pi$. Thus the map $X \to B(O_k)$ classifying ν is a lift of the fixed map τ_X. More precisely, it follows that

Corollary 7.8.2 *There is a natural bijection between normal cobordism classes of normal maps $(f : M \to X, \nu^k, T)$ of degree 1 of smooth manifolds to X and homotopy classes of liftings of $\tau_X : X \to B(G_k)$ to $B(O_k)$.*

We write $\mathcal{T}(X, \nu)$ for the set of these homotopy classes of liftings, which can thus be identified with the subset of $\Omega_m(X, \nu)$ of classes of degree 1: we can regard these as tangential structures on X. Observe that a choice of lifting induces a bijection of the set $\mathcal{T}(X, \nu)$ to $[X : G_k/O_k]$.

We now restrict to the simply-connected case and also suppose X has formal dimension $m \geq 5$. We can summarise Theorems 7.5.2, 7.5.4, and 7.6.1 as stating that surgery on a normal map of degree 1 to obtain a homotopy equivalence to X^m is possible if and only if an obstruction belonging to L_m vanishes, where we define the surgery group L_m ad hoc by

$$L_{4k} = \mathbb{Z},\ \ L_{4k+1} = 0,\ \ L_{4k+2} = \mathbb{Z}_2,\ \ L_{4k+3} = 0.$$

The isomorphism of L_{4k} on $\mathbb{Z}$ is given by the signature divided by 8 (the 8 comes from Proposition 7.3.3) and of L_{4k+2} on $\mathbb{Z}_2$ by the Kervaire invariant.

More precisely, if m is odd, by Theorem 7.6.1 we can perform the desired surgery, so the above necessary condition is sufficient. If $m = 4p$ is divisible by 4, by Theorem 7.5.2, surgery to obtain a homotopy equivalence is possible if and only if $\sigma(M) = \sigma(X)$. Here $\sigma(X)$ is determined by the homology of X, but $\sigma(M)$ depends on the choice of lift. By Hirzebruch's signature theorem 8.6.7, there is a polynomial L_m in Pontrjagin classes such that if we take the classes of $\mathbb{T}(M)$, cap product with the fundamental class gives $\sigma(M) = \langle L_m(\mathbb{T}(M)), [M]\rangle$. We can choose a bundle τ over X such that $\nu \oplus \tau$ is trivial. Then $f^*\tau$ is stably equivalent to $\mathbb{T}(M)$, so $L_m(\mathbb{T}(M)) = f^*L_p(\tau)$. Since f has degree 1, $f_*[M] = [X]$, hence $\langle L_m(\mathbb{T}(M)), [M]\rangle = \langle L_m(\tau), [X]\rangle$. The desired equality of signatures thus holds if and only if $\sigma(X) = \langle L_m(\tau), [X]\rangle$.

If $m \equiv 2 \pmod 4$, by Theorem 7.5.4 surgery to obtain a homotopy equivalence is possible if and only if the Kervaire invariant $\kappa := \mathrm{Kerv}(\phi : M \to X, T) \in \mathbb{Z}_2$ vanishes. Here the choice of a Wu orientation of ν induces Wu orientations of X and M, so by (7.7.3) above we have invariants η_X and η_M with $\eta_M = (-1)^\kappa \eta_X$: thus surgery is possible if and only if $\eta_M = \eta_X$.

To show that all elements of the groups L_m effectively arise as obstructions, we need the plumbing construction, which is best seen in the simplest case, when X is the Poincaré pair (D^m, S^{m-1}) and we study normal maps $f : (M, \partial M) \to (D^m, S^{m-1})$ inducing a homotopy equivalence $\partial M \to S^{m-1}$. Here ν is necessarily a trivial bundle ϵ^r, so T is a framing of $\mathbb{T}(M) \oplus \epsilon^r$.

Proposition 7.8.3 *(i) There exists a framed manifold Z^{4k}, which is a handlebody obtained by adding eight $(2k)$-handles to D^{4k}, such that the intersection matrix on $H_{2k}(Z)$ is the E_8 matrix (7.3.4).*

(ii) There exists a framed manifold Z^{4k+2}, which is a handlebody obtained by adding two $(2k+1)$-handles to D^{4k+2}, such that the normal map to D^{4k+2} has Kervaire invariant 1.

Proof (i) Write A for the tangent D^{2k} bundle of S^{2k}: this has Euler class 2, so the self-intersection of the zero cross-section is 2. Since $\mathbb{T}(S^{2k}) \oplus \epsilon^1$ is trivial, there is a framing of $\mathbb{T}(A) \oplus \epsilon^2$.

The E_8 matrix P is a positive definite symmetric 8×8 matrix of determinant 1, with entries 2 on the diagonal and 0 or -1 elsewhere. Take 8 copies A_i of A indexed by the rows of P, and for each non-zero entry $p_{i,j}$ $(i < j)$ of P, choose

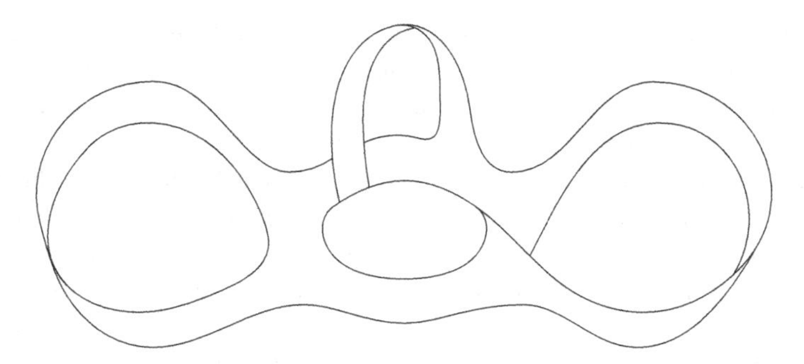

Figure 7.1 Plumbing

$(2k)$-discs $D_{i,j} \subset S_i^{2k}$, $D_{j,i} \subset S_j^{2k}$, ensuring that any two such discs in the same sphere S_i^{2k} are disjoint. The part of A_i lying over $D_{i,j}$ is a product bundle, so can be identified with a product $D_{i,j} \times D^{2k} \cong D_{i,j} \times D_{j,i}$. For each pair (i, j) we now identify $D_{i,j} \times D_{j,i}$ with $D_{j,i} \times D_{i,j}$ by the map interchanging the factors. This yields a manifold, containing copies of the A_i, but in which we have introduced intersections of the spheres so that the intersection matrix is 7.3.4 up to signs, but we can choose orientations of the basis elements to change all the signs to -1. The construction gives a manifold whose boundary has re-entrant corners, but these can be smoothed by the same techniques as in §2.6. The plumbing construction (but not this example) is illustrated in Figure 7.1.

(ii) Here the construction is simpler: we take two copies of the tangent disc bundle of S^{2k+1} and perform plumbing just once, so that the intersection matrix is just $\begin{pmatrix} 0 & 1 \\ -1 & 0 \end{pmatrix}$. We need to choose a framing so that μ takes the value 1 on each basis element.

We recall the definition of μ for a $(2k+1)$-connected normal map $f : (M, \partial M) \to (D^{4k+2}, S^{4k+1})$. By Theorem 7.1.1, any $\xi \in \pi_{2k+2}(f) \cong K_{2k+1}$ determines a regular homotopy class of immersions $\phi : S^{2k+1} \times D^{2k+1} \to M$ with homology class x. By §6.3, there are two regular homotopy classes of immersions in each homotopy class of maps $S^{2k+1} \to M^{4k+2}$, which are distinguished by the parity of the number $I(\phi)$ of self-intersection points of ϕ. In Lemma 7.5.1, we defined $\mu(x)$ to be $I(\phi)$ mod 2.

Now in Proposition 6.3.3 we constructed an immersion $j : S^{2k+1} \to \mathbb{R}^{4k+2}$ with a single transverse self-intersection and with normal bundle $\mathbb{T}(S^{2k+1})$. Thus pulling back the standard framing of $\mathbb{R}^{4k+2}$ by j gives a framing of $\mathbb{T}(S^{2k+1})$ such that the preferred regular homotopy class of immersions indeed has $I(\phi) = 1$. □

Since the matrices used have determinant ± 1, the manifolds constructed in Proposition 7.8.3 have boundaries with the homology of a sphere. It follows, except in low dimensions, that the boundary is simply-connected, hence is homotopy equivalent to a sphere. In fact it follows that the manifold Z^2 constructed above has boundary S^1; the boundary of Z^4 can be shown to be homeomorphic to Poincaré's dodecahedral space. In higher dimensions the boundary is a homotopy sphere, hence by Corollary 5.6.4 is homeomorphic to a sphere.

Write P_m for the set of normal cobordism classes of normal maps $(f : (M^m, \partial M) \to (D^m, S^{m-1}), T)$ with $f|_{\partial M} : \partial M \to S^{m-1}$ a homotopy equivalence. We observe that the structure of normal map amounts to giving a manifold M, a stable framing T of $\mathbb{T}(M)$, and the homotopy equivalence $\partial M \to S^{m-1}$.

Proposition 7.8.4 *If $m > 5$, the surgery obstruction gives a bijection $\beta : P_m \to L_m$.*

Proof Given a normal map, the surgery obstruction is well defined and belongs to L_m, so we have a map β.

Given two normal maps f_1, f_2 defining elements of P_m, we can form the boundary sum $M_1 + M_2$ and extend the map and framing. We claim that the surgery obstruction of the sum is the sum of the surgery obstructions. If m is odd, there is nothing to prove; if $m = 2k$, we first perform surgery below the middle dimension on each of f_1 and f_2: this induces surgeries on $f_1 + f_2$. We now have $K_k(M_1 + M_2) = K_k(M_1) \oplus K_k(M_2)$, and λ and μ split in a natural way. Since both σ and Arf are additive on direct sums, the claim follows. Similarly, we can define a normal map $\overline{f}$ by changing orientation, and $\beta(\overline{f}) = -\beta(f)$.

If $\beta(f_1) = \beta(f_2)$, we form $f_1 + \overline{f_2}$: by what we have just seen, $\beta(f_1 + \overline{f_2}) = 0$. We may thus perform surgery (keeping the boundary fixed) to construct a normal cobordism N^{m+1} of $M_1 + \overline{M_2}$ to a disc D^m.

The boundary sum may also be constructed as the union of $M_1, D^{m-1} \times I$ and M_2, since attaching $D^{m-1} \times [0, \frac{1}{2}]$ to M_1 by a collar on part of its boundary does not change the diffeomorphism class. Thus we write N as having $\partial_- N = M_1 \cup (D^{m-1} \times I) \cup M_2$ and $\partial_+ N = D^m$. Now adjusting corners we may rewrite N as N' with $\partial_- N' = M_2$, $\partial_+ N' = M_1$ and $\partial_c N' = D^m \cup \partial_c N \cup (D^{m-1} \times I)$. Since $\partial_c N \cong (S^{m-1} \times I)$, it follows that the same holds for $\partial_c N'$.

So we have a normal cobordism (keeping the boundary fixed) of M_1 to M_2. Thus β is injective.

It follows from Proposition 7.8.3 that the image of β contains a generator of L_m; now using sums and change of orientation as above it follows that β is surjective. □

Corollary 7.8.5 *Let Σ^{n-1} be a homotopy sphere which bounds a framed manifold N^n with $n \geq 6$. Then if n is odd, $\Sigma \cong S^{n-1}$; if $n \equiv 0$ (mod 4), Σ is determined up to diffeomorphism by $\sigma(N)$; if $n \equiv 2$ (mod 4), there are at most two diffeomorphism classes of such Σ.*

We return in Proposition 8.8.6 to the case $n \equiv 0$ (mod 4), and following that discuss the delicate question, the 'Kervaire invariant problem', of deciding for which $n \equiv 2$ (mod 4) Σ is unique (and so is diffeomorphic to S^{n-1}).

As well as seeking existence of a smooth manifold homotopy equivalent to X, we can investigate uniqueness by the same method. Consider pairs (M, f) with M a smooth manifold and $f : M \to X$ a homotopy equivalence, and let $(M, f) \sim (M', f')$ if there is a diffeomorphism $h : M \to M'$ with $f' \circ h \simeq f$. Write $\mathcal{S}(X)$ for the set of equivalence classes, which we may consider as smooth manifold structures on the homotopy type of X. There is a natural map $\mathcal{S}(X) \to \mathcal{T}(X)$.

Theorem 7.8.6 *For X simply-connected and $m \geq 5$ there is a sequence*

$$L_{m+1} \to \mathcal{S}(X) \to \mathcal{T}(X) \to L_m$$

which is 'exact': the image of $\mathcal{S}(X)$ in $\mathcal{T}(X)$ is the preimage of $0 \in L_m$, and the group L_{m+1} acts on $\mathcal{S}(X)$ and the orbits are the fibres of $\mathcal{S}(X) \to \mathcal{T}(X)$.

Proof The two latter maps, and exactness at $\mathcal{T}(X)$, are given by the above discussion.

Given an element $\alpha \in L_{m+1}$ and an element of $\mathcal{S}(X)$ represented by $f : M \to X$, by Proposition 7.8.4, α corresponds to an element of P_{m+1}, which we can represent by a normal map $g : (N, \partial N) \to (D^{m+1}, S^m)$ which defines a homotopy equivalence of the boundary. Choose embeddings of D^m in M and in ∂N (essentially unique by Theorem 2.5.6), and use them to glue N to $M \times I$ to give N', say. A retraction of N on D^m induces a map $G : N' \to M \times I \to X \times I$, and the restriction of G to $\partial_+ N'$ is a homotopy equivalence, so defines an element of $\mathcal{S}(X)$. Since N' inherits from N the structure of normal map, it gives a normal bordism between the two elements of $\mathcal{S}(X)$, so they map to the same in $\mathcal{T}(X)$. Any other choice of g representing α is normally cobordant to g; following this through gives an h-cobordism between the two choices for $\partial_+ N'$, so the element of $\mathcal{S}(X)$ is uniquely determined by α.

Conversely, given two elements of $\mathcal{S}(X)$ with the same image in $\mathcal{T}(X)$, there exists a normal bordism $G : N' \to X \times I$. If m is even, we can perform surgery (keeping the boundary fixed) to obtain a homotopy equivalence, and so have an h-cobordism: it follows that the two elements are equal. If $m = 2k - 1$, we can perform surgery to make the map G k-connected. It follows that N' can be obtained from $M := \partial_- N'$ by attaching k-handles. Moreover, the a-spheres of the handles are nullhomotopic in X and so in M. Since $k \geq 3$ we can perform a diffeotopy to take all of these attaching maps inside the disc D^m. But now N' is the boundary sum of $M \times I$ and a manifold defining an element of P_{m+1}. $\square$

We have studied framed manifolds with homotopy sphere boundaries; if we weaken the 'homotopy sphere' hypothesis slightly, we still obtain strong results. Suppose M^{2n-1} an $(n-2)$-connected manifold which bounds a framed manifold N^{2n} with $n \geq 3$. Again we can do surgery to make N $(n-1)$-connected. It follows from the homology exact sequence that $H_r(N; \mathbb{Z}) \cong H^{2n-r}(N, M; \mathbb{Z})$ vanishes for $2n - r \leq k$, i.e. for $r \geq n+1$. The same holds for other coefficient groups; hence $H_n(N; \mathbb{Z})$ is free abelian. By Lemma 5.6.10, N is a handlebody, so by Theorem 5.6.12 is determined up to diffeomorphism by (H, λ, α) where $H := H_n(N; \mathbb{Z})$, λ is the $(-1)^n$-symmetric bilinear map $H \times H \to \mathbb{Z}$ given by intersection numbers, and α is a map $H \to \pi_{n-1}(SO_n)$, satisfying

(i) $\lambda(x, x) = \pi(\alpha(x))$ for $x \in H$, and

(ii) $\alpha(x + y) = \alpha(x) + \alpha(y) + \lambda(x, y)(\partial\iota_n)$ for $x, y \in H$.

Moreover since N is framed, $\alpha(x)$ maps to 0 in $\pi_{n-1}(SO)$. Thus if n is even, $\alpha(x)$ is determined by $\pi(\alpha(x))$, so the classification of N reduces to that of the symmetric bilinear form λ, which is even since $\pi(\alpha(x))$ is even.

In the case $n = 3$, as $\pi_2(SO_3)$ vanishes, N is determined up to diffeomorphism by the skew-symmetric bilinear form λ on the free abelian group $H_3(N; \mathbb{Z})$; hence, by Proposition 7.3.5, by the set of integers $\{a_i\}$. It follows from the construction that N is the boundary sum of terms of the form of handlebodies of two types:

(i) N_k say, having two handles with intersection number k, and

(ii) diffeomorphic to $S^3 \times D^3$, having just one handle.

This can be extended to a complete diffeomorphism classification of closed, simply connected 5-manifolds M. In general, the tangent group of M is stably trivial provided obstructions in $H^r(M; \pi_{r-1}(SO))$ vanish; in the present case all these groups vanish except $H^2(M; \mathbb{Z}_2)$, and the obstruction here is the Stiefel–Whitney class $w_2(M)$. If $w_2(M) = 0$, M is stably framed, so by Proposition 8.1.4, determines a class in π_5^S and bounds a framed manifold N if and only if this class vanishes, which it does since (see §B.3(x)), $\pi_5^S = 0$. (Alternatively we can argue that M has a spinor structure, and since the cobordism group $\Omega_5^{Spin} = 0$, M bounds a spinor manifold.) Now perform surgery on N; by Corollary 7.2.2, we may suppose N 2-connected. It follows that N is a handlebody. Hence M is a connected sum of manifolds M_k (the boundary of N_k) and $S^3 \times S^2$. This argument is due to Smale [140].

The case $w_2(M) \neq 0$ is more complicated. The invariants of M consist of a triple (H, b, w), where

$H = H_2(M; \mathbb{Z})$ is a finitely generated abelian group,

b is the linking form – a nonsingular skew-symmetric bilinear self-pairing of the torsion subgroup T of H to $\mathbb{Q}/\mathbb{Z}$ – and

$w : H \to \mathbb{Z}_2$ is the homomorphism given by cap product with $w_2(M)$; moreover for $x \in T$, $w(x) = b(x, x)$.

The invariants determine $w_2(M) \in H^2(M; \mathbb{Z}_2)$, hence also $w_3 = Sq^1 w_2$ and the Stiefel–Whitney number $w_2 w_3[M]$. The contributions to $w_3 \in H^3(M; \mathbb{Z}_2)$ come only from summands of H of order 2, and it follows that $w_2 w_3[M]$ is equal to the number (modulo 2) of such summands. The oriented cobordism group Ω_5^{SO} has order 2, and the class of M in it is determined by $w_2 w_3[M]$.

Theorem 7.8.7 *[16] Any system of invariants as above is the set of invariants of a simply-connected 5-manifold, and two such manifolds with isomorphic invariants are diffeomorphic.*

We will not give the full proof, but merely an outline of the argument. For existence, first recall that by Proposition 7.3.9, the triple (H, b, w) is a direct sum of triples with G either $\mathbb{Z}$, $\mathbb{Z}_k \oplus \mathbb{Z}_k$ or $\mathbb{Z}_2$; we can construct manifolds as connected sums correspondingly.

For $G = \mathbb{Z}$, we take M as an S^3 bundle over S^2: the product $S^3 \times S^2$ if $w = 0$, and the non-trivial bundle $S^3 \tilde{\times} S^2$ if not.

For $G = \mathbb{Z}_k \oplus \mathbb{Z}_k$, if $w = 0$, we have the manifold M_k constructed above. This is diffeomorphic to the manifold obtained from $S^2 \times S^3$ by surgery on a sphere representing kz, where the class of z generates $H_2(S^3 \times S^2)$; now replacing $S^3 \times S^2$ by $S^3 \tilde{\times} S^2$ gives a suitable manifold in the case $w \neq 0$.

The case $G = \mathbb{Z}_2$ is trickier, since here we need a manifold which is not a boundary. Begin with the Hopf bundle $S^1 \to S^3 \to S^2$: this bounds a bundle B_1 with fibre D^2, which is the tubular neighbourhood of the 2-sphere which is the zero cross-section, and hence diffeomorphic to the complement of a disc D^4 in $P^2(\mathbb{C})$. The associated bundle B_2 with fibre S^2 is thus split by the original copy of S^3 into two parts, each diffeomorphic to B_1, but with one diffeomorphism reversing the orientation; in turn, B_2 is the boundary of the associated bundle B_3 with fibre D^3. Now $P^2(\mathbb{C})$ admits a diffeomorphism φ_1 given by complex conjugation. This induces -1 on $H^2(P^2(\mathbb{C}); \mathbb{Z})$, but is orientation preserving, hence (by the disc theorem) isotopic to a diffeomorphism leaving a disc D^4 pointwise fixed, hence giving a diffeomorphism φ_1' of B_1. There is thus a diffeomorphism φ_2 of B_2 given by φ_1' on one copy of B_1 and by the identity on the other. Finally, define M_q by using the diffeomorphism φ_2 of $B_2 = \partial B_3$ to glue two copies of B_3 together. A short calculation shows that indeed $H_2(M_q) \cong \mathbb{Z}_2$.

To establish uniqueness, since we dealt above with the case $w_2 = 0$, one can suppose w_2 non-zero. Given manifolds M, M' and an isomorphism $\alpha : H_2(M) \to H_2(M')$ compatible with the pairings b and maps w, we know that there is an oriented cobordism W of M to M'. By Corollary 7.2.2, we can perform surgery on W to make the map $W \to B(SO)$ 2-connected, and so $H_2(W) \cong \mathbb{Z}_2$. The main part of Barden's argument now involves surgery on 3-spheres embedded in W to convert W to an h-cobordism. The result follows by Theorem 5.5.6.

7.9 Notes on Chapter 7

§7.1 The useful terminology of 'normal maps' is due to Browder [31]. In an early paper, Milnor thanks Thom for having described the technique of surgery to him.

§7.2 This account of surgery below the middle dimension follows that in my book [167]. As with handlebody theory, the idea is to copy for manifolds what happens for CW complexes.

The proof of Corollary 7.2.4 fails in lower dimensions. The problem of embeddings of S^2 in S^4 is much more delicate, and no simple result is known. For knots in S^3 it follows from Thurston's geometrisation principle that unless the knot is a torus knot or a companion knot, the fundamental group of the complement is isomorphic to a subgroup of $SL_2(\mathbb{C})$.

§7.3 The results for forms over $\mathbb{Z}$ are classical. A nice survey of nonsingular quadratic forms was given by Milnor [94].

A convenient reference for forms over finite groups is my paper [161], but there is a substantial literature; many of the results are older. The general concept of quadratic form is discussed in [165].

The Arf invariant was first introduced in [11]. Invariants of a quadratic form q on a vector space V over a field k can be extracted from its Clifford algebra $C(q)$. This admits a mod 2 grading, and if V has even dimension, the centre Z of the even Clifford algebra is a quadratic extension of k. Except in characteristic 2, we can write Z in the form $k[z]/\langle z^2 = a\rangle$, and the class of a in $k^\times/(k^\times)^2$ is the discriminant of q. In characteristic 2, define $\wp(x) := x^2 + x$: then $Z = k[z]/\langle \wp(z) = a\rangle$ (where, if the associated bilinear form to q has a symplectic basis $\{e_i, f_i\}$, we may take $z = \sum_i e_i f_i$), and the class of a in $k^+/\wp(k^+)$ is Arf's invariant in general.

§7.4 Poincaré complexes were first defined in [164]. The definition in the general case is a little more elaborate than for simply connected spaces. In this section we only need immediate consequences of duality and properties of maps of degree 1 following [167].

The self-pairing of the torsion subgroup is traditionally called the linking pairing, and was first introduced by Seifert [133].

In [167], the result corresponding to Proposition 7.4.3 was formulated in homotopy terms, and used to prove that surgery is always possible for Poincaré pairs (Y, X) such that the map $\pi_1(X) \to \pi_1(Y)$ is an isomorphism – the so-called '$\pi - \pi$ Theorem'.

§7.5, §7.6 Milnor's exciting paper [92] constructing differentiable manifolds homeomorphic but not diffeomorphic to S^7 aroused great interest in this area. His talk [102] at the 1958 International Congress exhibited interrelations of relevant homotopy groups. This was followed by a preprint of Milnor in 1959 introducing a programme: introduce the group Θ_n of homotopy spheres, then show any homotopy sphere is stably framed, then study the obstruction to bounding a framed manifold, then study the case when it does. There were preliminary publications [95] and [97]. These included the calculation of P_{4k}, and essentially that of P_{4k+2}. The final proof that $P_{2r+1} = 0$ was accomplished by myself [159] and in the full account by Milnor and Kervaire [79].

The idea that the method extended to arbitrary simply-connected manifolds was due to Novikov [114] and Browder in 1962. Fuller accounts appeared in [115], [163], and [30]; both Browder and I gave talks at the international congress in 1966, and wrote books [31] and [167].

§7.7 Spivak's 'homotopy normal bundle' brought clarity to several previous results of this nature.

S-duality was introduced and developed in [141].

After the introduction of the Kervaire invariant in 1960 in [78], progress was made successively in 1966 by Brown and Peterson, then in 1969 by Browder [30] (his book [31] appeared in 1972). Browder starts from a normal map and uses the Spanier–Whitehead dual map $\psi : T_{Y/X}(\nu) \to T_{N/M}(\nu)$. For each cohomology class $x \in H^k(N, M)$ with $\psi^*(x) = 0$, write h for the composite map $S^s x \circ \psi : T_{Y/X}(\nu) \to T_{N/M}(\nu) \to S^s K(\mathbb{Z}_2, k)$. Since $h^*\iota = Sq^{k+1}\iota = 0$, one can form the functional Steenrod square (see §B.4) $Sq_h^{k+1}(S^s\iota) \in H^{2k+s}(T_{Y/X}(\nu))$, and evaluate this on the fundamental class to obtain $\mu(x) \in \mathbb{Z}_2$. This gives a definition of the map μ and hence of its Arf invariant independent of any preliminary surgeries.

A comprehensive study was made in 1972 by Brown [34], which has the advantage of defining an invariant for Poincaré complexes (with a Wu orientation) rather than for normal maps. Our account is a simplified version of this.

§7.8 We have given the general form of the reduction of diffeomorphism classification of (simply-connected) smooth manifolds to homotopy problems: the account follows the one I gave in [167].

The plumbing construction seems to have been first introduced by Milnor [95].

For simply-connected 4-manifolds M, it was observed by Milnor [94] that the homotopy type is determined by the intersection form on $H_2(M)$. A further step was taken by the author [162] showing that if M and M' are homotopy equivalent, then they are h-cobordant, and deducing that they can be made diffeomorphic by taking connected sums with a number of copies of $S^2 \times S^2$. This is far from establishing diffeomorphism: at the time of writing, no criterion is known for proving two 4-manifolds diffeomorphic. Another tantalising problem is finding which quadratic forms appear. For spinor manifolds, these forms must be even, and it follows from the calculation of spinor cobordism that the signature is divisible by 16. This value is realised by so-called K3 surfaces (for example, nonsingular quartic surfaces in $P^3(\mathbb{C})$), but for such a surface M $H_2(M)$ has rank 22 and it is not known whether there exist surfaces with $\sigma = 16$ but lower rank. See §5.7 for a fuller discussion.

When we drop the hypothesis of simple connectivity, it is necessary, as in §5.7, to replace the coefficient group $\mathbb{Z}$ by $\mathbb{Z}[\pi]$. This leads to surgery obstruction groups $L_m(\pi)$, generalising the above group L_m which is $L_m(1)$. The exact sequence of Theorem 7.8.6 holds in general with L_m replaced by $L_m(\pi_1(X))$. There is, however, no direct analogue of Proposition 7.8.4.

The groups $L_m(\pi)$ can be defined in an abstract way. When $m = 2k$ is even, they can be interpreted by equivalence classes of $(-1)^k$-hermitian forms over $\mathbb{Z}[\pi]$, by a relatively minor modification of the geometry of the simply connected case, using the results of §6.3 on embeddings of m-spheres in $2m$-manifolds. The odd dimensional case requires a different approach, and the surgery groups can be interpreted as quotients of the stable unitary group of such forms. Some calculations of these groups can be made: if π is finite, by methods of algebraic number theory, and for some infinite groups π using geometrical arguments. A first version of all this was given in [167].

This theory was re-worked in a more satisfactory way by Ranicki in a series of papers from 1973 on. We refer to his book [128].

8

Cobordism

We have already defined the word ‘cobordism’ in §5.1: recall that if W is a manifold, and $\partial_- W$ and $\partial_+ W$ are disjoint manifolds with union ∂W, we call the pair $(W, \partial_- W)$ a cobordism and the pair $(W, \partial_+ W)$ the dual cobordism; and also call W a cobordism of $\partial_- W$ to $\partial_+ W$ and say that $\partial_- W$, $\partial_+ W$ are cobordant.

In the earlier chapter, we were concerned with the geometry of a particular cobordism. We now observe that being cobordant is an equivalence relation amongst diffeomorphism classes of manifolds. For $M \times I$ is a cobordism of M to itself; if W is a cobordism from M_0 to M_1 then the same manifold, but with $\partial_\pm W$ interchanged, is a cobordism from M_1 to M_0; and if W_0 is a cobordism from M_0 to M_1 and W_1 is a cobordism from M_1 to M_2, then glueing W_0 to W_1 along M_1 gives a cobordism from M_0 to M_2. For this relation not to be vacuous, we insist throughout that the manifolds W in question be compact: otherwise the product $M \times [0, 1)$ would give a cobordism of any manifold M to the empty set.

The simple definition just given already leads to interesting results, but the concept of cobordism lends itself to a wide variety of possible generalisations and restrictions, and these lead to a flexible tool in the study of manifolds.

For example, we may choose to restrict the manifolds (*and* cobordisms) to be oriented, weakly complex, or k-connected (for a fixed k); we may add the structure of a map to a fixed space X; if X is a manifold, we may further require this map to be an embedding, or an immersion. We may consider pairs (M, V) with V a submanifold of M and then cobordisms (N, W) with W a submanifold of N (and $\partial_- W = V$, $\partial_- N = M$), where we may also fix the group of the normal bundle.

Next we consider pairs (M, φ), where M is a manifold and $\varphi : M \times G \to M$ defines a smooth action of the compact Lie group G on M. We may also restrict the orbit types of the action to lie in an assigned closed set of orbit types - an extreme example is the class of fixed-point-free actions.

We may also allow M to be a manifold with boundary – and then a cobordism is a manifold W with $\partial W = \partial_- W \cup \partial_c W \cup \partial_+ W$ – and impose one restriction on M and W and another on ∂M and $\partial_c W$. These variants of the definition may now be combined ad lib.

Lemma 8.0.1 *Disjoint union defines an addition which turns the set of cobordism classes (of a given dimension) into an abelian group.*

Proof The other kinds of structure pass at once to the disjoint union. Union is compatible with cobordism: if V, W are cobordisms of $\partial_- V$ to $\partial_+ V$, $\partial_- W$ to $\partial_+ W$, then the disjoint union $V \cup W$ is a cobordism of $\partial_- V \cup \partial_- W$ to $\partial_+ V \cup \partial_+ W$. Thus we have a binary operation on the set of cobordism classes, which is commutative and associative since disjoint unions are. The empty manifold acts as zero.

We obtain an inverse to W whenever $M \times I$ may be regarded as a cobordism of the disjoint union $(M \times 0) \cup (M \times 1)$ to the empty set (the induced structure on $M \times 0$ must coincide with that on M: on $M \times 1$ it can be different). □

For k-connected cobordism, we show in Lemma 8.8.1 that disjoint union can be replaced by connected sum.

In this chapter, vector bundles will be denoted by lower case Greek letters, so we write τ_M for the tangent bundle of M in place of $\mathbb{T}(M)$; normal bundles will usually be denoted by ν; and the trivial bundle of fibre dimension r by ε^r.

In the first section, we describe the basic Thom construction, leading to a bijection between certain sets of homotopy classes and certain bordism sets, and give an application to the problem of realising homology classes by submanifolds. Then we focus on the structure group on the normal bundle, and stabilisation, and define cobordism groups and rings.

The framework of cobordism lends itself to the construction of exact sequences, and we next describe this technique, which we will use many times. Then we treat cobordism of pairs; this leads to an interpretation of some relative groups.

The next section treats bordism as a homology theory, checks the axioms, introduces spectra, and dual notions of bordism and cobordism.

We then discuss equivariant cobordism, and show how the techniques of the preceding sections yield methods of calculation of the equivariant cobordism groups.

After a brief review of homology of classifying spaces, we describe the calculations of the unoriented bordism ring, and the unitary bordism ring. We hope to provide enough detail for the reader to follow the ideas, but refer to the original papers for details of calculations. We then attempt the same for oriented bordism and SU-bordism, with a detour to obtain the Hirzebruch signature theorem. We discuss k-connected cobordism, and then pull many results together in final calculations of groups of homotopy spheres and of knots.

Part of the use of cobordism theory is to make calculations, and in this chapter we will assume significantly more knowledge of homotopy theory than in earlier chapters. We attempt to provide enough background definitions and results for this discussion in Appendix B.

8.1 The Thom construction

We introduce the main tool in cobordism theory by considering the example which occurs in the earliest work on the subject: the study of submanifolds M^m of a fixed ambient manifold E^{m+k}.

Let ξ be an orthogonal vector bundle. As in §7.7, write V_ξ for the total space and B_ξ for the base; A_ξ for the subspace of V_ξ of all vectors of length ≤ 1 and S_ξ for its boundary, consisting of vectors of length 1. The Thom space $T(\xi)$ of ξ is obtained from A_ξ by identifying S_ξ to a point (denoted ∞): thus $T(\xi) = A_\xi / S_\xi$. We may identify B_ξ with the zero cross-section of the bundle, and hence with a subspace of $T(\xi)$. In the same section we met a special case of the Thom construction. Also Proposition 7.8.1 gave a preview of the next result.

If B_ξ is a smooth manifold, we can give ξ the structure of smooth vector bundle, and V_ξ and $T(\xi) \setminus \{\infty\}$ then also acquire the structure of smooth manifolds. Note that if B_ξ is a finite CW complex, so is $T(\xi)$; more precisely, if ξ has fibre dimension k, over each r-cell e^r of B_ξ we have a $(k+r)$-cell in A_ξ part of whose boundary lies over ∂e^r and part in S_ξ, so this gives a $(k+r)$-cell of $T(\xi)$, and all cells outside ∞ arise in this way.

Now let M^m be a submanifold of the compact manifold E^{m+k}, ν be its normal bundle. By Theorem 2.3.8 we can find an imbedding $h : A_\nu \to E$ defining a tubular neighbourhood of M in V.

The collapsing map $A_\nu \to T(\nu)$ defines a map $h(A_\nu) \to T(\nu)$ which extends to a continuous map $c_M : E \to T(\nu)$ which takes everything outside the tubular neighbourhood to ∞. This idea is due to Thom [150], and is called the *Thom construction*. Observe that if B_ν is identified with the zero cross-section of ν, we have $M = c_M^{-1}(B_\nu)$.

We introduce one more ingredient. Let $M^m \subset E^{n+k}$ have normal bundle ν, let ξ be a bundle whose base space B_ξ is a smooth manifold, and let $\phi : \nu \to \xi$ be a map of (orthogonal) vector bundles, hence inducing maps $B_\phi : B_\nu \to B_\xi$ and similarly for A, S and T. As above, identify A_ν with a tubular neighbourhood of M; write $c_M : E \to T(\nu)$ for the collapsing map, and form the composite $F_M := T_\phi \circ c_M : E \to T(\xi)$.

The first significant result in cobordism theory is that the Thom construction can, in a sense, be reversed. Define a cobordism of submanifolds of E to be a smooth compact submanifold W of $E \times I$, with $M_0 = \partial_- W = W \cap (E \times \{0\})$ and $M_1 = \partial_+ W = W \cap (E \times \{1\})$.

Proposition 8.1.1 *The Thom construction induces a bijection τ from the set of cobordism classes of submanifolds $M \subset E$ with normal bundle induced from ξ to the set of homotopy classes of maps $E \to T(\xi)$.*

Proof The construction takes a submanifold M with a bundle map $\phi_M : \nu_M \to \xi$ and gives a map $F_M := T_{\phi_M} \circ c_M : E \to T(\xi)$. To show we have a well-defined map τ we must show that cobordant submanifolds give rise to homotopic maps. Let $W^{m+1} \subset (E \times I)$ be a cobordism, with normal bundle ν_W induced via $\phi_W : \nu_W \to \xi$, and suppose the construction already performed for M_0 and M_1. It follows from the tubular neighbourhood theorem 2.5.5 that the chosen tubular neighbourhoods of M_0 and M_1 can be extended to a tubular neighbourhood of W in $E \times I$. Thus the collapsing map c_W for this neighbourhood extends those on the boundary. Hence $T_{\phi_W} \circ c_W$ is a homotopy between the maps obtained from M_0 and M_1. We thus have a well-defined map τ from cobordism classes to homotopy classes.

To show τ is surjective, suppose given a map $F : E \to T(\xi)$. Since $T(\xi) \setminus \{\infty\}$ is a smooth manifold, it follows from Proposition 2.3.4 that we can approximate F by a map F' agreeing with F on $F^{-1}(\infty)$ and which is smooth on a neighbourhood of $F^{-1}(B_\xi)$. If the approximation is close enough, $F' \simeq F$. Next by Theorem 4.5.6, we can further approximate F' by a map F'' transverse to B_ξ, and also suppose $F'' \simeq F'$. Now set $M := F'^{-1}(B_\xi)$. By Lemma 4.5.1, the normal bundle of M is induced from ξ by a map $\phi_M : \nu_M \to \xi$. If we now perform the Thom construction on M, the resulting $h : E^{m+k} \to T(\xi)$ agrees with F'', together with its first derivatives, on M^m. After a small homotopy, then, we can suppose $F'' = h$ on a neighbourhood of M. But the complement of such a neighbourhood is mapped, both by F'' and by h, to $T(\xi) \setminus B_\xi$, which is contractible. It follows that $h \simeq F'' \simeq F$, as desired.

That τ is injective follows by relativising the same arguments. Suppose given $M_0 \subset E \times 0, M_1 \subset E \times 1$ giving rise by the Thom construction to maps $f_0, f_1 : E \to T(\xi)$, and a homotopy $F : E \times I \to T(\xi)$ between f_0 and f_1. As above, we can replace F (keeping it fixed on $E \times \partial I$) by a homotopy F' of f_0 to f_1, which is smooth and transverse to B_ξ. Then $W := F'^{-1}(B_\xi)$ is a submanifold of $E \times I$, and provides a cobordism of M_0 to M_1; moreover the normal bundle of N is induced from ξ. □

A key point in the above arguments is where we first approximate the map F by a smooth map, and then make it transverse to a smooth submanifold of the target. Since we will use this idea several times below, in future we omit the references to Proposition 2.3.4 and Theorem 4.5.6.

We have described the Thom construction directly in a geometric context. We next relax the condition that B_ξ be a smooth manifold. There is a space $B(O_k)$ and a vector bundle γ_k over $B(O_k)$ such that, for any k-vector bundle ξ over a space X, there is a map $p : X \to B(O_k)$ such that ξ is equivalent to $p^*\gamma_k$, and this induces a bijection between isomorphism classes of vector bundles ξ and homotopy classes of maps p. (See §8.6 for more about classifying spaces). Moreover, we may construct $B(O_k)$ as the union of Grassmann manifolds $Gr_{m,k}$, and the map $Gr_{m,k} \to B(O_k)$ is m-connected. The bundle γ_k has associated disc bundle AO_k, say, and Thom space $T(O_k)$.

Lemma 8.1.2 *The Thom construction gives a bijection between cobordism classes of submanifolds $M^m \subset E^{m+k}$ and homotopy classes of maps $E \to T(O_k)$.*

Proof We apply Proposition 8.1.1 taking $Gr_{m,k}$ in place of B_ξ. For any submanifold M^m of E^{m+k}, the normal bundle is induced by a map to the classifying space $B(O_k)$, but we may replace these by maps to $G_{m,k}$. Since the map $G_{m,k} \to B(O_k)$ is m-connected, up to homotopy we can obtain $B(O_k)$ from $Gr_{m,k}$ by attaching cells of dimension $> m$. It follows that up to homotopy we can obtain $T(O_k)$ from the Thom space of $Gr_{m,k}$ by attaching cells of dimension $> m + k$. The result follows. □

We next replace O_k by an arbitrary structure group J (for example, J could be a Lie group), furnished with a homomorphism $J \to O_k$. There is (again see §8.6) a classifying space $B(J)$, and isomorphism classes of (vector) bundles over a space X with structure group J correspond bijectively to homotopy classes of maps $X \to B(J)$. There is an induced map $B(J) \to B(O_k)$ of classifying spaces. There is a universal bundle ξ_J over $B(J)$ and a map $f : X \to B(J)$ corresponds to the bundle $f^*\xi_J$. We write $A(J)$ for the disc bundle, $S(J)$ for its boundary sphere bundle and $T(J)$ for the Thom space $A(J)/S(J)$.

In fact we do not need J at all: only a space X playing the role of $B(J)$, and a map $X \to B(O_k)$ (here we can interpret the loop space $\Omega(X)$ as playing the role of J). However we adhere to the notation with J.

As in the case $J = O_k$, although $B(J)$ is rarely itself a smooth manifold, we can find a sequence of smooth manifolds $B(J^{(r)})$ and r-connected maps

$B(J^{(r)}) \to B(J^{(r+1)}) \to B(J)$. The same argument as for Lemma 8.1.2 now yields

Theorem 8.1.3 *The Thom construction induces a bijective map of the set of cobordism classes of pairs (E^{m+k}, M^m), with E fixed and J as structure group of the normal bundle, onto the set of homotopy classes $[E : T(J)]$.*

We can state this as a slogan: the extra structure defined on E by a submanifold whose normal bundle has group J is equivalent to the extra structure consisting of a map to $T(J)$.

Taking $J = SO_k$ in particular yields a natural bijection between cobordism classes of submanifolds $M^m \subset E^{m+k}$ with oriented normal bundle and homotopy classes of maps $E \to T(SO_k)$. Even more simply, taking J to be the trivial group gives

Proposition 8.1.4 *There is a natural bijection between cobordism classes of submanifolds $M^m \subset E^{m+k}$ with framed normal bundle and homotopy classes of maps $E \to S^k$.*

One application of Theorem 8.1.3 is to the problem of representing homology classes by manifolds. It seems that this problem, raised by Steenrod, was part of what led Thom to introduce the notion of cobordism. A first formulation is: let X be a space and $x \in H_n(X; \mathbb{Z})$: do there exist a closed oriented manifold M^n and a map $f : M \to X$ such that $f_*[M] = x$? We can vary this by using $\mathbb{Z}_2$ as coefficient group and not having an orientation. We can also take X as a manifold and require f to be an embedding: by general position results, this makes no difference if dim $M > 2n$. An affirmative result for manifolds implies one for spaces, since we can replace X by a manifold E homotopy equivalent to it; since we can apply such a result to the double $D(E)$ of E, it will suffice to consider the case of closed manifolds.

Given an oriented orthogonal vector bundle ξ over E with fibre dimension k, we have the Thom class $U \in H^k(T(\xi); \mathbb{Z})$.

Proposition 8.1.5 *Suppose E^{n+k} a closed oriented manifold. Then given a class $x \in H_k(E; \mathbb{Z})$, there is an oriented submanifold M^k of E whose fundamental homology class maps to x if and only if there is a map $F : E \to T(SO_n)$ with F^*U the Poincaré dual of x.*

Proof Since E is oriented, orientations of a submanifold M correspond to orientations of its normal bundle. We already know the correspondence between submanifolds of E and maps $F : E \to B(SO_k)$. It will thus suffice to show that $F^*U = [E] \cap x$.

In the situation of the Thom construction we have a commutative diagram

$$\begin{array}{ccc} H^k(T,\partial T) & \longleftarrow & H^k(TSO_k)\ , \\ \downarrow & & \downarrow \\ H^0(M) & \longleftarrow & H^0(BSO_k) \end{array}$$

so F^*U is the image in $H^k(E)$ of the class in $H^k(T, \partial T)$ corresponding to $1 \in H^0(M)$. Since the Thom class in $H^k(T, \partial T)$ is dual to the image of $[M]$ in $H_m(T)$, F^*U is indeed the cohomology class dual to the image of $[M]$ in $H_m(E)$. □

In the cases $k = 1, 2$ the condition is automatically satisfied: SO_1 is trivial and $T(SO_1)$ is D^1 with the boundary collapsed to a point, so can be identified with S^1, of type $K(\mathbb{Z}, 1)$; similarly $T(SO_2)$ and $K(\mathbb{Z}, 2)$ can both be identified with infinite complex projective space $P^\infty(\mathbb{C})$.

In his paper [149] Thom used his results on cobordism to prove that for any homology class $x \in H_n(X; \mathbb{Z}_2)$, there exist a closed manifold M^n and a map $f : M \to X$ such that $f_*[M] = x$. However for integer coefficients, while any $x \in H_n(X : \mathbb{Z})$ with $n \leq 6$ is the image of the fundamental class of a closed orientable manifold, this fails for any $n \geq 7$: there is an obstruction, obtained using the Steenrod reduced cube $\mathcal{P}^1$ (see §B.4).

8.2 Cobordism groups and rings

If we are interested in the manifold M^m but not the embedding in an ambient manifold E, it is natural to take E to be Euclidean space of large dimension $m + k$: by Whitney's embedding Theorem 4.2.2 we know that for $k > m + 1$ such embeddings exist and are unique up to diffeotopy. To apply the preceding section, we need E to be compact. Since embeddings in S^{m+k} yield ones in $\mathbb{R}^{m+k}$ by deforming M away from the point at infinity, we can take E as S^{m+k}.

Identifying $\mathbb{R}^{m+k}$ as a hyperplane in $\mathbb{R}^{m+k+1}$ leads us to identify S^{m+k} with a great sphere in S^{m+k+1}, and use the composite embedding $M^m \to S^{m+k} \to S^{m+k+1}$ to obtain independence of k. We may thus calculate the set Ω_m^O of cobordism classes of closed m dimensional manifolds by applying the theory of the preceding section to manifolds contained in spheres of large enough dimension.

We must also discuss the normal bundles. If ν^k is the normal bundle of M^m in S^{m+k}, the normal bundle in S^{m+k+1} is $\nu^k \oplus \varepsilon^1$. Before developing the theory more fully we present axioms for the 'stable groups' which will play the role of structure groups of the normal bundles.

A *stable group* **J** is given by a commutative diagram of groups and homomorphisms

$$\begin{array}{ccccccccc} \cdots \longrightarrow & J_{n-1} & \xrightarrow{i_{n-1}} & J_n & \xrightarrow{i_n} & J_{n+1} & \longrightarrow \cdots \\ & \downarrow{\scriptstyle \varphi_{n-1}} & & \downarrow{\scriptstyle \varphi_n} & & \downarrow{\scriptstyle \varphi_{n+1}} & \\ \cdots \longrightarrow & O_{n-1} & \longrightarrow & O_n & \longrightarrow & O_{n+1} & \longrightarrow \cdots \end{array} \tag{8.2.1}$$

where the inclusions in the lower row are natural. We impose the stability condition

(S): There is a function q_n of n, increasing (in the weak sense) and tending to infinity, such that i_n is q_n-connected.

We also need products and impose the following further conditions.

(M): We have a family of maps $\psi_{m,n} : J_m \times J_n \to J_{m+n}$ such that the following diagrams commute up to conjugating by an element in the component of the identity:

$$\begin{array}{ccccc} J_m \times J_n & \xrightarrow{\psi_{m,n}} & J_{m+n} & \xleftarrow{\psi_{m,n}} & J_m \times J_n \\ \downarrow{\scriptstyle i_m \times 1} & & \downarrow{\scriptstyle i_{m+n}} & & \downarrow{\scriptstyle 1 \times i_n} \\ J_{m+1} \times J_n & \xrightarrow{\psi_{m+1,n}} & J_{m+n+1} & \xleftarrow{\psi_{m,n+1}} & J_m \times J_{n+1} \end{array} \tag{8.2.2}$$

$$\begin{array}{ccc} J_m \times J_n & \xrightarrow{\psi_{m,n}} & J_{m+n} \\ \downarrow{\scriptstyle \varphi_m \times \varphi_n} & & \downarrow{\scriptstyle \varphi_{m+n}} \\ O_m \times O_n & \longrightarrow & O_{m+n} \end{array} \tag{8.2.3}$$

(A): The following diagram also commutes (in the same sense)

$$\begin{array}{ccc} J_l \times J_m \times J_n & \xrightarrow{\psi_{l,m} \times 1} & J_{l+m} \times J_n \\ \downarrow{\scriptstyle 1 \times \psi_{m,n}} & & \downarrow{\scriptstyle \psi_{l+m,n}} \\ J_l \times J_{m+n} & \xrightarrow{\psi_{l,m+n}} & J_{l+m+n} \end{array} \tag{8.2.4}$$

(C): We have commutativity in the diagram

$$\begin{array}{ccc} J_m \times J_n & \xrightarrow{T} & J_n \times J_m \\ \downarrow{\scriptstyle \varphi_{m+n}\psi_{m,n}} & & \downarrow{\scriptstyle \varphi_{m+n}\psi_{n,m}} \\ O_{m+n} & \xrightarrow{T'} & O_{m+n} \end{array} \tag{8.2.5}$$

where T is the natural interchange of factors, and T' means conjugation by an element whose determinant has sign $(-1)^{mn}$.

The important examples of stable groups **J** are the classical groups **O**, **SO**, **Spin**, **U**, **SU** and **Sp**, and the trivial group $\{1\}$. The above properties are immediate in these cases. Of interest also are the groups **Spin**^{c}, **Pin**, and **Pin**^{c} of [15, pp. 7–10]; however **Pin** fails to satisfy (M). Further examples can easily be constructed: for example, products of the above with each other or with any group of linear operators on a finite dimensional vector space.

We have presented the axioms in a geometrical setting, but note here that it would in fact suffice to have maps of classifying spaces $B(J_k)$ throughout; the map $B(J_k) \to B(O_k)$ induces an orthogonal vector bundle γ^k over $B(J_k)$ which is all we will need for our constructions.

An embedding $M^m \to S^{m+k}$ with J_k as structure group of the normal bundle now gives an embedding $M^m \to S^{m+k+1}$ with normal bundle with group J_{k+1}. It is however more natural to consider the tangent bundle. A *weak* **J**-*structure* on M^m is prescribed by choosing an integer r and reduction (e, f) of the group of $\tau_M \oplus \varepsilon^r$ to J_{m+r}; (r, e, f) and (r', e', f') are *equivalent* if the reductions (e, f) and (e', f') of $\tau_M \oplus \varepsilon^s$ are so for some $s \geqq r, r'$. When $J = U$ we call this a weakly complex structure.

We now show that if (S) holds, we can pass between the structure group on the stable tangent bundle and the structure group on a normal bundle. This fails for Pin: if M has a Pin normal bundle, the tangent bundle is not necessarily Pin: we have $\overline{w_2} = 0$ but $w_2 = w_1^2$.

Lemma 8.2.6 *Suppose in the diagram (8.2.3) that the map $\psi_o : J_r \to J_{r+s}$ induced by $\psi_{r,s}$ is c-connected. Let K be a CW complex of dimension $d \leqq$* $\min(c, r-2)$, *and ξ^r, η^s vector bundles over K, with a J_s-structure on η^s. Then the function f induce by ψ from J_r-structures on ξ^r to J_{r+s}-structures on $\xi^r \oplus \eta^s$ is bijective.*

Proof Let X_i be the classifying space for $J_i (i = r, s$ or $r + s)$; E_i the total space of the principal bundle with fibre O_i induced over X_i by φ_i. Write E_ξ, E_η, $E_{\xi\oplus\eta}$ for the spaces of the corresponding principal bundles over K. Then J_r-structures of ξ correspond to sections of the bundle over K with total space $E_\xi \times_{O_r} E_r$; similarly for $\xi \oplus \eta$. But the J_s-structure of η induces a fibrewise map

$$E_\xi \times_{O_r} E_r \to E_{\xi\oplus\eta} \times_{O_{r+s}} E_{r+s} \tag{8.2.7}$$

and the induced map $E_r \to E_{r+s}$ of fibres is at least $\min(c+1, r-1)$-connected since $X_r \to X_{r+s}$ is $(c+1)$-connected and $O_r \to O_{r+s}$ is $(r-1)$-connected. Thus (8.2.7) is at least $(d+1)$-connected, so any map of K to the second term can be factorised (up to homotopy) through the first, and f is surjective; moreover, the result is unique up to homotopy, so f is bijective. (It follows from the CHP that sections of a bundle are homotopic only if they are homotopic through sections.) □

Corollary 8.2.8 *Let $M^m \subset \mathbb{R}^{m+N}$ have a weak* **J***-structure, where the stable group* **J** *satisfies (S), and $q_N \geqq m$. Then the normal bundle has a J_N-structure; conversely, this implies a weak* **J***-structure on M.*

Proof In this case, $\xi \oplus \eta$ has a standard framing, and hence J-structure. We use (A) only to identify the ψ_0 of the lemma with a composite of maps i_n. □

The definition of a cobordism W of manifolds with weak **J**-structures demands a reduction of the structure group of τ_W. Now $\tau_W|\partial W \cong \tau_{\partial W} \oplus \varepsilon^1$, so the induced structure of the boundary is a reduction of the group of $\tau_{\partial W} \oplus \varepsilon^1$ rather than of $\tau_{\partial W}$ itself. Here we make the convention (necessary to obtain an equivalence relation) that the positive vector ε^1 is to be identified with the inward normal to $\partial_- W$ in W, but with the outward normal on $\partial_+ W$. Now a weak **J**-structure on a cobordism W induces weak **J**-structures on $\partial_- W$, $\partial_+ W$: we call it a cobordism between these manifolds with the induced structures.

We denote by Ω^J_m the set of cobordism classes of m-manifolds with a weak **J**-structure.

Lemma 8.2.9 *If* **J** *satisfies (S), and $N \geq m+2$, $q_N \geqq m+1$, there is a natural bijection of Ω^J_m to the set of cobordism classes of manifolds $M^m \subset S^{m+N}$ with J_N as group of the normal bundle, and hence to $\pi_{m+N}(T(J_N))$.*

Proof The first statement follows from Corollary 8.2.8, and the second from the results in the preceding section. □

Now let **J** be a stable group, with γ^k the universal bundle over $B(J_k)$. The inclusion $i_k : J_k \to J_{k+1}$ induces a bundle map $\phi_k : \gamma^k \oplus \varepsilon^1 \to \gamma^{k+1}$ over $Bi_k : B(J_k) \to B(J_{k+1})$. Write $B(\mathbf{J})$ for the limit of this sequence (we can regard the Bi_k as inclusions and form the union). In view of the identification

$T(\xi \oplus \varepsilon^1) = T(\xi) \wedge S^1$, the bundle map ϕ_k lifts to a map $h_k : ST(J_k) \to T(J_{k+1})$. The sequence of maps $h_k : ST(J_k) \to T(J_{k+1})$ defines a spectrum, which we denote by $\mathbb{TJ}$.

Theorem 8.2.10 *For* **J** *a stable group we have a bijection*

$$\Omega^J_m \cong \lim_{k\to\infty} \pi_{m+k}(T(J_k)) = \pi^S_m(\mathbb{TJ}).$$

Proof An embedding $i : M^m \to \mathbb{R}^{m+k} \subset S^{m+k}$ can be regarded as lying in a hyperplane (or great sphere) giving an embedding $i_1 : M^k \to \mathbb{R}^{m+k+1} \subset S^{m+k+1}$. Applying the Thom construction to the first gives a map $F : S^{m+k} \to T(J_k)$, and to the second gives its suspension $SF : S^{m+k+1} \to T(J_{k+1})$.

By definition, possession of a **J**-structure is equivalent to having a normal J_k-structure in S^{m+k} for some k. If we fix k, then by Lemma 8.2.9 we obtain the group $\pi_{m+k}(T(J_k))$. The desired group is the direct limit of these under the natural injection maps. □

If **J** satisfies (S), the suspension map $\pi_{m+k}(T(J_k)) \to \pi_{m+k+1}(T(J_{k+1}))$ is an isomorphism for $k > m + q_m$, so no limiting process is necessary.

The cobordism set Ω^J_m has a natural group structure: the sum of the classes of disjoint manifolds M, M' is defined to be the class of $M \cup M'$. Any M' is diffeomorphic (hence cobordant) to a manifold disjoint from M. The sum is well defined since the disjoint union of cobordisms of M with N and of M' with N' is a cobordism of $M \cup M'$ to $N \cup N'$. Commutativity and associativity are immediate. Since $\partial(M \times I) = (M \times \{0\}) \cup (M \times \{1\})$, we have an inverse (note that the normal bundle is different in the two cases).

The bijections of Lemma 8.2.9 and Theorem 8.2.10 are group isomorphisms since both are induced by the Thom construction. We can take manifolds M and M' to lie in distinct discs in S^{m+k}. The map given by the Thom construction takes the boundaries of these discs to ∞. If we then remove discs, and glue the two spheres together, we obtain the usual sum of homotopy classes.

Products are compatible with cobordism: if W is a cobordism from $\partial_- W$ to $\partial_+ W$, then $W \times M$ is a cobordism from $\partial_- W \times M$ to $\partial_+ W \times M$. Also, products are associative, and distributive over disjoint union, and there is a natural diffeomorphism of $M' \times M$ on $M \times M'$, which gives rise to a form of commutativity of multiplication.

If G, H are groups of orthogonal operators on $\mathbb{R}^q$, $\mathbb{R}^r$, then $B(G) \times B(H)$ is a classifying space for $G \times H$, and $\xi_G \times \xi_H$ is a universal bundle. As observed above, $T(G \times H) = T(G) \wedge T(H)$.

If $\mathbf{J}$ satisfies (M), the products $\psi_{m,n} : J_m \times J_n \to J_{m+n}$ induce maps $\psi'_{m,n} : T(J_m) \wedge T(J_n) \to T(J_{m+n})$, and if (A) holds, these associate up to homotopy. This provides $\mathbb{TJ}$ with the structure of a ring spectrum.

Theorem 8.2.11 *If* $\mathbf{J}$ *is a stable group satisfying (M), we have a bilinear product* $\Omega^J_m \times \Omega^J_n \to \Omega^J_{m+n}$ *which corresponds to the pairing in homotopy groups induced by the maps* $T(J_m) \wedge T(J_n) \to T(J_{m+n})$*. The product is associative if* $\mathbf{J}$ *satisfies (A), and defines a commutative graded ring if* $\mathbf{J}$ *satisfies (C).*

Proof The product of submanifolds $M \subset V$ and $N \subset W$ gives a submanifold $M \times N \subset V \times W$. Using the Thom construction as in Theorem 8.1.3 these determine elements of $[V : T(J_m)]$ and $[W : T(J_n)]$ and the product is given by

$$[V : T(J_m)] \times [W : T(J_n)] \to [V \times W : T(J_m \times J_n)].$$

The conclusion follows by taking V and W to be Euclidean spaces (or rather spheres) and stabilising. □

A case of particular simplicity is $\mathbf{J} = \{1\}$: each J_k consists only of the unit element, so we can take $B(J_k)$ to be a point; then $T(J_k) = S^k$. For each bundle occurring we have a specified isomorphism with a trivial bundle, i.e. a framing, and for clarity write Ω^{fr} for the cobordism group.

Corollary 8.2.12 *Framed cobordism groups are isomorphic to stable homotopy groups of spheres:* $\Omega^{fr}_n \cong \lim_{k\to\infty} \pi_{n+k}(S^k)$.

This, due to Pontrjagin [123], was the first theorem in the subject.

8.3 Techniques of bordism theory

In this section we introduce a couple of techniques, variants of which will often be used below. The first is a general method of constructing exact sequences. Recall from §5.1 that a cobordism of the bounded manifolds M_0 and M_1 is a manifold W with corner $\angle W$ which divides ∂W into three parts, with disjoint interiors: $M_0 = \partial_- W$, $\partial_c W$ and $M_1 = \partial_+ W$, with M_0 and M_1 disjoint. Thus $\partial_c W$ is a cobordism of ∂M_0 to ∂M_1.

By itself, this definition gives nothing: any manifold M with boundary is cobordant to the empty set by the manifold W obtained from $M \times I$ by rounding corners at $M \times \{1\}$. The interesting cases are those in which an extra condition is imposed on the cobordism $\partial_c W$.

Suppose two kinds of structure specified, which we call an α-structure and a β-structure, with the latter stronger than the former. For example, we may consider structure groups G_1 and $G_2 \subset G_1$, or maps to spaces X_1 and $X_2 \subset X_1$, or actions of groups H_1 and $H_2 \supset H_1$, or k_1-connectivity and $k_2(> k_1)$-connectivity.

By Ω_n^α and Ω_n^β we denote the cobordism groups of manifolds with α- (resp. β-) structure; and by $\Omega_n^{\alpha,\beta}$ the cobordism group of bounded manifolds with α-structure, whose boundaries have a β-structure including the given α-structure. We suppose there are natural group structures, though all we need is the zero element provided by the empty manifold.

Lemma 8.3.1 *There is an exact sequence* $\ldots \to \Omega_n^\beta \xrightarrow{i_n} \Omega_n^\alpha \xrightarrow{j_n} \Omega_n^{\alpha,\beta} \xrightarrow{\partial_n} \Omega_{n-1}^\beta \xrightarrow{i_{n-1}} \Omega_{n-1}^\alpha \to \ldots$

Proof The maps i_n and j_n are the natural ones; ∂_n is induced by taking the boundary.

Exactness at Ω_{n-1}^β follows since a β-manifold M^{n-1} represents an element z of Ker i_{n-1} if and only if, as α-manifold, it bounds some N^n. But then N represents an element $y \in \Omega_n^{\alpha,\beta}$ with $\partial_n y = z$.

We have $\partial_n \circ j_n = 0$ since an element of Ω_n^α is represented by a manifold with empty boundary. Now suppose N^n represents $y \in \Omega_n^{\alpha,\beta}$ with $\partial_n y = 0$. Then ∂N bounds a β-manifold N'. We form a closed manifold N'' by glueing N to N' along their common boundary. The α-structures on N and N' induce a α-structure on the union N''. We now define a cobordism W by taking $N'' \times I$ and introducing a corner along $\partial N \times \{0\}$, so that $\partial_- W = N \times \{0\}$, $\partial_c W = N' \times \{0\}$ and $\partial_+ W = N'' \times \{1\}$. Here $\partial_c W$ has a β-structure, so $N'' = \partial_+ W$ also represents y and is in the image of j_n.

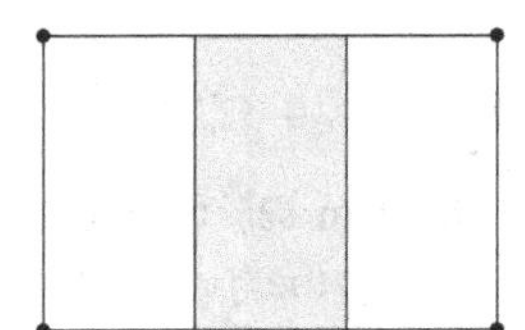
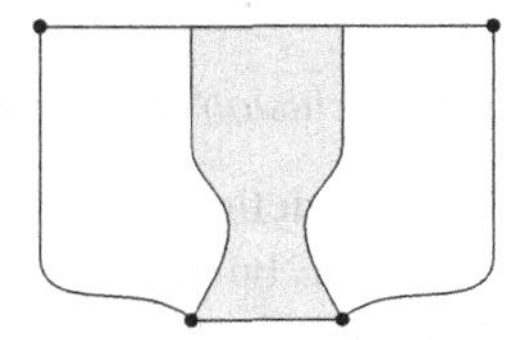

Figure 8.1 A new cobordism obtained by changing the corner

We have $j_n \circ i_n = 0$ since for any β-manifold M, we can interpret $N = M \times I$ as an (α, β) cobordism to the empty set by setting $\partial_-(M \times I) = M \times \{0\}$ and $\partial_c(M \times I) = M \times \{1\}$. Finally, if the closed α-manifold M has class x with $j_n x = 0$, there is a (α, β)-cobordism W of M to the empty set. Thus $\partial_- W = M$, $\partial_+ W = \varnothing$, and $N := \partial_c W$ is a closed β-manifold. Now letting V be the cobordism, diffeomorphic to W but with $\partial_- V = M$ and $\partial_+ V = N$, we see N is α-cobordant to M, so if N has class z we have $i_n z = x$. □

This procedure of changing the corner is itself a useful technique: we met it in §7.4 and will encounter it again.

The following addendum is easily proved by the same method.

Lemma 8.3.2 *Suppose given three kinds of structure: an α-structure, a β-structure, and a γ-structure, with γ stronger than β which in turn is stronger than α. Then there is a commutative diagram including the exact sequences corresponding to the three inclusions and one with the relative terms.*

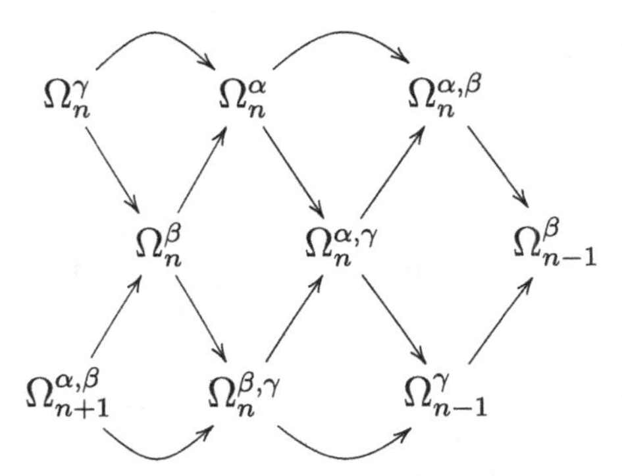

Lemma 8.3.1 is often applied together with a method of calculating $\Omega_n^{\alpha,\beta}$. To illustrate this, suppose any manifold W^n with α-structure has an induced β-structure except on a closed submanifold M^m, and define γ to be the type of structure induced on M by an α structure on a tubular neighbourhood V of M in W. This is imprecise; the details need to be clarified in each case where this is applied.

Lemma 8.3.3 *Inclusion induces an isomorphism $\Omega_m^\gamma \to \Omega_n^{\alpha,\beta}$.*

Proof The map is defined by taking the class of M in Ω_m^γ to that of $(V, \partial V)$ where V is the disc bundle over M which is part of the γ structure: by the definition of γ, this pair has an (α, β)-structure.

To prove the map surjective, take any $(W, \partial W)$ with (α, β)-structure, and construct M, V as above. Then $(V, \partial V)$ has a (α, β)-structure since ∂V is disjoint from M so we have an induced β structure on it. An (α, β)-cobordism X from $(V, \partial V)$ to $(W, \partial W)$ is obtained from $W \times I$ by rounding the corner at $\partial W \times \{0\}$ (using Proposition 2.6.2), and introducing a corner at $\partial V \times \{0\}$ (using Lemma 2.6.3) as in Figure 8.1: thus $\partial_c X = ((W \setminus \mathring{V} \times \{0\}) \cup (\partial W \times I)$.

Now suppose M such that $(V, \partial V)$ is (α, β)-cobordant to the empty set: write X for a cobordism. The α-structure on X gives a β-structure except on a compact submanifold L with boundary M. Then L has an induced γ-structure, so the class of M in Ω_m^γ is zero. □

Let V be a submanifold of M; then we call (M, V) a pair. If (N, W) is a pair of manifolds with boundary, and N is a cobordism of $\partial_- N$ to $\partial_+ N$, we set $\partial_- W = W \cap \partial_- N$, $\partial_+ W = W \cap \partial_+ N$. Our definition of submanifold then implies that W is a cobordism of $\partial_- W$ to $\partial_+ W$, and we shall call the pair (N, W) a cobordism of the pair $(\partial_- N, \partial_- N)$ to the pair $(\partial_+ N, \partial_+ W)$. Rather than restrict the structure groups of the stable tangent bundles of M and V independently; we usually restrict the structure group of the normal bundle of V in M: here there is no need to speak of weak structures.

We study cobordism of pairs by establishing a principle of 'extension of cobordism' (analogous to homotopy extension). This is illustrated in the next lemma.

Consider pairs (M^{v+q}, V^v), where M has a weak **J**-structure and the normal bundle an H_q-structure; more generally, consider $V^v \subset M^{v+q} \subset S^{v+q+r}$, where the structure groups of the normal bundles are H_q and J_r. Then the normal bundle $V^v \subset S^{v+q+r}$ has an $H_q \times J_r$-structure. Here we only consider the stable case $r > v + q + 1$ where the imbedding of M in S is irrelevant, so may replace J_r by **J**.

Let **J** be a stable group, and H_q a group mapping to O_q. Then setting $(\mathbf{J} \times H_q)_n = J_{n-q} \times H_q$ defines a stable group $\mathbf{J} \times H_q$, which satisfies **(S)** if J does.

Lemma 8.3.4 *The pair (M^{v+q}, V^v) is $(\mathbf{J}, H_q)$-cobordant to the empty pair if and only if M^{v+q} is $\mathbf{J}$-cobordant to zero and V^v is $\mathbf{J} \times H_q$-cobordant to zero.*

Proof The necessity of the condition is evident. To prove sufficiency we give a construction to extend a $\mathbf{J} \times H_q$-cobordism of V^v to the empty set to a $\mathbf{J} \times H_q$-cobordism of (M, V) to a pair (M', ψ). Since cobordism is an equivalence relation, it follows that M' is **J**-cobordant to φ, say by N'; then (N', φ) is the required $(\mathbf{J}, H_q)$-cobordism of (M', ψ) to (ψ, φ).

Let W^{v+1} be the given $\mathbf{J} \times H_q$-cobordism of V to φ: then there is an induced bundle over W with fibre D^q, whose total space we denote by L^{v+q+1}. Note that the restriction to V of this bundle is the normal bundle of V in M; hence we can identify a tubular neighbourhood of V in M with part of the boundary of L. We form $M \times I$, and attach L to $M \times 1$ by this identification, giving N. Since L and $M \times I$ have **J**-structures, which agree (by hypothesis, W is a cobordism of V with the $\mathbf{J} \times H_q$-structure induced from M) on the pair identified, N^{v+q+1}

has a weak **J**-structure. Also, $V \times I \cup W = W'$ is a submanifold whose normal bundle has group H_q.

Set $M \times 0 = \partial_- N$. Then (N, W') is a $\mathbf{J} \times H_q$-cobordism, and $W' \cap \partial_+ N = \varnothing$. This completes the proof of the lemma. □

Lemma 8.3.5 *The cobordism group of pairs (M^{v+q}, V^v), where M has a weak* **J***-structure and the normal bundle an H_q-structure is isomorphic to $\Omega^J_{v+q} \oplus \Omega^{J\times H_q}_v$.*

Proof In Lemma 8.3.4 we defined a map to the direct sum, and proved it a monomorphism; it clearly respects additive structure. The map to $\Omega^{J\times H_q}_v$ is onto, for given a $(J \times H_q)$-manifold V^v, we construct as above a bundle over V with fibre D^q, and can take M as the double of this manifold. Finally, the image contains $\Omega^J_{v+q} \oplus 0$: we need only consider pairs with V empty. □

8.4 Bordism as a homology theory

For **J** a stable group, and X any space, we denote by $\Omega^J_m(X)$ the cobordism group of (closed) manifolds M^m together with a weak **J**-structure and a map to X. The arguments of the preceding section generalise easily to this situation.

In fact we go further: given a pair of spaces $Y \subseteq X$ we define $\Omega^J_m(X, Y)$ to be the set of cobordism classes of (compact) manifolds M^m with a weak **J**-structure and a map $f : M \to X$ with $f(\partial M) \subset Y$. The definition of the cobordism relation is implicit in the above: a cobordism is a compact manifold W with corner, with a weak **J**-structure inducing the given weak **J**-structures on $\partial_\pm W$ (with the above convention), together with a map $g : (W, \partial_c W) \to (X, Y)$. Generalising Theorem 8.1.3, we have

Theorem 8.4.1 *If* **J** *is a stable group, the Thom construction induces isomorphisms*

$$\Omega^J_m(X, Y) \cong \lim_{k\to\infty} \pi_{m+k}((X^+ \wedge T(J_k), Y^+ \wedge T(J_k))).$$

Proof Given M^m, it follows from Whitney's embedding theorem and the refinements of Theorem 4.7.3 that for $k > m$ we can find embeddings of $(M, \partial M)$ as a submanifold of $(D^{m+k}, \partial D^{m+k})$, and that for $k > m + 1$ any two such embeddings are diffeotopic. It follows from Lemma 8.2.6 that for k large enough the weak **J**-structure on M induces a J_k-structure on the normal bundle ν of the embedding. Write ∂A_ν for the part of the disc bundle A_ν lying over ∂M. Then we have a map $A_\nu \to A(J_k) \to T(J_k)$ and also maps $(A_\nu, \partial A_\nu) \to$

$(M, \partial M) \to (X, Y)$. Taking the product, we have a map $(A_\nu, \partial A_\nu) \to (X \times T(J_k), Y \times T(J_k))$, which takes $(S_\nu, \partial S_\nu)$ to $(X \times \{\infty\}, Y \times \{\infty\})$.

Now collapse everything in D^{m+k} outside A_ν to a point, giving a map of $(D^{m+k}, \partial D^{m+k})$ to $(T(\nu), \partial T(\nu))$, and hence to

$$(X \times T(J_k)/X \times \{\infty\}, Y \times T(J_k)/Y \times \{\infty\}).$$

Recalling that $X \times T(J_k)/X \times \{\infty\} = X^+ \wedge T(J_k)$, we see as in Theorem 8.1.3, that this construction define a map

$$\Omega^J_m(X, Y) \to \lim_{k\to\infty} \pi_{m+k}((X^+ \wedge T(J_k), Y^+ \wedge T(J_k))).$$

The proof of the result now also closely follows that of Theorem 8.1.3. To establish surjectivity, we start with a map f and let K be the inverse image of ∞. Then f defines a map of $D^{m+k} \setminus K$ to $A(J_k) \times X$. We alter the first component by a small homotopy, to make it smooth and transverse to $B(J_k)$. This defines also a homotopy of f, say to f'. Now set $M^m = f'^{-1}(B(J_k) \times X)$; then f' induces a map $(M^m, \partial M^m) \to (X, Y)$, and the normal bundle of M has group reduced to J_k. It follows that the bordism class defined by M maps to the homotopy class of f.

Again, injectivity follows by a similar but simpler argument, and the proof that the bijection preserves group structures and the passage to the limit work as before. □

In particular we have an isomorphism $\Omega^J_m(X) \cong \lim_{k\to\infty} \pi_{m+k}(X^+ \wedge T(J_k))$. It follows as in Corollary 8.2.12 that

Corollary 8.4.2 *Under the isomorphism of Theorem 8.4.1, the external products $\Omega^J_m(X) \times \Omega^J_n(Y) \to \Omega^J_{m+n}(X \times Y)$ correspond to the homotopy pairings induced by $(X^+ \wedge T(J_k)) \wedge (Y^+ \wedge T(J_l)) \to (X^+ \wedge Y^+) \wedge T(J_{k+l})$.*

The maps $T(J_k) \wedge T(J_l) \to T(J_{k+l})$ give the limit $\mathbb{TJ}$ the structure of a ring spectrum (see §B.4). We can immediately extend Theorem 8.2.11 to

Theorem 8.4.3 *The Thom construction induces a natural equivalence between the functor Ω^J_* and homology theory with coefficients in the spectrum $\mathbb{TJ}$; this respects products in the multiplicative case.*

We have shown that Ω^J_* defines a homology theory. We prefer to present also a direct proof of this fact.

Theorem 8.4.4 *The groups $\Omega^J_*(X)$, $\Omega^J_*(X, Y)$ satisfy the Eilenberg–Steenrod axioms [50] for a homology theory.*

We first recall these axioms:

I, II: Ω_*^J is a functor from the category of pairs of spaces (X, Y) and continuous maps to the category of graded abelian groups. We denote $\Omega_*(X, \varnothing)$ by $\Omega_*(X)$.

III: For any pair (X, Y) there is a map $\partial : \Omega_m(X, Y) \to \Omega_{m-1}(Y)$ which is natural for maps of pairs.

IV: For any pair (X, Y), if $i : Y \to X$ and $j : (X, \varnothing) \to (X, Y)$ denote the inclusions, we have an exact sequence

$$\cdots \to \Omega_m^J(Y) \xrightarrow{i_*} \Omega_m^J(X) \xrightarrow{j_*} \Omega_m^J(X, Y) \xrightarrow{\partial_m} \Omega_{m-1}^J(Y) \to \cdots$$

V: Homotopic maps φ_0 and $\varphi_1 : (X_1, Y_1) \to (X_2, Y_2)$ induce the same map in bordism: $\varphi_{0,*} = \varphi_{1,*} : \Omega_*^J(X_1, Y_1) \to \Omega_*^J(X_2, Y_2)$.

VI: If $U \subset X$ has its closure in the interior of Y, then inclusion induces an isomorphism $\Omega_*^J(X \setminus U, Y \setminus U) \cong \Omega_*^J(X, Y)$.

Proof I, II: If $\varphi : (X_1, Y_1) \to (X_2, Y_2)$ is a map, M has a weak **J**-structure, and $f : (M, \partial M) \to (X_1, Y_1)$ represents a class $z \in \Omega_m^J(X_1, Y_1)$, then $\varphi \circ f$ represents $\varphi_*(z)$. This is well defined since if F defines a cobordism of f then $\varphi \circ F$ defines a cobordism of F. It is clear that the construction respects unions, so the map is additive.

III. If $f : (M, \partial M) \to (X, Y)$ gives a bordism class of (X, Y), then $f|\partial M$ gives a bordism class of Y. If $F : (W, \partial_c W) \to (X, Y)$ is a cobordism, then $F|\partial_c W$ is a cobordism between the boundary maps of $F|\partial_- W$ and $F|\partial_+ W$: thus restriction induces a map $\partial_m : \Omega_m^J(X, Y) \to \Omega_{m-1}^J(Y)$ which is compatible with disjoint union and hence a homomorphism. It is immediate that the construction is natural for maps of pairs.

IV. This is our first illustration of Lemma 8.3.1: here all manifolds have weak **J**-structures, and an α-structure consists of a map to X and a β-structure of a map to $Y \subset X$. Observe that in this case, if we form N'' by glueing manifolds N, N' with α-structure along their common boundary, both the weak **J**-structures and the maps to X fit to define a α-structure on N''.

V: If $\Phi : \varphi_0 \simeq \varphi_1$, then for any $f : (M, \partial M) \to (X, Y)$, we can regard $\Phi \circ f$ as defining a cobordism between $\varphi_0 \circ f$ and $\varphi_1 \circ f$.

VI: To prove surjectivity, we let $f : (M, \partial M) \to (X, Y)$ represent an element of $\Omega_m^J(X, Y)$. It is convenient first to alter f (if necessary) by a homotopy on a collar neighbourhood of ∂M so that some smaller neighbourhood is mapped into Y. Then $A = f^{-1}(X \setminus Y)$ and $B = \partial M \cup f^{-1}(U)$ have disjoint closures, so (see §A.2) we can find a continuous map $s : M \to I$ with $s(A) = 0$ and $s(B) = 1$.

We now approximate s by a smooth map, and make it transverse to $\frac{1}{2}$. Then $N := s^{-1}[-\frac{1}{2}, \frac{1}{2}]$ is a smooth submanifold of M, and $f|N$ determines an element of $\Omega^J_m(X \setminus U, Y \setminus U)$. But N and M determine the same class in $\Omega^J_m(X, Y)$. For a cobordism W, we use $f \times 1_I : M \times I \to X$ with a corner introduced at $\partial N \times 0$ and the corner at $\partial M \times 0$ rounded (as in the proof of Lemma 8.3.3). Since $(M \setminus N) \subset s^{-1}[1/2, 1]$, it is disjoint from A, and $f(M \setminus N) \subset Y$, so we can safely adjoin $(M \setminus N) \times 0$ to $\partial_c W$.

The proof of injectivity is similar. If $f : (W, \partial_c W) \to (X, Y)$ is a cobordism of $f|\partial_- W : (\partial_- W, \angle_- W) \to (X \setminus U, Y \setminus U)$ to $\partial_+ W = \varnothing$, we first adjust f so that $A = f^{-1}(X \setminus Y)$ and $B = \partial_c W \cup f^{-1}(U)$ have disjoint closures. Next choose a smooth $s : (W, A, B) \to (I, 0, 1)$, transverse to $\frac{1}{2}$, and set $V = s^{-1}[0, \frac{1}{2}]$. Then V is a cobordism of $\partial_- V$ to zero in $\Omega^J_m(X \setminus U, Y \setminus U)$: a cobordism of $\partial_- V$ to $\partial_- W$ is obtained exactly as above. This completes the proof of the theorem. □

Various standard properties of homology now follow.

Proposition 8.4.5 *(i) For any non-empty X, the maps $\{*\} \to X \to \{*\}$ induce a direct sum split $\Omega^J_*(X) \cong \Omega^J_* \oplus \tilde{\Omega}^J_*(X)$, where $\tilde{\Omega}^J_m(X) \cong \lim_{k\to\infty} \pi_{m+k}(X \wedge T(J_k))$. If CY is the cone on Y, $\tilde{\Omega}^J_*(CY) = 0$, and $\partial : \Omega^J_m(CY, Y) \cong \tilde{\Omega}^J_{m-1}(Y)$.*

(ii) If (X, Y) is a CW pair, or more generally if it has the homotopy extension property (HEP), $\Omega^J_*(X, Y) \cong \Omega^J_*(X/Y, pt) \cong \tilde{\Omega}^J_*(X/Y)$.

(iii) If $X \supset Y \supset Z$ is a triple, we have an exact sequence

$$\cdots \to \Omega^J_m(Y, Z) \to \Omega^J_m(X, Z) \to \Omega^J_m(X, Y) \to \Omega^J_{m-1}(Y, Z) \to \cdots$$

(iv) $\tilde{\Omega}^J_m(S^p) \cong \Omega^J_{m-p}$.

(v) Let $X = Y_1 \cup Y_2$, $Z = Y_1 \cap Y_2$, and suppose inclusion induces isomorphisms $\Omega^J_*(Y_i, Z) \cong \Omega^J_*(X, Y_{1-i})$ (by (i), this holds if the pairs (Y_i, Z) have the HEP). Then we have the exact sequences

$$\cdots \to \Omega^J_m(Z) \to \Omega^J_m(Y_1) \oplus \Omega^J_m(Y_2) \to \Omega^J_m(X) \to \Omega^J_{m-1}(Z) \to \cdots$$

$$\cdots \to \Omega^J_m(Z) \to \Omega^J_m(X) \to \Omega^J_m(X, Y_1) \oplus \Omega^J_m(X, Y_2) \to \Omega^J_{m-1}(Z) \to \cdots$$

(vi) $\Omega^J_*(X \cup Y) \cong \Omega^J_*(X) \oplus \Omega^J_*(Y)$ for disjoint union; $\tilde{\Omega}^J_*(X \vee Y) \cong \tilde{\Omega}^J_*(X) \oplus \tilde{\Omega}^J_*(Y)$.

(vii) If (X, Y) is a CW pair, $\Omega^J_m(X^p \cup Y, X^{p-1} \cup Y) \cong C_p(X, Y; \Omega^J_{m-p})$.

Proof (i) The splitting follows as we have an additive functor; the isomorphism follows as we have the same split on both sides of the equation. The next assertion follows from the homotopy axiom, the final one from the exact sequence.

(ii) Under the hypothesis, X/Y has the homotopy type of X with a cone on Y attached; by excision, this modulo the cone has the same groups as X modulo Y.

(iii) This is a standard exercise in diagram chasing.

(iv) Follows by induction from (ii) and (iii).

(v) These follow by another standard argument (the same for both).

(vi) Here $X \vee Y$ denotes the union of spaces X and Y with a single common point. Since $Z = \varnothing$, we can apply (v).

(vii) By (i), $\Omega^J_m(X^p \cup Y, X^{p-1} \cup Y) \cong \Omega^J_m(X^p/(X^{p-1} \cup (X^p \cap Y)))$. But $X^p/(X^{p-1} \cup (X^p \cap Y))$ is a wedge of p-spheres. Now apply (iv) and (vi). □

These results all illustrate how we can begin to calculate the groups $\Omega^J_m(X, Y)$ in terms of the Ω^J_m. We can formalise this process as a spectral sequence.

Theorem 8.4.6 *Let (X, Y) be a CW pair. Then there is a first quadrant Ω^J_*-module spectral sequence, converging strongly to $\Omega^J_*(X, Y)$, which starts with $E^2_{pq} = H_p(X, Y; \Omega^J_q)$.*

Proof By Proposition 8.4.5 (iii), the triple $(X^p \cup Y, X^q \cup Y, X^r \cup Y)$ $(r < q < p)$ has an exact bordism sequence. All the maps are induced by inclusions and boundary homomorphisms, so all expected diagrams commute. Such a collection of exact sequences defines a spectral sequence. We write $X^\infty = X$, $X^{-\infty} = \varnothing$: then the limit term is $\Omega^J_*(X, Y)$. The module structure is induced by natural products $\Omega^J_m \times \Omega^J_n(X^p \cup Y, X^q \cup Y) \to \Omega^J_{m+n}(X^p \cup Y, X^q \cup Y)$: if M^m is a closed manifold, and $f : (N, \partial N) \to (X^p \cup Y, X^q \cup Y)$, then we use the manifold $M \times N$ (with induced J-structure) and the map induced by first projecting on N.

By Proposition 8.4.5 (vii), the E^1 term is

$$E^1_{pq} = \Omega^J_{p+q}(X^p \cup Y, X^{p-1} \cup Y) \cong C_p(X, Y; \Omega^J_q).$$

The boundary d^1 is induced by taking the boundary of a manifold: it is easy to verify that this coincides with the usual boundary in the chain complex of (X, Y). It follows that $E^2_{pq} = H_p(X, Y; \Omega^J_q)$ and hence (since $\Omega^J_q = 0$ for $q < 0$) we have a first quadrant spectral sequence.

As to convergence, we note that

$$\begin{aligned} \Omega^J_n(X^{-\infty} \cup Y) &= \Omega^J_n(X^p \cup Y) \quad \text{for all } p < 0 \\ \Omega^J_n(X^p \cup Y) &= \Omega^J_n(X^\infty \cup Y) \quad \text{for all } p > n, \end{aligned}$$

the first since $X^{-1} = \varnothing = X^{-\infty}$ and the second since (by the cellular approximation theorem) any map of an n-manifold into X is homotopic to a map into X^n. These two isomorphisms imply strong convergence of the sequence. □

We have now discussed the homology theory associated with the spectrum $\mathbb{TJ}$ and so with the stable group **J**. There is also an associated cohomology theory, defined by

$$\Omega_J^n(X) := H^n(X; \mathbb{TJ}) := \lim_{N\to\infty} [S^N X : T(J_{N+n})].$$

The geometric content of this definition arises again by Theorem 8.4.1. If M is a closed smooth manifold, the suspension $S^N M$ is obtained from $\mathbb{R}^N \times M$ by adding a single point ∞. It follows from the Theorem that $[S^N M : T(J_{N+n})]$ corresponds bijectively to cobordism classes of submanifolds of $\mathbb{R}^N \times M$ whose normal bundles have group reduced to J_{N+n}.

Theorem 8.4.7 *Let* **J** *satisfy (S). Let M^m have a weak* **J***-structure. Then* $\Omega_J^n(M) \cong \Omega_{m-n}^J(M, \partial M)$.

Proof In this case, $\mathbb{R}^N \times M^m$ also has a weak **J**-structure. By Corollary 8.2.8, a J_{N+n}-structure on the normal bundle of V^{m-n} in $\mathbb{R}^N \times M^m$ induces a weak **J**-structure on the tangent bundle of V, and conversely if N is large enough. We thus have a bijective correspondence between $\Omega_J^n(M)$ and cobordism classes of manifolds V^{m-n} with weak **J**-structure and an imbedding in $\mathbb{R}^N \times M^m$, for large enough N. But if N is large, any map to $\mathbb{R}^N \times M^m$ is homotopic to an embedding, homotopic embeddings are diffeotopic, and a diffeotopy gives a cobordism. Hence specifying an imbedding in $\mathbb{R}^N \times M^m$ up to diffeotopy is equivalent to specifying a map to $\mathbb{R}^N \times M^m$, or indeed to M^m, up to homotopy. It remains only to note that if M has boundary, ∂V is imbedded in $\mathbb{R}^N \times \partial M$, so we must insist that it be mapped to ∂M. □

This result shows that a manifold with weak **J**-structure is orientable for the homology theory Ω_*^J. We also have a form of the Gysin isomorphism theorem.

Theorem 8.4.8 *Let* **J** *be a stable group satisfying (S),* (**M**)*; X a topological space, ξ a J_k-bundle over X. Then* $\Omega_n^J(X) \cong \tilde{\Omega}_{n+k}^J(T(\xi))$.

Proof Let $f : X \to B(J_k)$ classify ξ, and f_N denote the composite

$$B(J_N) \times X \xrightarrow{1\times f} B(J_N) \times B(J_k) \xrightarrow{B\psi_{N,k}} B(J_{N+k}).$$

Write F_N for the map $B(J_N) \times X \to B(J_{N+k}) \times X$ whose components are f_N and projection on the second factor. F_N is covered by a bundle map of $\gamma^N \oplus \xi$ to γ^{N+k}. Also, $B(J_N)$ is mapped by the natural injection i to $B(J_{N+k})$, and we have a commutative exact diagram

$$\begin{array}{ccccccccc} 0 & \to & \pi_r(B(J_N)) & \to & \pi_r(B(J_N)\times X) & \to & \pi_r(X) & \to & 0 \\ & & \downarrow i_* & & \downarrow F_{N*} & & \| & & \\ 0 & \to & \pi_r(B(J_{N+k})) & \to & \pi_r(B(J_{N+k})\times X) & \to & \pi_r(X) & \to & 0. \end{array}$$

Thus F_{N*} is an isomorphism in the limit as $N \to \infty$. We have an induced map of Thom spaces

$$T(J_N) \wedge T(\xi) \to T(J_{N+k}) \wedge X^+,$$

which then also in the limit gives homotopy isomorphisms. Thus

$$\begin{aligned} \Omega^J_n(X) &\cong \lim_{N\to\infty} \pi_{N+n+k}(T(J_{N+k}) \wedge X^+) \\ &\cong \lim_{N\to\infty} \pi_{N+n+k}(T(J_N) \wedge T(\xi)) \\ &\cong \tilde{\Omega}^J_{n+k}(T(\xi)). \end{aligned}$$ □

The calculation in Lemma 8.3.5 of cobordism of pairs involves the groups $\Omega^{J\times H_q}_*$, which admit a natural Ω^J_*-module structure. We can now 'compute' them directly using bordism groups.

Lemma 8.4.9 *We have* $\Omega^{J\times H_q}_n \cong \Omega^J_{n+q}(T(H_q))$, *and more generally*

$$\Omega^{J\times H_q}_n(X) \cong \Omega^J_{n+q}(T(H_q) \wedge X^+).$$

Proof By Theorem 8.4.1, we have

$$\begin{aligned} \Omega^{J\times H_q}_n(X) &= \lim_{N\to\infty} \pi_{n+N}(T(J \times H_q)_N \wedge X^+) \\ &= \lim_{N\to\infty} \pi_{n+N}(T(J_{N-q} \times H_q) \wedge X^+) \\ &= \lim_{N\to\infty} \pi_{n+N}(T(J_{N-q}) \wedge T(H_q) \wedge X^+) \\ &= \Omega^J_{n+q}(T(H_q) \wedge X^+). \end{aligned}$$ □

As with any homology theory, we can define bordism theory with coefficients. If $n > 1$ and $r > 1$ are natural numbers, write e^r_n for a space obtained from S^n by attaching an $(n+1)$-cell by a map $S^n \to S^n$ of degree r; thus $\tilde{H}_N(e^r_n; \mathbb{Z})$ is isomorphic to $\mathbb{Z}_r$ if $N = n$ and is zero otherwise.

We can now define $\Omega^J_N(X; \mathbb{Z}_r) := \tilde{\Omega}^J_{N+n}(X \wedge e^r_n)$. Elementary properties of this definition are easily deduced from homotopy properties of the spaces e^r_n. We will not go into further details.

The following construction is also sometimes useful. Let $\mathbf{J}$ be a stable group and H be any topological group. Then we can define a stable group $\mathbf{J} \rhd H$ by setting $(\mathbf{J} \rhd H)_n := J_n \times H$, operating on $\mathbb{R}^n$ via its projection on J_n. (This is *not* the same as the $\mathbf{J} \times H$ defined above.)

We have $B(J_n \rhd H) = B(J_n) \times B(H)$ and $T(J_n \rhd H) = T(J_n) \wedge B(H)$. In particular, if X is any CW complex, the loop space ΩX is equivalent to a topological group, and we have

$$\begin{aligned}\Omega_n^{J\rhd\Omega X} &= \lim_{N\to\infty} \pi_{n+N}(T(J_{n+N} \times \Omega X)) \\ &= \lim_{N\to\infty} \pi_{n+N}(T(J_{n+N} \wedge X)) = \tilde{\Omega}_n^J(X).\end{aligned}$$

Thus the groups $\tilde{\Omega}_*^J(X)$ may be considered as the coefficient groups of the homology theory $\Omega_*^{J\rhd\Omega X}$.

8.5 Equivariant cobordism

The object of this section is to give a method for reducing the calculation of equivariant cobordism groups to that of the bordism groups of certain classifying spaces.

We begin by formulating the definitions of equivariant cobordism groups. First define $I_m^O(G)$ to be the cobordism group of manifolds with a smooth action of the compact Lie group G. Next, let A be a closed collection of orbit types (in the sense of §3.5), and write $I_m^O(G; A)$ for the cobordism group of those actions such that all orbit types belong to A. Here we identify a type defined by a pair (H, E) with the type defined by $(H, E \oplus \mathbb{R})$, where H acts trivially on $\mathbb{R}$, to be able to use the same list of types for manifolds and for cobordisms.

We also wish to incorporate a structure group. Let J be a stable group satisfying (**M**), (**A**) and (**S**), and M have a J-structure (on its stable tangent bundle). We say that a smooth action of G on M *respects the J-structure* if the following condition is satisfied. For some n, we are given an action of G on a principal J_n-bundle P which defines the J-structure, lifting the given action of G on M. This defines actions of G on the associated bundles; in particular, on the principal J_{n+1}-bundle, so the condition is independent of n. To avoid technicalities we restrict to three cases: J may be O, SO, or U: in the second case all the bundles are orientable; in the third all are unitary, in particular E is acted on by a unitary group which we denote $U(E)$.

Write $I_m^J(G; A)$ for the group of cobordism classes of manifolds M^m with J-structure and a smooth G-action which respects it, and such that each orbit type belongs to A.

First consider the case of free actions: we have the single orbit type when H is trivial; denote it by 'free'. In this case, the projection $M \to G\backslash M$ is a fibre bundle with group G. Such bundles over X are classified by (homotopy classes of) maps $X \to B(G)$, where $B(G)$ is the classifying space of G.

Lemma 8.5.1 *The cobordism group $I_m^J(G; free)$ is isomorphic to the bordism group $\Omega_{m-g}^J(B(G))$, where $g = \dim G$.*

Proof We have just observed that a free action on M leads to a map $G\backslash M \to B(G)$; the converse is also immediate. The same remarks apply to manifolds with boundary, so the correspondence goes over to cobordism. □

Now let A be a closed set of orbit types, $\alpha \in A$ a maximal element, and write $A' := A \setminus \{\alpha\}$: since α is maximal, this too is closed. There is a natural map $I_m^J(G; A') \to I_m^J(G; A)$.

Lemma 8.5.2 *There is a natural exact sequence*

$$\ldots I_m^J(G; A') \to I_m^J(G; A) \to I_m^J(G; (A, A')) \to I_{m-1}^J(G; A') \to I_{m-1}^J(G; A) \ldots$$

Proof This is a direct application of the general principle of Lemma 8.3.1. Here the third term is defined as the group of cobordism classes of cobordisms W^m with J-structure and a smooth G-action which respects it, and such that each orbit type belongs to A and those on ∂W to A'. □

This formal result is only of value once we have a way to compute the third term.

We recall that an orbit type α is associated to a subgroup H^α of G and a representation of H^α on a Euclidean space E^α (both defined up to conjugacy). Since α is maximal in A, W^α is a closed submanifold of W, and by hypothesis is disjoint from ∂W. By Theorem 3.5.8, a neighbourhood of W^α in M is equivariantly diffeomorphic to a bundle over $X^\alpha = W^\alpha/G$ with fibre $G \times_{H^\alpha} E^\alpha$. Note that $\dim X^\alpha = \dim W^\alpha - \dim G + \dim H^\alpha$, and that $\dim M - \dim W^\alpha = \dim E^\alpha$.

According to Lemma 8.3.3, the third term $I_m^J(G; (A, A'))$ is isomorphic to the bordism group of G-manifolds W^α together with a G-bundle $\pi : N^\alpha \to W^\alpha$ with fibre E^α on which H^α acts as indicated. To proceed, we let P be the principal $J(E^\alpha)$-bundle associated to π: P is the set of isometries of E^α on fibres of π. On P we have the natural (right) action of $J(E^\alpha)$, also an induced (left) action of G which commutes with it, hence an action of $G \times J(E^\alpha)$: this action has only a single orbit type. The isotropy group is the set of elements

$$H^* = \{(h^{-1}, \rho(h)) \mid h \in H^\alpha\} \subseteq G \times J(E^\alpha).$$

Recalling the discussion in §3.5 of the structure of a G-manifold with just one orbit type, we now consider the submanifold P^{H^*} and the induced action on it of

$$N(H^*) = \{(g, r) \in G \times J(E^\alpha) : \rho(g^{-1}hg) = r^{-1}\rho(h)r \text{ for all } h \in H^\alpha\}.$$

Denote by L^α the quotient group $N(H^*)/H^*$. Then P^{H^*} is a principal L^α-bundle over X. The mechanism of classifying spaces tells us that such bundles correspond to homotopy classes of maps $X \to BL^\alpha$.

Theorem 8.5.3 *We have*

$$I^J_m(G; (A, A')) \cong \Omega^J_{m-c}(BL^\alpha),$$

where $c = \dim G - \dim H^\alpha + \dim E^\alpha$.

Proof As in the proof of Lemma 8.5.1, the homotopy class of $X \to BL^\alpha$ determines the isomorphism class of X and hence W with all its structure. Since the same applies to bounded manifolds, we can pass to cobordism classes. □

This argument extends trivially to the case when A has two maximal elements α, β such that neither $\alpha \prec \beta$ nor $\beta \prec \alpha$ (for example, the subgroups H^α and H^β are conjugate), then if $A'' := A \setminus \{\alpha, \beta\}$ then

$$I^J_m(G; (A, A'')) \cong I^J_m(G; (A'' \cup \{\alpha\}, A'')) \oplus I^J_m(G; (A'' \cup \{\beta\}, A'')),$$

for the orbit types M^α and M^β have disjoint closures. Similarly we can deal with further such summands.

The simplest example is the group $G = \mathbb{Z}_2$ with $J = O$. Any action is semi-free; the possible non-trivial orbit types have $H = \mathbb{Z}_2$ and $E = \mathbb{R}^k$ for some k, with the antipodal action. In this case, H^* has order 2 and is central in $G \times O_k$, so its normaliser is the whole group and $L = N(H^*)/H^*$ is isomorphic to O_k. Denote by A the set of all orbit types. It follows from Theorem 8.5.3, together with the remark following it, that we have

$$I^O_m(\mathbb{Z}_2; (A, free) \cong \bigoplus_k \Omega^O_{m-k}(B(O_k)).$$

Theorem 8.5.4 *There is a split short exact sequence*

$$0 \to I^O_m(\mathbb{Z}_2; A) \to \oplus_k \Omega^O_{m-k}(B(O_k)) \to \Omega_{m-1}(B\mathbb{Z}_2) \to 0.$$

Proof By Lemma 8.5.2 we have an exact sequence

$$I^O_m(\mathbb{Z}_2; free) \to I^O_m(\mathbb{Z}_2; A) \to I^O_m(\mathbb{Z}_2; (A, free)) \to I^O_{m-1}(\mathbb{Z}_2; free) \to I^O_{m-1}(\mathbb{Z}_2; A).$$

We claim that the map $I^O_m(\mathbb{Z}_2; free) \to I^O_m(\mathbb{Z}_2; A)$ is trivial. Indeed, given a free action of $\mathbb{Z}_2$ on M, the mapping cylinder of the projection $M \to \mathbb{Z}_2\backslash M$ can be identified with a bundle over $\mathbb{Z}_2\backslash M$ with fibre the interval $[-1, 1]$. This is a smooth manifold, with a $\mathbb{Z}_2$-action given by -1 in each fibre, and has boundary M.

The long sequence thus breaks into short exact sequences, and we can substitute $I^O_{m-1}(BZ_2; free) \cong \Omega_{m-1}(B\mathbb{Z}_2)$ by Lemma 8.5.1 and the value of

$I_m^O(\mathbb{Z}_2; (A, free))$ from the calculation preceding the theorem. The fact that the sequence splits follows since the middle group has exponent 2. □

The same approach may be applied to the case $G = \mathbb{Z}_p$, with p an odd prime, but here the details do not simplify. Here it is more natural to take the structure group J as SO or U. It is still true that any action is semifree, and we have a calculation of the bordism group of free actions; although the calculation of $\Omega^{SO}(B\mathbb{Z}_p)$ and $\Omega^U(B\mathbb{Z}_p)$ is less easy than for $\Omega^O(B\mathbb{Z}_2)$, this may be explicitly done, using the results quoted in the following section. A non-free orbit type has $H = \mathbb{Z}_p$, but to describe the isotropy action we must specify for each $r = 1, 2, \ldots, p-1$ a multiplicity $a_r \geq 0$ and then have an action on $\mathbb{C}^{n_\alpha}$ with $n_\alpha = \sum_r a_r$ where the generator t of G is represented by the diagonal action with the eigenvalue ζ_p^r repeated a_r times. In each case, H^* is isomorphic to $\mathbb{Z}_p$, but the calculation of its normaliser $N(H^*)$ depends very much on α; only in the case when all a_r except one vanish is the corresponding group H^* central. Nor is there any reason for the map $I_m^{SO}(\mathbb{Z}_p; free) \to I_m^{SO}(\mathbb{Z}_p; A)$ to be trivial.

8.6 Classifying spaces, Ω_*^O, Ω_*^U

We first describe the cohomology of the classifying spaces, and begin with the unitary group, where the structure is simplest.

The group U_n has a subgroup consisting of diagonal matrices; this is a torus T^n, a product of n copies of the circle group $U_1 = S^1$. The classifying space $B(S^1)$ can be taken to be infinite complex projective space $P^\infty(\mathbb{C})$ and $H^*(B(S^1); \mathbb{Z})$ is the polynomial ring $\mathbb{Z}[t]$ on a single generator $t \in H^2(B(S^1); \mathbb{Z})$. Thus $H^*(B(T^n); \mathbb{Z})$ is the polynomial ring $\mathbb{Z}[t_1, \ldots, t_n]$.

The inclusion induces maps $B(T^n) \to B(U_n)$ and $H^*(B(U_n); \mathbb{Z}) \to H^*(B(T^n); \mathbb{Z})$. It is well known that the map $H^*(B(U_n); \mathbb{Z}) \to H^*(B(T^n); \mathbb{Z})$ is injective, and that its image is the subring of polynomials invariant under the action of the Weyl group W. The Weyl group W of U_n is the symmetric group, and acts by permutations: the invariants form the ring of symmetric functions in the t_i. We can identify this with the polynomial ring generated by the elementary symmetric functions c_i $(1 \leq i \leq n)$, which can be defined by the formal identity $\prod_1^n(x - t_i) = x^n + \sum_1^n(-1)^r x^{n-r} c_r$. The class c_i is known as the *Chern class*. An additive basis of $H^*(B(U_n); \mathbb{Z})$ is given by the elements $s_{i_1,i_2,\ldots}$ $(i_1 \geq i_2 \ldots)$, defined as the sum of all the *distinct* monomials formed from $t_1^{i_1} t_2^{i_2} \cdots$ by permuting the variables. To distinguish these from similar calculations below, we sometimes write $s_I(c)$ for emphasis.

Taking the limit as $n \to \infty$ gives $H^*(B(U); \mathbb{Z})$ as the polynomial ring in an infinite sequence of variables $c_1, c_2, \ldots$. There is an additive isomorphism of $H^*(B(U); \mathbb{Z})$ to the cohomology of the Thom spectrum $\mathbb{TU}$.

The multiplicative structure for $\mathbb{TU}$ appears in homology, and is induced by direct sum, which gives maps $B(U_m) \times B(U_n) \to B(U_{m+n})$ and $TU_m \wedge TU_n \to TU_{m+n}$. To evaluate this in cohomology, observe that it comes from the identity $T^m \times T^n = T^{m+n}$ and hence $H^*(T^m; \mathbb{Z}) \otimes H^*(T^n; \mathbb{Z}) \cong H^*(T^{m+n}; \mathbb{Z})$; we can identify these as polynomial rings with generators, say, $t_1, \ldots, t_m, u_1, \ldots, u_n$. The induced map $\nabla : H^*(B(U_{m+n}); \mathbb{Z}) \to H^*(B(U_m); \mathbb{Z}) \otimes H^*(B(U_n); \mathbb{Z})$ is given by $\nabla(s_I) = \sum s_{I_1} \otimes s_{I_2}$, where the sum is extended over all partitions of the set $I = \{i_1, i_2, \ldots\}$ as a *disjoint* union $I = I_1 \cup I_2$. This is compatible with the inclusion maps which increase m and n, so we can pass to the limit, giving a diagonal map $\nabla : H^*(B(U); \mathbb{Z}) \to H^*(B(U); \mathbb{Z}) \otimes H^*(B(U); \mathbb{Z})$.

Dualising gives an algebra structure on $H_*(B(U); \mathbb{Z})$. If we define $\{\tau_I\}$ to be the dual basis to $\{s_I\}$ it follows that if I_1 and I_2 are disjoint we have $\tau_{I_1}\tau_{I_2} = \tau_{I_1, I_2}$ and hence $\tau_I = \tau_1^{i_1}\tau_2^{i_2}\cdots$. Thus $H_*(B(U); \mathbb{Z})$ is a polynomial ring with the τ_r as generators.

It follows from the Gysin isomorphism theorem that we have an isomorphism $H_*(B(U_m); \mathbb{Z}) \to \tilde{H}_*(TU_m : \mathbb{Z})$ of degree m. Since the diagram

$$\begin{array}{ccccc} H_*(B(U_m); \mathbb{Z}) \otimes H_*(B(U_n); \mathbb{Z}) & \cong & H_*(B(U_m) \times B(U_n); \mathbb{Z}) & \to & H_*(B(U_{m+n}); \mathbb{Z}) \\ \downarrow \wr & & \downarrow \wr & & \downarrow \wr \\ \tilde{H}_*(TU_m; \mathbb{Z}) \otimes \tilde{H}_*(TU_n; \mathbb{Z}) & \cong & \tilde{H}_*(TU_m \wedge TU_n; \mathbb{Z}) & \to & \tilde{H}_*(TU_{m+n}; \mathbb{Z}) \end{array}$$

is commutative, we have an induced isomorphism of $H_*(B(U); \mathbb{Z})$ on the stable homology ring $\tilde{H}_*(TU; \mathbb{Z})$.

The structure for the orthogonal group O_n is very similar. The subgroup X_n of diagonal matrices is a product of n copies of $O_1 \cong S^0$: it is a maximal elementary 2-subgroup. The classifying space $B(O_1)$ can be taken to be infinite real projective space $P^\infty(\mathbb{R})$ and $H^*(B(O_1); \mathbb{Z}_2)$ is the polynomial ring $\mathbb{Z}_2[t]$ on a single generator $t \in H^1(B(O_1); \mathbb{Z}_2)$. Thus $H^*(B(X_n); \mathbb{Z}_2)$ is the polynomial ring $\mathbb{Z}_2[t_1, \ldots, t_n]$. The inclusion induces maps $B(X_n) \to B(O_n)$ and $H^*(B(O_n); \mathbb{Z}_2) \to H^*(B(X_n); \mathbb{Z}_2)$; the image is the subring of polynomials invariant under the action of the group of permutations of the t_i, so is the ring of symmetric functions, and hence the polynomial ring generated by the elementary symmetric functions. In this case, the class defined by the ith elementary symmetric function is known as the *Stiefel–Whitney class*, and is denoted w_i, and we write $s_I(w)$ for the symmetric functions defined as above. We refer to [103] for a good general introduction to Stiefel–Whitney and other characteristic classes.

Thus $H^*(B(O); \mathbb{Z}_2)$ is the polynomial algebra in classes w_i. It is additively isomorphic to $\tilde{H}^*(\mathbb{TO}; \mathbb{Z}_2)$. Direct sum of vector bundles induces a diagonal map for these, hence a multiplication on $\tilde{H}_*(\mathbb{TO}; \mathbb{Z}_2)$, given by essentially the same formulae as in the unitary case.

Returning to the unitary case and applying Proposition B.4.1, we obtain our first calculation.

Proposition 8.6.1 *The ring $\Omega_*^U \otimes \mathbb{Q}$ is a polynomial ring with one generator in each even dimension, and each group Ω_n^U is finitely generated.*

Since we had dual bases above, s_n is orthogonal to all the τ_I such that I has more than one part, and hence to all decomposable classes in $H_*(B(U); \mathbb{Z})$ (i.e. classes which can be expressed as sums of products of classes of lower degree). Thus $z \in H_n(B(U); \mathbb{Z})$ is decomposable if and only if $\langle z, s_n(c)\rangle = 0$.

Since $H_*(B(U); \mathbb{Z})$ and $\tilde{H}_*(TU; \mathbb{Z})$ are polynomial rings, any ring homomorphism $\tilde{H}_*(TU; \mathbb{Z}) \to \mathbb{Z}$ is determined by its values on the generators τ_r, and these values may be chosen arbitrarily. A corresponding statement holds with coefficients $\mathbb{Q}$ in place of $\mathbb{Z}$. We seek a formula to express this.

Any additive homomorphism $\phi : H_*(TU; \mathbb{Q}) \to \mathbb{Q}$ is given by taking inner product with an element Φ of the direct product $H^{**}(TU; \mathbb{Q}) \cong H^{**}(B(U); \mathbb{Q})$ of the groups $H^n(B(U); \mathbb{Q})$. Dualising, it follows that ϕ is a ring homomorphism if and only if $\nabla(\Phi) = \Phi \otimes \Phi$. As above, it is convenient to consider Φ as a symmetric element of the power series ring in infinitely many variables t_i. Since $\nabla(t_i^r) = t_i^r \otimes 1 + 1 \otimes t_i^r$, we see that for any coefficients a_r, the infinite product

$$\Phi := \prod_i \left(1 + \sum_{r=1}^{\infty} a_r t_i^r\right)$$

has the desired property. Since this formula allows one arbitrary coefficient at each stage, it allows independent choices for the $\phi(\tau_r)$, so Φ is necessarily of this form.

We have seen that if M^{2n} has a weak **U**-structure given by a lift $f : M \to B(U)$ of the map inducing τ_M, the class of M in $\Omega_*^U \otimes \mathbb{Q}$ is decomposable if and only if $s_n(M) := \langle M, f^* s_n(c)\rangle = 0$. It will be useful to have some calculations of these numbers. Denote by $Y_{m,n}^{\mathbb{C}}$ a nonsingular hypersurface of degree $(1, 1)$ in $P^m(\mathbb{C}) \times P^n(\mathbb{C})$, for example, that given by $\sum_{i=0}^{min(m,n)} x_i y_i = 0$. These examples were introduced by Milnor [96]. Write also $Y_{m,n}^{\mathbb{R}}$ for a nonsingular hypersurface of degree $(1, 1)$ in $P^m(\mathbb{R}) \times P^n(\mathbb{R})$.

Proposition 8.6.2 *We have $s_n(P^n(\mathbb{C})) = n + 1$ and $s_{m+n-1}(Y_{m,n}^{\mathbb{C}}) = -\binom{m+n}{m}$.*

Proof Write λ for the canonical line bundle over $P^n(\mathbb{C})$. Then $\tau_{P^n(\mathbb{C})} \oplus \varepsilon \equiv (n+1)\lambda$, so the characteristic classes of $\tau_{P^n(\mathbb{C})}$ agree with those of $(n+1)\lambda$, so are induced by the map $P^n(\mathbb{C}) \to B(T^{n+1}) \to B(U_{n+1})$.

Since $c_1(\lambda)$ is the generator x of $H^2(P^n(\mathbb{C}); \mathbb{Z})$, the generators t_i of the cohomology groups $H^2(; \mathbb{Z})$ of the factors BT of BT^{n+1} all map to x. Thus s_n, which is the image of $\sum t_i^n$, maps to $(n+1)x^n$, and evaluating this on $[P^n(\mathbb{C})]$ gives $(n+1)$.

Set $M := P^m(\mathbb{C}) \times P^n(\mathbb{C})$ and write λ_1, λ_2 for the line bundles induced from the two factors. We have $H^2(M; \mathbb{Z}) \cong \mathbb{Z} \oplus \mathbb{Z}$: write x_1, x_2 for the generators coming from the two factors. Thus $\tau_M \oplus 2\varepsilon \equiv (m+1)\lambda_1 \oplus (n+1)\lambda_2$ and in calculating characteristic classes we may take $(m+1)$ of the t_i equal to x_1 and $(n+1)$ equal to x_2. These all pull back by the inclusion $i : Y_{m,n}^{\mathbb{C}} \subset M$, and the normal bundle ν_Y of Y in M is the pullback of $\lambda_1 \otimes \lambda_2$, with first Chern class $i^*(x_1 + x_2)$.

Since s_n is defined as a sum of contributions coming from summands, we have in general $s_n(\xi \oplus \eta) = s_n(\xi) + s_n(\eta)$. Now as

$$\tau_Y \oplus \nu_Y \oplus 2\varepsilon \equiv (m+1)i^*\lambda_1 \oplus (n+1)i^*\lambda_2,$$

and as $i^*x_1^{m+n-1}[Y] = 0$, it follows that

$$s_n(Y) = -s_n(\nu_Y)[Y] = -i^*(x_1 + x_2)^{m+n-1}[Y] = -(x_1 + x_2)^{m+n-1}[i_*Y]$$
$$= -(x_1 + x_2)^{m+n}[M] = -\binom{m+n}{m}.$$

□

The same calculations yield

Corollary 8.6.3 *We have* $s_n(w)(P^n(\mathbb{R})) \equiv n+1$ *(mod 2) and* $s_{m+n-1}(w)(Y_{m,n}^{\mathbb{R}}) \equiv \binom{m+n}{m}$ *(mod 2).*

The calculations for (special) orthogonal and symplectic groups are similar to the unitary case, provided for the orthogonal group we localise away from the prime 2. The groups SO_{2n+1} and Sp_{2n} each contain a maximal torus T^n, but in these cases the action of the Weyl group includes, as well as permutations, the inversions in each factor. (For SO_{2n} we only allow an even number of inversions.) The ring of invariants thus consists of the symmetric functions in the variables t_i^2 (also for SO_{2n} the product $\prod t_i$). The class defined by the ith elementary symmetric function is known as the *Pontrjagin class*, and is denoted p_{4i}. The same arguments apply here to calculate the dual. It now follows as before from Proposition B.4.1 that

Proposition 8.6.4 *The rings* $\Omega_*^{SO} \otimes \mathbb{Q}$ *and* $\Omega_*^{Sp} \otimes \mathbb{Q}$ *are polynomial rings with one generator in each dimension divisible by 4.*

The unitary structure on $P^n(\mathbb{C})$ induces an SO-structure, and the dummy variables t_i play the same role as before. As s_n is given by $\sum t_i^n$, we see that if n is even, the formula for s_n in terms of Chern classes is the same as the formula in terms of Pontrjagin classes, thus it follows from Proposition 8.6.2 that

Lemma 8.6.5 *We have* $s_{2n}(p)(P^{2n}(\mathbb{C})) = 2n + 1$.

Thus we can take the manifolds $P^{2k}(\mathbb{C})$ as generators of $\Omega_*^{SO} \otimes \mathbb{Q}$.

It follows from Proposition 8.6.4 that any ring homomorphism $\Omega_*^{SO} \otimes \mathbb{Q} \to \mathbb{Q}$ is determined by its values on a list of generators. The most celebrated example of this is the signature. We defined the signature $\sigma(M)$ of an oriented manifold M of dimension $4k$ in §7.5 as the signature of the quadratic form given by intersection numbers on $H_{2k}(M; \mathbb{R})$. We saw in that section that $\sigma(M)$ vanishes if M is an oriented boundary, and σ is clearly additive on disjoint unions, hence defines an additive homomorphism $\sigma : \Omega_*^{SO} \to \mathbb{Z}$.

Lemma 8.6.6 *The signature* σ *is multiplicative for products, hence defines a ring homomorphism* $\sigma : \Omega_*^{SO} \to \mathbb{Z}$.

Proof Consider the product $M^m \times N^{4k-m}$ of two oriented manifolds. We have $H^{2k}(M \times N) = \oplus_i H^i(M) \otimes H^{2k-i}(N)$. Under cup product the term $H^i(M) \otimes H^{2k-i}(N)$ is dually paired with $H^{m-i}(M) \otimes H^{2k-m+i}(N)$, so only the term $m = 2i$ can contribute to the signature. The self-pairing of $H^i(M) \otimes H^{2k-i}(N)$ with itself to $H^m(M) \otimes H^{4k-m}(N) \cong \mathbb{R}$ is the tensor product of the self-pairings of $H^i(M)$ and $H^{2k-i}(N)$ to $\mathbb{R}$. If i is odd, there is a Lagrangian subspace K of $H^i(M)$, so $K \otimes H^{2k-i}(N)$ is a Lagrangian subspace of $H^i(M) \otimes H^{2k-i}(N)$ and this has signature zero. If i is even, we can diagonalise the quadratic forms on $H^i(M)$ and $H^{2k-i}(N)$, and the calculation is trivial. □

The value is given by Hirzebruch's signature theorem [74]. First recall the expansion

$$\frac{t}{tanh(t)} = 1 + \sum_{k=1}^{\infty} (-1)^{k-1} \frac{2^{2k}}{(2k)!} B_k t^{2k},$$

which we may use to define the Bernoulli numbers B_k

$$B_1 = \frac{1}{6}, \quad B_2 = \frac{1}{30}, \quad B_3 = \frac{1}{42}, \quad B_4 = \frac{1}{30}, \quad B_5 = \frac{5}{66},$$
$$B_6 = \frac{691}{2730}, \quad B_7 = \frac{7}{6}, \ldots$$

let us write this as $t/tanh(t) = 1 + \sum_{k=1}^{\infty} \beta_k t^{2k}$. Now define the class $L_* \in H^{**}(B(O); \mathbb{Q})$ by the formula $L_* := \prod_i (1 + \sum_{k=1}^{\infty} \beta_k t_i^{2k})$, where the t_i are the auxiliary variables introduced above.

Theorem 8.6.7 *For M oriented, the signature is given by* $\sigma(M) = f^* L_m[M]$, *where* $f : M \to B(SO)$ *induces* τ_M.

Proof We have seen that we can take the manifolds $P^{2k}(\mathbb{C})$ as generators of $\Omega_*^{SO} \otimes \mathbb{Q}$. Thus two ring homomorphisms agreeing on these coincide, so it suffices to verify the formula for $M = P^{2k}(\mathbb{C})$. We see at once that each manifold $P^{2k}(\mathbb{C})$ has signature 1.

Since $p(\tau_M) = (1+\alpha^2)^{2k+1}$, we have

$$\begin{aligned} L_*[P^{2k}(\mathbb{C})] &= \left(\frac{\alpha}{tanh\,\alpha}\right)^{2k+1}[P^{2k}(\mathbb{C})] \\ &= coefficient\ of\ \alpha^{2k}\ in\ \left(\frac{\alpha}{tanh\,\alpha}\right)^{2k+1} \\ &= \frac{1}{2\pi i}\oint \frac{dz}{(tanh\,z)^{2k+1}} \\ &= \frac{1}{2\pi i}\oint \frac{du}{u^{2k+1}(1-u^2)} \quad (substituting\ u = tanh(z)) \\ &= Res_0(1+u^2+u^4+\ldots)/u^{2k+1} \\ &= 1. \end{aligned}$$

□

Explicit formulae for the L classes may be calculated: for example,

$$L_1 = \frac{1}{3}p_1,\ L_2 = \frac{7p_2 - p_1^2}{45},\ L_3 = \frac{62p_3 - 13p_1p_2 + 2p_1^3}{945}.$$

We will use the below formula for the leading coefficient (see, for example, [103]).

Lemma 8.6.8 *The coefficient of p_k in L_k is $2^{2k}(2^{2k-1}-1)B_k/(2k)!$.*

For A_* a graded vector space over a field F, we count the dimensions by the *Poincaré series*

$$P(A_*; F)(t) := \sum_0^\infty \dim_F(A_n)t^n.$$

Thus if A is a polynomial algebra with the degrees of generators in a set S, we have $P(A)(t) = \prod_{i\in S}(1-t^i)^{-1}$. In particular, by Proposition 8.6.1,

$$P(\Omega_*^U \otimes \mathbb{Q})(t) = P(H_*(U;\mathbb{Q}))(t) = \prod_{i=1}^\infty (1-t^2)^{-1},$$

and by Proposition 8.6.4,

$$P(\Omega_*^{SO} \otimes \mathbb{Q})(t) = P(H_*(SO;\mathbb{Q}))(t) = \prod_{i=1}^\infty (1-t^4)^{-1}.$$

Thom's great achievement [150] was the calculation

Theorem 8.6.9 *The ring Ω_*^O is a polynomial ring over $\mathbb{Z}_2$ with one generator in each dimension* not *of the form $2^k - 1$. The bordism class of a manifold M^m*

is determined by the Stiefel–Whitney numbers of M. Moreover M qualifies as a generator if and only if $s_m(w)[M] \neq 0$.

We outline the steps in the proof. Since $M \times I$ can be regarded as a cobordism of the union of two copies of M to the empty set, any element of Ω_*^O has order 2. It thus suffices to perform calculations in mod 2 cohomology.

The next step is to calculate the action of the Steenrod algebra $\mathcal{S}_2$ on $H^*(B(O); \mathbb{Z}_2)$ and hence that on $H^*(\mathbb{TO}; \mathbb{Z}_2)$. This shows that the latter is a free $\mathcal{S}_2$-module, and hence that there is a map $\psi : \mathbb{TO} \to \prod \mathbb{K}(\mathbb{Z}_2, n)$ to a product of Eilenberg–MacLane spectra which induces a cohomology isomorphism, hence is a (stable) homotopy equivalence, so induces an isomorphism of Ω_*^O to a sum of copies of $\mathbb{Z}_2$. This can be formulated as follows. For any M^m and cohomology class $k \in H^m(B(O); \mathbb{Z}_2)$, the classifying map $\phi_M : M \to B(O)$ of τ_M induces $\phi_M^* k \in H^m(M; \mathbb{Z}_2)$ and hence a number $\phi_M^* k[M] \in \mathbb{Z}_2$, called a *Stiefel–Whitney number* of M. The result implies that these numbers determine the class of M in Ω_m^O.

More generally, ψ induces, for any X, an isomorphism $\Omega_*^O(X) \to \prod H_n(X; \mathbb{Z}_2)$: given a map $f : M \to X$ and a cohomology class $k \in H^n(B(O); \mathbb{Z}_2)$, the map ϕ_M induces $\phi_M^* k \in H^n(M; \mathbb{Z}_2)$, hence a dual homology class $[M] \cap \phi_M^* k \in H_{m-n}(M; \mathbb{Z}_2)$ and a class $f_*([M] \cap \phi_M^* k) \in H_{m-n}(X; \mathbb{Z}_2)$. Now the composed map $\Omega_m^O(X) \to \oplus_n H_{m-n}(X; \mathbb{Z}_2)$ is a natural isomorphism.

Further, $H_*(\mathbb{TO}; \mathbb{Z}_2)$ is a free comodule over the dual $\mathcal{S}^2$ of $\mathcal{S}_2$, and this is a polynomial ring with one generator in each dimension of the form $2^k - 1$. Thus

$$P(\mathcal{S}_2)(t) = \prod_k (1 - t^{2^k - 1})^{-1}.$$

It follows that

$$P(\Omega_*^O; \mathbb{Z}_2)(t) = \prod_{i \text{ not of form } 2^n - 1} (1 - t^i)^{-1}.$$

For the multiplicative structure we can argue abstractly using the fact that $H_*(\mathbb{TO}; \mathbb{Z}_2)$ is a polynomial ring, or we can argue as follows.

If M is such that $s_m(w)[M^m] \neq 0$, the class of M in $H_m(\mathbb{TO}; \mathbb{Z}_2)$ is indecomposable, hence so is the class of M in Ω_m^O. If m is even, we can take M as $P^m(\mathbb{R})$; otherwise if $m + 1$ is not a power of 2, write $m + 1 = 2^{r-1}(2s + 1)$ with $s > 0$, then by Corollary 8.6.3 we can take $M = Y^{\mathbb{R}}_{2^{r-1}, 2^r s}$.

Since we have exhibited manifolds M^m with $s_m(w)[M] \neq 0$ for each m not of the form $2^k - 1$, these indecomposables generate a polynomial ring, and the above counting argument shows that this is the whole of Ω_*^O.

Alternative choices are as follows. Write $P(m, n)$ for the bundle over $P^m(\mathbb{R})$ with fibre $P^n(\mathbb{C})$ where the structure group $\mathbb{Z}_2$ acts by complex conjugation.

Lemma 8.6.10 *If $N = 2^{r-1}(2s+1)$ with $s > 0$, set $V^{2N-1} := P(2^r - 1, 2^r s)$. Then $s_{2N-1}(w)[V^{2N-1}] = 1$.*

We omit the proof (an elementary calculation) which, like the construction of the V^{2N-1}, is due to Dold [41].

A similar, but more elaborate argument gives the result in the unitary case, which is due to Milnor [96] and Novikov [113]. Introduce the notation r_n by

$$\begin{aligned} r_n :&= p \text{ if } n \text{ is a power of the prime } p, \\ &= 1 \text{ if } n \text{ is not a prime power.} \end{aligned}$$

Theorem 8.6.11 *The ring Ω_*^U is a polynomial ring with one generator in each even dimension. The bordism class of a manifold M^{2m} with weak* **U***-structure is determined by the Chern numbers of M. Moreover M qualifies as a generator if and only if $s_m(c)[M] = \pm r_{m+1}$.*

The argument includes the same steps, but encounters additional technical difficulties. One must analyse the Steenrod algebras $\mathcal{S}_p$ for each prime p and the corresponding actions on $H^*(B(U); \mathbb{Z}_p)$ and hence on $H^*(\mathbb{TU}; \mathbb{Z}_p)$. This time the modules are not free, since the Bockstein β_p acts trivially, but are free over the quotient $\overline{\mathcal{S}}_p$ of S_p by the ideal generated by β_p. It follows that, for each p, there is a map of $\mathbb{TU}$ to a product of Eilenberg–MacLane spectra $\mathbb{K}(\mathbb{Z}, n)$ which induces an isomorphism of (mod p) cohomology.

For the multiplicative structure we find that $H_*(\mathbb{TU}; \mathbb{Z}_p)$ is a polynomial ring, it is a free comodule over the dual $\overline{\mathcal{S}}^p$ of $\overline{\mathcal{S}}_p$, and that this is a polynomial ring with one generator in each dimension of the form $2(p^k - 1)$.

Additional calculations are needed first, to ensure that we can fit these together for all primes p to obtain a map which is a stable homotopy equivalence, and then to make an analysis of the multiplicative structure.

Again some of this can be bypassed using explicit constructions of manifolds. By Proposition 8.6.2 we have $s_{m+n-1}(Y_{m,n}^{\mathbb{C}}) = -\binom{m+n}{m}$. Thus for manifolds of dimension N we have values of $s_N[M]$ taking all values $-\binom{N+1}{m}$ with $1 \le m \le N$. The highest common factor of these is just r_{N+1}.

8.7 Calculation of Ω_*^{SO} and Ω_*^{SU}

We consider two cases:

$\mathbf{J} = \mathbf{O}$, $\mathbf{SJ} = \mathbf{SO}$, $\mathbf{J}/\mathbf{SJ} = \{\pm 1\}$, $d = 1$, $\mathbb{K} = \mathbb{R}$,

$\mathbf{J} = \mathbf{U}$, $\mathbf{SJ} = \mathbf{SU}$, $\mathbf{J}/\mathbf{SJ} = S^1$, $d = 2$, $\mathbb{K} = \mathbb{C}$;

we will present the two theories in parallel as far as possible. We will omit many details (the account of these results occupies the whole of the memoir

[39] and 140 pp. of Stong's book [147]) but aim to describe all the geometrical ideas involved.

We will focus on geometrical arguments, and begin with certain exact sequences. Some of the arguments will apply to other cases satisfying similar conditions: for example, taking $(\mathbf{J}, \mathbf{SJ})$ to be $(\mathbf{Pin}, \mathbf{Spin})$ or $(\mathbf{Spin^c}, \mathbf{Spin})$ or $(\mathbf{U} \times H, \mathbf{SU} \times H)$ with H compact.

In each case, we have $J_n/SJ_n \cong \mathbf{J}/\mathbf{SJ} \cong S^{d-1}$. We will write P for $P^\infty(\mathbb{K})$ and P^k for $P^k(\mathbb{K})$. We write γ^n for the standard vector bundle over $B(J_n)$ or $B(SJ_n)$ and η for the standard line bundle over P^k. We regard P as the classifying space for the group S^{d-1}, so the map $\mathbf{J} \to \mathbf{J}/\mathbf{SJ} \cong S^{d-1}$ induces $\pi : BJ \to P$.

Lemma 8.3.1 gives us an exact sequence in which the third term is the cobordism group $\Omega_m^{J,SJ}$ of bounded J-manifolds with a weak SJ-structure on the boundary. We first interpret this relative term using Lemma 8.3.3.

Theorem 8.7.1 *We have a natural isomorphism* $\Omega_m^{J,SJ} \cong \Omega_{m-d}^{SJ}(P)$.

Proof This is an instance of the general method of Lemma 8.3.3, but there are many details to clarify.

We will specify the J-structure of a manifold M by the classifying map of its stable normal bundle, $\nu_M : M \to BJ$. We have a fibration $B(SJ) \to B(J) \xrightarrow{\pi} P$, and an SJ-structure of M is determined by a nullhomotopy of $\pi \circ \nu_M$ which is thus covered by a homotopy of ν_M to a map into $B(SJ)$.

The standard line bundle over P has a J-structure, classified by $P \xrightarrow{\eta} BJ_d \xrightarrow{\iota} BJ$; we may assume that $\pi \circ \iota \circ \eta$ is the identity map 1_P of P. The section $\iota \circ \eta$, together with the group action, shows that the fibration $B(SJ) \to B(J) \to P$ is trivial.

Write $(-1)_P : P \to P$ for the negative of the identity: this is given in the real case by the identity, and in the complex case by complex conjugation. Moreover P is an H-space, and the diagram

$$\begin{array}{ccc} BJ \times BJ & \xrightarrow{B\psi} & BJ \\ \downarrow \pi \times \pi & & \downarrow \pi \\ P \times P & \to & P \end{array}$$

is homotopy commutative; we may choose our model of BJ to make it commutative.

Now suppose M^m a J-manifold such that ∂M is an SJ-manifold. Consider the map $\pi_M := \pi \circ \nu_M : M \to P$; up to homotopy, we may suppose that this maps M to a finite dimensional projective subspace P^k. We can make this map smooth and transverse to the submanifold P^{k-1}, whose preimage will then be a smooth submanifold V^{m-d} of M^m, with normal bundle induced from η. As ∂M

has an SJ-structure, $\pi \circ \nu_M$ is trivial on ∂M (which has trivial normal bundle in M), so may be assumed to avoid P^{k-1}. Thus V lies in the interior $\mathring{M}$ of M, and is closed.

The stable normal bundle ν_V of V is the sum of the bundles induced from ν_M and from η. We give the second summand *minus* the obvious structure. So the normal bundle ν_V is induced by

$$V \xrightarrow{\nu_M|V} BJ \xrightarrow{1\times\pi} BJ \times P \xrightarrow{1\times -1} BJ \times P \xrightarrow{1\times\eta} BJ \times BJ \xrightarrow{B\psi} BJ.$$

The composite $\pi \circ \nu_V$ is thus induced by

$$V \xrightarrow{\nu_M|V} BJ \xrightarrow{\pi} P \xrightarrow{(1,-1)} P \times P \to P.$$

Thus a null-homotopy of the composite map $P \to P$ defines one for $\pi \circ \nu_V$, and hence an SJ-structure for V.

Now choose a tubular neighbourhood W of V in M: this is a bundle over V, with fibre D^d, associated to $(\pi \circ \nu|V)^*\eta$. It follows as in Lemma 8.3.3 that $(M, \partial M)$ is (J, SJ)-cobordant to $(W, \partial W)$. We need to verify that the SJ-structure on ∂M extends to $(M \setminus \mathring{W})$: this follows since $\pi \circ \nu_M$ takes $(M \setminus \mathring{W})$ to the contractible set $(P^k \setminus P^{k-1})$.

Thus the (J, SJ)-cobordism class of $(M, \partial M)$ agrees with that of $(W, \partial W)$, hence is determined by the class of $(V, \pi \circ \nu|V)$ in $\Omega_{m-d}^{SJ}(P)$. The formula which determines it is as follows. Let η' be the bundle induced from η. Then $\nu_V = \nu_M + \bar{\eta}'$, where the bar recalls the sign change above. Thus $\nu_V + \eta' = \nu_M + \bar{\eta}' + \eta' = \nu_M + \varepsilon^2$.

Conversely, given any element of $\Omega_{m-d}^{SJ}(P)$, represented say by (V, f), we can take the bundle E with fibre D^d associated to $f^*\eta$ and give it a J-structure. The stable normal bundle $\nu_{\partial E}$ of the boundary ∂E is the restriction of ν_E. But $\pi \circ \nu_E$ is essentially f, by definition, and is covered by a bundle map over V of E to the disc bundle $D(\eta)$ associated to η, and hence of ∂E to the corresponding sphere bundle $S(\eta)$. But $S(\eta)$ is contractible, so we have a null-homotopy of $\partial E \to S(\eta) \to P$, and so an SJ-structure on ∂E. Since all our constructions carry over to cobordisms, we have indeed an isomorphism $\Omega_m^{J,SJ} \cong \Omega_{m-d}^{SJ}(P)$. □

We remarked above that for cobordism theory, the extra structure provided by a submanifold is equivalent to the extra structure provided by a map to its Thom space. Moreover P is homeomorphic to the Thom space of η. This leads to

Theorem 8.7.2 *We have a natural isomorphism* $\Omega_{m-d}^{SJ}(P) \cong \Omega_{m-d}^{SJ} \oplus \Omega_{m-2d}^{J}$.

Proof Given an SJ-manifold V^{m-d} and a map $\chi_V : V \to P$, we may suppose χ_V maps V into P^{k-1}. We make this transverse to P^{k-2}, and write $B = \chi_V^{-1}(P^{k-2})$.

Then $\nu_B = \nu_V|B + (\chi_V|B)^*\eta$, and we use this formula to give B a J-structure. Since the cobordism class of (V, f) determines the cobordism classes of V and B, we have a homomorphism $\Omega^{SJ}_{m-d}(P) \to \Omega^{SJ}_{m-d} \oplus \Omega^{J}_{m-2d}$.

Conversely, the class of (V, f) is determined by those of V, B, and the map $B \to P$ inducing the normal bundle of B in V. By Lemma 8.3.5 we can separate the contributions of B and V, provided the stable normal bundle of B is induced by $B \to B(SJ) \times P$. But since the fibration $B(SJ) \to B(J) \to P$ is trivial, $B(SJ) \times P$ is homotopy equivalent to $B(J)$. □

By Lemma 8.3.1 we have an exact sequence

$$\ldots \Omega^{SJ}_m \to \Omega^{J}_m \to \Omega^{J,SJ}_m \to \Omega^{SJ}_{m-1} \cdots$$

and combining Theorem 8.7.1 and Theorem 8.7.2 gives a natural isomorphism $\Omega^{J,SJ}_m \cong \Omega^{SJ}_{m-d} \oplus \Omega^{J}_{m-2d}$. We now study the maps in the sequence obtained by making this substitution.

Theorem 8.7.3 *There is an exact sequence*

$$\Omega^{SJ}_n \xrightarrow{i_S} \Omega^{J}_n \xrightarrow{(d_1,d_2)} \Omega^{SJ}_{n-d} \oplus \Omega^{J}_{n-2d} \xrightarrow{\binom{\times\alpha}{0}} \Omega^{SJ}_{n-1} \to \cdots$$

where α is the class of S^{d-1} with a twisted framing. Also, there exists s_2 : $\Omega^{J}_{n-2d} \to \Omega^{J}_n$ with $(d_1, d_2) \circ s_2 = (0, 1)$.

Proof Write (d_1, d_2) for the components of the map $\Omega^{J}_m \to \Omega^{SJ}_{m-d} \oplus \Omega^{J}_{m-2d}$, so that the image of the class of M by d_1 (resp. d_2) is determined by V (resp. B) in the notation above; also write (q_1, q_2) for the components of the map $\Omega^{SJ}_{m-d} \oplus \Omega^{J}_{m-2d} \to \Omega^{SJ}_{m-1}$.

As to q_1, we can suppose B empty and χ_V trivial. Then the disc bundle defining the class in $\Omega^{J,SJ}_m$ is trivial, and has boundary $V \times S^{d-1}$. Since we have a product bundle, we obtain multiplication by the class, α say, of S^{d-1} with appropriate SJ-structure. To determine this, we can take V to be a point and M a disc D^d. Recall that V was constructed from M by making π_M transverse to P^{k-1}. Now π_M maps $\partial M = S^{d-1}$ to a point, so induces a map of $S^d = M/\partial M$ which meets P^{k-1} transversely in just one point. This coincides (up to homotopy) with the inclusion of a projective line P^1. So α is the class of S^{d-1}, with SJ-structure defined by a framing of the normal bundle, twisted in this way.

We now construct a map $s_2 : \Omega^{J}_{m-2d} \to \Omega^{J}_m$ and show that $d_1 \circ s_2 = 0$ and $d_2 \circ s_2 = id$. From this, and the exactness of the sequence it follows that $q_2 = 0$.

To define s_2, suppose that B^{m-2d} is a J-manifold, and form $(\pi \circ \nu_B)$, which we may suppose a map $B \to P^k$ for some k. Write Q_{k+2} for the projective bundle over P^k associated to $\eta \oplus \varepsilon^2$. Let M^m be the induced bundle over B, V^{m-d} the sub-bundle corresponding to $\eta \oplus \varepsilon^1$, and identify B itself with the sub-bundle of V corresponding to η.

Write f for the map $M \to B$ and g for the composite map $M \to B \to P^k$. Then there is a natural splitting $\tau_M \oplus g^*\eta \cong f^*\tau_B \oplus g^*(\eta \oplus \varepsilon^1) \oplus (-1 \circ g)^*\varepsilon^1$. We give all these bundles the induced J-structures. Since the construction passes to cobordisms, we have a well-defined map of cobordism classes with $s_2(B) := (M)$.

Consider the decomposition $\mathbb{K}^{k+3} = \mathbb{K}^{k+1} \oplus \mathbb{K}^2$. Then P^k is the projective space of $\mathbb{K}^{k+1}$, so a point $x \in P^k$ corresponds to a line $\ell_x \subset \mathbb{K}^{k+1}$ which we can identify with the fibre of η over x. We now identify the fibre of $\eta \oplus \varepsilon^2$ over x with $\ell_x \oplus \mathbb{K}^2$, and so Q_{k+2} with the subspace of $P^k \times P^{k+2}$ of pairs (x, y) with $\ell_y \subset \ell_x \oplus \mathbb{K}^2$.

Now define $\zeta : Q_{k+2} \to P^{k+2}$ to be the map induced by projection on P^{k+2}. Then $\zeta^{-1}(P^{k+1})$ is the set of pairs (x, y) with $\ell_y \subset \ell_x \oplus \mathbb{K} \oplus 0$ and $\zeta^{-1}(P^k)$ is the set of pairs (x, x) with $x \in P^k$. Thus $\zeta : Q_{k+2} \to P^{k+2}$ is transverse to P^{k+1} and P^k, and these have preimages the sub-bundles associated to $\eta \oplus \varepsilon^1$ and η.

We claim that $\zeta \circ \beta \simeq \pi \circ \nu_M$. Since the target of these maps is the Eilenberg–MacLane space P, this only needs checking on the level of the cohomology class. It follows that this map is transverse to P^{k+1} and P^k, and these have preimages V and B. Hence we have $d_2 \circ s_2 = id$.

To see $d_1 \circ s_2 = 0$, we must find an SJ-manifold with boundary V. But V is a $P^1(= S^d)$-bundle over B, with structure group $Z(= S^{d-1})$, so bounds the associated disc bundle, which is topologically the product by I of the mapping cylinder of the principal bundle. Since the principal bundle was obtained from $\pi \circ \nu_B$, this has an SJ-structure. □

When $d = 1$ we have $S^{d-1} = S^0$, but each point has the positive orientation: this twists the standard framing of ∂D^1 by changing a sign. In this case $J = O$ and the map $\Omega_{m-1}^{SO} \to \Omega_{m-1}^{SO}$ is multiplication by 2. If $d = 2$, we have $S^{d-1} = S^1$, and the twisted framing differs from the standard one. Here (see §B.3(x)) homotopy theory tells us that $\alpha \in \pi_1^S$ is the non-zero element η_2, and $2\eta_2 = 0$.

We now define **RJ** as the stable group given by the pullback diagram

$$\begin{array}{ccccc} B(SJ) & \to & B(RJ) & \to & B(J) \\ \downarrow & & \downarrow & & \downarrow \\ P^0 & \to & P^1 & \to & P \end{array} .$$

Proposition 8.7.4 *(i) There is a split short exact sequence*

$$0 \to \Omega_n^{RJ} \xrightarrow{i_R} \Omega_n^J \xrightarrow{d_2} \Omega_{n-2d}^J \to 0 \tag{8.7.5}$$

split by a map $s_0 : \Omega_n^J \to \Omega_n^{RJ}$ *with* $i_R \circ s_0 = 1$.

(ii) The following sequence also is exact:

$$\Omega_n^{SJ} \xrightarrow{i_{SR}} \Omega_n^{RJ} \xrightarrow{d_1} \Omega_{n-d}^{SJ} \xrightarrow{\times\alpha} \Omega_{n-1}^{SJ} \to \cdots \tag{8.7.6}$$

Here i_{SR}, i_R and $i_S = i_R \circ i_{SR}$ are the maps induced by the natural inclusions $SJ \subset RJ \subset J$.

Proof (i) It follows from the definition of RJ that $d_2 \circ i_R = 0$, and we have already proved d_2 surjective. For exactness at Ω_J^n, suppose M defines an element of $\mathrm{Ker} d_2$. Thus we may take $\pi \circ \nu_M$ as a map to P^k, make it transverse to P^{k-2}, and write B for the preimage: then by hypothesis B is cobordant to the empty set. As in Lemma 8.3.5 we may extend this cobordism to one of $(M, \pi \circ \nu_M)$, and thus suppose $\pi \circ \nu_M$ a map to $P^k \setminus P^{k-2}$. But this is homotopic to a map into P^1. Thus M defines a class in Ω_n^{RJ}.

It remains to define s_0 and prove $i_R \circ s_0$ the identity. We begin as usual with $M \xrightarrow{\nu_M} BJ \xrightarrow{\pi} P$; again as usual we may replace the target by P^k. We thus have a map $M \times P^1 \to P^k \times P^1 \to P^{2k+1}$, where the final map is the Veronese embedding

$$((x_0, \ldots, x_k), (y_0, y_1)) \to (x_0y_0, \ldots, x_ky_0, x_0y_1, \ldots, x_ky_1).$$

The composite is transverse to a generic linear subspace L, say given by $x_0y_1 = x_1y_0$, of P^{2k} (at a point where transversality failed we would have $y_0 = y_1 = 0$), and we define $s_0[M]$ to be the class of the preimage M' of L.

Adapting the above proof that $\pi \circ \nu_V$ is nullhomotopic shows in this case (where an extra factor P^1 appears) that $\pi \circ \nu_{M'}$ is homotopic to a map to P^1, so M' defines a class in Ω_n^{RJ}. Finally, if M itself defines such a class, we may take $k = 1$ above, so that the projection of M' on M is a diffeomorphism. □

Since $d_2 \circ s_2 = i$, $1 - s_2d_2$ retracts Ω_n^J on the kernel of d_2, which we can now identify with Ω_n^{RJ}. We denote this map by $\rho : \Omega_n^J \to \Omega_n^{RJ}$.

There are alternative presentations of the above material. One can *define* Ω_n^{RJ} as the kernel of d_2. One can also show (cf. Theorem 8.7.1) that there is a natural isomorphism $\Omega_m^{J,RJ} \cong \Omega_{m-2d}^J$. It can also be shown that $\Omega_n^{RJ} \cong \bar{\Omega}_{n+d}^{SJ}(P^2)$.

We observe that $\Omega_*^J(P)$ is a free Ω_*^J-module with base the classes x_j defined by the inclusions of P^j in P.

We define a module endomorphism Δ of $\Omega_*^J(P)$ as follows. Given a class represented by $f : M \to P^k \subset P$, we make f transverse to P^{k-1}, set $L := f^{-1}(P^{k-1})$ and define $\Delta(M, f) := (L, f|L)$. It follows that $\Delta(x_j) = x_{j-1}$.

Write $\varepsilon : \Omega_*^J(P) \to \Omega_*^J$ for the augmentation and $\mu : \Omega_*^J \to \Omega_*^J(P)$ for the map sending $[M]$ to the class of $(M, \pi \circ \nu_M)$. The map $P \times P \to P$ which classifies the tensor product of line bundles induces a multiplication in $\Omega_*^J(P)$ with respect to which μ is a ring homomorphism.

We observe that $\varepsilon \circ \Delta \circ \mu \circ i_R = i_S \circ d_1$ and that $\varepsilon \circ \Delta^2 \circ \mu = d_2$. The retraction s_0 is given in this notation by $s_0(z) = \varepsilon\Delta(\mu(z).x_1)$ for any $z \in \Omega_*^J$.

We find that Ω_*^{RJ} is a subring of Ω_*^J if $J = O$ but not if $J = U$, so we define a multiplication on Ω_*^{RJ} by

$$x * y := s_0(xy).$$

We also define

$$\partial := i_{SR} \circ d_1 : \Omega_*^{RJ} \to \Omega_*^{RJ}.$$

Since $d_1 \circ i_{SR} = 0$ in the exact sequence (8.7.6), we have $\partial^2 = 0$.

For any class $a \in \Omega_*^{RJ}$, $\mu(a)$ is in the image of $\Omega_*^J(P^1)$, so can be written $\mu(a) = \alpha x_0 + \alpha' x_1$ with $\alpha,\ \alpha' \in \Omega_*^J$. Then $a = \varepsilon\mu(a) = \alpha + \alpha'\varepsilon(x_1)$ and we have $\partial a = \varepsilon\Delta\mu a = \varepsilon(\alpha' x_0) = \alpha'\varepsilon(x_0) = \alpha'$.

Lemma 8.7.7 *(i)* $a * b = a.b + 2w_2.\partial a.\partial b$.
(ii) $\partial(a.b) = a.\partial b + \partial a.b - \varepsilon(x_1).\partial a.\partial b$.
(iii) $\partial(x.\partial(y)) = \partial x.\partial y$.
(iv) $\partial(a * b) = a.\partial b + \partial a.b + w_1.\partial a.\partial b$.

Proof (i) Let $a,\ b \in \Omega_*^{RJ}$, and write $\mu(a) = \alpha x_0 + \alpha' x_1$, $\mu(b) = \beta x_0 + \beta' x_1$, so $s_0(a.b) = \varepsilon\Delta(\alpha\beta x_1 + (\alpha\beta' + \alpha'\beta)x_1^2 + \alpha'\beta' x_1^3)$. Calculations give $\varepsilon\Delta(x_1) = \varepsilon(x_0)$, $\varepsilon\Delta(x_1^2) = \varepsilon(x_1)$ and $\varepsilon\Delta(x_1^3) = 3\varepsilon(x_1^2) - 2\varepsilon(x_2)$; thus

$$\begin{aligned} s_0(a.b) &= \alpha\beta\varepsilon(x_0) + (\alpha\beta' + \alpha'\beta)\varepsilon(x_1) + \alpha'\beta'(3\varepsilon(x_1^2) - 2\varepsilon(x_2)) \\ &= (\alpha + \alpha'\varepsilon(x_1))(\beta + \beta'\varepsilon(x_1)) + 2\alpha'\beta'(\varepsilon(x_1^2) - \varepsilon(x_2)) \\ &= a.b + 2\partial a.\partial b.(\varepsilon(x_1^2) - \varepsilon(x_2)). \end{aligned}$$

(ii) With $a,\ b$ as above, we have

$$\begin{aligned} \partial(a.b) &= \varepsilon\Delta\mu(a.b) \\ &= \varepsilon\Delta(\alpha\beta + (\alpha\beta' + \alpha'\beta)x_1 + \alpha'\beta' x_1^2) \\ &= \varepsilon((\alpha\beta' + \alpha'\beta)x_0 + \alpha'\beta' x_1) \\ &= (\alpha + \alpha'\varepsilon(x_1))\beta' + \alpha'(\beta + \beta'\varepsilon(x_1)) - \alpha'\beta'\varepsilon(x_1) \\ &= a.\partial b + \partial a.b - (\varepsilon(x_1)).\partial a.\partial b. \end{aligned}$$

(iii) follows from (ii) since $\partial^2 = 0$. (iv) now follows from (i)–(iii). □

Here w_2 is given by $\varepsilon(x_1^2) - \varepsilon(x_2)$, and $w_1 = 2\partial(w_2) - \varepsilon(x_1)$. If $d = 2$ we have $w_2 := [(P^1)^2] - [P^2]$.

Since $\partial^2 = 0$, $(\Omega_*^{RJ}, \partial)$ defines a chain complex: denote its homology by H_*^J. Explicitly,

$$H_n^J := \frac{\operatorname{Ker}(\partial : \Omega_n^{RJ} \to \Omega_{n-d}^{RJ})}{\operatorname{Im}(\partial : \Omega_{n+d}^{RJ} \to \Omega_n^{RJ})}.$$

Although, in the case $d = 2$, ∂ is not a derivation, it follows from Lemma 8.7.7 that Ker ∂ is a subring of Ω_*^{RU} and that *Im* ∂ is an ideal in it, so that the quotient H_*^J is a ring.

We next want an exact sequence derived from (8.7.6). The general procedure, due to Massey [87], is as follows. Suppose given an exact sequence

$$\ldots P \xrightarrow{a} P \xrightarrow{b} Q \xrightarrow{c} P \xrightarrow{a} P \ldots$$

(he calls this an exact couple). Then $d := b \circ c$ has $d^2 = 0$ since $c \circ b = 0$, so we can form the homology H of Q with respect to d. Set $A := a(P) \subset P$.

Lemma 8.7.8 *There is an exact sequence*

$$\ldots A \xrightarrow{a_1} A \xrightarrow{b_1} H \xrightarrow{c_1} A \ldots$$

If a, b, c have respective degrees d_a, d_b, d_c then a_1, b_1, c_1 have degrees d_a, $d_b - d_a$, d_c.

Observe that this gives another exact couple, called the derived couple.

Proof Define a_1 as the restriction of a. For $y = a(x) \in A$ define $b_1(y)$ as the class of $b(x)$: we have $db(x) = bcb(x) = 0$ so do get a class in H. And for an element $\zeta \in H$ represented by $z \in Q$ with $bc(z) = 0$ define $c_1(\zeta) = c(z)$: this is indeed in Ker$(b) =$ Im(a). Any other representative is of form $z + bc(w)$ and $c(z + bc(w)) = c(z)$.

Composites vanish since $b_1(a_1(ax))$ is the class of $b(a(x)) = 0$; $c_1(b_1(ax))$ is represented by $c(b(x)) = 0$; and $a_1(c_1(\zeta)) = a(c(z)) = 0$.

If $y = a(w)$ and $b_1(y) = 0$, then $b(w) = d(x) = b(c(x))$ for some x so $w - c(x) \in$ Ker$(b) =$ Im(a); $w = c(x) + a(v)$ so $y = a^2(v) \in$ Im(a_1). If $c_1(\zeta) = 0$, then $z \in$ Ker$(c) =$ Im(b): set $z = b(y)$: then $\zeta = b_1(ay)$. If $a_1(x) = 0$, then $x \in$ Ker$(a) =$ Im(c). This proves exactness. □

Write A_n^J for the image of $\theta : \Omega_{n-d+1}^{SJ} \to \Omega_n^{SJ}$. Applying Lemma 8.7.8 to (8.7.6) gives the exact sequence

$$\ldots \to A_n^J \to A_{n+d-1}^J \to H_n^J \to A_{n-d}^J \to A_{n-1}^J \to \ldots \qquad (8.7.9)$$

As in the preceding section, the completion of the calculation of the cobordism rings depends on exhibiting particular examples. We will give these, but omit the detailed calculations, some of which yield

Lemma 8.7.10 *We have* $s_{n+2}(s_2(M^{d(n+2)})) \equiv (n+1)c_1^n[M^{dn}]$ *(mod 2)* $c_1^{n+2}[s_2(M^{dn})] = -c_1^n[M^{dn}]$.

As in the proof of Theorem 8.6.9, for n not a power of 2, choose integers r, s with $r+s=n$ and $\binom{n}{r}$ odd: for example, write $n=2^p(2q+1)$ (so $q\geq 1$) and set $r=2^{p+1}q$, $s=2^p$. Define cobordism classes by

$z_{2dn} := \rho(P^{2r}\times P^{2s}) \in \Omega_{2dn}^{RJ}$,

$z_{2dn-d} := d_1(P^{2r}\times P^{2s}) \in \Omega_{2dn-d}^{SJ}$.

It follows using Lemma 8.7.10 that $s_m(z_{dm})$ is odd in both cases, and from the formulae relating the maps that $d_1 z_{2dn} = z_{2dn-d}$.

In case $n=2^j>1$ is a power of 2, first set $z_{2dn} := P^n \times P^n$. In the case $d=1$, since $P^n(\mathbb{R})\times P^n(\mathbb{R}) \sim P^n(\mathbb{C})$, which is orientable, this gives $d_1 z_{2n} = 0$. If $d=2$ we set $z_{2dn-d} := d_1(P^n\times P^n)$.

In the preceding section we defined Poincaré series and calculated the series for Ω_*^U and Ω_*^{SO} over $\mathbb{Q}$ and for Ω_*^O over $\mathbb{Z}_2$. It now follows from the exact sequence (8.7.5) that

$$P(\Omega_*^{RO};\mathbb{Z}_2)(t) = (1-t^2)\prod_{i \text{ not of form } 2^n-1}(1-t^i)^{-1}.$$

Theorem 8.7.11 *(i) $\Omega_*^{SO}/Tors$ is a polynomial ring.*

(ii) All torsion in Ω_*^{SO} has order 2.

(iii) Classes in Ω_*^{SO} are detected by Stiefel–Whitney numbers and Pontrjagin numbers.

(iv) The image of Ω_*^{SO} in $\Omega_*^{RO}\subset\Omega_*^O$ is Ker ∂; the image of Tors Ω_*^{SO} is Im ∂.

A presentation of Ω_*^{SO} by generators and relations is not convenient: (iv) gives a better description.

Proof (i) By Proposition 8.6.4, $\Omega_*^{SO}\otimes\mathbb{Q}$ is a polynomial ring, with one generator in each dimension divisible by 4. It follows (again from [96] or [113]) that $\Omega_*^{SO}\otimes\mathbb{Z}[\frac{1}{2}]$ is a polynomial ring. We next claim that Ω_*^{SO}/Tors is a polynomial ring: it suffices to observe that since by Lemma 8.6.5 $s_n(p)[P^{2n}(\mathbb{C})] = 2n+1\neq 0$, so the classes of the $P^{2n}(\mathbb{C})$ are polynomial generators of $\Omega_*^{SO}\otimes\mathbb{Q}$ and since also these numbers are odd, their images in Ω_*^O generate a polynomial algebra.

Next observe that Ω_*^{RO} is a subring of Ω_*^O. This follows from Lemma 8.7.7 (i) and the fact that Ω_*^O has exponent 2, or more simply from the fact that $P^1=S^1$ can be regarded as a subgroup of P in this case. Next, the map ∂ is a derivation: this follows from (ii) of the same Lemma, and the fact that $\epsilon(x_1)$ is a class of dimension 1, hence is zero as $\Omega_1^O=0$.

We defined classes $z_n\in\Omega_n^{RO}$ above, for n, $n+1$ not powers of 2, and showed that in each case, $s_n(z_n)=1$. If $n=2^j\geq 2$, the class x_n of $P^n(\mathbb{R})$ has $s_n(x_n)=1$ by Proposition 8.6.3 above. Hence Ω_*^O is the polynomial ring in these generators. Now $P^{2^i}(\mathbb{R})^2$ is cobordant to $P^{2^i}(\mathbb{C})$, which is orientable. The classes z_n

and x_n^2 thus all belong to Ω_*^{RO} and generate a polynomial ring. Since this subring has the same Poincaré polynomial as Ω_*^{RO}, it is the whole ring.

We have now calculated the derivation ∂ on all generators of Ω_*^{RO}, for as $P^{2^i}(\mathbb{C})$ is orientable, $\partial(x_n^2) = 0$. Since we can regard Ω_*^{RO} as the tensor product of the algebras $\mathbb{Z}_2[z_m, z_{m+1}]$ with $m+1 = 2^p(2q+1)$, $p \geq 1$, $q \geq 1$ and $\mathbb{Z}_2[x_n^2]$, the ring $H_*(\Omega_*^{RO}; \partial)$ is the tensor product of the homologies of these subalgebras, which is the polynomial algebra in the z_{m+1}^2 and x_n^2. Thus $H_*^O := H_*(\Omega_*^{RO}; \partial)$ is a polynomial ring over $\mathbb{Z}_2$, with one generator in each dimension divisible by 4.

We now recall the exact sequence given by (8.7.9) with $J = O$:

$$\ldots \to A_n^O \xrightarrow{a_1} A_n^O \xrightarrow{b_1} H_n^O \xrightarrow{c_1} A_{n-1}^O \to A_{n-1}^O \to \ldots$$

Since a_1 is induced from α it is multiplication by 2. Thus $A_n^O = 2\Omega_n^{SO}$. The torsion-free rank of A_{4n}^O is equal to the number $p(n)$ of partitions of n, so the image of b_1, isomorphic to $A_{4n}^O/2A_{4n}^O$ has rank over $\mathbb{Z}_2$ at least this. Hence b_1 is surjective in these, hence in all degrees. By exactness, the kernel of multiplication by 2 on A_n^O vanishes. Thus A_n^O is torsion-free. This proves (ii).

By (8.7.6) the kernel of $\Omega^{SO} \to \Omega^{RO} \subset \Omega^O$ is the image of multiplication by 2, so is torsion-free. An element on which all Stiefel–Whitney numbers vanish is in this kernel, hence of infinite order, hence by Proposition 8.6.4 is detected by Pontrjagin numbers; thus (iii) holds.

The first assertion of (iv) follows from Proposition 8.7.4 (ii); the second now follows from the above calculation that $\Omega_*^{SO}/\text{Tors}$ and Ker $\partial/Im\,\partial$ have the same Poincaré polynomial. □

Calculations in homology lead to the further results, completing the above.

Lemma 8.7.12 *(i) As $\mathcal{S}_2$-module, $H^*(\mathbb{TSO}; \mathbb{Z}_2)$ is the direct sum of a free module and copies of $\mathcal{S}_2/Sq^1.\mathcal{S}_2$.*

(ii) The spectrum $\mathbb{TSO}$ is homotopy equivalent to a wedge of spectra $\mathbb{K}(\mathbb{Z}_2, n)$ and $\mathbb{K}(\mathbb{Z}, n)$.

(iii) An element of Ω_*^O is in the image of Ω_*^{SO} (resp. of Ω_*^{RO}) if and only if all Stiefel–Whitney numbers with w_1 (resp. w_1^2) as a factor vanish.

We turn to Ω_*^{SU}. It again follows from Proposition B.4.1 that

$$P(\Omega_*^{RU}; \mathbb{Q})(t) = P(H_*(B(RU) : \mathbb{Q}))(t) = (1-t^2)^{-1}\prod_{i=3}^{\infty}(1-t^{2i})^{-1},$$

$$P(\Omega_*^{SU}; \mathbb{Q})(t) = P(H_*(B(SU) : \mathbb{Q}))(t) = \prod_{i=2}^{\infty}(1-t^{2i})^{-1}.$$

Since Ω_*^{RU} is a direct summand of Ω_*^U, it is torsion-free. Write $\overline{\Omega}_*^{SU}$ for the pure subgroup $(\Omega_*^{SU} \otimes \mathbb{Q}) \cap \Omega_*^U$ generated by Ω_*^{SU}. It follows by comparing Poincaré series that an element of Ω_*^U is in the image of $\overline{\Omega}_*^{SU}$ (resp. of Ω_*^{RU}) if and only if all Chern numbers with c_1 (resp. c_1^2) as a factor vanish.

Calculations at odd primes were made by Novikov [113]. The following is analogous to Theorem 8.6.11 for Ω_*^U.

Theorem 8.7.13 *The ring $\Omega_*^{SU} \otimes \mathbb{Z}[\frac{1}{2}]$ is a polynomial algebra with one generator in each even dimension $\neq 2$. The class of a manifold M^{2m} is determined by Chern numbers, and M^{2m} qualifies as a generator if and only if $s_m(c)[M]$ is $\pm r_m r_{m+1}$ times a power of 2.*

The structure at the prime 2 is not simple: a precise description of the torsion-free quotient is given by [147, p. 265]. The torsion subgroup is described by

Theorem 8.7.14 *(i) All torsion in Ω_*^{SU} has order 2. We have*

$$P(Tors\ \Omega_*^{SU}; \mathbb{Z}_2)(t) = (t + t^2) \prod_{i=1}^{\infty} (1 - t^{8i})^{-1}.$$

(ii) The image of $\Omega_{2j}^{SU} \to \Omega_{2j}^{RU}$ is $Ker\, \partial$ if $2j \not\equiv 4 \pmod 8$ and is $Im\, \partial$ if $2j \equiv 4 \pmod 8$.

Proof We outline the main arguments involved in the proof. First recall that by Proposition 8.7.4, Ω_*^{RU} maps injectively to Ω_*^U, so is torsion-free.

Since $2\alpha = 0$, in the sequence $\Omega_*^{SU} \xrightarrow{\alpha} \Omega_*^{SU} \to \Omega_*^{RU}$ the image of the first map has exponent 2; the quotient by it embeds in a free group, so there is no further torsion. More precisely, as Ω_*^{RU} vanishes in odd dimensions, the sequence (8.7.6) reduces to

$$0 \to \Omega_{2k-1}^{SU} \xrightarrow{\alpha} \Omega_{2k}^{SU} \to \Omega_{2k}^{RU} \to \Omega_{2k-2}^{SU} \xrightarrow{\alpha} \Omega_{2k-1}^{SU} \to 0. \qquad (8.7.15)$$

To calculate the 2-torsion, we again use the derived couple (8.7.9): here $d = 2, J = U$, so we have

$$\ldots \to A_n^U \to A_{n+1}^U \to H_n^U \to A_{n-2}^U \to A_{n-1}^U \to \ldots$$

Also, $A_n^U := \theta(\Omega_{n-1}^{SU}) \subset \Omega_n^{SU}$ has exponent 2. Since Ω_*^{RU} vanishes in odd dimensions, so does H_*^U: it follows that the map $\alpha : A_{2k-1}^U \to A_{2k}^U$ is an isomorphism. It thus follows from (8.7.15) that $A_{2k}^U = \text{Tors}(\Omega_{2k}^{SU})$ and $A_{2k-1}^U \cong \Omega_{2k-1}^{SU}$.

We now claim that the map $\alpha : A_{2k-2}^U \to A_{2k-1}^U$ vanishes. For $A_{2k-2}^U = \theta(\Omega_{2k-3}^{SU})$ and Ω_{2k-3}^{SU} is in the image of the map θ given by multiplication by η_2. Thus the image of α is contained in the image of θ^3. However by §B.3(x), we have $\eta_2^3 = 0$.

The sequence (8.7.9) thus reduces to split short exact sequences

$$0 \to A^U_{2k+1} \to H^U_{2k} \to A^U_{2k-2} \to 0. \qquad (8.7.16)$$

To calculate H^U_* we use another sequence. Since Ω^{RU}_* is torsion-free we have a short exact sequence $0 \to \Omega^{RU}_* \xrightarrow{2} \Omega^{RU}_* \to \Omega^{RU}_* \otimes \mathbb{Z}_2 \to 0$, which we regard as an exact sequence of chain complexes with ∂ as differential. There is thus an exact homology sequence, which we denote

$$\ldots \to H^U_* \xrightarrow{2} H^U_* \to H^V_* \xrightarrow{\partial} H^U_* \to \ldots.$$

Now the groups H^U_n have exponent 2 since, for each RU-manifold M, we have $\partial[P^1(\mathbb{C}) \times M] = 2[M]$. Thus the map $2 : H^U_* \to H^U_*$ is zero. The groups H^U_n and H^V_n vanish in odd degrees, and ∂ has degree -2, so the sequence reduces to

$$0 \to H^U_{2k} \to H^V_{2k} \to H^U_{2k-2} \to 0. \qquad (8.7.17)$$

We next compute H^V_*. We think of Ω^{RU}_* as a polynomial subring of Ω^U_*, and ∂ as a derivation: a correct formulation is given in Lemma 8.7.7.

We have defined elements $z_{2n} \in \Omega^{RU}_{2n}$ for n, $n+1$ not powers of 2 such that $s_{2n}(z_{2n})$ is odd. For $m = 2^j \geq 2$, define

$x_{4m} := \rho(P^m(\mathbb{C}) \times P^m(\mathbb{C}))$;

$x_{4m-2} := d_1(P^m(\mathbb{C}) \times P^m(\mathbb{C}))$.

Calculations similar to those in the preceding case yield

$s_{m,m}(c)(x_{4m}) \equiv 1 \pmod 2$ (the class s_{2m} does not suffice here), and

$s_{2m-1}(c)(x_{4m-2}) \equiv 2 \pmod 4$.

It follows that the z_{2n}, x_{4m-2} and x_{4m} give polynomial generators of $\Omega^U_* \otimes \mathbb{Z}[\frac{1}{2}]$ and $\Omega^U_* \otimes \mathbb{Z}_2$ in all dimensions except 2 and 4. Since all are in Ω^{RU}_*, we only need to add the class x_2 of $P^1(\mathbb{C})$ to obtain a complete set of generators of $\Omega^{RU}_* \otimes \mathbb{Z}_2$.

We have $\partial z_{4n} = z_{4n-2}$, $\partial x_{4m} = x_{4m-2}$ and $\partial x_2 = 0$, so if ∂ were a true derivation, we would have H^V_* polynomial with generators x_2 and z^2_{4n} in each dimension divisible by 8. In fact it follows from Lemma 8.7.7 that the elements $h_2 = z_2$ and $h_{8n} := z^2_{4n} + z_2 z_{4n-2} z_{4n}$ $(n \geq 2)$ are cycles. It follows that H^V_* is a polynomial algebra with their classes as generators.

Further calculations using Lemma 8.7.7 exhibit elements of H^U_* mapping to h_2^2 and the h_{8k}. Thus $H^U_{4n} \to H^V_{4n}$ is surjective, so by the exact sequence (8.7.17), H^U_{4n-2} vanishes and the maps $H^V_{4n+2} \to H^U_{4n} \to H^V_{4n}$ are isomorphisms; moreover, H^U_* is a polynomial algebra with the classes of h_2^2 and the h_{8k} as generators. The Poincaré series of H^U_* is thus given by $P(H^U_*; t) = (1-t^4)^{-1} \prod_{k=2}^{\infty}(1-t^{8k})^{-1}$.

Since H_n^U vanishes unless n is divisible by 4, $A_n^U = 0$ unless $n \equiv 1$ or $n \equiv 2$ (mod 4). It now follows from the exact sequence (8.7.16) and the isomorphism $A_{2k-1}^U \to A_{2k}^U$ that if $PB(t)$ denotes the Poincaré series of the even part of A_*^U, then $P(H_*^U; t) = (t^2 + t^{-2})PB(t)$; thus $PB(t) = t^2 \prod_{k=1}^{\infty}(1 - t^{8k})^{-1}$. Hence $P(A_*^U; t) = (t + t^2) \prod_{k=1}^{\infty}(1 - t^{8k})^{-1}$,and the rank of Tors Ω_n^{SU} is as stated. □

Further calculations yield more detailed results.

Theorem 8.7.18 [9] *(i) Write* $\mathcal{S}_2^* := \mathcal{S}_2/\langle Sq^1\rangle$. *Then as* $\mathcal{S}_2$*-module,* $H^*(\mathbb{TSU}; \mathbb{Z}_2)$ *is a sum of copies of* $\mathcal{S}_2^*$ *and* $\mathcal{S}_2^*/\mathcal{S}_2^*.Sq^2$.

(ii) The spectrum $\mathbb{TSU}$ *is homotopy equivalent to a wedge of copies of spectra* $\mathbb{K}(\mathbb{Z}, n)$ *and spectra* $\mathbb{BO}\langle k\rangle$.

(iii) An SU-manifold bounds in Ω_*^{SU} *if and only if all its Chern numbers and KO characteristic numbers vanish.*

8.8 Groups of knots and homotopy spheres

We first consider k-connected cobordism, where the manifolds M and cobordisms W are to be k-connected for some integer $k \geqq 1$. In this case, M is orientable: we make the further convention that M is oriented.

Since the set of k-connected manifolds is not closed under disjoint union, we define an addition on the set of cobordism classes using connected sum. We remark that in general, the disjoint union and connected sum of two manifolds are cobordant: a cobordism of $M \cup M'$ to $M \,\sharp\, M'$ is given by taking $(M \times I) \cup (M' \times I)$ and attaching a 1-handle to join $M \times 1$ and $M' \times 1$.

Lemma 8.8.1 *Connected sum of k-connected manifolds of a given dimension* $n > 2$ *is a commutative associative operation with unit, compatible with cobordism. The set of equivalence classes thus acquires the structure of an abelian group* $\Omega_n\langle k\rangle$.

Proof The operation is well-defined by Theorem 2.7.4 (with the remark following dealing with orientation); by Proposition 2.7.6, it is commutative and associative, and the sphere S^n acts as unit. That the connected sum $M \,\sharp\, M'$ is k-connected if M and M' are follows if $k = 0$ from the definition, if $k = 1$ from the fact that for $n > 2$ removing a point does not introduce a fundamental group, and if $k > 1$ from the fact that removing a point does not change homology in dimension $< n$.

We must next check that the operation is compatible with cobordism. Let V and W be connected cobordisms, of dimension $n + 1$, and $f_- : D^n \to \partial_- V$, $f_+ : D^n \to \partial_+ V$, $g_- : D^n \to \partial_- V$, and $g_+ : D^n \to \partial_+ V$ be used to define the connected sums $\partial_- V \natural \partial_- W$ and $\partial_+ V \natural \partial_+ W$. Join $f_-(0)$ to $f_+(0)$ by an arc α in V: a tubular neighbourhood of the arc gives an imbedding $F : D^n \times I \to V$ with $f_- = F|D^n \times 0$ and $f_+ = F|D^n \times 1$. Similarly define $G : D^n \times I \to W$. Now delete the interiors of the images of F and G and glue the boundaries, and we have a cobordism of $\partial_- V \natural \partial_- W$ to $\partial_+ V \natural \partial_+ W$.

The inverse $\overline{M}$ of M is as usual obtained by change of orientation. We can regard $M \times I$ as a cobordism of $M \cup \overline{M}$ to the empty set. Attaching $D^n \times I$ with one end in M and one in $\overline{M}$ gives W with $\partial W = M \natural \overline{M}$. Now remove a disc from the interior of W to obtain a cobordism of $M \natural \overline{M}$ to S^m. □

For 0-connected cobordism (where we do not assume M oriented), we noted above that disjoint union is cobordant to connected sum, so that the map $\Omega_n\langle 0\rangle \to \Omega_n^O$ is surjective for $n \geqq 1$; it is easily seen to be bijective.

For k-connected cobordism, we need the connective covers of groups and classifying spaces. For any X we denote by $X^{\langle k\rangle}$ the $(k-1)$-connected cover of X: thus the map $\pi_r(X^{\langle k\rangle}) \to \pi_r(X)$ is zero for $r < k$ and an isomorphism for $r \geq k$. Observe that $B(J^{\langle k-1\rangle}) = (B(J))^{\langle k\rangle}$: we will write $BJ^{\langle k\rangle}$ for $B(J^{\langle k\rangle})$, which is k-connected.

The classifying map $\tau_M : M \to B(O)$ of its normal bundle lifts to a map $\tau_M^k : M \to BO^{\langle k\rangle}$ if M is k-connected, and the lift is unique up to homotopy if M is $(k+1)$-connected. We now claim

Theorem 8.8.2 *If $m > 2k+2$, there is a natural isomorphism $\Omega_m\langle k\rangle \to \pi_m^S(T(O^{\langle k\rangle}))$.*

Proof It follows from the remark preceding the theorem that there is a natural map $\psi_m^k : \Omega_m\langle k\rangle \to \pi_m^S(T(O^{\langle k\rangle}))$.

By Theorem 8.1.3 the Thom construction induces a bijection from the set of cobordism classes of m-manifolds whose stable normal bundle is induced from $BO^{\langle k\rangle}$ with the set $\pi_m^S(T(O^{\langle k\rangle}))$.

By Theorem 7.2.1, if X is a finite CW-complex and $m \geq 2r$, any normal map $(f : M \to X, \nu, T)$ is normally cobordant to a normal map $(f' : M' \to X, \nu, T')$ such that f' is r-connected. Applying this with X (a high enough skeleton of) $BO^{\langle k\rangle}$ and $r = k+1$, we see that if $m > 2k+2$ any element of $\pi_m^S(T(O^{\langle k\rangle}))$ is represented by a manifold M^m with $f : M \to BO^{\langle k\rangle}$ $(k+1)$-connected. It thus follows from the exact sequence

$$\pi_{k+1}(M) \to \pi_{k+1}(BO^{\langle k\rangle}) \to \pi_{k+1}(f) \to \pi_k(M) \to \pi_k(BO^{\langle k\rangle}) \to \dots$$

that M is k-connected, so the map ψ_m^k is surjective.

Similarly, given a cobordism W between two k-connected M, M' defining the same element of $\Omega_m\langle k\rangle$, provided $m+1 > 2k+2$ we can perform surgery on W, eventually making $W \to BO^{\langle k\rangle}$ $(k+1)$-connected and hence W k-connected. Hence ψ^k_m is injective. □

Corollary 8.8.3 *Excluding small m, we have isomorphisms* $\Omega_m\langle 1\rangle \cong \Omega^{SO}_m$, $\Omega_m\langle 2\rangle \cong \Omega_m\langle 3\rangle \cong \Omega^{Spin}_m$.

For we have $BO^{\langle 1\rangle} = B(SO)$, $BO^{\langle 2\rangle} = BO^{\langle 3\rangle} = B(Spin)$.

The above argument deals with the cases $m > 2k+2$ using surgery below the middle dimension. The cases $m = 2k$ and $m = 2k+1$ are of special interest. The case $m = 2k$ was discussed in Theorem 5.6.12.

If $m < 2k$ a k-connected m-manifold is a homotopy sphere (terminology introduced in §5.6), and the value of k is irrelevant to further study. From now on we focus on homotopy spheres. We begin our treatment by deriving a number of exact sequences: in all cases exactness will follow from the general principle of Lemma 8.3.1. First, however, we list the types of cobordism to be considered. In each case there is a natural definition of addition by connected sum, which gives the set a group structure. This is treated in Lemma 8.8.1 for k-connected cobordism and is similar in other cases. At the centre of our interest are groups of homotopy spheres:

(θ) Submanifolds $\Sigma^m \subset S^{m+k}$ with a homotopy equivalence $\Sigma^m \to S^m$. We denote the set of cobordism classes by Θ^k_m. We also consider

($f\theta$) Submanifolds $\Sigma^m \subset S^{m+k}$ with a homotopy equivalence $\Sigma^m \to S^m$ and a framing of the normal bundle (with compatible orientation class). Here we denote the set of cobordism classes by $F\Theta^k_m$. In parallel with these we consider

(*so*) The standard submanifold $S^m \subset S^{m+k}$ and a framing of the normal bundle (with compatible orientation class). Framings are classified up to homotopy by $\pi_m(SO_k)$, and we can identify this with the cobordism group.

(sph) Submanifolds $M^m \subset S^{m+k}$ with a framing of the normal bundle.

For a cobordism we must have a submanifold W^{m+1} of $S^{m+k} \times I$ (equal in case (*so*) to $S^m \times I$) together in cases (θ) and ($f\theta$) with a homotopy equivalence $W^{m+1} \to S^m$, in cases (*so*), ($f\theta$) and (sph) with a framing of the normal bundle of W^{m+1} in $S^{m+k} \times I$; in each case inducing the given structures on $\partial_- W$ and $\partial_+ W$.

A structure of type (*so*) is stronger than one of type ($f\theta$) which in turn is stronger than one of type (sph). Each of these three inclusions induces by Lemma 8.3.1 an exact sequence, and by the Corollary to that Lemma we also have an exact sequence of the three relative groups. We now reinterpret these.

As in §7.8, write $B(G_n)$ for the classifying space for spherical fibrations with fibre S^{n-1} and G_n for the monoid of maps of S^{n-1} to itself of degree ± 1, with multiplication given by composition of maps. Fixing the orientation gives a submonoid SG_n and a classifying space $B(SG_n)$. We write $F_n \subset G_{n+1}$ for the set of base-point preserving maps $S^n \to S^n$ of degree ± 1, and SF_n for those of degree $+1$. The suspension of a self-map of S^{n-1} is a self-map of the same degree of S^n which fixes a base point; thus we have an inclusion $G_n \subset F_n$. There are corresponding classifying spaces $B(F_n)$ and $B(SF_n)$. Since all components of $\Omega^n S^n$, including SF_n, are homotopy equivalent, we have $\pi_r(F_n) \cong \pi_{r+n}(S^n)$. Further discussion is given in §B.2.

Now (sph) is the cobordism group of submanifolds $M^m \subset S^{m+k}$ with a framing of the normal bundle; by Proposition 8.1.4 the group of cobordism classes is identified with $\pi_{m+k}(S^k) = \pi_m(\Omega^k S^k) = \pi_m(SF_k)$. We now see that the (so, sph) sequence can be identified with the exact homotopy sequence of (SF_k, SO_k).

The relative term for the $(so, f\theta)$ sequence is represented by manifolds W with boundary, with an assigned embedding $W^{m+1} \subset D^{m+k+1}$, with $\partial W = S^m \subset S^{m+k}$, a framing of the normal bundle of W^{m+1} in D^{m+k+1}, and a homotopy equivalence of W with a point. Since W is contractible, the framing of its normal bundle is unique (up to homotopy) and can be ignored. We regard D^{m+k+1} as the upper hemisphere of S^{m+k+1} and complete W to a closed manifold $\overline{W} \subset S^{m+k+1}$ by attaching the standard disc $D^{m+1} \subset D^{m+k+1}$ in the lower hemisphere and rounding the corner. There is a natural homotopy equivalence of $\overline{W}$ with S^{m+1}. Conversely, given a homotopy sphere $\Sigma^{m+1} \subset S^{m+k+1}$ we have (by the Disc Theorem 2.5.6) an essentially unique embedding $D^{m+1} \to \Sigma^{m+1}$; its neighbourhood in S^{m+k+1} may be identified with a disc D^{m+k+1}, and the whole construction can be reversed. The relative group is thus identified with Θ^k_{m+1}.

The relative term for the $(f\theta, sph)$ sequence is represented by manifolds W with boundary, with $W \subset D^{m+k+1}$, framed normal bundle, and a homotopy equivalence $\partial W \to S^m$. We denote the corresponding group of cobordism classes by P^{*k}_{m+1}.

By Lemma 8.3.2 we now have

Proposition 8.8.4 *We have a commutative braid of long exact sequences.*

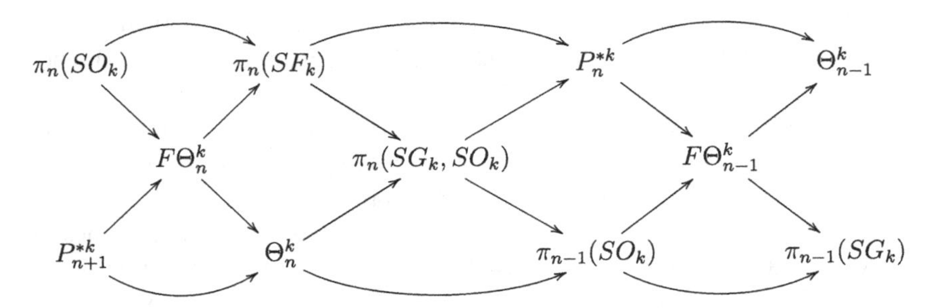

For each of the above 6 sequences of groups, the natural inclusions $S^{n+k} \subset S^{n+k+1}$ and $D^{n+k+1} \subset D^{n+k+2}$ induce maps which increase k by 1. In each case we see that for k large enough ($k > n+1$ suffices), these maps are isomorphisms and the groups stabilise. We denote the limiting groups by omitting k from the notation (and also the asterisk from P^*). All sequences of Proposition 8.8.4 thus remain exact when we omit the affix k. We may identify $\pi_n(SG)$ with the stable homotopy group π_n^S and the map $\pi_n(SO) \to \pi_n(SG)$ with the classical J-homomorphism $J_n : \pi_n(SO) \to \pi_n^S$.

We now give calculations for the stabilised groups. By §B.3(xi) $\pi_n(SO)$ is isomorphic to $\mathbb{Z}$ for $n \equiv -1 \pmod 4$, to $\mathbb{Z}_2$ for $n \equiv 0$ or $n \equiv 1 \pmod 8$, and is trivial otherwise. We proved in Proposition 7.8.4 by surgery that P_n is isomorphic to $\mathbb{Z}$ for $n \equiv 0 \pmod 4$, to $\mathbb{Z}_2$ for $n \equiv 2 \pmod 4$, and is trivial otherwise (provided $n > 5$). It follows from the stabilised braid (8.8.4) that the groups Θ_n are closely related to the stable homotopy groups π_n^S. A first deduction is

Proposition 8.8.5 *All the groups in the* stabilised *diagram of (8.8.4) with $n \geq 5$ are finitely generated abelian groups, and all are finite with the exceptions of $\pi_{4r-1}(SO)$, $\pi_{4r}(SF, SO)$, P_{4r} and $F\Theta_{4r-1}$, which have rank 1.*

For the case $n = 4s-1$ we first consider an element y of the group $\pi_{4s}(SF, SO)$ of cobordism classes of framed manifolds N with boundary diffeomorphic to S^{4s-1}. The boundary $x \in \pi_{4s-1}(SO)$ induces an orthogonal bundle $\xi(x)$ over S^{4s}: we can then form the Pontrjagin class $p_s(\xi(x))$ and evaluate on the fundamental class $[S^{4s}]$ giving an integer $p_s(x)$, say. Additivity properties of bundles and classes show that we have a homomorphism $p_s : \pi_{4s-1}(SO) \to \mathbb{Z}$. According to [22], if x_0 generates $\pi_{4s-1}(SO)$ then (up to sign) $p_s(x_0) = a_s(2s-1)!$, where we set $a_s = 2$ if s is odd and $a_s = 1$ if s is even. Thus the image of y in $\pi_{4s-1}(SO)$ is $p_s(x)/a_s(2s-1)!$ times a generator.

On the other hand, attaching a disc to the boundary of N yields a closed manifold M. The normal bundle of M is trivial except on the disc, so is induced from a bundle over S^{4s} which we can identify with the above bundle $\xi(x)$. According to the signature theorem 8.6.7, the signature of M is given by $L_s(\nu_M)[M]$. Since all the intermediate Pontrjagin classes of ν_M vanish, it follows by Lemma 8.6.8 that

$$L_s(\nu_M) = 2^{2s}(2^{2s-1} - 1)B_s p_s(x)/(2s)!,$$

so the image of y in P_{4s} is $2^{2s-3}(2^{2s-1} - 1)B_s p_s(x)/(2s)!$ times a generator.

Thus in some sense the generators in $\pi_{4s-1}(SO)$ and P_{4s} differ by a factor $a_s 2^{2s-2}(2^{2s-1} - 1)B_s/4s$. It now follows from exactness of the braid that

Proposition 8.8.6 *We have* $|\Theta_{4s-1}| = a_s 2^{2s-2}(2^{2s-1} - 1)B_s|\pi^S_{4s-1}|/4s$.

More precisely, according to Adams [5] (see also §B.3(xviii)), $\mathrm{Ker} J_{4s-1}$ is a subgroup of $\pi_{4s-1}(SO)$ of index $\mathrm{den}(B_s/4s)$. Here, if $z \in \mathbb{Q}$ is expressed as a fraction p/q with $p, q \in \mathbb{Z}$ as small as possible, we write $p := \mathrm{num}(z)$ and $q := \mathrm{den}(z)$ for the numerator and denominator of z. Thus $|\pi^S_{4s-1}| = \mathrm{den}(B_s/4s)|\mathrm{Coker}\, J_{4s-1}|$. It also follows that $p_s(\mathrm{Ker}\, J_{4s-1}) = a_s(2s-1)!\mathrm{den}(B_s/4s)$, hence the signatures of the manifolds M obtained by closing elements of $\pi_{4s}(SF, SO)$ form the group of multiples of $a_s 2^{2s+1}(2^{2s-1} - 1)\mathrm{num}(B_s/4s)$.

The integer $m(2s) := \mathrm{den}(B_s/4s)$ is given by the following formula (due to Milnor and Kervaire [102], see also Adams [4]). For n an integer and p a prime, denote by $\nu_p(n)$ the greatest integer r such that p^r divides n. Then

For p odd, $\nu_p(m(t)) = 1 + \nu_p(t)$ if $t \equiv 0 \pmod{(p-1)}$, and $= 0$ if not.

For $p = 2$, $\nu_p(m(t)) = 2 + \nu_2(t)$ if t is even, and $= 1$ if t is odd.

Since $P_{4s} \cong \mathbb{Z}$, with the isomorphism given by $\sigma/8$, it follows that the image of $\pi_{4s}(SF, SO)$ in P_{4s}, which is the kernel of $P_{4s} \to \Theta_{4s-1}$ is a subgroup of index $a_s 2^{2s-2}(2^{2s-1} - 1)\mathrm{num}(B_s/4s)$, so this number is the order of the group traditionally denoted bP_{4s}, which is the kernel of the epimorphism $\Theta_{4s-1} \to \pi_{4s-1}(SF, SO)$, the latter group having order $|\mathrm{Coker}\, J_{4s-1}|$.

For other values of n, we compare Θ_n with π^S_n via the intermediary $\pi_n(SF, SO)$ (or, if $n \equiv 0 \pmod 4$, its torsion subgroup). It was shown by Adams [5] that J_n is a (split) monomorphism if $n \equiv 0$ or $1 \pmod 8$. Thus if $n \not\equiv -1 \pmod 4$ $\mathrm{Tors}\, \pi_n(SF, SO)$ is the cokernel of J_n. Moreover Θ_n maps onto $\mathrm{Tors}\, \pi_n(SF, SO)$ except perhaps when $n \equiv 2 \pmod 4$. In this last case, π^S_n maps onto $\pi_n(SF, SO)$ and we have the map $K_n : \pi^S_n \to P_n \cong \mathbb{Z}_2$, defining the Kervaire invariant of framed manifolds. Thus if $n \equiv 0 \pmod 4$, Θ_n is isomorphic to $\mathrm{Tors}\, \pi_n(SF, SO)$; if either $n \equiv 1 \pmod 4$ or $n \equiv 2 \pmod 4$ and K_n vanishes, Θ_n maps isomorphically to $\pi_n(SF, SO)$.

The delicate question of deciding for which values of n K_n is zero, is known as the 'Kervaire invariant problem'. It was shown by Browder [30] that K_n vanishes unless $n + 2 = 2^{k+2}$ is a power of 2. There are simple constructions showing K_n non-zero if $k = 0$, 1 or 2 (the classical framings of the tangent bundles of S^1, S^3 and S^7 induce framings of the projective spaces, and one uses $P^1(\mathbb{R}) \times P^1(\mathbb{R})$, $P^3(\mathbb{R}) \times P^3(\mathbb{R})$, and $P^7(\mathbb{R}) \times P^7(\mathbb{R})$); there is a somewhat less simple example for $k = 3$, and a proof by strenuous calculations [17] if $k = 4$. Recently it was shown by Hill, Hopkins, and Ravenel [69] that K_n vanishes for all $k \geq 6$, leaving only the case $k = 5$ (dimension 126) open.

A modified version of the braid (8.8.4) turns out to have better properties: we will replace the term SF_k by SG_k. Now $\pi_n(SG_k)$ is the group of homotopy classes of maps $S^n \times S^{k-1} \to S^{k-1}$ with the restriction to S^{k-1} homotopic to the identity. By the Thom construction, we can identify this with cobordism classes of framed manifolds $M^n \subset S^n \times S^{k-1}$ such that M^n has intersection number 1 with $* \times S^{k-1}$, or equivalently, the projection of M^n on S^n has degree 1.

Correspondingly, we can interpret $\pi_n(SG_k, SO_k)$ as the group of cobordism classes of framed manifolds $(M^n, \partial M) \subset (D^n, \partial D^n) \times S^{k-1}$ such that the projection $\partial M \subset S^{n-1} \times S^{k-1} \to S^{n-1}$ is a diffeomorphism and the projection $S^{n-1} \to S^{k-1}$ is induced by the framing. We have thus interpreted the exact homotopy sequence of (SG_k, SO_k) as an exact cobordism sequence.

We now need a replacement for the group denoted P_n^{*k} above. Write P_n^k for the group of cobordism classes of framed manifolds $(M^n, \partial M) \subset (D^n, \partial D^n) \times S^{k-1}$ such that the projection $\partial M \subset S^{n-1} \times S^{k-1} \to S^{n-1}$ is a homotopy equivalence and the projection $S^{n-1} \to S^{k-1}$ is induced by the framing. We now claim

Proposition 8.8.7 *We have a commutative braid of long exact sequences.*

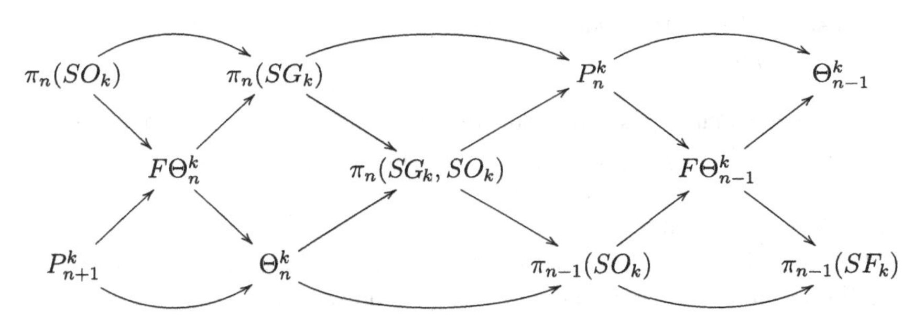

Proof The exact homotopy sequence of (SG_k, SO_k) was described above, and the exact sequence $\pi_n(SO_k) \to F\Theta_n^k \to \Theta_n^k$ is as before. We next describe the remaining maps.

First consider $F\Theta_n^k \to \pi_n(SG_k)$. Given a framed homotopy sphere $\Sigma_n \subset S^{n+k}$, take a tubular neighbourhood T. By Proposition 5.6.6, there is a diffeomorphism $h : \partial T \to S^n \times \partial D^k$: choose h such that the standard framing of $S^n \times \partial D^k$ pulls back to the framing of ∂T induced from that on S^{n+k}. Now the first vector of the normal framing of Σ^n induces a map $f : \Sigma^n \to \partial T$, and we take $h(f(\Sigma^n)) \subset S^n \times S^{k-1}$, with the framing induced by the remaining vectors of the normal framing of Σ^n.

The map $\Theta_n^k \to \pi_n(SG_k, SO_k)$ is defined similarly. We identify Θ_n^k with the group of framed homotopy discs in D^{n+k} with boundary the standard S^{n-1}. Now follow through the same steps as above.

The map $\pi_n(SG_k, SO_k) \to P^k_n$ is a forgetful map, defined by weakening the structure.

To define $P^k_{n+1} \to F\Theta^k_n$, we start with a framed manifold $(M^{n+1}, \partial M) \subset (D^{n+1}, \partial D^{n+1}) \times S^{k-1}$ such that the projection $\partial M \subset S^n \times S^{k-1} \to S^n$ is a homotopy equivalence. Map this to $\partial M \subset S^n \times S^{k-1} \subset S^{n+k}$ with the given framing extended by the normal vector to $S^n \times S^{k-1}$ in S^{n+k}.

The maps so far defined form a commutative diagram, and we define the remaining maps $P^k_{n+1} \to \Theta^k_n$ and $\pi_n(SG_k) \to P^k_n$ as the composites in the diagram. It follow easily that all four sequences have order 2. The exactness of the two remaining sequences follows again (with a little care) from Lemma 8.3.1. □

Since $G_n \subset F_n \subset G_{n+1}$, the stabilisations as $n \to \infty$ have the same homotopy groups; it follows that the same goes for the diagrams (8.8.4) and (8.8.7).

The reason why the second braid is an improvement on the first is the following.

Proposition 8.8.8 *The natural map $P^k_m \to P_m$ is surjective for $k \geq 2$ and an isomorphism for $k \geq 3$.*

Proof Recall that P^k_m is the group of cobordism classes of framed manifolds $(M^m, \partial M) \subset (D^m, \partial D^m) \times S^{k-1}$ such that the projection $\partial M \subset S^{m-1} \times S^{k-1} \to S^{m-1}$ is a homotopy equivalence and the projection $\partial M \to S^{k-1}$ is induced by the framing.

For surjectivity, since $P_{2n+1} = 0$, it suffices to consider the case $m = 2n$ even. By Proposition 7.8.3, generators of P_{2n} are represented by framed manifolds M constructed by attaching n-handles to D^{2n}. Since changing orientation and forming boundary sums respect this description, it follows that all elements of P_{2n} are so represented.

Write $e_i : S^{n-1} \times D^n \to S^{2n-1}$ for the attaching maps of the handles. Since all embeddings of S^{n-1} in S^{2n-1} are isotopic, there is a diffeomorphism of the image of e_i to the submanifold obtained from $\partial D^n \times D^n \subset \partial(D^n \times D^n)$ by rounding the corner. Thus e_i extends to an embedding $f_i : (D^n, \partial D^n) \times D^n \to (D^{2n}, \partial D^{2n})$ diffeomorphic to that induced by the map $D^n \times D^n \to D^{2n}$ rounding the corner.

We seek to construct a smooth embedding $F : (M, \partial M) \to (D^{2n}, \partial D^{2n}) \times S^1$ such that the trivial normal bundle agrees with the given stable framing; in fact we replace S^1 by I and then wrap round by $t \to e^{2\pi it}$. Choose distinct points $t_i \in I$ and a smooth map $\phi : S^{2n-1} \to I$ such that the image of $\phi \circ e_i$ is the point

t_i. We first define a continuous map F_0: it is given on D^{2n} by $F_0(z) = (z, \phi(z))$ and on the handle h_i (identified with $D^n \times D^n$) by $F_0(x, y) = (f_i(x, y), t_i)$.

The map F_0 is not injective: each handle overlaps the core. Write $\nu : (D^n, \partial D^n) \to ([0, 1], 0)$ for the map given by $(1 - \|x\|^2)$, and deform the map of the handle to $F_1(x, y) = (f_i(x, y), t_i + \epsilon\nu(x))$, where ϵ is small enough that the handles remain disjoint. Then F_1 is injective, but has a corner along each copy of $S^{n-1} \times S^{n-1}$. We define a map F_2 by rounding these corners. This has the desired effect of deforming the interior part $\mathring{D}^n \times D^n$ of each handle into the interior of $D^{2n} \times I$, and gives the desired smooth embedding in $(D^{2n}, \partial D^{2n}) \times S^1$.

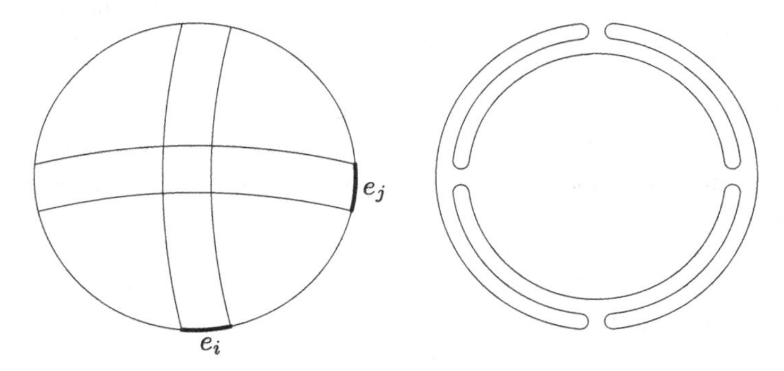

Figure 8.2 Embedding a plumbed manifold

We attempt to illustrate this in Figure 8.2: here the first figure represents a disc with two handles, pictured as a basket suspended by a couple of handles; the second figure indicates how these fit at the boundary. To prove injectivity, suppose given a framed manifold $(M^m, \partial M) \subset (V, \partial V)$ with $(V, \partial V) = (D^m, \partial D^m) \times S^{k-1}$ and $\partial M \subset S^{m-1} \times S^{k-1} \to S^{m-1}$ a homotopy equivalence, such that M represents 0 in the stabilised group P_m. Then there is a cobordism W of M to a disc: we seek to extend the embedding of $M = \partial_- W$ to $(W, \partial_c W) \to (V, \partial V) \times I$, ideally such that on $\partial_c W$ we have a product embedding. It is enough to consider a single r-handle attached to (the interior of) M. In view of the clause in Theorems 7.5.2, 7.5.4 (m even), and 7.6.1 (m odd) stating that for (simply-connected) surgery on manifolds of dimension $2n$ or $2n + 1$ it is sufficient to perform surgery on spheres S^r with $r \leq n$, we may suppose here that $2r \leq m$, and that M is $(r - 1)$-connected.

Using the first vector of the framing, we extend the a-sphere of the handle to an embedding $\phi : S^r \times I \to V$ such that $\phi(S^r \times \{0\})$ is the a-sphere and the rest of the image is disjoint from M. We next show that $\phi(S^r \times \{1\})$ is nullhomotopic in the complement $V \setminus M$ of M. Since M has codimension greater than 2, the complement is 1-connected.

We now use the hypothesis that $H_m(M, \partial M) \to H_m(V, \partial V)$ is surjective. It follows that $H_i(V, M \cup \partial V) = 0$ with possible exceptions $i = k, i = m + k - 1$, $r + 1 \leq i \leq m + 1 - r$. By the universal coefficient theorem, the same holds for $H^i(V, M \cup \partial V)$. By duality, $H_i(V \setminus M) = 0$ except perhaps for $i = m - 1$, $i = 0$, $k + r - 2 \leq i \leq m + k - r - 1$. Since $k \geq 2$, $V \setminus M$ is r-connected, so our r-sphere is indeed nullhomotopic in $V \setminus M$.

We can thus extend the map ϕ on a collar neighbourhood of the boundary to a map $\psi : (D^{r+1}, \partial D^{r+1}) \to (V, M)$ with $\psi^{-1}(M) = \partial D^{r+1}$.

The map ψ is covered by a stable normal framing of the handle. As in the proof of Theorem 7.1.1, this framing determines a regular homotopy class of immersions $D^{r+1} \times D^{m+k-r-2} \to V$. We wish the immersion to restrict to the given embedding $S^r \times D^{m-r} \to M$. Since $m \geq 2r$ and $k \geq 3$, we have $m + k - r - 2 \geq r + 1$. Thus $\pi_r(SO_{m+k-r-2})$ maps onto $\pi_r(SO)$, so the stable framing induces a normal framing. It follows by Theorem 6.2.1 that the map ψ is homotopic (relative to its boundary) to an immersion.

If $m + k - 1 > 2(r + 1)$ putting this map in general position makes it an embedding; in the critical case $m = 2r$ and $k = 3$, we can use the Whitney trick (see Theorem 6.3.4 but allow boundaries) to obtain an embedding. Now using the normal framing on this handle allows us to extend the embedding of M to the desired embedding of M with the handle. □

Inserting this result in the braid diagram (8.8.7) together with results in § B.3 (ix) on homotopy groups of spheres, (xiv) on homotopy groups of orthogonal groups and (xix) on $\pi_r(SO_k) \to \pi_r(SG_k)$, it follows that

Theorem 8.8.9 *All groups in the diagram*

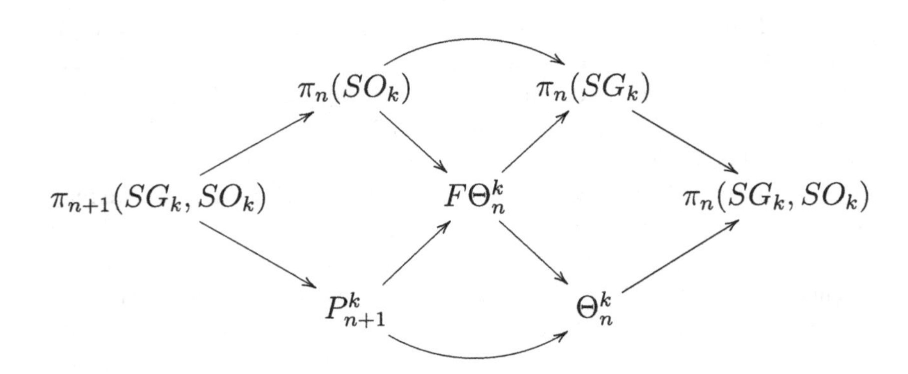

for $k \geq 3$ *are finite* except *for*
$(A): n = 4s+1,\ k = 4s+2$*:* $\pi(SO) \to F\Theta \to \pi(SG)$*, of rank 1,*

$$(S): n = 4s-1,\ k > 2s+1: \quad \begin{array}{ccc} \pi(SG, SO) & \to & \pi(SO) \\ \downarrow & & \downarrow \\ P & \to & F\Theta \end{array} \quad \textit{of rank 1,}$$

$(O): n = 4s-1,\ k \leq 2s$*:* $P \to F\Theta \to \Theta$*, of rank 1,*
$(B): n = 4s-1,\ k = 4s$*: 'the direct sum of the diagrams (A) and (S)',*
$(C): n = 4s-1,\ k = 2s+1$*: 'the direct sum of the diagrams (A) and (O)'.*

In particular, Θ^k_{4s-1} has rank 1 if $k \leq 2s+1$, and otherwise Θ^k_n is finite.

We can use the above results to investigate groups of embeddings of spheres in spheres. Denote by Σ^k_m the set of diffeotopy classes of embeddings $S^m \to S^{m+k}$. Since by Lemma 2.5.11 orientation-preserving embeddings $(D^{m+k}, D^m) \to (S^{m+k}, S^m)$ are unique up to diffeotopy, we can define a connected sum of two embeddings by removing an embedded disc-pair from each, and glueing along the boundary (with an orientation reversal). It follows that Σ^k_m acquires the structure of a group.

Since diffeotopic embeddings are cobordant, there is a natural forgetful map $\sigma : \Sigma^k_m \to \Theta^k_m$. By Lemma 8.3.1 the map σ lies in an exact sequence $\ldots \to R^k_{m+1} \to \Sigma^k_m \to \Theta^k_m \to R^k_m \to \ldots$ The relative term R^k_{m+1} is the set of cobordism classes of homotopy discs $\Delta^{m+1} \subset D^{m+k+1}$ together with a diffeomorphism $S^m \to \partial\Delta^{m+1}$. It follows from Corollary 5.6.3 that for $m \geq 5$ Δ^{m+1} is diffeomorphic to D^{m+1}. If also $k \geq 3$, it now follows from Theorem 5.6.7 (i) that $\Delta^{m+1} \subset D^{m+k+1}$ is diffeomorphic to the standard pair. Thus R^k_{m+1} is the cobordism group of standard pairs together with a diffeomorphism of S^m on the boundary. The embedding now plays no part, thus for $m \geq 5$, $k \geq 3$ the map $R^k_{m+1} \to R_{m+1}$ is an isomorphism. Hence $R^k_{m+1} \cong R_{m+1} \cong \Theta_{m+1}$. This proves

Proposition 8.8.10 *There is an exact sequence* $\ldots \to \Theta_{m+1} \to \Sigma^k_m \to \Theta^k_m \to \Theta_m \to \ldots$

Since the groups Θ_m are all finite, it follows that the rank of Σ^k_m is the same as that of Θ^k_m: thus is 1 if $m = 4s-1$ and $k \leq 2s+1$, and zero otherwise.

It follows from the Whitney embedding theorem that $\Sigma^k_m = 0$ for k large. More precisely, by Theorem 6.4.11, any two embeddings of S^m in S^{m+k} are isotopic (and hence Σ^k_m vanishes) provided $2k > m+3$. However in the limiting case $2k = m+3$ the group does not vanish: if also k is odd, it is infinite by the above; more precisely, by [61], we have $\Sigma^{2s+1}_{4s-1} \cong \mathbb{Z}$. It was shown in [64] that in the other critical case, $\Sigma^{2s}_{4s-3} \cong \mathbb{Z}_2$.

8.9 Notes on Chapter 8

§8.1 The first result in this area is due to Pontrjagin [123], who succeeded in relating framed bordism to homotopy groups of spheres. Thom's paper [150], as well as formally introducing the construction, obtained a transversality theorem.

§8.2 I also believe that at least part of his motivation was the problem of representing homology classes by embedded submanifolds.

§8.3 I do not know where it was first observed that the definition of bordism naturally leads to exact sequences. The second technique was formally introduced in [158].

§8.4 In his paper [12], Atiyah introduced bordism as a homology theory, showed that smooth oriented manifolds are also orientable for this theory, and made applications to bordism groups.

There are other abstract structures using bordism. Graeme Segal defined in [134] axioms for quantum field theory, which we can summarise as follows. A cobordism category is a category with objects (diffeomorphism classes of) closed manifolds (of a given dimension) and morphisms (diffeotopy classes of) bordisms: to obtain interesting examples one usually imposes extra structure: for example, an embedded submanifold of codimension 2.

A 'topological field theory' is then a functor ϕ from such a category to, for example, the category of vector spaces over $\mathbb{C}$ and maps: it is required also to take disjoint unions to tensor products. Since the empty manifold is mapped to $\mathbb{C}$, if M is a closed manifold, and so a cobordism from the empty set to itself, $\phi(M)$ is a linear map $\mathbb{C} \to \mathbb{C}$: multiplication by a number, giving an invariant $\alpha(M) \in \mathbb{C}$. Non-trivial examples are not easy to construct.

§8.5 The main reference for this section is the book [38], which has a wealth of information about actions of finite cyclic groups. Chapter IV of that book contains the calculation of equivariant bordism groups of $\mathbb{Z}_2$-actions. The $\mathbb{Z}_p$-actions are discussed in Chapter VII: the results are, of course, not complete. However many geometrical consequences of their calculations are given throughout the book.

§8.6 In his original 1954 paper [150], as well as introducing transversality and using it to reduce the calculation of cobordism groups to a homotopy problem, Thom was able to give the full calculation of Ω_*^O, using Serre's calculation [135] of cohomology of Eilenberg–MacLane spaces, and to calculate $\Omega_*^{SO} \otimes \mathbb{Q}$ using Proposition B.4.1. Milnor's paper [96] followed in 1960 and Novikov's [113] appeared in 1962. Milnor's book [103] gives an alternative introduction to characteristic classes, the calculation of the cohomology of classifying spaces, cobordism and the calculation of cobordism rings, including the Hirzebruch signature theorem.

Unitary bordism has more structure than the calculation in Theorem 8.6.11 shows. One aspect of this is:

Theorem 8.9.1 *There is an isomorphism of the universal formal group over* $\mathbb{Z}$ *on* Ω_*^U.

We explain this statement. If T is a connected 1-dimensional analytic Lie group with multiplication $\mu : T \times T \to T$, and x a local coordinate at the unit, we can expand $\mu \circ x$ as a power series $F(x, y)$ with $F \in \mathbb{R}[[x, y]]$. The group properties are reflected in the identities

$$F(x, 0) = F(0, x) = x, \quad F(x, y) = F(y, x), \quad F(F(x, y), z) = F(x, F(y, z)).$$

One thus defines a formal group over a ring R as a formal power series in 2 variables $F(x, y)$, with constant term 0, satisfying these rules. The simplest examples are $F_r(x, y) = x + y + rxy$ for $r \in R$.

Now consider $P := P^\infty(\mathbb{C}) \cong B(U_1)$. There is a multiplication map $\mu : P \times P \to P$ induced, for example, by tensor product of line bundles.

Since P has a cell structure with one cell in each even dimension we can identify $\Omega_*^U(P)$ with $\Omega_*^U[[z]]$, with a generator $z \in \Omega_2^U(B(U_1))$ which can be taken as defined by the inclusion $P^1(\mathbb{C}) \subset P^\infty(\mathbb{C})$. Now $\mu^*(z) \in \Omega_*^U[[x, y]]$ defines a formal group.

For the proof of Theorem 8.9.1 we refer to Quillen [127]. This result is the jumping off point for the use of complex cobordism theory as a tool for elaborate calculations in homotopy theory. It is used to set up the so-called Adams–Novikov spectral sequence. One can localise Ω_*^U homology theory at a prime p; it then splits into the so-called BP-theories with much smaller coefficient group (polynomial with generators only in dimensions $p^r(2p - 2)$). We refer to [129] for an introduction to this area.

§8.7 Certain exact sequences were devised by the author [157] to relate $\Omega_*^{\mathbf{O}}$ and $\Omega_*^{\mathbf{SO}}$, as a means of calculating the latter. A more abstract proof was found by Atiyah [12] (who invented bordism theory for the purpose). My original insight was that the apparently complicated structure of Ω_*^{SO} might be the similar to the structure of $H^*(X; \mathbb{Z})$ for a space X such that each of $H^*(X : \mathbb{Q})$ and $H^*(X; \mathbb{Z}_2)$ is a polynomial ring.

The original exact sequences were extended by Conner and Floyd to the case of $\Omega_*^{\mathbf{U}}$ and $\Omega_*^{\mathbf{SU}}$, and used in the calculations of the latter, with details in [39].

Further calculations of Ω^{SU} were obtained by Anderson, Brown, and Peterson [9].

The groups Ω^{Spin} were calculated, also by Anderson, Brown, and Peterson, in [10]. They first determine the structure of $H^*(\mathbb{T}\mathbb{S}pin; \mathbb{Z}_2)$ as a module over the Steenrod algebra: it is a sum of copies of $\mathcal{S}_2$, $\mathcal{S}_2/\mathcal{S}_2(Sq^3)$ and $\mathcal{S}_2/\mathcal{S}_2(Sq^1, Sq^2)$. They deduce that the Thom spectrum is homotopy equivalent to a wedge of spectra of type $\mathbb{K}(\mathbb{Z}_2, n)$ and $\mathbb{BO}\langle n\rangle$; and thence that cobordism class in Ω_*^{Spin} is determines by Stiefel–Whitney and KO-characteristic numbers.

Complete results are also available for $Spin^c$: here cobordism class is determined by Stiefel–Whitney numbers and characteristic numbers in $\mathbb{Q}$: calculations for this case can be reduced to those for *Spin* in view of the isomorphism $\Omega_n^{Spin^c} \cong \tilde{\Omega}_{n-2}^{Spin}(P^\infty(\mathbb{C}))$.

In addition to the original references, Stong's book [147] aims to give complete details of all the calculations involved in determining the cobordism groups mentioned above, and their interrelations with each other and with framed bordism.

For Ω_*^{Sp}, it was again shown in [113] that the tensor product by $\mathbb{Z}[\frac{1}{2}]$ is a polynomial algebra. Extensive calculations have been made by Kochman [80].

§8.8 The sequences 8.8.4 were extracted from the methods introduced by Milnor and Kervaire [79] for calculating the groups Θ_n. Our account follows the presentation by Levine [85], which in turn combined the earlier work of Milnor and Kervaire (see, for example, [79]) with ideas of Haefliger [61].

Milnor's discovery [92] of non-diffeomorphic differential structures on the topological manifold S^7 was a great surprise: up to then, though smooth and piecewise linear (PL) structures were used, the philosophy was that one was really studying problems in pure topology. Likewise the existence of non-trivial embeddings of spheres in spheres contrasts with the theorem of Stallings [143] (in the topological category) and Zeeman [183] (in the piecewise linear category) that embeddings of spheres in spheres, in codimension at least 3, are topologically unknotted. It is thus possible to regard all the results about embeddings of spheres in spheres as a manifestation of smoothing theory.

Explicit results of this kind were obtained by Rourke and Sanderson. In the first of the three papers [130] they set out to construct a theory of neighbourhoods of locally flat submanifolds of PL manifolds to play the role in PL topology of the tubular neighbourhoods in differential topology. By introducing a notion of 'block bundles' they constructed a (simplicial) space $B\widetilde{PL}_k$ such that for any PL manifold M^m the set of isomorphism classes of regular neighbourhoods of M embedded locally flatly in PL $(m+k)$-manifolds maps bijectively to the homotopy set $[M : B\widetilde{PL}_k]$.

In the third paper, after defining various simplicial spaces, in particular a piecewise differentiable version $\widetilde{BPD}_k$ of $\widetilde{BPL}_k$ which is homotopy equivalent to it, they interpret the braid (8.8.7) as the homotopy braid coming from the inclusions $B(SO_k) \subset B\widetilde{SPL}_k \subset B(SG_k)$.

In the subsequent paper [131], Rourke and Sanderson construct a theory of neighbourhoods of locally flat submanifolds of topological manifolds. A starting point is the notion of microbundle introduced by Milnor [99]. Following a subsequent idea of Haefliger, they consider a microbundle with fibre dimension $(n+k)$ together with a submicrobundle with fibre dimension n. From these they form a (simplicial) classifying space $BTop^n_{n+k}$ and establish the existence of a (Kan) fibration $Top^n_{n+k} \to Top_n$, whose fibre is denoted $Top_{n+k,n}$. An $(n+k)$ dimensional neighbourhood of a manifold N^n induces a lift of $N \to BTop_n$ to a cross-section of the induced fibration.

They then establish that if $i \leq k$ and either $n \leq 2$ or $n + k \geq 5$ the map $\pi_i(Top_{n+k,k}) \to \pi_i(Top_n)$ is an isomorphism. It follows that with this dimension restriction, neighbourhoods of N are classified by maps $N \to BTop_k$. This leads to obstruction theories to the existence of normal microbundles or block bundles with fibre D^k or $\mathbb{R}^k$.

It also follows that the above results in the PL case carry over to the Top case. Thus one can identify $F\Theta^k_n$ with $\pi_n(STop_k)$, Θ^k_n with $\pi_n(STop_k, SO_k)$, and P^k_n with $\pi_n(SG_k, STop_k)$. Thus the stability theorem Proposition 8.8.8 establishes a homotopy pullback diagram

$$\begin{array}{ccc} STop_k & \to & SG_k \\ \downarrow & & \downarrow \\ STop & \to & SG \end{array},$$

and the exact sequence of Proposition 8.8.10 interprets Σ^k_n as the homotopy group of the diagram

$$\begin{array}{ccc} SO_k & \to & STop_k \\ \downarrow & & \downarrow \\ SO & \to & STop \end{array},$$

and hence of the diagram

$$\begin{array}{ccc} SO_k & \to & SG_k \\ \downarrow & & \downarrow \\ SO & \to & SG \end{array}.$$

This final result had been obtained by Haefliger in [64].

Appendix A
Topology

A.1 Definitions

A *topology* on a set X is a collection $\mathcal{U}$ of subsets, called *open sets*, such that $X \in \mathcal{U}$, the union of any subfamily of $\mathcal{U}$ belongs to $\mathcal{U}$, and the intersection of two elements of $\mathcal{U}$ also belongs to $\mathcal{U}$. A topology can be defined by prescribing a set $\mathcal{V}$ of subsets of X to be a 'subbase' of open sets: then define $\mathcal{U}$ to consist of arbitrary unions of finite intersections of elements of $\mathcal{V}$. A set $\mathcal{W}$ is a *base* of open sets if every open set is a union of elements of $\mathcal{W}$.

A subset F of X is *closed* if its complement $X \setminus F$ is open. If A is any subset of X (in particular, if A is a point) a subset V of X is a *neighbourhood* of A if there is an open set U with $A \subseteq U \subseteq V$.

If $Y \subset X$ is a subset of a space X with a topology $\mathcal{U}$, the subspace topology on Y is given by taking as open sets the $U \cap Y$ with $U \in \mathcal{U}$.

A topology is said to be *Hausdorff* if for any $x_1 \neq x_2 \in X$ we can find $U_1, U_2 \in \mathcal{U}$ with $x_1 \in U_1$, $x_2 \in U_2$ and U_1, U_2 disjoint, i.e. $U_1 \cap U_2 = \emptyset$. This is a rather weak condition, and all spaces we will consider are Hausdorff. In a Hausdorff space, each point is a closed set. There are also stricter separation conditions (which hold for smooth manifolds): a topology is *completely regular* if any point x and closed set F not containing it are contained in disjoint open sets, and *normal* if disjoint closed sets F_1, F_2 are contained in disjoint open sets $U_1, U_2 \in \mathcal{U}$.

A mapping $f : X \to Y$ between two topological spaces is *continuous* if whenever V is open in Y, $f^{-1}(V)$ is open in X. It is a *homeomorphism* if f is bijective and both f and f^{-1} are continuous. We call f an *embedding* if it is injective and gives a homeomorphism between X and $f(X)$ with the subspace topology.

An important condition on a topology is the existence of a countable base of open sets. This holds for $\mathbb{R}^n$ since we can take the balls with rational radii and centres having rational coordinates.

A set $\mathcal{U} = \{U_\alpha \mid \alpha \in A\}$ of subsets of X is a *covering* if $\bigcup_{\alpha \in A} U_\alpha = X$; it is an *open covering* if each U_α is open in X, and it is *locally finite* if each point of X has a neighbourhood intersecting only a finite number of the U_α. A covering $\mathcal{V} = \{V_\beta \mid \beta \in B\}$ of X refines $\mathcal{U}$ if for each β there is an α such that $V_\beta \subseteq U_\alpha$.

The space X is *compact* if for every open covering $\mathcal{U}$, a finite subset of $\mathcal{U}$ already covers X. It is *locally compact* if every neighbourhood of a point contains a compact neighbourhood. Since any point has a neighbourhood which is a disc, any manifold is locally compact. A space is *paracompact* if every open covering has a locally finite refinement by an open covering.

Any compact subset K of a Hausdorff space X is closed. For if $x \notin K$, then for each $k \in K$, x and k have disjoint neighbourhoods U_x, V_x. The $K \cap V_x$ form an open cover of K, so there is a finite subcover. The intersection of the corresponding U_x is an open neighbourhood of x disjoint from K.

If $\{U_a\}$ is a locally finite family of subsets of X and $K \subset X$ is compact, then K has a neighbourhood intersecting only finitely many of the U_a. For each point $k \in K$ has such an open neighbourhood N_k; we may choose a finite subset of the N_k which cover K, and their union is a neighbourhood of K with the desired property.

If $f : X \to Y$ is continuous and $K \subset X$ is compact, the image $f(K)$ is compact. For if $\{U_\alpha\}$ is an open cover of $f(K)$ we can write $U_\alpha = f(K) \cap V_\alpha$ with V_α open in Y. Since f is continuous, $f^{-1}(V_\alpha)$ is open in X, and these give an open covering of K. Taking a finite subcovering here gives a finite subcover of $\{U_\alpha\}$.

Thus if K is a compact space and $f : K \to Y$ is continuous, f takes closed sets to closed sets, so if f is bijective it is a homeomorphism; if f is injective, it is an embedding.

Lemma A.1.1 *If X is a locally compact space any neighbourhood of a compact set $K \subset X$ contains a compact neighbourhood of K.*

Proof Let U be the given neighbourhood of K: then U is a neighbourhood of each $x \in K$, so we can find neighbourhoods A_x, B_x, C_x of x in X with $C_x \subset B_x \subset A_x \subset U$ and A_x, C_x open and B_x compact. Since the open sets C_x cover the compact set K, there is a finite subcover $\{C_{x_n}\}$. The (finite) union of the B_{x_n} is compact and contains the open neighbourhood $\bigcup_n C_{x_n}$ of K. □

Taking K as a point $x \in X$, any open neighbourhood A_x of $x \in X$ contains a compact neighbourhood B_x, which contains an open neighbourhood C_x: and so on.

The product $\prod A_i$ of a family of spaces has a topology defined by the subbase consisting of products $\prod U_i$ with U_i open in A_i for each i and $U_i = A_i$ for all but finitely many. If each A_i is compact, so is $\prod_i A_i$.

The *inverse limit* $\varprojlim A_i$ of a sequence $A_{i+1} \xrightarrow{\alpha_i} A_i$ $(i \geq 1)$ is defined to be the subset of the product $\prod A_i$ with $\alpha_i(x_{i+1}) = x_i$ for each i. If the A_i are topological spaces, it inherits a topology as a subspace of the product.

A.2 Topology of metric spaces

A *metric* on a set X is a mapping $\rho : X \times X \to \mathbb{R}$ such that $\rho(x, y) \geq 0$ for all $x, y \in X$, $\rho(x, y) = 0$ if and only if $x = y$, and $\rho(x, z) \leq \rho(x, y) + \rho(y, z)$ for all x, y and $z \in X$. This defines a topology with a base consisting of the sets $\{x \mid \rho(x, y) < d\}$ for all $y \in X$, $d > 0$. Equivalently, a subset $U \subseteq X$ is open if, for each $x \in U$, there exists $\varepsilon > 0$ such that $\rho(x, y) < \varepsilon$ implies $y \in U$.

We have seen in Theorem 2.1.1 that smooth manifolds are metric as topological spaces.

The prime example of a metric space is $\mathbb{R}^n$, with points $x = (x_1, \ldots, x_n)$ and distance function $\rho(x, y) = \|x - y\| = \sqrt{\sum_1^n (x_i - y_i)^2}$. The basic examples of topological spaces are subsets of $\mathbb{R}^n$ with the topology given by the induced metric. We are not concerned with arbitrary subsets: more typical are polyhedra, or subsets defined by vanishing of a certain number of polynomial functions. However, we will need the general terminology as we will also need to consider spaces of mappings.

In a metric space X, we define a sequence $\{x_n\}$ of points to converge to a limit x_∞ if $\rho(x_n, x_\infty) \to 0$ as $n \to \infty$. The limit, if it exists, is unique, since if y were another limit we would have $\rho(y, x_\infty) = 0$. We call a metric space X *complete* if it satisfies Cauchy's convergence condition, namely that for any sequence $x_n \in X$ such that $\rho(x_m, x_n) \to 0$ as $m, n \to \infty$ there exists a limit point $x_\infty \in X$ such that $\rho(x_n, x_\infty) \to 0$ as $n \to \infty$.

For metric spaces X, topological conditions can be expressed in terms of convergence of sequences; for example, $f : X \to Y$ is continuous iff for all $x_i \to x \in x$ we have $f(x_i) \to f(x)$.

If X is a metric space, $x \in X$, $F \subseteq X$ is closed, and $x \notin F$, then x has a neighbourhood disjoint from F, so there exists $\varepsilon > 0$ such that $y \in F$ implies $\rho(x, y) \geq \varepsilon$, so $\rho(x, F) := \inf\{\rho(x, y) \mid y \in F\}$ is strictly positive. For any $A \subset$

X, $\rho(x, A) = 0$ if and only if x is in the closure of A if and only if there is a sequence $a_i \in A$ with $a_i \to x$.

Clearly $|\rho(x, F) - \rho(y, F)| \leq \rho(x, y)$, so the map $x \mapsto \rho(x, F)$ is continuous. If F and F' are disjoint closed sets, there are disjoint open neighbourhoods $G := \{x \mid \rho(x, F) < \rho(x, F')\}$ of F and similarly for G'. Hence any metric space is normal. It may be that $\rho(F, F') = 0$: for example, consider $F = \{(x, y) \in \mathbb{R}^2 \mid xy = 1\}$ and $F' = \{(x, y) \in \mathbb{R}^2 \mid y = 0\}$. However if K is compact and disjoint from F we have $\rho(F, K) > 0$, for the image of K by the continuous map $x \mapsto \rho(F, x)$ is a closed subset of $\mathbb{R}$ not containing $\{0\}$. Even if $\rho(F, F') = 0$, the formula $s(P) := \rho(P, F)/(\rho(P, F) + \rho(P, F'))$ defines a continuous map $s : X \to I$ with $s(F) = 0$ and $s(F') = 1$.

A metric space K is compact if and only if every sequence has a convergent subsequence. To see this, first observe that if $x_i \to y$, then the set whose elements are the x_i and y is compact, for given any open cover, one of the open sets of the cover contains y, hence all but finitely many of the x_i. Now if $\{x_i\}$ has no convergent subsequence, the set $\bigcup_i \{x_i\}$ is closed, its complement U is open, and $\{U \cup \{x_i\}\}$ is an open cover of K with no finite subcover. Conversely, if there is a cover with no finite subcover, there is a countable one $\{U_r\}$ and if we choose $x_n \notin \bigcup_{r \leq n} U_r$ if a subsequence converged to $y \in K$ we would have $y \in U_n$ for some n and then U_n would contain all but finitely many of the subsequence.

From this, or directly, it follows that the direct product of two, or indeed of any family of compact spaces is compact.

We will call a sequence $\{x_n\}$ with no convergent subsequence *discrete*. If $\{x_n\}$ is a discrete sequence, the set having these as elements is a closed set.

Lemma A.2.1 *Let $f : A \times B \to C$ be a continuous map of compact metric spaces. Then for any $\varepsilon > 0$ there exists $\delta > 0$ such that $\rho(b, b') < \delta$ implies that $\rho(f(a, b), f(a, b')) < \varepsilon$ for all $a \in A$.*

Proof Suppose not. Then there exist $\varepsilon > 0$ and sequences b_n, $b'_n \in B$ with $\rho(b_n, b'_n) < \frac{1}{n}$ and $a_n \in A$ with $\rho(f(a_n, b_n), f(a_n, b'_n)) \geq \varepsilon$. In view of compactness, these all have convergent subsequences; passing to these, we may suppose $b_n \to b$, $b'_n \to b'$ and $a_n \to a$. It follows that $\rho(b, b') = 0$, so $b = b'$ and by continuity that $\rho(f(a, b), f(a, b')) \geq \varepsilon$, a contradiction. □

The notion of compactness for spaces is accompanied by the important notion of properness for maps.

Lemma A.2.2 *The following conditions on a map $f : X \to Y$ of metric spaces are equivalent:*

(i) f is closed and for each $y \in Y$, $f^{-1}(y)$ is compact;

(ii) every sequence $x_i \in X$ such that $f(x_i)$ converges has a convergent subsequence;

(iii) for each compact subset K of Y, $f^{-1}(K)$ is compact.

A map is said to be *proper* if it satisfies these conditions.

Proof *(i)* $\Rightarrow$ *(ii)* Suppose (i) holds, that $\{x_n\}$ is discrete, but that $f(x_n)$ converges to a limit y. Since $C = \{x_n \mid n \in \mathbb{N}\}$ is closed, so is $f(C)$, and since $f(x_n) \to y$, $y \in C$. The same argument shows that for any subsequence $\{x_{n_k}\}$ of $\{x_n\}$ we have $y = f(x_{n_k})$ for some k. Thus $y = f(x_n)$ for all but finitely many n; hence $f^{-1}(y)$ contains a discrete sequence, contradicting its compactness.

(ii) $\Rightarrow$ *(iii)* Suppose (ii) holds, that $K \subset Y$ is compact, and that $f^{-1}(K)$ is not. Then $f^{-1}(K)$ contains a discrete sequence $\{x_n\}$. Since $\{f(x_n)\}$ lies in the compact set K, it has a convergent subsequence. It follows from (ii) that $\{x_n\}$ has a convergent subsequence, so is not discrete.

(iii) $\Rightarrow$ *(i)* It follows at once from (iii) that preimages of points are compact. Let C be closed in X and $f(x_n)$ be a sequence of points of $f(C)$ converging to a limit y. Then the set K consisting of y and the points $f(x_n)$ is compact, so by (iii) $f^{-1}(K)$ is compact. The sequence x_n of points in this compact set has a convergent subsequence x_{n_k} with limit x, say; as C is closed, $x \in C$. Thus $f(x_{n_k}) \to f(x)$; hence $y = f(x) \in f(C)$. □

It follows from the characterisation (iii) that the composite of two proper maps is proper. Also since the product of compact spaces is compact, for any X and compact K, the projection $K \times X \to X$ is proper. Since every closed subset of a compact space is compact, any continuous map $f : K \to Y$ with K compact is proper.

Lemma A.2.3 *A proper injective map $f : X \to Y$ of Hausdorff spaces is an embedding.*

Proof Replacing Y by $f(X)$, we may suppose f bijective. But now f takes closed sets to closed sets, hence also open sets to open sets, so is a homeomorphism. □

We now give some results for metric spaces which are useful for proving existence of embeddings when we weaken the requirement of compactness.

Lemma A.2.4 *(i) Let Y be a metric space, X a closed subset. For any open neighbourhood U of X in Y, there is a positive continuous function f on X such that if $x \in X$ and $\rho(x, y) < f(x)$, we have $y \in U$.*

(ii) If X is a compact subset of the metric space Y, any open neighbourhood U of X in Y contains an ε-neighbourhood for some $\varepsilon > 0$.

Proof (i) Define $f(x) = \rho(x, Y \setminus U)$: then $|f(x) - f(x')| \leq \rho(x, x')$, so f is continuous: it is non-zero and satisfies the condition.

(ii) Take $\varepsilon = \inf f$, where f is given by (i). □

We may apply this result in particular when $Y = X \times X$ with X embedded as the diagonal $\Delta(X)$. Thus if X is compact, there exists $\varepsilon > 0$ such that $\rho(x, y) < \varepsilon \Rightarrow (x, y) \in U$. Combining these ideas gives

Lemma A.2.5 *If X is a compact subset of the metric space Y, and U an open neighbourhood of $X \times X$ in $Y \times Y$, then for some $\varepsilon > 0$, if V is the ε-neighbourhood of X in Y, U contains $V \times V$.*

Proof Take $\varepsilon = \frac{1}{2}\rho(X \times X, (Y \times Y \setminus U))$. Then if $\rho(v_1, X) < \varepsilon$, $\rho(v_2, X) < \varepsilon$ we have $\rho((v_1, v_2), X \times X) < 2\varepsilon = \rho(X \times X, (Y \times Y \setminus U))$, so (v_1, v_2) does not lie in $Y \times Y \setminus U$. □

Corollary A.2.6 *Let Y be a metric space, $f : Y \to Z$ a map such that each $P \in Y$ has a neighbourhood U_P with $f|U_P$ an embedding, and $X \subset Y$ such that $f|X$ is injective. Then X has a neighbourhood V in Y such that $f|V$ is injective.*

If also each $f(U_P)$ is open, $f|V$ is an embedding.

Proof Let $D = \{(y_1, y_2) : y_1 \neq y_2, f(y_1) = f(y_2)\} \subset Y \times Y$. Since $f|X$ is injective, D is disjoint from $X \times X$. The closure $\bar{D}$ is contained in the closed subset defined by $f(y_1) = f(y_2)$, which is equal to $D \cup \Delta(Y)$. But by hypothesis, each point (P, P) has a neighbourhood $U_P \times U_P$ disjoint from D. Thus $\bar{D}$ is disjoint from $\Delta(Y)$, so D is closed. Now apply Lemma A.2.5, taking $U = Y \times Y \setminus D$: this gives a neighbourhood V of X such that $V \times V$ does not meet D, so $f|V$ is injective.

As each $f|U_P$ is an embedding, f induces a homeomorphism between U_P and $f(U_P)$ with the subspace topology. Thus the inverse map is continuous on $f(U_P)$, which is open in $f(V)$. Thus it is continuous at each point of $f(V)$. □

The following can be used to replace Theorem 1.1.4, which we proved for smooth manifolds.

Proposition A.2.7 *Suppose X locally compact and a countable union of compact subsets. Then there exist coverings by sets $F_a \subset G_a$ with each F_a compact, each G_a open, $\{G_a\}$ locally finite, and $\bigcup_a F_a = X$.*

Proof By Proposition 1.1.3, we can find compact subsets C_n and open subsets $B_{n+\frac{1}{2}}$ such that $X = \bigcup_n C_n$ and for all $n \geq 1$, $C_n \subset B_{n+\frac{1}{2}} \subset C_{n+1}$. It now suffices to set $F_n := C_{n+1} \setminus B_{n-\frac{1}{2}}$ and $G_n := B_{n+\frac{3}{2}} \setminus C_{n-1}$: these are locally finite since any $x \in X$ belongs to some $C_n \setminus C_{n-1}$, so the open set $B_{n+\frac{1}{2}} \setminus C_{n-1}$ is a neighbourhood of x, and meets G_N only if $n-2 \leq N \leq n+1$. □

The relation between paracompactness and countability is given by

Proposition A.2.8 *(i) If each component of X is open, X is paracompact if and only if each component is.*

(ii) A connected locally compact space X is paracompact if and only if it is a countable union of compact subsets.

Proof (i) is immediate since an open cover of X induces (and is induced by) open covers of each of its components.

(ii) If X is paracompact, the open covering by neighbourhoods of points with compact closures has a locally finite refinement. Since these sets have compact closures, each meets only finitely many others. Starting with one such set U_0, only finitely many others meet it; only finitely many meet one of the above, and so on. But since X is connected, each U_α is connected to U_0 by a finite chain. Thus there are only countably many U_α, and X is the union of their (compact) closures.

Conversely if $X = \bigcup_{n \geq 0} U_n$ is a countable union, setting $V_n := \bigcup_{0 \leq i \leq n} U_i$, we may assume the sequence V_n increasing. Any compact subset is covered by the U_n, hence by a finite subset, hence is contained in some V_n. Each point of V_n has a compact neighbourhood; V_n is covered by these neighbourhoods, hence by finitely many. Their union is compact, so is contained in some V_m. Thus, passing to a subsequence, we may suppose that V_{n+1} contains an open neighbourhood of V_n. Now any open cover of X induces one of the compact set $V_{n+1} \setminus Int V_n$, which has a finite refinement. The union of all these refines the given cover, covers all of X, and is locally finite since any point is in some $V_{n+1} \setminus Int V_n$, so has a neighbourhood contained in V_{n+2} and disjoint from V_{n-1}. □

The following useful result has a different nature.

Proposition A.2.9 *If X is a finite dimensional metric space, any open covering $\{U_\alpha\}$ has a finite dimensional refinement. More precisely, there exist an open covering $\{S_j \mid j \in J\}$ of X, with each S_j contained in U_α for some α, and a map $d : J \to \{0, \ldots, N\}$ such that if $d(j) = d(j')$, $j \neq j'$ then $\bar{S}_j \cap \bar{S}_{j'} = \emptyset$.*

We omit the proof, which is given by Hurewicz and Wallman on [76, p. 54]. To understand the result, the reader should consider the picture of a simplicial complex K of dimension N: each simplex of dimension r admits coordinates $\{x_0, \ldots, x_r\}$ with $x_i \geq 0$, $\sum_i x_i = 1$, and $S_r(K)$ is a union of sets contained in the interior of each r-simplex. Replace this simplicial complex K by its barycentric subdivision K': each vertex V of this is labelled by the dimension $d(V)$ of the simplex of which V is the barycentre. Now map each point of K' to the nearest vertex: more precisely, define an open neighbourhood of the vertex V to be

$$N(V) := \{x \in K' \mid (\forall W \neq V)\rho(x, V) \geq \rho(x, W) - 2^{-N}\},$$

where W runs over the vertices of K'. Now set $S_r(K) := \bigcup\{N(V) \mid d(V) = r\}$: a disjoint union of the neighbourhoods $N(V)$ with $d(V) = r$. Now if $f : X \to K$, define $S_r(X) := f^{-1}(S_r(K))$ to obtain subsets with the desired properties.

Notes on this section. The results on compactness and proper maps can be extended to general (not metric) topological spaces (see [24, §12]).

A closer study of the notion of properness is also given in [47, §3.2], using the following concept. For any map $f : X \to Y$, define the *improper set* $Z(f)$ as the set of $y \in Y$ such that there is a discrete sequence $\{x_n \mid n \in \mathbb{N}\}$ on X with $f(x_n) \to y$. This is the smallest closed subset of Y such that the restriction of f to a map $X \setminus f^{-1}(Z) \to Y \setminus Z$ is proper: thus is empty if and only if f is proper.

A.3 Proper group actions

A (left) *action* of a group G on a set X is a map $\phi : G \times X \to X$ such that $\phi(1, x) = x$ for all $x \in X$ and $\phi(g, \phi(h, x)) = \phi(gh, x)$ for all $x \in X$ and $g, h \in G$. We usually denote $\phi(g, x)$ by $g.x$. We are really only interested in smooth group actions, so X will be a Hausdorff space throughout.

Given an action ϕ, the *isotropy group* of $x \in X$ is $G_x := \{g \in G \mid g.x = x\}$. The *orbit* of x is $G.x := \{g.x \mid g \in G\}$. The action induces a bijection $G/G_x \to G.x$ since

$$g.x = h.x \Leftrightarrow h^{-1}g.x = x \Leftrightarrow h^{-1}g \in G_x \Leftrightarrow hG_x = gG_x.$$

Equivalently, the map $\phi_x : G \to X$ defined by $\phi_x(g) := g.x$ induces an injection of G/G_x into X.

Given a left group action, we denote the set of orbits by $G\backslash X$ and the projection by $q : X \to G\backslash X$. We give $G\backslash X$ the quotient topology and call it the *orbit space*. The map q is open, for if U is open in X, $q^{-1}(q(U)) = \bigcup_{g\in G} g.U$, a union of open sets, hence open; by the definition of quotient topology, $q(U)$ is open.

Proposition A.3.1 *Let $\phi : G \times X \to X$ be a group action. Then the following are equivalent:*

(i) The map $(\phi, \pi) : G \times X \to X \times X$ *(where* π *denotes the projection) is a proper map;*

(ii) (ϕ, π) *is closed and all isotropy groups* G_x *are compact;*

(iii) for any compact subsets $K, L \subseteq X$, $T_{K,L} := \{g \in G \mid g.K \cap L \neq \emptyset\}$ *is compact.*

Proof $(i) \Rightarrow (ii)$ since $G_x \times \{x\}$ is the preimage of (x, x) under (ϕ, π).

$(ii) \Rightarrow (i)$ since the preimage of (y, x) is empty if $y \notin G.x$, and if $y = g.x$ is the coset gG_x, homeomorphic to G_x.

Now by Lemma A.2.2, (i) is equivalent to the condition that for any compact subset of $X \times X$, its preimage under (ϕ, π) is compact. It is sufficient to consider subsets of the form $L \times K$, where K and L are compact subsets of X. We have $(\phi, \pi)^{-1}(L \times K) = \{(g, x) \mid g.x \in L, x \in K\} \subseteq T_{K,L} \times K$. Thus if $T_{K,L}$ is compact, so is this (closed) subset of it; and if this set is compact, so is its projection on the first factor, which is $T_{K,L}$. □

A group action will be called *proper* if it satisfies the equivalent conditions of Proposition A.3.1. It is not true that for any proper group action ϕ itself is a closed map: consider, for example, $G = X = \mathbb{R}$ with action by translation.

Lemma A.3.2 *(i) A group action of a compact group is proper.*

(ii) Given two Lie subgroups H, K *of* G *with* K *compact, the natural action of* H *on the coset space* G/K *is proper.*

Proof (i) It will suffice to show that the preimage of a compact $C \subset X \times X$ is compact. The second projection C_2 of C is compact, and the preimage of C is a closed subset of the compact set $G \times C_2$.

(ii) It is enough to show that the action of G on G/K is proper. Any compact subset C of $G/K \times G/K$ is a subset of some $C_1 \times C_2$ with each C_i compact, and the preimage of C_i in G is a compact set B_i. The image of $B_1 \times B_2$ by the map $(x, y) \to xy^{-1}$ is a compact set B. Now the preimage of C in $G \times G/K$ is a closed subset of the compact set $B \times C_2$. □

Proposition A.3.3 *Let* $\phi : G \times X \to X$ *be a proper group action and* $x \in X$. *Then*

(i) the isotropy group G_x *is compact;*

(ii) the map $\phi_x : G \to X$ *given by* $\phi_x(g) = g.x$ *is proper;*

(iii) the orbit $G.x$ *is a closed subset of* X*;*

(iv) the induced map $G/G_x \to G.x$ *is a homeomorphism.*

Proof (i) $G_x \times \{x\}$ is the preimage of the point (x, x) (a compact set) under the proper map (ϕ, π).

(ii) This is a closed map as it is the restriction of (ϕ, π) to the closed subset $G \times \{x\}$. The preimage of a compact set K is the preimage of $K \times \{x\}$ under (ϕ, π), so is compact.

(iii) It is the image of G under the proper, hence closed map ϕ_x.

(iv) This map is bijective by construction, continuous since ϕ_x is, and by the definition of the quotient topology, and closed since ϕ_x is. □

Proposition A.3.4 *Let $\phi : G \times X \to X$ be a smooth proper group action. Then the quotient space $G\backslash X$ is Hausdorff, locally compact, and paracompact.*

Proof Write Δ for the diagonal in $G\backslash X \times G\backslash X$. Since (ϕ, π) is closed, $C := \{(x, g.x) \mid x \in X\}$ is closed in $X \times X$. Now $C = (q, q)^{-1}(\Delta)$, and since $G\backslash X \times G\backslash X$ has the quotient topology, it follows that Δ is closed in $G\backslash X \times G\backslash X$. Thus $G\backslash X$ is Hausdorff.

By Theorem 3.3.5, any point of X has an invariant neighbourhood of the form $j(G \times_H V)$ with $H \subseteq G$ a compact subgroup and V a disc on which H acts orthogonally. Thus any point of $G\backslash X$ has a neighbourhood of the form $H\backslash V$, which is compact. So $G\backslash X$ is locally compact.

By Proposition A.2.8, paracompactness will follow provided $G\backslash X$ is a countable union of compact subsets. But this follows since X is such a union, and the image of a compact set is compact. □

For the special case when G is compact, we have

Proposition A.3.5 *If $\phi : G \times X \to X$ is a group action with G compact, then*

(i) the map ϕ is a proper map;

(ii) the action is proper;

(iii) the map $q : X \to X/G$ is proper;

(iv) for any $Y \subset X$, any neighbourhood of Y contains a G-invariant neighbourhood.

Proof (i) Suppose F a closed subset of $G \times X$: we want to prove that any limit point x of $\phi(F)$ belongs to $\phi(F)$. Suppose $(g_i, x_i) \in F$ and $g_i.x_i \to x$. Since G is compact, $\{g_i\}$ has a convergent subsequence. Passing to this subsequence, we may write $g_i \to g$. Then $x_i = g_i^{-1}.(g_i.x_i) \to y := g^{-1}.x$. Thus $(g_i, x_i) \to (g, y)$, so $(g, y) \in F$ and $x = g.y \in \phi(F)$.

(ii) Similarly if $(g_i, x_i) \in F$ and $(g_i.x_i, x_i) \to (y, x)$ we have $x_i \to x$ and may suppose $g_i \to g$; thus $g.x = y$, $(g, x) \in F$ and $(y, x) = (\phi, \pi)(g, x)$.

(iii) The preimage under q of a point $q(x)$ is the orbit $G.x$, which is compact since G is. Now suppose F is closed in X: then $G \times F$ is closed in $G \times X$; since by (i) ϕ is proper, $G.F = \phi(G \times F)$ is closed in X. Now by the definition of quotient topology, $q(F)$ is closed in X/G since $q^{-1}(q(F)) = G.F$ is closed in X.

(iv) Let U be an open set containing Y. Then $W := X \setminus q^{-1}q(X \setminus U)$ is G-invariant and contained in U. Since q is proper $X \setminus U$ is closed, thus W is open. □

A similar argument shows that in general if $K \subset G$ is compact then the restriction of ϕ to $K \times X \to X$ is proper, and hence if $A \subseteq X$ is closed (compact) so is $K.A$.

Proposition A.3.6 *Let G act properly on M and ρ be a G-invariant metric on M. Define $\overline{\rho} : G\backslash M \times G\backslash M \to \mathbb{R}$ by $\overline{\rho}(G.x, G.y) := inf_{g \in G}\, \rho(x, g.y)$. Then $\overline{\rho}$ is a metric on $G\backslash M$.*

Proof Since the action is proper, the orbit $G.y$ is closed. Thus if $x \notin G.y$, $\rho(x, G.y) > 0$, i.e. $G.x \neq G.y$ implies $\overline{\rho}(G.x, G.y) \neq 0$.

For any $x, y, z \in M$ and any $\varepsilon > 0$ we can choose $g, g' \in G$ with $\rho(x, g.y) < \overline{\rho}(G.x, G.y) + \varepsilon$ and $\rho(y, g'.z) < \overline{\rho}(G.y, G.z) + \varepsilon$. Thus

$\overline{\rho}(G.x, G.z) \leq \rho(x, gg'.z) \leq \rho(x, g.y) + \rho(g.y, gg'.z)$,

and this is equal to

$\rho(x, g.y) + \rho(y, g'.z) < \overline{\rho}(G.x, G.y) + \overline{\rho}(G.y, G.z) + 2\varepsilon$.

Since this holds for any $\varepsilon > 0$, we have $\overline{\rho}(G.x, G.z) \leq \overline{\rho}(G.x, G.y) + \overline{\rho}(G.y, G.z)$, so the triangle inequality holds. □

Note As for the definition of proper maps, one can define and study a 'bad set'. If G is a locally compact group acting on a Hausdorff space X, then $x \in X$ is a *wandering point* if it has a neighbourhood V_x such that $\{g \in G \mid V_x.g \cap V_x \neq \varnothing\}$ has compact closure, or equivalently, if there exists a compact subset $K \subset G$ such that $g \notin K$ implies $V_x.g \cap V_x = \varnothing$. The set $\Omega(X)$ of all wandering points is open, and the action of G on $\Omega(X)$ is proper; the action on X is proper if and only if $\Omega(X) = X$. In the case when G is a discrete group, the term 'properly discontinuous' is often used instead of 'proper'.

A.4 Mapping spaces

We begin by discussing topologies on the set $C^0(X, Y)$ of continuous maps between two topological spaces X and Y. We are only interested here in the case when X and Y are manifolds, and hence metrisable.

Perhaps the most commonly used topology on function spaces is the so-called *compact-open topology*, which we call the C^0 topology. This is the topology on $C^0(X, Y)$ defined by taking the sets

$$A(K, U) := \{f \mid f(K) \subset U\} \quad \text{with } K \subset X \text{ compact, } U \subset Y \text{ open}$$

as a sub-base of open sets. It can be described as the topology of uniform convergence of f on compact sets.

There is also the *fine topology* (or fine C^0 topology), which we define by taking the

$$B(U) := \{f \mid (1 \times f)(X) \subset U\} \quad \text{with } U \text{ open in } X \times Y$$

as a base of open sets.

Lemma A.4.1 *(i) The sets* $I(\{K_\alpha, U_\alpha\}) := \bigcap_\alpha A(K_\alpha, U_\alpha)$, *with* $K_\alpha \subset X$ *compact,* $U_\alpha \subset Y$ *open,* $\{K_\alpha\}$ *locally finite, are a subbase for the fine topology.*

(ii) For $f \in C^0(X, Y)$ *and* ρ *a metric on* Y, *the sets*

$$J(f, k) := \{g \in C^0(X, Y) \mid (\forall x \in X)\, \rho(f(x), g(x)) < k(x)\},$$

with $k \in C^0(X, \mathbb{R}_{>0})$, *are a base of neighbourhoods of* f *in the fine topology.*

Proof We have $J(f, k) = B(U)$, where $U = \{(x, y) \in X \times Y \mid \rho(y, f(x)) < k(x)\}$, hence $J(f, k)$ is open. That these give a base of neighbourhoods of f follows by applying Lemma A.2.4 to neighbourhoods of the graph of f in $X \times Y$.

The set $A(K_\alpha, U_\alpha)$ is the preimage by $1 \times f$ of the open subset

$$((X \setminus K_\alpha) \times Y) \cup (X \times U_\alpha)$$

of $X \times Y$. Any finite intersection of these subsets is thus also open. But by hypothesis, any $x \in X$ has an open neighbourhood U_x intersecting K_α for only finitely many α. Thus the intersection of $U_x \times Y$ with $I(\{K_\alpha, U_\alpha\})$ is equal to its intersection with a finite number of the $A(K_\alpha, U_\alpha)$ and hence is open. It follows that $I(\{K_\alpha, U_\alpha\})$ is open.

For the converse, it will suffice to check that any neighbourhood $J(f, k)$ of f contains one of the form $I(\{K_\alpha, U_\alpha\})$. It will suffice if the K_α cover X and $z \in K_\alpha$, $y \in U_\alpha$ implies $\rho(f(z), y) < k(z)$. For each x, set $U_x := \{y \mid \rho(y, f(x)) < \frac{1}{3}k(x)\}$, choose a compact neighbourhood $K_x \subset f^{-1}(U_x) \cap \{z \mid k(z) > \frac{2}{3}k(x)\}$ now let $\{K_\alpha\}$ be a locally finite subcover of the sets K_x. □

If X *is compact, the fine topology, the* C^0 *topology and the topology of uniform convergence are the same.*

For when X is compact, the functions k in $J(f, k)$ have positive lower bounds, so the base of neighbourhoods $J(f, k)$ is equivalent to the base of neighbourhoods $J(f, c)$ (c constant), which defines the uniform topology.

Both topologies are Hausdorff; indeed completely regular.

We will shortly see that the C^0 topology is metrisable, hence Hausdorff and normal.

For the fine topology, any closed set C not containing f is disjoint from some $J(f, k)$, so $J(f, \frac{1}{2}k)$ is an open set containing f disjoint from the open set $\{g \in C^0(X, Y) \mid (\forall x \in X)\, \rho(f(x), g(x)) > \frac{1}{2}k(x)\}$ which contains C.

If X is not compact, the fine topology is very large, and the two topologies are distinct.

Proposition A.4.2 *(i) The space $C^0(X, Y)$ with the C^0 topology has a complete metric.*

(ii) A sequence of maps which converges in the fine topology is eventually constant outside a compact set.

(iii) If X is not compact, the fine topology on $C^0(X, Y)$ is not metrisable, and does not admit a countable base, even locally.

Proof (i) First suppose X compact, then choose a complete metric ρ on Y and take the uniform metric $\rho(f, g) = \sup_{x \in X} \rho(f(x), g(x))$. This is complete since if $\{f_n\}$ is a Cauchy sequence, so is each $\{f_n(x)\}$, which thus converges to a limit $f(x)$, and f is continuous as the uniform limit of $\{f_n\}$.

For X not compact, write $X = \bigcup_{i=1}^{\infty} X_i$ as a countable union of compact subsets. Then the topology for $C^0(X_i, Y)$ is defined by a complete metric ρ_i, hence also by the bounded metric $\rho'_i(f, g) := \min(\rho_i(f, g), 2^{-i})$. The metric $\rho := \sum_{i=1}^{\infty} \rho'_i$ defines the product topology on $\Pi_i C^0(X_i, Y)$, and hence the required topology on the subset $C^0(X, Y)$. Moreover, $C^0(X, Y)$ is a closed subset of the complete $\Pi_i C^0(X_i, Y)$ and is thus also complete.

(ii) Assume $f_n \to f$ and that for no compact $K \subset X$ is the sequence f_n eventually constant outside K. Choose an increasing sequence $\{K_n\}$ of compact subsets of X with union X. By hypothesis, there exist $x_n \in (X \setminus K_n)$ and $i_n > n$ with $f_{i_n}(x_n) \neq f(x_n)$. Set $\delta_n := \rho(f_{i_n}(x_n), f(x_n))$. Since the sequence x_n diverges, we can find a positive continuous function k on X such that $k(x_n) = \frac{1}{2}\delta_n$ for infinitely many n. For none of these n is $f_{i_n} \in J(f, k)$, contradicting the assumption that $f_n \to f$.

(iii) follows from (ii). □

Not only compactness of spaces, but properness of maps is important in discussing these topologies, and we have

Lemma A.4.3 *If Y is a locally compact, paracompact metric space, the set $C^0_{pr}(X, Y)$ of proper maps is open in $C^0(X, Y)$ in the fine topology.*

Proof By Proposition A.2.7, there exist coverings of Y by sets $F_a \subset G_a$ with each F_a compact, each G_a open, $\{G_a\}$ locally finite, and $\bigcup_a F_a = Y$.

For $f : X \to Y$ a proper map, the sets $K_a := f^{-1}(F_a)$ are compact. They are locally finite, since for any $x \in X$, $f(x)$ has a neighbourhood U_x meeting only finitely many of the G_a, so $f^{-1}(U_x)$ is a neighbourhood of x meeting only finitely many of the K_a. Hence by Lemma A.4.1, $I(\{K_a\}, \{G_a\})$ is a neighbourhood of f in the fine topology.

We claim that any $g \in I(\{K_a\}, \{G_a\})$ is proper. For any compact subset $L \subset Y$ meets only finitely many G_a, so $g^{-1}(L)$ is contained in the union of the corresponding K_a, so is compact. □

We turn to the question of continuity of the composition map.

Proposition A.4.4 *(i) The composition map $C^0(X, Y) \times C^0(Y, Z) \to C^0(X, Z)$ is continuous for the C^0 topologies.*

(ii) The map $C^0(Y, Z) \to C^0(X, Z)$ defined by composition with a continuous map f is continuous for the fine topologies if and only if f is proper.

(iii) The composition map $C^0_{pr}(X, Y) \times C^0(Y, Z) \to C^0(X, Z)$ is continuous for the fine C^0 topologies.

Proof (i) It will suffice to show that the preimage of a subbasic open set $A(K_X, U_Z)$ is open, and thus to show that if $g \circ f \in A(K_X, U_Z)$, it contains a neighbourhood of (f, g).

Since $g \circ f \in A(K_X, U_Z)$ and f is proper, $f(K_X)$ is a compact subset of Y and $g^{-1}(U_Z)$ is an open neighbourhood of it. By Lemma A.1.1, this contains a compact neighbourhood, so we can find a compact K_Y and an open U_Y with $f(K_X) \subset U_Y \subset K_Y \subset g^{-1}(U_Z)$.

It follows that the preimage of $A(K_X, U_Z)$ contains the open neighbourhood $A(K_X, U_Y) \times A(K_Y, U_Z)$ of (f, g).

(ii) If f is not proper, there is a discrete sequence $x_n \in X$ such that $f(x_n)$ converges to a limit $y_0 \in Y$. Let $g : Y \to Z$ be continuous, and consider a neighbourhood $J(g \circ f, k)$ of $g \circ f$. We want to show that for some k, $f^*J(g \circ f, k)$ is not open, in fact does not contain a neighbourhood $J(g, \ell)$ of g. For if it does, $\rho(g(y), h(y)) < \ell(y)$ for all y implies $\rho(g(f(x)), h(f(x))) < k(x)$ for all x.

Since x_n is discrete, we can choose k with $k(x_n) = n^{-1}$ for all n. Now as $f(x_n) \to y_0$, if $\rho(g(f(x_n)), h(f(x_n))) < k(x_n) = n^{-1}$ for all n, it follows that $\rho(g(y_0), h(y_0)) = 0$. Thus we do not have a neighbourhood of g.

(iii) We copy (i); so start with neighbourhood $I(\{K^X_\alpha, U^Z_\alpha\})$ of $g \circ f$: here as well as the U^Z_α being open, the K^X_α are locally finite. Any y has a compact neighbourhood B_y: then $f^{-1}(B_y)$ is compact (as f is proper), so meets only finitely many of the K^X_α. Hence the $f(K^X_\alpha)$ are locally finite.

We now have a locally finite family of compact sets $f(K_\alpha^X)$ with neighbourhoods $g^{-1}(U_\alpha^Z)$ and seek $f(K_\alpha^X) \subset U_\alpha^Y \subset K_\alpha^Y \subset g^{-1}(U_\alpha^Z)$ with the K's compact, the U's open and the K_α^Y locally finite. We restate the problem. First use countable compactness to say the set of α is countable. We have a locally finite family of compact sets A_n with open neighbourhoods D_n and seek $A_n \subset B_n \subset C_n \subset D_n$ with the C_n compact, the B_n open and the C_n locally finite.

Shrinking the D_n, we may suppose each meets only finitely many of the A_i. Now by Lemma A.1.1 we can find B_n and C_n as above, but have yet to make the C_n locally finite. Set

$$C'_n := C_n \setminus \bigcup \{D_r \mid r < n,\ D_r \cap A_n = \emptyset\}.$$

Since $A_n \subset C_n$ and we have only removed subsets disjoint from A_n, we have $A_n \subset C'_n$. If C'_r meets C'_n with $r < n$ then C'_n meets D_r so $D_r \cap A_n \neq \emptyset$. For each r this holds for finitely many n, and there are only finitely many $n < r$, so C'_r meets only finitely many C'_n. It remains only to take B'_n as a neighbourhood of A_n contained in C'_n.

In the original notation, it follows that the preimage of $I(\{K_\alpha^X, U_\alpha^Z\})$ contains $I(\{K_\alpha^X, U_\alpha^Y\}) \times I(\{K_\alpha^Y, U_\alpha^Z\})$, a product of neighbourhoods of f and g. □

We next discuss the Baire property, which is important for many of our applications.

Theorem A.4.5 (Baire's Theorem) *Let X be a complete metric space. The intersection of a countable family of dense open subsets of X is dense.*

Proof Let the given subsets be $\{U_i\}$, and let V be any non-empty open set. Then $V \cap U_1$ is non-empty and open, and so contains a metric neighbourhood $U(x_1, \varepsilon_1)$, say. Next, $U_2 \cap U(x_1, \varepsilon_{\frac{1}{2}})$ is non-empty and open, so contains some $U(x_2, \varepsilon_2)$. We can thus construct a decreasing sequence of neighbourhoods $U(x_i, \varepsilon_i)$ and have $\varepsilon_i \to 0$. Then $\{x_i\}$ is a Cauchy sequence, so has a limit point x, which lies in each $\bar{U}(x_i, \varepsilon_i)$ (since the later x_j do) and so in each U_i and in V. □

This result shows that any complete metric space has the Baire property. It follows from Proposition A.4.2 that $C^0(X, Y)$ with the C^0 topology has the Baire property. For the fine topology, we have to work harder.

Theorem A.4.6 *If X is paracompact and Y a complete metric space, then $C^0(X, Y)$ with the fine topology is a Baire space.*

Further, if $Q \subset C^0(X, Y)$ is closed in the C^0 topology, then Q with the fine topology is a Baire space.

Proof Let $\{U_i\}$ be a countable sequence of open dense sets and V a further open set. Choose $f_0 \in V$ and a neighbourhood $J(f_0, k_0)$ of f_0 with closure contained in V.

Now suppose inductively chosen functions $f_0, \dots, f_r$ and neighbourhoods $J(f_i, k_i)$ $(0 \le i \le r)$ such that $f_i \in V$, $f_i \in J(f_j, k_j)$, $k_i < 2^{-i}$ and $\overline{J(f_i, k_i)} \subset U_i$ for $j \le i \le r$. Since U_{r+1} is dense, it meets the open set $\bigcap_{i=0}^{r} J(f_i, k_i)$: choose f_{r+1} in the intersection, and choose a neighbourhood $J(f_{r+1}, k_{r+1})$ with closure contained in it and with $k_{r+1} < 2^{-(r+1)}$.

Since ρ is a complete metric, and the sequence f_r converges uniformly, we can define f to be its limit. Since all f_i with $i > r$ belong to $J(f_r, k_r)$, f belongs to its closure, which is contained in U_r $(r > 0)$ or V $(r = 0)$. Thus $V \cap \bigcap_r U_r$ is non-empty, as required.

Given a countable sequence of open dense subsets W_i of Q, we can take $U_i := W_i \cup (C^0(X, Y) \setminus Q)$ and argue as above. We only need to note that since $Q \subset C^0(X, Y)$ is closed in the C^0 topology, the uniform limit f of the maps $f_i \in Q$ also belongs to Q. □

For smooth manifolds V^v and M^m, write $C^r(V, M)$ for the set of maps $V \to M$ whose restrictions in any local coordinates have continuous partial derivatives of all orders $\le r$; in particular, $C^\infty(V, M)$ is the set of smooth maps of V to M. Taking r-jets gives an injective map $j^r : C^r(V, M) \to C^0(V, J^r(V, M))$. The topology on $C^r(V, M)$ induced by regarding it as a subspace of $C^0(V, J^r(V, M))$ with the compact-open topology is called the C^r *topology*, and the topology induced from the fine topology is the *fine* C^r *topology*. The image of $j^r : C^r(M, N) \to C^0(M, J^r(M, n))$ is closed in the C^r topology.

The inclusion of $C^\infty(V, M)$ in $C^r(V, M)$ induces topologies on it, and we define the C^∞ *topology* to be the union of the C^r topologies, in the sense that a set is open if it is open in one of these topologies. Correspondingly, the fine C^∞ topology, which we christen the W^∞ *topology*, is the union of the fine C^r topologies.

The properties of these metrics are similar to those for the case $r = 0$, and the proofs run in parallel, though with complications of detail (the case $r = \infty$ requiring a little more effort), so we omit most of them. The discussion extends to manifolds with boundaries, corners, etc. The following statements hold for all $r \le \infty$: it is the case $r = \infty$ which is of prime interest to us.

We have equivalent characterisations of the fine C^r topology if the above conditions on the images of the maps are replaced by conditions on the r-jets. However if in the C^∞ version of $I(\{K_\alpha, U_\alpha\})$ we allow the U_α to be open in jet spaces $J^r(V, M)$ for varying values of r we obtain a new topology, the *very strong topology*, which we do not discuss further in this book.

Both topologies on $C^\infty(V, M)$ are completely regular. They agree if V is compact.

For the W^∞ topology, a convergent sequence of maps is eventually constant outside a compact set; hence the topology is neither metrisable nor even locally countable.

Theorem A.4.7 *With the C^∞ topology, $C^\infty(V, M)$ is a complete metric space.*

Proof First suppose V is compact. Each jet space $J^r(V, M)$ is a smooth manifold, and admits a complete Riemannian metric ρ^r, say. The distance function $\rho^r(f, g) := \sup_{P\in V} \rho^r(j^r f(P), j^r g(P))$ is well defined since V is compact, and defines the C^r topology on $C^\infty(V, M)$.

The same topology on $J^r(V, M)$ is given by the non-Riemannian metric $\rho'^r = \inf(\rho^r, 1)$, and the metric $\rho(f, g) = \sum_r 2^{-r} \rho'^r(f, g)$ defines the C^∞ topology on $C^\infty(V, M)$.

A Cauchy sequence $\{f_i\}$ in $C^\infty(V, M)$ must *à fortiori* be Cauchy with the metric ρ^r. Since $J^r(V, M)$ is complete, the maps $j^r f_i$ converge to a limit g^r, which is continuous, since the convergence was uniform.

The coordinates $u_{\omega,j}$ of $j^r f_i$ are the partial derivatives of the $u_{0,j}$. Let ω' be derived from ω by increasing ω_i by unity, and $|\omega'| \leq r$: then $u_{\omega',j} = \partial u_{\omega,j}/\partial x_i$ and so $u_{\omega,j}$ is the indefinite integral with respect to x_i of $u_{\omega',j}$. Integration commutes with uniform limits, so the relation $u_{\omega',j} = \partial u_{\omega,j}/\partial x_i$ also holds for g^r. Thus the $u_{0,j} = y_j$ are r-times continuously differentiable, g^r is the r-jet of a C^r function g, independent of r, so g is smooth, and is the limit of the sequence.

For V not compact, write $V = \bigcup_{i=1}^\infty V_i$ as a countable union of compact submanifolds (with boundary). Then the topology for $C^\infty(V_i, M)$ is defined by a metric ρ_i, bounded by 1. Hence the metric $\rho = \sum_{i=1}^\infty 2^{-i} \rho_i$ defines the product topology on $\Pi_i C^\infty(V_i, M)$, and hence the required topology on the subset $C^\infty(V, M)$.

Now $C^\infty(V, M)$ is a closed subset of the complete $\prod_i C^\infty(V_i, M)$ which is thus also complete. □

Lemma A.4.8 *The set $C^\infty_{pr}(V, M)$ of proper smooth maps is open in $C^\infty(V, M)$ in the W^∞ topology.*

Proof Let $f : V \to M$ be a proper map, and $\{\varphi_\alpha : U_\alpha \to \mathring{D}^m(3)\}$ a locally finite open cover of M as in Theorem 1.1.4, so that M is covered by the compact sets $K_\alpha := \varphi_\alpha^{-1}(D^m(2))$. Since f is proper, $F_\alpha := f^{-1}(K_\alpha)$ is compact. Then $\mathcal{W} := \{g \mid \forall\alpha\ g(F_\alpha) \subset U_\alpha\}$ is an open neighbourhood of f. For any $g \in \mathcal{W}$ and any compact $L \subset M$, L meets only finitely many U_α, so $g^{-1}(L)$ is contained in the union of the corresponding F_α, so is compact. □

The composition map $C^\infty(V, M) \times C^\infty(M, N) \to C^\infty(V, N)$ is continuous for the C^∞ topologies; however for the W^∞ topologies this fails unless V is compact: more precisely, $C^\infty_{pr}(V, M) \times C^\infty(M, N) \to C^\infty(V, N)$ is continuous, and the map $C^\infty(M, N) \to C^\infty(V, N)$ defined by composition with $f : V \to M$ is continuous if and only if f is proper.

Theorem A.4.9 (see, for example, [73, 2.4.4], [57, 3.4]) *If F is any subspace of $C^\infty(V, M)$ which is closed in the C^∞ topology, then F (with either the C^∞ topology or the W^∞ topology) has the Baire property.*

For example, if $f \in C^\infty(V, M)$ and K is a closed subset of V, we can take $F = \{g \in C^\infty(V, M) \mid g|K = f|K\}$.

We also have

Theorem A.4.10 *If W is open in $C^\infty(V, M)$ with the C^∞ or W^∞ topology, then W has the Baire property.*

Proof Since $C^\infty(V, M)$ is completely regular, for any $f \in W$ we can choose a neighbourhood U of f whose closure $F \subset W$. If now the U_i are dense open subsets of W, the $U_i \cup (W \setminus F)$ are dense open subsets of X, hence their intersection $\bigcap U_i \cup (W \setminus F)$ is dense in X, and hence intersects U. Thus U meets $\bigcap U_i$, and since this holds for any neighbourhood of f contained in U, f is in the closure of $\bigcap U_i$. As this holds for all $f \in W$, $\bigcap U_i$ is dense in W. $\square$

Appendix B

Homotopy theory

I do not know any book on homotopy theory which covers all the material to which I need to refer, but one useful introduction is May's book [89].

B.1 Definitions and basic properties

A continuous map $X \times I \to Y$ is said to be a *homotopy* between the maps $X \to Y$ given by its restrictions to $X \times \{0\}$ and $X \times \{1\}$. The relation of homotopy between maps is an equivalence relation. A major concern of homotopy theory is the set of homotopy equivalence classes of maps $X \to Y$, which in this appendix we denote by $[X : Y]$. Unless otherwise stated we fix base points in X and Y and require maps and homotopies to respect the base point. The base point is usually denoted $*$, but is often suppressed from the notation. A map $X \to Y$ homotopic to the constant map $X \to *$ is said to be nullhomotopic. We write X^+ for the disjoint union of X and a point, taken as base point.

An important type of homotopy occurs when $B \subset A$, $h : A \times I \to A$ satisfies $h(x, 0) = x$ for all $x \in A$, $h(x, t) = x$ for all $x \in B$, $t \in I$ and $h(A \times \{1\}) = B$: B is then called a *deformation retract* of A and h is a *deformation retraction*. A simple example is when A is a square and B the union of three sides.

Two spaces X, X' are said to be *homotopy equivalent* if there are maps $f : X \to X'$ and $f' : X' \to X$ such that each composite $f \circ f'$, $f' \circ f$ is homotopic to the identity map.

If $f : S^{n-1} \to X$ is a continuous map, we define a space $X \cup_f e^n$: as a set, we have the disjoint union of X and $\mathring{D}^n$; the map $g : D^n \to X \cup_f e^n$ is given by the identity on $\mathring{D}^n$ and by f on S^{n-1}; and we declare a subset to be open if its preimages by both g and the inclusion of X are open. This process is called attaching an n-cell to X. We can allow $n = 0$: S^{-1} is the empty set, so $X \cup_f e^0 = X^+$ is the disjoint union of X and a point.

A space obtained by attaching a finite number of cells to the empty set is a cell complex. A *CW-complex* is obtained by a (possibly infinite) sequence of attachments of cells to $\emptyset$, subject to the condition that each attaching map has image in a finite subcomplex, and that the topology is given by declaring a set to be open if its intersection with each finite subcomplex is. A *CW-pair* (K, L) consists of a CW-complex L and a CW-complex K obtained from L by attaching cells. We are mainly interested in finite CW-complexes and pairs, or at worst those with a finite number of cells of each dimension.

Given a CW-complex (or pair) we can change the attaching maps by homotopies (and K by a homotopy equivalence) to ensure that cells are attached in order of increasing dimension: the argument parallels that of §5.2, which is modelled on the CW case. The space obtained at the intermediate stage when all cells of dimension $\leq n$ have been attached, is called the *n-skeleton* of K and denoted $K^{(n)}$.

In general, we use the term 'space' for a topological space homotopy equivalent to a CW-complex. This class of objects is closed under various natural constructions, including fibrations and formation of function spaces (with the compact-open topology).

For any space X and $n \geq 1$, the set $[S^n : X]$ has the structure of a group and is denoted $\pi_n(X)$. The group is abelian if $n \geq 2$; if X is connected, it is independent of the base point. The group $\pi_1(X)$ is called the fundamental group of X.

Given a space Y and subspace X, we can similarly define $\pi_n(Y, X)$ using maps $f : D^n \to Y$ with $f(S^{n-1}) \subset X$; more generally given any map $j : X \to Y$ we define $\pi_n(j)$. There is an exact sequence

$$\ldots \pi_n(X) \xrightarrow{j_*} \pi_n(Y) \to \pi_n(j) \to \pi_{n-1}(X) \ldots.$$

Going one further, given a commutative diagram

$$\begin{array}{ccc} A & \xrightarrow{p} & B \\ \Phi : q \downarrow & & r \downarrow, \\ C & \xrightarrow{s} & D \end{array}$$

we can define $\pi_n(\Phi)$ by homotopy classes of commutative diagrams of maps of an n-sphere, the upper and lower hemispheres of its boundary, and the equator into Φ: this is a group for $n \geq 3$. There are exact sequences

$$\ldots \pi_n(p) \to \pi_n(s) \to \pi_n(\Phi) \to \pi_{n-1}(p),$$
$$\ldots \pi_n(q) \to \pi_n(r) \to \pi_n(\Phi) \to \pi_{n-1}(q).$$

A space X is *contractible* if it is homotopy equivalent to a point. It is *weakly contractible* if any map $K \to X$, with K a finite CW-complex, is homotopic to a constant map. It is sufficient to check this for K a sphere, i.e. that $\pi_n(X)$ is trivial for all $i \geq 0$.

If we merely suppose that every map $K \to X$, with K a finite CW-complex of dimension $\leq n$, is homotopic to a constant map, X is called *n-connected*. For this, it is sufficient that $\pi_n(X)$ is trivial for all $0 \leq i \leq n$.

Recall that a map $f : X \to Y$ is said to be a *weak homotopy equivalence* if, for any CW-pair (K, L) and maps $a : L \to X$ and $b : K \to Y$ with $b \,|\, L = f \circ a$ there exists $c : K \to X$ with $c \,|\, L = a$ and $f \circ c$ homotopic to b keeping L fixed.

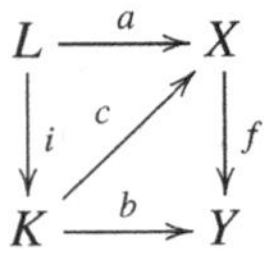

$$\begin{array}{ccc} L & \xrightarrow{a} & X \\ {\scriptstyle i}\downarrow & \overset{c}{\nearrow} & \downarrow{\scriptstyle f} \\ K & \xrightarrow{b} & Y \end{array}$$

For this it suffices to consider pairs $S^{k-1} \subset D^k$ instead of $L \subset K$; thus for X connected it suffices if f induces isomorphisms $f_* : \pi_r(X) \to \pi_r(Y)$ of homotopy groups.

The map $f : X \to Y$ is said to be n-connected if this condition holds for all (K, L) with K of dimension $\leq n$. If f is the inclusion of a subset, we say that the pair (Y, X) is *n-connected*. For this it is sufficient that $\pi_n(Y, X)$ is trivial for all $0 \leq i \leq n$: equivalently (if $n \geq 2$) that X and Y are connected, the map $f_* : \pi_r(X) \to \pi_r(Y)$ is an isomorphism for $r < n$ and surjective for $r = n$.

For any K, we define the *cylinder* on K to be the product $K \times I$, the *cone* CK on K to be obtained from $K \times I$ by identifying the subspace $K \times \{0\}$ to a point (so there is an inclusion $K \to CK$ with $x \mapsto (x, 1)$), and the *suspension* SK to be obtained by further identifying $(* \times I) \cup (K \times \{1\})$ to a point. More generally, for any map $f : K \to L$ we define the *mapping cone* $L \cup_f CK$ to be obtained from the disjoint union $L \cup CK$ by identifying, for each $x \in K$, the point $(x, 1) \in CK$ with $f(x) \in L$: this generalises the procedure of attaching a cell to L using a map $f : S^{n-1} \to L$. We also define the *mapping cylinder* $Cyl(f) := L \cup_f (K \times I)$ to be obtained from the disjoint union $L \cup (K \times I)$ by identifying, for each $x \in K$, the point $(x, 1) \in (K \times I)$ with $f(x) \in L$: this contains $K \times \{0\}$ as a subspace, and has L as a deformation retract.

The *join* of two spaces K and L is the space $K * L$ obtained from $K \times L \times I$ by identifying each $\{k\} \times L \times \{0\}$ to $k \in K$ and each $K \times \{l\} \times \{1\}$ to $l \in L$. The *smash product* of spaces K and L is defined to be

$$K \wedge L := (K \times L)/(K \times \{*\} \cup \{*\} \times L).$$

In particular, the suspension $SK = S^1 \wedge K$.

A map $i : K \to L$ is said to have the *homotopy extension property* (HEP) if given any map $f : L \to Y$ and homotopy $g : K \times I \to Y$ such that $g(x, 0) = f(i(x))$ for each $x \in K$ there is a homotopy $h : L \times I \to Y$ such that $h(i(x, t)) = g(x, t)$ for each $(x, t) \in K \times I$ and $h \circ (i \times 1_I) = g$. This is a typical property of inclusion maps: the inclusion of a subcomplex L in a CW-pair (K, L) has the HEP. Any map $f : K \to L$ is homotopy equivalent to the inclusion $K \to Cyl(f) = L \cup_f (K \times I)$, which has the HEP. If $i : K \to L$ has the HEP, identifying CK to a point gives a homotopy equivalence $L \cup_i CK \to L/K$: to obtain a homotopy inverse, extend the homotopy of CK which shrinks the cone to its vertex to a homotopy of the identity map of $L \cup_i CK$: at the end of the homotopy is a map sending CK to a point, hence factoring through L/K.

For any $f : K \to L$ and any X, the sequence

$$[K : X] \leftarrow [L : X] \leftarrow [L \cup_f CK : X]$$

is exact, for a map $L \to X$ extends to $L \cup_f CK$ if and only if its restriction to K is nullhomotopic. For any $f : K \to L$, denote by Af the inclusion $L \to L \cup_f CK$. Since Af has the HEP, $(L \cup_f CK) \cup_g CL$ is homotopy equivalent to $CL/(L \cup_f CK) = SK$, so up to homotopy $A^2 f$ is a map $L \cup_f CK \to SK$. Iterating once more gives a map $A^3 f : SK \to SL$ which differs from the suspension Sf by reversing orientation in I. Thus the sequence $A^r f$ of maps induces, for any X, an exact sequence

$$[K : X] \leftarrow [L : X] \leftarrow [L \cup_f K : X] \leftarrow [SK : X] \leftarrow [SL : X] \ldots$$

Each set $[SK : X]$ admits a natural group structure, and $[S^2 K : X]$ is abelian.

A map $p : X \to Y$ is said to be have the *covering homotopy property (CHP)* if given a space K, a map $a : K \to X$ and a homotopy $b : K \times I \to Y$ such that $b \,|\, (K \times 0) = p \circ a$, there exists a homotopy $c : K \times I \to X$ such that $a = c \,|\, (K \times 0)$ and $b = p \circ c$.

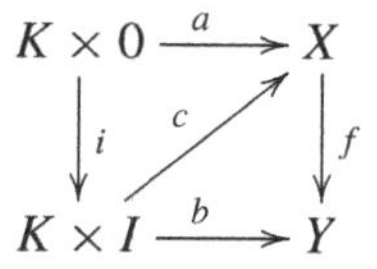

$$\begin{array}{ccc} K \times 0 & \xrightarrow{\ a\ } & X \\ \downarrow{\scriptstyle i} & \overset{c}{\nearrow} & \downarrow{\scriptstyle f} \\ K \times I & \xrightarrow{\ b\ } & Y \end{array}$$

If this holds for K a finite CW-complex, it follows for any CW-complex; it also follows if (K, L) is a CW-pair that c can be chosen to extend a lift already given on $L \times I$. It suffices to require this condition for pairs $(K, L) = (D^n, S^{n-1})$. We may regard the CHP as a sort of dual notion to the HEP.

We recall from §1.3 that if G is a Lie group acting on a smooth manifold F, a map $\pi : E \to B$ is the projection of a fibre bundle (with base space B, total space E, and fibre F) if B can be covered by open sets U_α such that

(i) There are homeomorphisms $\varphi_\alpha : U_\alpha \times F \to \pi^{-1}(U_\alpha)$ such that for all $m \in U_\alpha, x \in F, \pi \varphi_\alpha(m, x) = m$.

(ii) For each pair (α, β) there is a continuous map $g_{\alpha\beta} : U_\alpha \cap U_\beta \to G$ such that for $m \in U_\alpha \cap U_\beta, x \in F, \varphi_\beta(m, x) = \varphi_\alpha(m, g_{\alpha\beta}(m).x)$.

Lemma B.1.1 *The projection map $\pi : E \to B$ of a fibre bundle has the CHP.*

This is trivial if π is the projection of a product $B \times F \to B$, thus we can lift a homotopy whose image is contained in some U_α; and now the result is proved by subdividing $K \times I$ into small pieces.

This result motivates the definition that a map $\pi : E \to B$ is a *fibration* if it has the CHP. Given a fibration, write F for the fibre $F := \pi^{-1}(*)$. Then for any space X, the sequence $[X : F] \to [X : E] \to [X : B]$ is exact, for given a map $f : X \to B$ with $\pi \circ f$ homotopic to the map to $*$, we can lift the homotopy to give a homotopy of f to a map into F.

Now let X be a connected space and consider the space EX of continuous maps $\alpha : I \to X$. There are two projections $p_0, p_1 : EX \to X$ given by $p_0(\alpha) = \alpha(0)$ and $p_1(\alpha) = \alpha(1)$: each has the CHP. The map p_0 is a homotopy equivalence: a homotopy inverse is given by constant maps $c : X \to EX$ with $c(x)(t) = x$; the map $h : EX \times I \to EX$ given by $h(\alpha, t) = \alpha_t$ with $\alpha_t(u) = \alpha(min(t, u))$ is a homotopy of $c \circ p_0$ to the identity. Thus $PX := p_0^{-1}(*)$ is contractible. The restriction $q_1 := p_1 \mid PX$ also has the CHP, and $\Omega X := q_1^{-1}(*)$ is called the *loop space* of X.

For any map $f : K \to L$ we form the pullback

$$X := \{(k, \alpha) \in K \times EL \mid f(k) = \alpha(0)\};$$

write $i = (i_1, i_2)$ for the inclusion of X in $K \times EL$. Since p_0 is a homotopy equivalence, so is the projection $i_1 : X \to K$. The composite $f \circ i_1 = p_0 \circ i_2 : X \to L$ is homotopic to the map $\pi : p_1 \circ i_2$.

Lemma B.1.2 *The projection $\pi : X \to L$ defined above has the CHP.*

Proof Given $g : Y \to X$ and a homotopy $G : Y \times I \to L$ such that $G \mid Y \times \{0\} = \pi \circ g$ we need to construct $h : Y \times I \to X$ with $h \mid Y \times \{0\} = g$ and $\pi \circ h = g$. To this end, write $i \circ g = (g_1, g_2)$, $i \circ h = (h_1, h_2)$; use t as parameter for paths belonging to EL and s as the homotopy parameter in I; thus write h_2 as $h_2(y, t, s) \in L$.

Then the conditions that (g_1, g_2) and (h_1, h_2) factor through X are

$$f(g_1(y)) = g_2(y, 0), f(h_1(y, s)) = h_2(y, 0, s);$$

that h extends g is

$$h_1(y, 0) = g_1(y), \quad h_2(y, t, 0) = g_2(y, t),$$

and that h lifts G is

$$h_2(y, 1, s) = G(y, s).$$

We take $h_1(y, s) = g_1(y)$, and then the equations define $h_2(y, t, s)$ if either $t = 0$, $s = 0$ or $t = 1$: moreover the two values for $h_2(y, 0, 0)$ agree since $f(h_1(y, 0)) = f(g_1(y)) = g_2(y, 0)$ and those for $h_2(y, 1, 0)$ do since $G(y) =$

$\pi(g(y)) = g_2(y, 1)$. Since the union of 3 sides of the square $I \times I$ is a retract of the whole square, we can extend these values to define h_2 for all values. □

The fibre of π is called the *mapping fibre* of f; we may denote it by M_f. Thus $M_f := \{(k, \alpha) \in K \times EL \mid f(k) = \alpha(0),\ \alpha(1) = *\}$. We have seen that if f has the HEP, $L \cup_f K \simeq L/K$. Dually, if $f : K \to L$ has the CHP, with fibre F, then F is homotopy equivalent to M_f. Let us write Bf for the map $M_f \to K$: up to homotopy, if f has the CHP, this agrees with the inclusion $F \subset K$. As π has the CHP, so does $M_f \to K$, and this has fibre ΩL, so $B^2 f : \Omega L \to M_f$. Analogously to the above discussion of Af, up to homotopy we can identify $B^3 f$ with $\Omega f : \Omega K \to \Omega L$. It follows that for any space X, there is an exact sequence

$$\ldots [X : \Omega K] \to [X : \Omega L] \to [X : M_f] \to [X : K] \to [X : L].$$

Composition of loops induces a group structure on the set $[X : \Omega K]$, and there is a natural bijection of this set on $[SX : K]$. In particular, $\pi_r(\Omega X) \cong \pi_{r+1}(X)$. Taking X a sphere in the exact sequence gives

$$\ldots \pi_n(K) \to \pi_n(L) \to \pi_{n-1}(M_f) \to \pi_{n-1}(K) \to \pi_{n-1}(L).$$

Here we may identify $\pi_{n-1}(M_f)$ with the group $\pi_n(f)$ and the sequence with the exact homotopy sequence described above. If also $f : K \to L$ has the CHP, with fibre F, then M_f is homotopy equivalent to F.

Lemma B.1.3 *Given a sequence* $A_{i+1} \xrightarrow{\alpha_i} A_i$ *where the maps* α_i *are fibrations, there are natural isomorphisms* $q_n : \pi_n(\varprojlim A_i) \cong \varprojlim \pi_n(A_i)$.

Given a sequence of maps $f_i : A_i \to B_i$ *between two sequences of fibrations, with each* f_i *a weak homotopy equivalence and* $f_i \circ \alpha_i = \beta_i \circ f_{i+1}$ *for each* i, *the induced map* $\varprojlim A_i \to \varprojlim B_i$ *is a weak homotopy equivalence.*

For a map $S^n \to \varprojlim A_i$ defines a sequence of maps $S^n \to A_i$, so we have a natural map q_n. Since α_i is a fibration, if the homotopy class of a map $S^n \to A_i$ lifts to that of a map to A_{i+1}, so does the map itself. It follows that q_n is surjective; injectivity follows similarly.

The second assertion now follows.

Many of the definitions and results in this section have a formal nature. A set of axioms for homotopy theory, with a development along these lines, was given by Quillen [126].

B.2 Groups and homogeneous spaces

We observed in §3.1 that for any Lie group G and Lie subgroup H, we have a fibre bundle with projection $G \to G/H$ and fibre H; and that if we have two

Lie subgroups $H_1 \subset H_2 \subset G$, the projection $G/H_2 \to G/H_1$ is that of a fibre bundle, with fibre H_2/H_1, so has the CHP.

The group $GL_n(\mathbb{R})$ acts transitively on the space P of positive definite quadratic forms on $\mathbb{R}^n$, and O_n is the isotropy group of the usual inner product, so we have an induced diffeomorphism of $GL_n(\mathbb{R})/O_n$ on P, and hence a fibre bundle $O_n \to GL_n(\mathbb{R}) \to P$. Since P is a convex subset of a Euclidean space, it is contractible. Thus $GL_n(\mathbb{R})$ is homotopy equivalent to O_n. It is usually more convenient to work with the compact group O_n.

Similarly, any Lie group G has maximal compact subgroups K, any two are conjugate, and G/K is contractible. Thus for homotopy purposes, we may replace G by K. In particular, we may replace $GL_n(\mathbb{C})$ by U_n.

Since O_n acts transitively on the Grassmann manifold $Gr_{n,k}$ of k-dimensional subspaces of $\mathbb{R}^n$, and the subgroup leaving $\mathbb{R}^k \oplus \{0\}$ can be identified with $O_k \times O_{n-k}$, we can identify $Gr_{n,k}$ with the coset space $O_n/(O_k \times O_{n-k})$. This is a smooth manifold, and there is a natural vector bundle $\gamma_{n,k}$ over $Gr_{n,k}$ whose fibre is the k dimensional linear subspace.

The space $V'_{n,k}$ of injective linear maps $\mathbb{R}^k \to \mathbb{R}^n$ is homotopy equivalent to the space of isometric linear embeddings $\mathbb{R}^k \to \mathbb{R}^n$. The latter is called the *Stiefel manifold*, and denoted $V_{n,k}$ (we call $V'_{n,k}$ the *weak Stiefel manifold*). It can be identified with O_n/O_{n-k}, hence with SO_n/SO_{n-k}. For any n-vector bundle $\xi : E \to B$ with group O_n there is an associated bundle with fibre $V_{n,k}$: a point in its total space can be interpreted as an isometry of $\mathbb{R}^k$ into some fibre of ξ.

For any Lie group G, there is a contractible space $E(G)$ admitting a free action of G. Write $B(G) := E(G)/G$ and $\pi_G : E(G) \to B(G)$ for the projection. Then this is a principal G-bundle, and for any principal G-bundle ξ over any space X there is a map $f : X \to B(G)$, unique up to homotopy, such that ξ is equivalent to $f^*\pi_G$. The bundle $\pi_G : E(G) \to B(G)$ is determined uniquely up to homotopy by this condition.

The space $B(G)$ is called a *classifying space* for G. Since $E(G)$ is contractible, it follows that G is homotopy equivalent to the loop space $\Omega B(G)$.

The classical construction of a classifying space is based on the Grassmann manifolds. The natural inclusion $Gr_{n,k} \subset Gr_{n+1,k}$, is $(n-k)$-connected, and the union $\bigcup_m Gr_{n,k}$ can be taken as a classifying space $B(O_k)$ for bundles with group O_k. This construction may be adapted for other Lie groups.

There is an alternative construction, due to Milnor [91], using the sequence of iterated joins $G * G * \ldots * G$ (on which G acts freely), and taking $E(G)$ as the union.

Yet another approach is axiomatic. The set $\mathcal{E}_G(X)$ of equivalence classes of bundles over X with a given structure group G is a contravariant functor of

X, and it is not difficult to verify the hypotheses of Brown's representability theorem [33]. This shows again that there exists a space $B(G)$ and a bundle ξ_G over it with structure group G such that taking a map $f : X \to B(G)$ to the bundle $f^*\xi_G$ induces a bijection of $[X : B(G)]$ on $\mathcal{E}_G(X)$.

In some sense, we can regard any space X as a classifying space for ΩX, which plays the part of the group, since we have a fibration $\Omega X \to PX \to X$ with PX contractible.

An $(n-1)$-spherical fibration consists of a fibration $\pi : F \to E \to X$ together with a homotopy equivalence $S^{n-1} \to F$. It follows from the axiomatic approach that there is a classifying space $B(G_n)$ for the set $\mathcal{E}^n_S(X)$ of homotopy equivalence classes of $(n-1)$-spherical fibrations over X and a fibration $\nu_n : S^{n-1} \to S(G_n) \to B(G_n)$, such that $f \mapsto f^*\nu_n$ gives a bijection $[X : B(G_n)] \to \mathcal{E}^n_S(X)$.

This notation goes with writing G_n for the set of maps of S^{n-1} to itself of degree ± 1, with the multiplication given by composition of maps. Although this is not a group, it can be treated as one for the purposes of homotopy theory. In particular we have a homotopy equivalence $G_n \to \Omega B(G_n)$. Restricting to maps of degree $+1$, or to fibrations with a fixed orientation of the fibre, gives a monoid SG_n and a classifying space $B(SG_n)$. The inclusion $O_n \subset G_n$ gives rise to a natural map $B(O_n) \to B(G_n)$.

We write $F_n \subset G_{n+1}$ for the set of base-point preserving maps $S^n \to S^n$ of degree ± 1, and SF_n for those of degree $+1$. The suspension of a self-map of S^{n-1} is a self-map of the same degree of S^n which fixes a base point; thus we also have an inclusion $G_n \subset F_n$. Since all components of $\Omega^n S^n$, including SF_n, are homotopy equivalent, we have $\pi_r(F_n) \cong \pi_{r+n}(S^n)$. We have a fibration $SF_{n-1} \to SG_n \to S^{n-1}$, and hence an exact sequence

$$\ldots \to \pi_{r+n-1}(S^{n-1}) \to \pi_r(G_n) \to \pi_r(S^{n-1}). \qquad \text{(B.2.1)}$$

The classifying spaces $B(G)$ are infinite dimensional, and not homotopy equivalent to finite dimensional spaces. They may, however, be approximated by smooth manifolds. Since the map $Gr_{m,k} \to B(O_k)$ is m-connected, for a manifold M of dimension at most m, the set of homotopy classes of maps $M \to Gr_{m,k}$ maps bijectively to that of maps $M \to B(O_k)$. In general, we first replace the original $B(G)$, or indeed any space X, by the $(N+1)$-skeleton X_1 of its singular complex. Next, provided the homotopy groups of X are countable, we can replace X_1 by a countable $(N+1)$-simplicial complex X_2; then by a locally finite complex X_3, and finally imbed X_3 properly in Euclidean $(2N+3)$-space and take an open neighbourhood X_4 of which it is a deformation retract.

In the construction of classifying spaces we have emphasised principal bundles. However, for any G-space L we can study bundles with group G and fibre L, and the classification is the same as for the associated principal bundles: they are induced from the universal bundle $E(G) \times_G L$. For example, using the action of $GL_n(\mathbb{R})$ on $\mathbb{R}^n$, we obtain a universal vector bundle over $B(GL_n(\mathbb{R}))$.

Likewise we have a universal orthogonal vector bundle γ_k over $B(O_k)$, whose total space contains the associated unit disc bundle $A(O_k)$. Writing $S(O_k)$ for its boundary sphere bundle, we have the Thom space $T(O_k) = A(O_k)/S(O_k)$. Thus for any group G with a given homomorphism $G \to O_k$ we have induced bundles $S(G) \subset A(G)$ and $T(G)$ is obtained from $A(G)$ by identifying $S(G)$ to a point.

More generally, since each sphere bundle is a spherical fibration, we have an inclusion $O_n \subset G_n$ and maps $B(O_n) \to B(G_n)$, $S(O_n) \to S(G_n)$. Here the role of $A(G_n)$ is played by the mapping cylinder $Cyl(\pi)$, where $\pi : S(G_n) \to B(G_n)$ denotes the projection, and we define $T(G_n)$ to be its mapping cone. Again, any map $X \to B(G_n)$ induces a spherical fibration ξ over X and we have a Thom space. In this situation there is still a natural isomorphism, called the Gysin isomorphism

$$H^r(X) \to H^{k+r}(A_\xi, S_\xi) \cong \tilde{H}^{k+r}(T(\xi)).$$

A summary of calculations of cohomology of classifying spaces is in §8.6.

In general, if $x \in H^n(B(G); A)$ is a cohomology class, and $\pi : E \to X$ is a G-bundle, π is induced by a map $f : X \to B(G)$, so we have a class $f^*x \in H^n(X; A)$. Such a class is called a *characteristic class* of the bundle π, and denoted $x(f)$. For example, we have $H^*(B(O_n) : \mathbb{Z}_2) \cong \mathbb{Z}_2[w_1, \ldots, w_n]$, so any polynomial in $w_1, \ldots, w_n$ defines a characteristic class for vector bundles of fibre dimension n.

If M is a smooth manifold, its tangent bundle $\mathbb{T}(M)$ is classified by a map $\phi : M \to B(O)$, so a class $x \in H^n(B(O); A)$ induces a characteristic class $x(M) := \phi^*(x) \in H^n(M; A)$. If $\mathbf{J}$ is a stable group and M has a J structure, we may replace O by $\mathbf{J}$ here.

If moreover M has the same dimension n, we have $\phi^*(x)[M] \in A$: this is called a *characteristic number* of M (if $A = \mathbb{Z}$ we do just have a number). If W is a cobordism of M to M', $x \in H^n(B(G) : A)$, and $\psi : W \to B(G)$ classifies a G-structure on W, then $\phi^*(x)[M] = \phi'^*(x)[M']$, since ψ restricts to ϕ and ϕ', and $\langle \psi^*(x), [M] - [M'] \rangle = 0$ since $[M] - [M'] = 0$ in homology, as the boundary of W. Thus characteristic numbers are cobordism invariants.

The same argument applies with any non-classical homology theory; for example, with KO-theory.

B.3 Homotopy calculations

In this section we summarise the results of a large number of homotopy calculations. We have included text intended to make the summary less unreadable, but make no attempt to give proofs. The results may be found in texts on homotopy theory, but the author has not discovered a convenient single reference for these results.

(i) There are natural maps $\pi_n(X) \to H_n(X; \mathbb{Z})$ and $\pi_n(X, Y) \to H_n(X, Y; \mathbb{Z})$. The *Hurewicz Isomorphism Theorem* states that if X is $(n-1)$-connected (and $n \geq 2$), the natural map $\pi_n(X) \to H_n(X; \mathbb{Z})$ is an isomorphism.

It follows that $\pi_r(S^n)$ is zero for $r < n$ and isomorphic to $\mathbb{Z}$ for $r = n$. We write ι_n for the class in $\pi_n(S^n)$ of the identity map.

The Hurewicz theorem has a relative version: if (K, L) is $(n-1)$-connected (and K, L are simply-connected), the natural map $\pi_n(K, L) \to H_n(K, L; \mathbb{Z})$ is an isomorphism. If we define the homology groups of a map $f : A \to B$ as those of the pair $(Cyl(f), A)$ we can write this as: if f is $(n-1)$-connected, $\pi_k(f) \to H_k(f; \mathbb{Z})$ is an isomorphism for $k \leq n$.

(ii) The group SU_2 is homeomorphic to the sphere S^3, and its action on $P^1(\mathbb{C}) \simeq S^2$ gives a fibre bundle map $\eta_2 : S^3 \to S^2$ called the *Hopf map*; similarly using quaternions or Cayley numbers gives maps $\eta_4 : S^7 \to S^4$ and $\eta_8 : S^{15} \to S^8$: using the real numbers gives $\eta_1 : S^1 \to S^1$ of degree 2, so homotopic to $2\iota_1$.

(iii) There is a natural homomorphism $H : \pi_{2n-1}(S^n) \to \mathbb{Z}$, called the *Hopf invariant*. Given $f : S^{2n-1} \to S^n$, form $X_f := S^n \cup_f e^{2n}$, then $H^n(X_f)$ and $H^{2n}(X_f)$ are infinite cyclic with preferred generators u, v, say, and we set $u^2 = H(f)v$. This invariant vanishes for n odd (the cup product is skew-symmetric here), and takes the value 1 for each of η_2, η_4, η_8.

One generalisation of H is defined as follows. The map $\pi_r(j_n)$ induced by the inclusion $j_n : S^n \vee S^n \to S^n \times S^n$ has a right inverse given by adding the maps induced by the two projections of $S^n \times S^n$. Then H is the composite

$$\pi_r(S^n) \to \pi_r(S^n \vee S^n) \to \pi_{r+1}(S^n \times S^n, S^n \vee S^n) \to \pi_{r+1}(S^{2n}),$$

where the first map is induced by collapsing the equator to a point, the second by the splitting in the exact homotopy sequence of $(S^n \times S^n, S^n \vee S^n)$ and the third by collapsing $S^n \vee S^n$ to a point.

(iv) Let $f : (D^m, S^{m-1}) \to (X, *)$ represent $\alpha \in \pi_m(X)$ and $g : (D^n, S^{n-1}) \to (X, *)$ represent $\beta \in \pi_n(X)$: then the *Whitehead product* $[\alpha, \beta] \in \pi_{m+n-1}(X)$ is the homotopy class of the map $F : \partial(D^m \times D^n) \to X$ given by $F(x, y) = f(x)$ if $y \in \partial D^n$ and $= g(y)$ if $x \in \partial D^m$.

We have $[\iota_n, \iota_n] \in \pi_{2n-1}(S^n)$, and $H([\iota_n, \iota_n])$ is 0 if n is odd, and 2 if n is even.

(v) The 'Hopf invariant 1' problem, the question whether $H : \pi_{2n-1}(S^n) \to \mathbb{Z}$ is surjective, was solved by Adams [3]: it is surjective only if n is 2, 4, or 8.

This is analogous to the Kervaire invariant problem.

(vi) A further relative version of the Hurewicz theorem is the *Blakers–Massey Theorem* [18]. Given a commutative square

$$\begin{array}{ccc} A & \xrightarrow{p} & B \\ \Phi : q\downarrow & & r\downarrow, \\ C & \xrightarrow{s} & D \end{array}$$

of simply-connected spaces, we can define $H_*(\Phi, \mathbb{Z})$ so that there are exact sequences $H_*(q; \mathbb{Z}) \to H_*(r; \mathbb{Z}) \to H_*(\Phi, \mathbb{Z}) \to H_{*-1}(q; \mathbb{Z})$. Then if p is $(r-1)$-connected, q is $(s-1)$-connected, and $H_*(\Phi, \mathbb{Z}) = 0$, $\pi_n(\Phi)$ vanishes for $n < r + s - 1$ and $\pi_{r+s-1}(\Phi) \cong H_r(p; \mathbb{Z}) \otimes H_s(q; \mathbb{Z})$.

(vii) We can apply (vi) to the square given by the inclusions of S^n in the two hemispheres E_-^{n+1} and E_+^{n+1} of S^{n+1} (these inclusions are n-connected), and theirs in S^{n+1}. This gives $\pi_r(\Phi) = 0$ for $r \leq 2n$ and $\pi_{2n+1}(\Phi) \cong \mathbb{Z}$. Since the hemispheres are contractible, the sequence $\pi_r(E_-^{n+1}, S^n) \to \pi_r(S^{n+1}, E_+^{n+1}) \to \pi_r(\Phi)$ becomes $\pi_{r-1}(S^n) \to \pi_r(S^{n+1}) \to \pi_r(\Phi)$.

The map $\pi_{r-1}(S^n) \to \pi_r(S^{n+1})$ is called the *suspension map*. It is thus an isomorphism for $r \leq 2n - 1$, so the groups $\pi_{n+k}(S^n)$ for $n \geq k + 2$ are all isomorphic; the limit value is denoted π_k^S. Also we have an exact sequence

$$\pi_{2n}(S^n) \to \pi_{2n+1}(S^{n+1}) \to \mathbb{Z} \to \pi_{2n-1}(S^n) \to \pi_{2n}(S^{n+1}) \to 0.$$

Here the second map is the Hopf invariant, and $1 \in \mathbb{Z}$ maps to $[\iota_n, \iota_n]$. It follows from the above that if n is even, the second map is zero so we have an exact sequence $0 \to \mathbb{Z} \to \pi_{2n-1}(S^n) \to \pi_{n-1}^S \to 0$; if $n \neq 1, 3, 7$ is odd we must replace $\mathbb{Z}$ by $\mathbb{Z}_2$ here.

(viii) For the groups $\pi_r(S^n)$ we have a range given by $r < n$ where the groups vanish, and a range $n \leq r < 2n - 1$ where they are stable. We get information in the next 'metastable' range $2n - 1 \leq r < 3n - 2$ as follows.

We use the isomorphism of $\pi_r(\Omega S^{n+1})$ on $\pi_{r+1}(S^{n+1})$. Up to homotopy, ΩS^{n+1} has a cell structure with one kn-cell for each $k \in \mathbb{N}$. Hence $(\Omega S^{n+1}, S^n)$ is $(2n-1)$-connected and, by the relative Hurewicz theorem, $\pi_{2n}(\Omega S^{n+1}, S^n) \cong \mathbb{Z}$. Now applying (vi) to the square

$$\begin{array}{ccc} S^n & \longrightarrow & \Omega S^{n+1} \\ \downarrow & & \downarrow \\ * & \longrightarrow & \Omega S^{n+1}/S^n \simeq S^{2n} \cup e^{3n} \ldots \end{array},$$

we find that $\pi_r(\Omega S^{n+1}, S^n) \to \pi_r(S^{2n})$ is an isomorphism for $r < 3n - 1$. This yields the so-called *EHP sequence*

$$\pi_{n+k}(S^n) \xrightarrow{E} \pi_{n+k+1}(S^{n+1}) \xrightarrow{H} \pi_{n+k+2}(S^{2n+2}) \xrightarrow{P} \pi_{n+k-1}(S^n) \xrightarrow{E} \ldots, \quad \text{(B.3.1)}$$

generalising the sequence (vii), and valid for a range $k < 2n - 1$. Here the map P agrees (up to suspension) with the Whitehead product with ι_n: $\pi_k(S^n) \to \pi_{n+k-1}(S^n)$.

A more general version can be obtained using the fibration $S^n \to \Omega S^{n+1} \to \Omega S^{2n+1}$ (after localisation at 2) constructed by James [1] and Toda [6].

(ix) The homotopy group $\pi_r(S^n)$ is finite for $r > n$ except if n is even and $r = 2n - 1$ when it is the direct sum of $\mathbb{Z}$ and a finite group.

(x) The calculation of the homotopy groups $\pi_r(S^n)$ is a massive enterprise: see [129] for the state of the art. The stable groups form a ring under composition; the first few, with generators (here we use the same notation η_2 for the class of the suspension in π_1^S of η_2), are given by

$$\pi_1^S \cong \mathbb{Z}_2[\eta_2],\ \pi_2^S \cong \mathbb{Z}_2[\eta_2^2],\ \pi_3^S \cong \mathbb{Z}_{24}[\eta_4],\ \pi_4^S = 0,\ \pi_5^S = 0.$$

We have $\eta_2^3 = 0 \in \pi_3^S$.

(xi) The group SO_n acts transitively on the unit sphere S^{n-1} in $\mathbb{R}^n$, and the stabiliser of the unit point on the x_n-axis is the subgroup SO_{n-1}. Thus there is a fibre bundle $SO_{n-1} \to SO_n \to S^{n-1}$, with an exact homotopy sequence. Since $\pi_i(S^n)$ vanishes for $i < n$, we have isomorphisms $\pi_r(SO_{n-1}) \to \pi_r(SO_n)$ for $r \leq n - 3$. More generally, if X has dimension $\leq r$, the suspension map $[X : BSO_n] \to [X : BSO_{n+1}]$ is bijective for $n \geq r + 1$, so stably isomorphic vector bundles over X of fibre dimension $\geq r + 1$ must be isomorphic.

Also all groups $\pi_r(SO_N)$ for $N \geq r + 2$ are isomorphic; the common value is denoted $\pi_r(SO)$.

(xii) It was proved by Bott [21] that $\pi_r(SO)$ is infinite cyclic if $r \equiv 3$ (mod 4), isomorphic to $\mathbb{Z}_2$ if $r \equiv 0$ or $r \equiv 1$ (mod 8), and zero otherwise. A good account of Bott's proof is given in [98].

(xiii) The exact sequence of the fibre bundle $SO_{n-1} \to SO_n \to S^{n-1}$ includes

$$\to \pi_{n-1}(SO_{n-1}) \xrightarrow{i_*} \pi_{n-1}(SO_n) \xrightarrow{\pi_*} \mathbb{Z} \xrightarrow{\partial} \pi_{n-2}(SO_{n-1}) \xrightarrow{i_*} \pi_{n-2}(SO) \to 0. \tag{B.3.2}$$

If $x \in \pi_{n-1}(SO_n)$ classifies a bundle ξ, then $\pi_* x$ can be identified with the Euler number of ξ. If $x = \partial\iota_n$, then ξ is the tangent bundle of S^n, so $\pi_*\partial\iota_n$ is 2 for n even, and 0 for n odd. The image of π_* is 0 for n odd, $\mathbb{Z}$ for $n = 2, 4, 8$ and $2\mathbb{Z}$ for n even otherwise.

(xiv) Using (ix) and (xi), we see inductively that each group $\pi_r(SO_k)$ is finitely generated; the rank is 0 except if

(a) $k = 2s + 1$, $r = 4i - 1$, $1 \leq i \leq s$, or

(b) $k = 2s + 2$, either $r = 4i - 1$ with $1 \leq i \leq s$ or $r = 2s + 1$.

In these cases the rank is 1 except if $k = 4s$ and $r = 4s - 1$ when the rank is 2.

(xv) The Stiefel manifolds $V_{n,k} = SO_n/SO_{n-k}$ occur in fibre bundles $SO_{n-k} \to SO_n \to V_{n,k}$ $(1 < k < n)$ and $V_{n-k,l-k} \to V_{n,l} \to V_{n,k}$ $(k < l < n)$, which give further exact homotopy sequences. It now follows that for $r \leq n-k-1$ we have $\pi_r(V_{n,k}) = 0$, i.e. $V_{n,k}$ is $(n-k-1)$-connected.

(xvi) The calculation (xiii), that the kernel of $\pi_{n-2}(SO_{n-1}) \to \pi_{n-2}(SO)$ is isomorphic to $\mathbb{Z}$ for n odd, and to $\mathbb{Z}_2$ for n even, now implies that the first non-vanishing homotopy group $\pi_{n-k}(V_{n,k})$ is isomorphic to $\mathbb{Z}$ if $(n-k)$ is even and to $\mathbb{Z}_2$ if $(n-k)$ is odd.

(xvii) There is a homomorphism $J : \pi_k(SO_n) \to \pi_{n+k}(S^n)$, called the *J-homomorphism*, defined as follows. An element $\phi \in \pi_k(SO_n)$ is represented by a map $f : S^k \times D^n \to D^n$. Write $c : D^n \to S^n$ for a map which collapses ∂D^n to $*$. Write S^{n+k} as the union of $S^k \times D^n$ and $D^{k+1} \times S^{n-1}$, and define $g : S^{n+k} \to S^n$ to map the first part by $c \circ f$ and the second to $*$. Then $J(\phi)$ is the class of g in $\pi_{n+k}(S^n)$. An equivalent definition in the language of cobordism is given in §8.8.

For $x \in \pi_k(SO_n)$, we have $H(J(x)) = S^n(\pi(x)) \in \pi_{n+k}(S^{2n-1})$. Taking $k = 2s-1$, $n = 2s$ and $x = \partial\iota_{2s}$, then since $\pi(x) = 2\iota_{2s-1}$ we deduce $H(J(x)) = 2$, so the homomorphism $J : \pi_{2s-1}(SO_{2s}) \to \pi_{4s-1}(S^{2s})$ has rank 1.

(xviii) The image of the stable J homomorphism $J_k : \pi_k(SO) \to \pi_k^S$ was determined after heroic calculations by Adams [5]; a simpler proof was found in joint work with Atiyah [8].

(a) If $k \equiv 0$ or $k \equiv 1 \pmod 8$, the map J_k is a split monomorphism.

(b) If $k = 4m-1$ the image of J_k has order equal to $\mathrm{den}(B_m/4m)$, and is a direct summand of π_S^k.

(xix) It follows from (vi) that $\pi_r(SF_n)$ is finite for $r > 0$ except if n is even and $r = n-1$ when it is the direct sum of $\mathbb{Z}$ and a finite group.

In the exact sequence (B.2.1)

$$\ldots \to \pi_{r+n-1}(S^{n-1}) \to \pi_r(SG_n) \to \pi_r(S^{n-1}) \to \pi_{r+n-2}(S^{n-1}),$$

the final map is the Whitehead product with ι_{n-1}, so has infinite image if and only if n is odd and $r = n-1$. Thus $\pi_r(SG_n)$ is infinite if and only if either $r = n-1$ and n is even or $r = 2n-3$ and n is odd. The image of the map $\pi_r(SO_n) \to \pi_r(SG_n)$ has infinite order in each of these cases.

To summarise: the homotopy groups are finite except as follows:

Case	r	n	$rank(\pi_r(SO_n))$	$rank(\pi_r(SG_n))$
A	$4s+1$	$4s+2$	1	1
B	$4s-1$	$4s$	2	1
S	$4s-1$	$2s+1 < n \neq 4s$	1	0
C	$4s-1$	$2s+1$	1	1

(xx) If we take the exact sequence (B.2.1), increase n by 1, replace r by k, and compare with (B.3.1), we see that if $\pi_k(S^n)$ is stable, i.e. $2n \geq k+2$, we have an isomorphism $\pi_k(SG_{n+1}) \cong \pi_k(SF_{n+1})$.

The calculations in (xiv) can be compared with Haefliger's result [64] $\pi_r(F_n, G_n) \cong \pi_{r-n+1}(SO, SO_{n-1})$ for $r \leq 3n - 6$, which he established by geometrical arguments.

B.4 Further techniques

We have defined CW complexes as built up from spheres by attaching cells. If these are attached in order of increasing dimension, a complex K has an n-skeleton $K^{(n)}$: the union of cells of dimension $\leq n$. The inclusion $i : K^{(n)} \to K$ has the HEP and is n-connected: the map $H_r(K^{(n)}) \to H_r(K)$ is an isomorphism for $r < n$ and an epimorphism for $r = n$; and the mapping cone $K \cup_i CK^{(n)}$ is n-connected.

There is also a dual approach. We may start with K, attach $(n+1)$-cells to K to kill $\pi_n(K)$; then $(n+2)$-cells to kill $\pi_{n+1}, \ldots$, obtaining eventually an inclusion $j : K \to K_{(n)}$ with $\pi_r(j)$ an isomorphism for $r \leq n-1$ and $\pi_r(K_{(n)}) = 0$ for $r \geq n$. Denote the mapping fibre of j by $p^n : K^{\langle n \rangle} \to K$: then $K^{\langle n \rangle}$ is $(n-1)$-connected and $\pi_r(p^n)$ is an isomorphism for $r \geq n$. The pair $(K^{\langle n \rangle}, p^n)$ is called the $(n-1)$-connected cover of K, and is determined up to homotopy by these conditions.

It follows that, up to homotopy, there is for each k a fibration $K^{\langle k-1 \rangle} \to K^{\langle k \rangle} \to K(k, \pi_k(K))$. For any Y we have an induced map $[Y : K^{\langle k \rangle}] \to [Y : K]$; this is surjective if Y is k-connected, and bijective if Y is $(k+1)$-connected. The sequence of maps $\ldots \to K^{\langle 2 \rangle} \to K^{\langle 1 \rangle} \to K$ is called the Postnikov tower of K.

Given CW complexes K, L and a map $f : K^{(k-1)} \to L$ of the $(k-1)$-skeleton, the obstruction to extending f over a k-cell of K is an element of $\pi_{k-1}(L)$; collecting these over all k-cells gives a cochain on K, which is necessarily a cocycle. Its class in $H^k(K; \pi_{k-1}(L))$ is the obstruction to extending the restriction of f to $K^{(k-2)}$ over $K^{(k)}$.

If this obstruction vanishes, we can seek to extend over the $(k+1)$-skeleton, and so on. However, the later obstructions will in general depend on choices made at earlier stages. If k is the *least* integer such that $H^k(K; \pi_{k-1}(L))$ is non-zero, the obstruction in this group depends on no choices, and is called the *primary obstruction.*

If ξ is a vector bundle, and $\xi\langle k \rangle$ the associated bundle with fibre $V_{n,k}$, the primary obstruction to finding a section of $\xi\langle k \rangle$ is denoted $W_{n-k}(\xi)$; it lies in $H^{n-k}(B; \pi_{n-k}(V_{n,k}))$. The reduction modulo 2 of $W_{n-k}(\xi)$ is equal to the Stiefel–Whitney class $w_{n-k}(\xi)$.

Given $n \geq 1$ and a group π, abelian if $n \geq 2$, spaces $K(\pi, n)$ were constructed by Eilenberg and MacLane [49], with the property that $\pi_r(K)$ vanishes for $r \neq n$ and that $\pi_n(K) \cong \pi$: this determines $K(\pi, n)$ up to homotopy equivalence. For π abelian, there is a natural isomorphism $[X : K(\pi, n)] \cong H^n(X; \pi)$. It follows that a map $K(\pi, m) \to K(\rho, n)$ determines a natural transformation $H^m(X; \pi) \to H^n(X : \rho)$. Such a transformation is called a *cohomology operation*.

In particular, $[K(\mathbb{Z}_p, n) : K(\mathbb{Z}_p, n+k)] \cong H^{n+k}(K(\mathbb{Z}_p, n); \mathbb{Z}_p)$. Composing with an element of this group gives a natural transformation from $H^n(X; \mathbb{Z}_p)$ to $H^{n+k}(X : \mathbb{Z}_p)$. There are maps

$$H^{n+k}(K(\mathbb{Z}_p, n); \mathbb{Z}_p) \to H^{n+k+1}(K(\mathbb{Z}_p, n+1); \mathbb{Z}_p),$$

which are isomorphisms for $n > k$, so the groups with $n > k$ have a common value $H^k(K(\mathbb{Z}_p); \mathbb{Z}_p)$: elements of this give stable operations. Composition endows the set of these operations with a natural ring structure; this ring is known as the *Steenrod algebra* and denoted $\mathcal{S}_p$. Particular such operations are the Bockstein $\beta_p : H^n(X; \mathbb{Z}_p) \to H^{n+1}(X : \mathbb{Z}_p)$ and Steenrod's squares $Sq^i : H^n(X; \mathbb{Z}_2) \to H^{n+i}(X; \mathbb{Z}_2)$ and reduced pth powers $\mathcal{P}^r : H^n(X; \mathbb{Z}_p) \to H^{n+2r(p-1)}(X : \mathbb{Z}_p)$. These operations generate $\mathcal{S}_p$ and formulae for their composites (the Adem relations) are well known. There are rules (Cartan formulae) for evaluating these operations on the cup product of two classes. These define a diagonal map which furnishes $\mathcal{S}_p$ with the structure of a Hopf algebra. It thus has a canonical anti-automorphism, which is denoted χ.

It was shown by Milnor [93] that the dual algebra $\mathcal{S}_p^\vee$ is a polynomial algebra on a 1-dimensional generator b_p and generators c_r $(r \geq 1)$ of degrees $2(p^r - 1)$. The quotient $\overline{\mathcal{S}}_p$ of $\mathcal{S}_p$ by the ideal generated by β_p has dual the polynomial algebra on the c_r. A careful and thorough account of this material is given in [145].

Steenrod squares are related to Stiefel–Whitney classes as follows. If ξ is a vector bundle, with projection $\pi : E \to B$ and Thom space $T(\xi)$, we have the Gysin isomorphism $\Phi : H^*(B : \mathbb{Z}_2) \to \tilde{H}^*(T(\xi); \mathbb{Z}_2)$, with $\Phi(1) = U$, say: then $Sq^i U = \Phi(w_i(\xi)) = w_i(\xi).U$. Classes $v_i \in H^i(B(O); \mathbb{Z}_2)$ are defined uniquely by the rule $w_i = v_i + \sum_{j=1}^{i-1} Sq^j v_{i-j}$, which may be written compactly as $w_* = Sq^* v_*$. In the special case of the tangent bundle of a manifold M^m, we have the formulae, known as Wu relations, $Sq^i x[M] = xv_i[M]$ for any $x \in H^{m-i}(M : \mathbb{Z}_2)$: these follow from the above and duality in M (see [103, IX, 5]).

As well as primary operations such as Steenrod squares there are secondary operations. The general idea is that if something vanishes for two independent reasons, this leads to a construction. Perhaps the simplest example: given maps $A_0 \xrightarrow{f_1} A_1 \xrightarrow{f_2} A_2 \xrightarrow{f_3} A_3$ such that $f_2 \circ f_1$ and $f_3 \circ f_2$ are nullhomotopic, choose homotopies $h_1 : A_0 \times I \to A_2$ and $h_2 : A_1 \times I \to A_3$: then both $h_1 \circ f_3$ and $h_2 \circ f_0$ are homotopies of $f_3 \circ f_2 \circ f_1 \circ f_0$ to a point. Glueing these together thus gives a map $SA_0 \to A_3$. This depends not only on the homotopy classes of the f_i but also on the choices of the homotopies, so (in the additive case) is unique up to adding elements of $f_3 \circ [SA_0 : A_2]$ and $[SA_1 : A_3] \circ Sf_1$.

For example, if $A_2 = K(G, m)$ and $A_3 = K(H, n)$, the map f_3 defines a cohomology operation $\phi : H^m(X; G) \to H^n(X; H)$. Thus if f_2 represents a class $\xi \in H^m(A_1; G)$ such that $f_1^*\xi = 0$ and $\phi(\xi) = 0$ we obtain an element of $H^n(SA_0; H) \cong H^{n-1}(A_0; H)$, which is denoted $\phi_{f_1}\xi$.

If p is a prime, we can localise a (finitely generated) abelian group A at p by forming the tensor product $A \otimes \mathbb{Z}_{(p)}$ with the group of integers localised at p (i.e. rational numbers with denominator prime to p). An Eilenberg–MacLane space $K(A, n)$ localises to $K(A \otimes \mathbb{Z}_{(p)}, n)$. Building up using fibrations, one can define the *localisation* $X_{(p)}$ at p of any simply-connected space X: it is unique up to homotopy, and $\pi_n(X_{(p)})$ is the localisation of $\pi_n(X)$ at p. See, for example, [23] for a textbook account. Similarly we can localise at any set S of primes. This permits calculations where we can ignore throughout the contribution of all primes not in S. This technique of 'mod $\mathcal{C}$' theory is due to Serre [136].

We define a *spectrum* $\mathbb{A}$ to be a sequence of (based) spaces A_n ($n \in \mathbb{Z}$) and maps $i_n : SA_n \to A_{n+1}$: equivalently, we may require maps $A_n \to \Omega A_{n+1}$. It is called an Ω-spectrum if the maps $A_n \to \Omega A_{n+1}$ are all homotopy equivalences.

The map i_n induces $\pi_{r+n}(A_n) \to \pi_{r+n+1}(SA_n) \to \pi_{r+n+1}(A_{n+1})$ and, for any C, $H_{r+n}(A_n; C) \to H_{r+n+1}(SA_n : C) \to H_{r+n+1}(A_{n+1}; C)$: the limits of these are defined to be $\pi_r(\mathbb{A})$ and $H_r(\mathbb{A}; C)$.

Proposition B.4.1 *Let $\mathbb{X}$ be a spectrum whose homology groups are finitely generated. Then the natural map $\pi_k^S(\mathbb{X}) \to H_k(\mathbb{X}; \mathbb{Z})$ has finite kernel and cokernel.*

This is proved using the methods of *mod $\mathcal{C}$* theory [136]. It is a very useful first step in calculation of bordism groups.

We give important examples of spectra. The *sphere spectrum* $\mathbb{S}$ is defined by the sequence S^n and $SS^n \simeq S^{n+1}$. The Eilenberg–MacLane spectrum $\mathbb{K}(A, k)$ is defined by the sequence $K(A, n+k)$ and the homotopy equivalences $K(A, n+k) \to \Omega K(A, n+k+1)$. The cohomology ring $H^*(\mathbb{K}(\mathbb{Z}_p, k); \mathbb{Z}_p)$ is free on one generator over $\mathcal{S}_p$; $H^*(\mathbb{K}(\mathbb{Z}, k); \mathbb{Z}_p)$ is free over $\overline{\mathcal{S}}_p$.

For J a stable group in the sense of §8.2, the sequence of maps $h_k : ST(J_k) \to T(J_{k+1})$ defines a spectrum, which we denote by $\mathbb{TJ}$.

A different example is obtained using the homotopy equivalence $\Omega U \to B(U)$ established by Bott: set $A_{2n} = B(U)$ and $A_{2n-1} = U$. This gives an Ω-spectrum $\mathbb{BU}$ with $\pi_{2k}(\mathbb{BU}) \cong \mathbb{Z}$ for *all* $n \in \mathbb{Z}$. Similarly using a homotopy equivalence $\Omega^8 O \to O$ we define a spectrum $\mathbb{BO}$. For any spectrum $\mathbb{A}$ we can define the $(k-1)$-connected cover $\mathbb{A}\langle k\rangle$: as for spaces, $\mathbb{A}^{\langle k\rangle}$ is $(k-1)$-connected and $\pi^k : \mathbb{A}^{\langle k\rangle} \to \mathbb{A}$ induces isomorphisms of the homotopy groups π_r for $r \geq k$. The spectrum $\mathbb{BO}\langle k\rangle$, which is a Ω-spectrum with 0-term $B(O)\langle k\rangle$, plays a role in Chapter 8.

A spectrum $\mathbb{A}$ is a ring spectrum if we are given a system of maps $A_m \wedge A_n \to A_{m+n}$ compatible with the i_n. There is a natural condition of associativity. For the above examples, $\mathbb{S}$ is a ring spectrum, a ring structure on A induces one on $\mathbb{K}(A, k)$, and $\mathbb{TJ}$ is a ring spectrum if (M) and (A) hold for $\mathbf{J}$.

Any spectrum $\mathbb{A} = \{A_n, i_n\}$ gives rise to a homology theory (satisfying the axioms discussed in §8.4) on defining

$$\begin{aligned}
H_N(X; \mathbb{A}) &= \lim_{N\to\infty} \pi_{n+N}(A_n \wedge X^+) \\
H_N(X, Y; \mathbb{A}) &= \lim_{N\to\infty} \pi_{n+N}(A_n \wedge X^+, A_n \wedge Y^+) \\
&= \lim_{N\to\infty} \pi_{n+N}(A_n \wedge X, A_n \wedge Y).
\end{aligned}$$

If $\mathbb{A}$ is a ring spectrum we obtain external products which are associative if the spectrum is.

References

[1] R. Abraham, Transversality in manifolds of mappings, *Bull. Amer. Math. Soc.* **69** (1963) 470–474.

[2] M. Adachi, *Embeddings and immersions*, Trans. Math. Monographs 124, Amer. Math. Soc. 1993. (Translated from the 1984 Japanese original by K. Hudson.)

[3] J. F. Adams, On the non-existence of elements of Hopf invariant one, *Ann. Math.* **72** (1960) 20–104.

[4] J. F. Adams, On the groups $J(X)$ II, *Topology* **3** (1965) 137–172.

[5] J. F. Adams, On the groups $J(X)$ IV, *Topology* **5** (1966) 21–71.

[6] J. F. Adams, *Lectures on Lie groups*, xii+182 pp, W. A. Benjamin Inc, 1969.

[7] J. F. Adams, *Stable homotopy and generalised homology*, x+373 pp, Chicago Lectures in Math., Univ. of Chicago Press, 1974.

[8] J. F. Adams and M. F. Atiyah, K theory and the Hopf invariant, *Quart. Jour. Math. (Oxford)* **17** (1966) 31–38.

[9] D. W. Anderson, E. H. Brown, Jr., and F. P. Peterson, *SU* cobordism, *KO*-characteristic numbers, and the Kervaire invariant, *Ann. Math.* **83** (1966) 54–67.

[10] D. W. Anderson, E. H. Brown, Jr., and F. P. Peterson, The structure of the spin cobordism ring, *Ann. Math.* **86** (1967) 271–298.

[11] C. Arf, Untersuchungen über quadratischen Formen in Körpern der Charakteristik 2, *Crelle's Journal* **183** (1941) 148–167.

[12] M. F. Atiyah, Bordism and cobordism, *Proc. Camb. Phil. Soc.* **57** (1961) 200–208.

[13] M. F. Atiyah, Immersions and embeddings of manifolds, *Topology* **1** (1961) 125–132.

[14] M. F. Atiyah, Thom complexes, *Proc. London Math. Soc.* **11** (1961) 291–310.

[15] M. F. Atiyah, R. Bott, and A. Shapiro, Clifford modules, *Topology* **3** supplement 1 (1964) 3–38.

[16] D. Barden, Simply-connected 5-manifolds, *Ann. Math.* **82** (1965) 365–385.

[17] M. G. Barratt, J. D. S. Jones, and M. E. Mahowald, Relations amongst Toda brackets and Kervaire invariant in dimension 62, *Jour. London Math. Soc.* **30** (1985) 533–550.

[18] A. L. Blakers and W. S. Massey, The homotopy groups of a triad I, II, *Ann. Math.* **53** (1951) 161–205, **55** (1952) 192–201.

[19] J. M. Boardman, Singularities of differentiable maps, *Publ. Math. IHES* **33** (1967) 21–57.

[20] A. Borel, *Seminar on transformation groups*, esp. Chapter VII (by G. E. Bredon) and Chapter VIII (by R. S. Palais), Ann. Math. study 46, Princeton Univ. Press, 1956.

[21] R. Bott, The stable homotopy of the classical groups, *Ann. Math.* **70** (1959) 313–337.

[22] R. Bott and J. Milnor, On the parallelizability of the spheres, *Bull. Amer. Math. Soc.* **64** (1958) 87–89.

[23] R. Bott and L. W. Tu, *Differential forms in algebraic topology*, xiv+331 pp, Springer Graduate Texts 82, Springer-Verlag, 1982.

[24] N. Bourbaki, *Elements de Mathematique, Partie I, Livre III: Topologie Générale*, 187 pp, Hermann, 1951.

[25] G. E. Bredon, Examples of differentiable group actions, *Topology* **3** (1965) 115–122.

[26] G. E. Bredon, Transformation groups on spheres with two types of orbits, *Topology* **3** (1965) 103–114.

[27] G. E. Bredon, *Introduction to compact transformation groups*, xiii+459 pp, Pure and Applied Math. 46, Academic Press, 1972.

[28] G. E. Bredon, *Topology and geometry*, xiv+557 pp, Springer Graduate Texts 139, Springer-Verlag, 1993.

[29] W. Browder, Fiberings of spheres and h-spaces which are rational homology spheres, *Bull. Amer. Math. Soc.* **68** (1962) 202–203.

[30] W. Browder, The Kervaire invariant of framed manifolds and its generalisations, *Ann. Math.* **90** (1969) 157–186.

[31] W. Browder, *Surgery on simply-connected manifolds*, x+132 pp, Springer-Verlag, 1972.

[32] A. B. Brown, Functional dependence, *Trans. Amer. Math. Soc.* **38** (1935) 379–394.

[33] E. H. Brown, Jr., Cohomology theories, *Ann. Math.* **75** (1962) 467–484.

[34] E. H. Brown, Jr., Generalizations of the Kervaire invariant, *Ann. Math.* **95** (1972) 368–383.

[35] H. Cartan and S. Eilenberg, *Homological algebra*, xv+390 pp, Princeton Univ. Press, 1956.

[36] J. Cerf, Topologie de certains espaces de plongements, *Bull. Soc. Math. France* **89** (1961) 227–380.

[37] C. Chevalley, *Theory of Lie groups, I*, viii+217 pp, Princeton Univ. Press, 1946.

[38] P. E. Conner and E. E. Floyd, *Differentiable periodic maps*, vii+148 pp, Springer-Verlag, 1964.

[39] P. E. Conner and E. E. Floyd, *Torsion in SU-bordism*, Memoirs Amer. Math. Soc. 60 (1966).

[40] J. Dieudonné, *Foundations of modern analysis*, xiv+361 pp, Academic Press, 1960.

[41] A. Dold, Erzeugende der Thomschen algebra $\mathfrak{N}$, *Math. Zeits.* **65** (1956) 25–35.

[42] S. K. Donaldson, An application of gauge theory to 4-dimensional topology, *Jour. Diff. Geom.* **18** (1983) 279–315.

[43] S. K. Donaldson, Irrationality and the h-cobordism conjecture, *Jour. Diff. Geom.* **26** (1987) 141–168.
[44] S. K. Donaldson, Polynomial invariants for smooth 4-manifolds, *Topology* **29** (1990) 257–316.
[45] S. K. Donaldson, The Seiberg-Witten equations and 4-manifold topology, *Bull. Amer. Math. Soc.* **33** (1996) 45–70.
[46] S. K. Donaldson and P. B. Kronheimer, *The geometry of four-manifolds*, x+440 pp, Oxford Math. Monographs, Oxford Univ. Press, 1990.
[47] A. A. du Plessis and C. T. C. Wall, *The geometry of topological stability* (London Math. Soc. monographs, new series, no. 9), viii+572 pp, Oxford Univ. Press, 1995.
[48] C. Ehresmann, Introduction à la théorie des structures infinitésimales et des pseudogroupes de Lie, 16 pp. in *Colloque de topologie et géométrie différentielle, Strasbourg, 1952*, Université de Strasbourg, 1953.
[49] S. Eilenberg and S. MacLane, Relations between homology and homotopy groups of spaces, *Ann. Math.* **46** (1945) 480–509.
[50] S. Eilenberg and N. E. Steenrod, *Foundations of algebraic topology*, xv+328 pp, Princeton Univ. Press, 1952.
[51] R. Fintushel and R. J. Stern, The blowup formula for Donaldson invariants, *Ann. Math.* **143** (1996) 529–546.
[52] T. M. Flett, *Mathematical analysis*, xiv+439 pp, McGraw-Hill, 1966.
[53] M. H. Freedman, The topology of four-dimensional manifolds, *Jour. Diff. Geom.* **17** (1982), 357–453.
[54] M. H. Freedman and F. Quinn, *Topology of 4-manifolds*, viii+259 pp, Princeton Univ. Press, 1990.
[55] M. Furuta, Monopole equations and the 11/8 conjecture, *Math. Res. Lett.* **8** (2001), 279–291.
[56] C. G. Gibson, K. Wirthmüller, A. A. du Plessis, and E. J. N. Looijenga, *Topological stability of smooth mappings*, Springer Lecture Notes in Math. 552, Springer-Verlag, 1976.
[57] M. Golubitsky and V. Guillemin, *Stable mappings and their singularities*, x+209 pp, Springer Graduate Texts 14, Springer-Verlag, 1973.
[58] M. L. Gromov, Stable mappings of foliations into manifolds (in Russian), *Izv. Akad. Nauk SSSR* **33** (1969) 707–734.
[59] M. L. Gromov, *Partial differential relations*, Ergebnisse 9, Springer-Verlag, 1986.
[60] A. Haefliger, Plongements différentiables de variétés dans variétés, *Comm. Math. Helv.* **36** (1961) 47–82.
[61] A. Haefliger, Knotted $(4k-1)$-spheres in $6k$-space, *Ann. Math.* **75** (1962) 452–466.
[62] A. Haefliger, Plongements de variétés dans le domaine stable, *Sém. Bourbaki* **245** (1962–63) 63–77.
[63] A. Haefliger, Plongements différentiables dans le domaine stable, *Comm. Math. Helv.* **37** (1963) 155–176.
[64] A. Haefliger, Differentiable embeddings of S^n in S^{n+q} for $q > 2$, *Ann. Math.* **83** (1966) 402–436.
[65] A. Haefliger, Lectures on the Theorem of Gromov, pp 128–141 in *Proceedings of Liverpool Singularities Symposium II*, Springer Lecture Notes in Math. 209, 1971.

[66] A. Haefliger and M. W. Hirsch, Immersions in the stable range, *Ann. Math.* **75** (1962) 231–241.

[67] A. Haefliger and M. W. Hirsch, On the existence and classification of differentiable embeddings, *Topology* **2** (1963) 129–135.

[68] E. Hewitt and K. A. Ross, *Abstract harmonic analysis*, viii+519 pp, Springer-Verlag, 1963.

[69] M. A. Hill, M. J. Hopkins, and D. C. Ravenel, On the nonexistence of elements of Kervaire invariant one, arXiv:0908.3724. See also H. Miller's Bourbaki seminar 1029 and Hill, Hopkins, and Ravenel, 'The Arf-Kervaire invariant problem in algebraic topology, introduction', pp 23–57 in *Current developments in mathematics, 2009*, Int. Press, 2010 and 'Sketch of the proof', pp 1–43 in *ibid, 2010*, Int. Press, 2011.

[70] M. W. Hirsch, Immersions of manifolds, *Trans. Amer. Math. Soc.* **93** (1959) 242–276.

[71] M. W. Hirsch, Smooth regular neighbourhoods, *Ann. Math.* **76** (1962) 524–530.

[72] M. W. Hirsch, Embeddings and compressions of polyhedra and smooth manifolds, *Topology* **4** (1966) 361–369.

[73] M. W. Hirsch, *Differential topology*, x+221 pp, Springer Graduate Texts 33, Springer-Verlag, 1976.

[74] F. E. P. Hirzebruch, *Neue topologische Methoden in der algebraischen Geometrie*, 165 pp, Springer-Verlag, 1956; translated by R. L. E. Schwarzenberger, with additions, as *Topological methods in algebraic geometry*, xii+234 pp, Classics in Mathematics, Springer-Verlag, 1978.

[75] W. Hurewicz, *Lectures on ordinary differential equations*, MIT Press, 1958.

[76] W. Hurewicz and H. Wallman, *Dimension theory*, 165 pp, Princeton Univ. Press, 1948.

[77] D. Husemoller, *Fiber bundles* (Third edition), xx+353 pp, Springer Graduate Texts 20, Springer-Verlag, 1994.

[78] M. Kervaire, A manifold which does not admit any differentiable structure, *Comm. Math. Helv.* **34** (1960) 257–270.

[79] M. Kervaire and J. W. Milnor, Groups of homotopy spheres I, *Ann. Math.* **77** (1963) 504–537.

[80] S. O. Kochman, *The symplectic cobordism ring I, II,* Memoirs Amer. Math. Soc. 24 (1980) ix, 206 pp; 42 (1982) vii, 170 pp.

[81] P. B. Kronheimer and T. Mrowka, Recurrence relations and asymptotics for 4-manifold invariants, *Bull. Amer. Math. Soc.* **30** (1994) 215–221.

[82] A. A. Kosinski, *Differential manifolds*, xvi+248 pp, Academic Press, 1993.

[83] S. Lang, *Introduction to differentiable manifolds*, x+126 pp, Interscience, 1962.

[84] S. Lefschetz, *Introduction to topology*, viii+219 pp, Princeton Univ. Press, 1949.

[85] J. Levine, A classification of differentiable knots, *Ann. Math.* **82** (1965) 15–50.

[86] B. Malgrange, *Ideals of differentiable functions*, 106 pp, Oxford Univ. Press, 1966.

[87] W. S. Massey, Exact couples in algebraic topology, *Ann. Math.* **56** (1952) 363–396.

[88] J. N. Mather, Stability of C^∞-mappings: I The division theorem, *Ann. Math.* **87** (1968) 89–104. II Infinitesimal stability implies stability, *Ann. Math.* **89** (1969) 254–291. III Finitely determined map-germs, *Publ. Math. IHES* **35** (1969)

127–156. IV Classification of stable germs by $\mathbb{R}$- algebras, *Publ. Math. IHES* **37** (1970) 223–248. V Transversality, *Adv. Math.* **4** (1970) 301–335. VI The nice dimensions, Springer Lecture Notes in Math. 192 (1971) 207–253.

[89] J. P. May, *A concise course in algebraic topology*, x+243 pp, Chicago Lectures in Math., Univ. of Chicago Press, 1999.

[90] B. Mazur, *Differential topology from the point of view of simple homotopy theory*, Publ. Math. I.H.E.S. 15 1963.

[91] J. W. Milnor, Construction of universal bundles I, *Ann. Math.* **63** (1956) 272–284.

[92] J. W. Milnor, On manifolds homeomorphic to the 7-sphere, *Ann. Math.* **64** (1956) 399–405.

[93] J. W. Milnor, The Steenrod algebra and its dual, *Ann. Math.* **67** (1958) 150–171.

[94] J. W. Milnor, On simply-connected 4-manifolds, pp 122–128 in *Symposium Internacional de Topologia Algebrica*, Mexico (1958).

[95] J. W. Milnor, Differentiable structures on spheres, *Amer. J. Math.* **81** (1959) 962–972.

[96] J. W. Milnor, On the cobordism ring Ω^* and a complex analogue I, *Amer. J. Math.* **82** (1960) 505–521.

[97] J. W. Milnor, A procedure for killing the homotopy groups of differentiable manifolds, *Amer. Math Soc. Symp in Pure Math.* **3** (1961) 39–55.

[98] J. W. Milnor, *Morse theory* (notes by M. Spivak and R. Wells), vi+153 pp, Ann. Math. study 51 Princeton Univ. Press, 1963.

[99] J. W. Milnor, Microbundles I, *Topology* **3**, supplement 1 (1964) 53–80.

[100] J. W. Milnor, *Topology from the differentiable viewpoint*, x+64 pp, Univ. Press of Virginia, 1965.

[101] J. W. Milnor, *Lectures on the h-cobordism theorem* (notes by L. Siebenmann and J. Sondow), v+116 pp, Princeton Univ. Press, 1965.

[102] J. W. Milnor and M. Kervaire, *Bernoulli numbers, homotopy groups and a theorem of Rohlin*, Proceedings of ICM (Edinburgh 1958), 1962.

[103] J. W. Milnor and J. Stasheff, *Characteristic classes*, 144 pp, lecture notes, 1957; reissued and amplified vii+331 pp, as Ann. Math. study 76, Princeton Univ. Press, 1974.

[104] D. Montgomery and C. T. Yang, The existence of a slice, *Ann. Math.* **65** (1957) 108–116.

[105] D. Montgomery and C. T. Yang, Orbits of highest dimension, *Trans. Amer. Math. Soc.* **87** (1958) 284–293.

[106] D. Montgomery and L. Zippin, *Topological transformation groups*, xi+289 pp, Interscience, 1955.

[107] J. Morgan and G. Tian, *The geometrization conjecture*, x+291 pp, Clay Math. Monographs 5, Amer. Math. Soc., 2014.

[108] M. Morse, Relations between the critical points of a function of n independent variables, *Trans. Amer. Math. Soc.* **27** (1925) 345–396.

[109] M. Morse, *The calculus of variations in the large*, Amer. Math. Soc. Colloq. publs. 18, 1934.

[110] P. S. Mostert, ed., *Proceedings of the conference on transformation groups (New Orleans 1967)*, xiii+457 pp, Springer-Verlag, 1968.

[111] G. D. Mostow, Equivariant embeddings in Euclidean space, *Ann. Math.* **65** (1957) 432–446.

[112] G. D. Mostow, On a conjecture of Montgomery. *Ann. Math.* **65** (1957) 513–516.

[113] S. P. Novikov, Homotopy properties of Thom complexes, *Mat. Sb. (N.S.)* **57** (1962) 407–442 (in Russian).

[114] S. P. Novikov, Diffeomorphisms of simply-connected manifolds, *Dokl. Akad. Nauk SSSR* **143** (1962) 1046–1049, translated as *Soviet Math. Doklady* **3** (1962) 540–543.

[115] S. P. Novikov, Homotopy equivalent smooth manifolds I, *Izv. Akad. Nauk SSSR. ser. mat.* **28** (1964) 365–474, translated as *AMS Translations* **48** (1965) 271–396.

[116] R. Oliver, *Whitehead groups of finite groups*, London Math. Soc. Lecture Notes 132, viii+349 pp, Cambridge Univ. Press, 1988.

[117] R. S. Palais, Imbedding of compact differentiable transformation groups in orthogonal representations, *Jour. Math. Mech.* **6** (1957) 673–678.

[118] R. S. Palais, Local triviality of the restriction map for embeddings, *Comm. Math. Helv.* **34** (1960) 305–312.

[119] R. S. Palais, On the existence of slices for actions of noncompact Lie groups, *Ann. Math.* **73** (1961) 285–323.

[120] G. Perelman, The entropy formula for the Ricci flow and its geometric applications, ArXiv math.DG/0211159; Ricci flow with surgery on 3-manifolds, ArXiv math.DG/0303109.

[121] A. A. du Plessis and C. T. C. Wall, *The geometry of topological stability* (London Math. Soc. monographs, new series, no. 9), viii+572 pp, Oxford Univ. Press, 1995.

[122] J. Poncet, Groupes de Lie compacts de transformations de l'espace euclidean et les sphères comme espaces homogènes, *Comm. Math. Helv.* **33** (1959) 109–120.

[123] L. S. Pontrjagin, Characteristic cycles on differentiable manifolds, *Math. Sb (NS 63)* **21** (1947) 233–284 (AMS translations ser. 1, 32).

[124] L. S. Pontrjagin, *Smooth manifolds and their applications in homotopy theory*, *Trudy Mat. Inst. Steklov* **45** (1955) Akad. Nauk SSSR (AMS translations ser. 2, **11**).

[125] C. Procesi, *Lie groups: an approach through invariants and representations*, xxiv+596 pp, Universitext, Springer-Verlag, 2007.

[126] D. G. Quillen, *Homotopical algebra*, Springer Lecture Notes in Math. 43, Springer-Verlag, 1967.

[127] D. G. Quillen, On the formal group laws of unoriented and complex cobordism theory, *Bull. Amer. Math. Soc.* **75** (1969) 1293–1298.

[128] A. A. Ranicki, *Algebraic L-theory and topological manifolds*, Tracts in Math. 102, Cambridge Univ. Press, 1992.

[129] D. C. Ravenel, *Complex cobordism and stable homotopy groups of spheres*, Pure and Applied Math. 121, Academic Press, 1986.

[130] C. P. Rourke and B. J. Sanderson, Block bundles I, II, III, *Ann. Math.* **87** (1968) 1–29, 256–278, 431–483.

[131] C. P. Rourke and B. J. Sanderson, On topological neighbourhoods, *Compositio Math.* **22** (1970) 387–424.

[132] A. Sard, The measure of the critical points of differentiable maps, *Bull. Amer. Math. Soc.* **48** (1942) 883–890.

[133] H. Seifert, Verschlingungsinvarianten, *Sitz. Preuss. Akad. Wiss. Berlin* **26** (1933) 811–828.
[134] G. Segal, Geometric aspects of quantum field theory, pp 1387–1396 in *Proceedings of the International Congress of Mathematicians, Vol. I, II (Kyoto, 1990)*, Math. Soc. Japan, 1991.
[135] J.-P. Serre, Cohomologie modulo 2 des complexes d'Eilenberg–MacLane, *Comm. Math. Helv.* **27** (1953) 198–231.
[136] J.-P. Serre, Groupes d'homotopie et classes de groupes abéliens, *Ann. Math.* **58** (1953) 258–294.
[137] S. Smale, The classification of immersions of spheres in Euclidean space, *Ann. Math.* **69** (1959) 327–344.
[138] S. Smale, Generalized Poincaré's conjecture in dimensions greater than four, *Ann. Math.* **74** (1961) 391–406.
[139] S. Smale, On the structure of manifolds, *Amer. Jour. Math.* **84** (1962) 387–399.
[140] S. Smale, On the structure of 5-manifolds, *Ann. Math.* **75** (1962) 38–46.
[141] E. H. Spanier and J. H. C. Whitehead, Duality in homotopy theory, *Mathematika* **2** (1955) 56–80.
[142] M. Spivak, Spaces satisfying Poincaré duality, *Topology* **6** (1967) 77–102.
[143] J. R. Stallings, On topologically unknotted spheres, *Ann. of Math.* **77** (1963) 490–503.
[144] N. E. Steenrod, *The topology of fibre bundles*, viii+224 pp, Princeton Univ. Press, 1951.
[145] N. E. Steenrod, *Cohomology operations: lectures by N. E. Steenrod written and revised by D. B. A. Epstein*, vii+139 pp, Ann of Math. study 50, Princeton Univ. Press, 1962.
[146] S. Sternberg, *Lectures on differential geometry*, xviii+390 pp, Prentice-Hall, 1964.
[147] R. R. Stong, *Notes on cobordism theory*, vi+404+viii pp, Princeton Univ. Press, 1968.
[148] T. Tao, *Hilbert's fifth problem and related topics*, xiv+338 pp, Graduate studies in math 153, Amer. Math. Soc., 2014.
[149] R. Thom, Sur un problème de Steenrod, *C. R. Acad. Sci. Paris* **236** (1953) 1128–1130.
[150] R. Thom, Quelques propriétés globales des variétés différentiables, *Comm. Math. Helv.* **29** (1954) 17–86.
[151] R. Thom, Les singularités des applications différentiables, *Ann. Inst. Fourier Grenoble* **6** (1955–56) 43–87.
[152] R. Thom, La classification des immersions d'après Smale, *Sém. Bourbaki* **157** (Dec. 1957).
[153] R. Thom and H. I. Levine, *Singularities of differentiable mappings*, Bonn Math. Schrift **6** (1959). Reprinted in Springer Lecture Notes in Math. 192 (1971) 1–89.
[154] W. P. Thurston, *Three-dimensional geometry and topology. Vol. 1* (ed. Silvio Levy), Princeton Mathematical Series 35, x+311 pp. Princeton Univ. Press, 1997.
[155] F. van der Blij, An invariant of quadratic forms mod 8, *Kon. Ned. Akad. van Wet. ser. A* **62** (1959) 291–293.

[156] O. Veblen and J. H. C. Whitehead, *The foundations of differential geometry*, Cambridge Univ. Press, 1932.
[157] C. T. C. Wall, Determination of the cobordism ring, *Ann. Math.* **72** (1960) 292–311.
[158] C. T. C. Wall, Cobordism of pairs, *Comm. Math. Helv.* **35** (1961) 136–145.
[159] C. T. C. Wall, Classification of $(n-1)$-connected $2n$-manifolds, *Ann. Math.* **75** (1962) 163–189.
[160] C. T. C. Wall, Classification problems in differential topology I: classification of handlebodies, *Topology* **2** (1963) 253–261.
[161] C. T. C. Wall, Quadratic forms on finite groups and related topics, *Topology* **2** (1963) 281–298.
[162] C. T. C. Wall, On simply-connected 4-manifolds, *Jour. London Math. Soc.* **39** (1964) 141–149.
[163] C. T. C. Wall, An extension of results of Novikov and Browder, *Amer. Jour. Math.* **88** (1966) 20–32.
[164] C. T. C. Wall, Poincaré complexes I, *Ann. Math.* **86** (1967) 213–245.
[165] C. T. C. Wall, On the axiomatic foundations of the theory of hermitian forms, *Proc. Camb. Phil. Soc.* **67** (1970) 243–250.
[166] C. T. C. Wall, ed., *Proceedings of Liverpool Singularities Symposium I*, Springer Lecture Notes in Math. 192, 1970.
[167] C. T. C. Wall, *Surgery on compact manifolds* (London Math. Soc. monographs 1), x+280 pp, Academic Press, 1970. (Second edition, with editorial comments by A. A. Ranicki), xvi, 302 pp, Amer. Math. Soc., 1999.
[168] C. T. C. Wall, Geometric properties of generic differentiable manifolds, pp 707–774 in *Geometry and topology: III Latin American school of mathematics* (eds. J. Palis and M. P. do Carmo), Springer Lecture Notes in Math. 597, 1977.
[169] C. T. C. Wall, Classification and stability of singularities of smooth maps, pp 920–952 in *Singularity theory* (eds. D. T. Lê, K. Saito, and B. Teissier), World Scientific, 1995.
[170] C. T. C. Wall, On the structure of finite groups with periodic cohomology, pp 381–413 in *Lie groups: structure, actions and representations; in honor of Joseph A. Wolf* (eds. A. Huckleberry, I. Penkov, and G. Zuckerman), Progress in Math. 306, Birkhäuser, 2013.
[171] A. H. Wallace, Modifications and cobounding manifolds, *Canad. Jour. Math.* **12** (1960) 503–528.
[172] H. Weyl, *Die idee der Riemannschen Fläche*, B. G. Teubner, 1913. (Reprinted, with supplements, 1997.)
[173] H. Whitney, Analytic extensions of differentiable functions defined on closed sets, *Trans. Amer. Math. Soc.* **36** (1934) 63–89.
[174] H. Whitney, Sphere spaces, *Proc. Nat. Acad. Sci. USA* **21** (1935) 464–468.
[175] H. Whitney, Differentiable manifolds, *Ann. Math.* **37** (1936) 645–680.
[176] H. Whitney, The general type of singularity of a set of $(2n-1)$ smooth functions of n variables, *Duke Math. Jour.* **10** (1943) 161–172.
[177] H. Whitney, The self-intersections of a smooth n-manifold in $2n$-space, *Ann. Math.* **45** (1944) 220–246.
[178] H. Whitney, The self-intersections of a smooth n-manifold in $(2n-1)$-space, *Ann. Math.* **45** (1944) 247–293.

[179] H. Whitney, On singularities of mappings of Euclidean spaces I, mappings of the plane into the plane, *Ann. Math.* **62** (1955) 374–410.
[180] H. Whitney, Tangents to an analytic variety, *Ann. Math.* **81** (1965) 469–549.
[181] E. Witten, Monopoles and 4-manifolds, *Math. Res. Lett.* **1** (1994) 769–796.
[182] J. A. Wolf, *Spaces of constant curvature*, McGraw-Hill, 1967.
[183] E. C. Zeeman, Unknotting combinatorial balls, *Ann. Math.* **78** (1963) 501–526.

Index of notations

Set theory

$\emptyset$ is the empty set.
$x \in X$ denotes that x is an element of the set X.
$X \subseteq Y$ means that X is a subset of Y.
$\bar{X}$ denotes the closure of X (in Y).
$\{x \mid A(x)\}$ is the set of objects x satisfying a condition $A(x)$.
$X \times Y$ is the set of pairs $\{(x, y) \mid x \in X, y \in Y\}$.
$\Delta(X)$ is the diagonal subset $\Delta(X) := \{(x, y) \in X \times X \mid x = y\}$ of $X \times X$.
$f : X \to Z$ means that f is a map from X to Z.
$\operatorname{Im} f$ denotes the image of f: $\{f(x) \mid x \in X\}$.
$|$ denotes restriction, for example, for $Y \subseteq X \xrightarrow{f} Z$, $f|Y$ is the restriction of f to Y.
$\circ$ denotes composite: the composite $g \circ f$ is given by $g \circ f(x) := g(f(x))$.
$\mathbb{R}$ is the set of real numbers,
$\mathbb{R}^n$ is the vector space of n-tuples $x = (x_1, \ldots, x_n)$ with each $x_k \in \mathbb{R}$,
$\|x\| := \surd(x_1^2 + \cdots + x_n^2)$.
$\mathbb{R}^n_+$ is the subset with $x_1 \geq 0$, $\mathbb{R}^n_{++}$ the subset with $x_1 \geq 0, x_2 \geq 0$.
$[a, b]$ is the closed interval $\{x \in \mathbb{R} \mid a \leq x \leq b\}$;
$[a, b)$ is the half-open interval $a \leq x < b$ (allowing $b = \infty$); similarly $(a, b]$.
$D^n_x(r)$ is the closed disc $\{y \in \mathbb{R}^n \mid \|x - y\| \leq r\}$,
$S^{n-1}_x(r)$ the sphere $\{y \in \mathbb{R}^n \mid \|x - y\| = r\}$,
$\mathring{D}^n_x(r)$ the open disc $\{y \in \mathbb{R}^n \mid \|x - y\| < r\}$,
$D^n_+(r) := D^n_x(r) \cap \mathbb{R}^n_+$ is the closed half-disc
$\mathring{D}^n_+(r) := \mathring{D}^n_x(r) \cap \mathbb{R}^n_+$ the open half-disc.

If x is omitted, the centre is the origin; if r is omitted, the radius is $r = 1$.
$D^k(a, b) := \{x \in \mathbb{R}^k \mid a \leq \|x\| \leq b\}$.

Thus $I := [0, 1] = D^1_+$ and $\mathbb{R}^+ := \mathbb{R}^1_+ = [0, \infty)$.
$u : \mathbb{R}^n \setminus \{0\} \to S^{n-1}$ is defined by $u(x) := x/\|x\|$. (§4.2).

Groups, fields, etc.

$\mathbb{Z}$ is the ring of integers.
$\mathbb{Z}_n$ is the additive group of integers modulo the natural number n.
$\mathbb{R}$ is the field of real numbers.
$\mathbb{Q}$ is the field of rational numbers.
$\mathbb{C}$ is the field of complex numbers.
$\sigma(q)$ is the signature of a quadratic form q defined over $\mathbb{R}$,
$\mathrm{Arf}(\mu)$ is the Arf invariant of a quadratic form μ defined over $\mathbb{Z}_2$.
$\mathfrak{G}(\mu)$ is the Gauss sum of the quadratic form μ on a finite group.
$P(A_*; F)(t) := \sum_0^\infty \dim_F(A_n)t^n$ is the Poincaré series of A_*.
$|G|$ is the order of the finite group G.
$\mathrm{Tors}(A)$ is the torsion subgroup of the abelian group A.
$A \oplus B$ is the direct sum of A and B;
$A \otimes B$ is the tensor product of A and B.
$G^\vee$ is the dual group to G.
$\mathrm{Ker}(\phi)$ is the kernel of the group homomorphism $\phi : A \to B$;
$\mathrm{Coker}(\phi)$ is the cokernel of $\phi : A \to B$.
G/H is the quotient (space) of (right cosets) of a group G by a subgroup H. If H is a normal subgroup of G, this is the quotient group.

$\mathbf{GL}_m(K)$ is the group of nonsingular $(m \times m)$ matrices over the field K.
$\mathbf{SL}_m(K)$ is the subgroup of matrices of determinant 1.
$\mathbf{GL}^+_m(\mathbb{R}) \subset \mathbf{GL}_m(\mathbb{R})$ is the subgroup of matrices with positive determinant.
$\mathbf{O}_m \subset \mathbf{GL}_m(\mathbb{R})$ is the orthogonal group, $\{A \in \mathbf{GL}_m(\mathbb{R}) \mid AA^t = I\}$.
$\mathbf{U}_m \subset \mathbf{GL}_m(\mathbb{C})$ is the unitary group, $\{A \in \mathbf{GL}_m(\mathbb{C}) \mid A\overline{A}^t = I\}$.
$\mathbf{SO}_m := \mathbf{O}_m \cap \mathbf{SL}_m(\mathbb{R})$.
$\mathbf{SU}_m := \mathbf{U}_m \cap \mathbf{SL}_m(\mathbb{C})$.
G_n is the monoid of maps of S^{n-1} to itself of degree ± 1.
$F_n \subset G_{n+1}$ is the set of base-point preserving maps $S^n \to S^n$.
Top_n is defined in §8.9.
SG_n, SF_n, $STop_n$ are the corresponding subsets of orientation-preserving maps.

For each of the above groups and monoids C_n,
$B(C_n)$ is the classifying space of C_n;
$B(G_n)$ is the classifying space for spherical fibrations with fibre S^{n-1};
C is the union of the C_n; and
$B(C)$ is the inductive limit of the sequence $B(C_n)$.

Manifolds, etc.

$Bp(x)$ is the bump function (§1.1).
T_PM is the tangent space at $P \in M$ to the smooth manifold M;
$T_P^\vee M$ is the dual vector space.
$\mathbb{T}(M)$ is the tangent vector bundle of M; the dual is $\mathbb{T}^\vee(M)$.
$\mathbb{T}^0(M)$ is the zero cross-section.
$\mathbb{N}(M/V)$ is the normal bundle of the smooth submanifold $V \subset M$.
∂M is the boundary of M: for example, $\partial D_x^n(r) = S_x^{n-1}(r)$.
$\angle M$ is the corner of M.
$\mathring{M} := M \setminus \partial M$ is the interior of M.
$\partial_- W$, $\partial_+ W$ and $\partial_c W$ are the lower, upper, and middle parts of the boundary ∂W of a cobordism W.
$D(M)$ is the double of M.
$M_1 \# M_2$ is the connected sum of manifolds M_1 and M_2.
$M_1 + M_2$ is the boundary sum of M_1 and M_2 (§2.7).
$P^m(\mathbb{R}) = P(\mathbb{R}^{m+1})$ is the set of lines through the origin in $\mathbb{R}^{m+1}$;
$P^m(\mathbb{C}) = P(\mathbb{C}^{m+1})$ the set of lines in $\mathbb{C}^{m+1}$.
$P^\infty(\mathbb{R}) := \bigcup_{n \in \mathbb{N}} P^n(\mathbb{R})$; $P^\infty(\mathbb{C}) := \bigcup_{n \in \mathbb{N}} P^n(\mathbb{C})$.
$Gr_{m,k}$ is the Grassmann manifold of k-dimensional subspaces of $\mathbb{R}^m$.
$V_{m,k} \cong O_m/O_k$ is the Stiefel manifold of isometric embeddings $\mathbb{R}^k \to \mathbb{R}^m$.
$V'_{m,k} \cong GL_m(\mathbb{R})/GL_k(\mathbb{R})$ is the set of linear embeddings $\mathbb{R}^k \to \mathbb{R}^m$.
$J^k(V, M)$ is the space of k-jets of maps $V \to M$;
$j^k f : V \to J^k(V, M)$ is the k-jet of the map $f : V \to M$;
$V^{(r)}$ is the subset of V^r consisting of r-tuples of *distinct* points of V.
${}_rJ^k(V, M)$ is the subset of $(J^k(V, M))^r$ lying over $V^{(r)}$.
${}_rj^k f : V^{(r)} \to {}_rJ^k(V, M)$ is the multijet of $f : V \to M$. §4.4.
$C^r(V, M)$ is the set of C^r maps $V \to M$ $(0 \leq r \leq \infty)$;
$C^r_{pr}(V, M)$ is the set of proper C^r maps.
$\text{Imm}(V, M)$ is the set of (smooth) immersions $V \to M$;
$\text{Emb}(V, M)$ is the set of (smooth) embeddings $V \to M$;
$\text{Diff}(M)$ is the set of diffeomorphisms of M.
Σ^i, $\Sigma^i(V, M) \subset J^1(V, M)$, $\Sigma^i f$ are Thom-Boardman sets: see §4.5.
$I(\phi)$ is the number of double points of an immersion $\phi : V^k \to M^{2k}$.

Cobordism theory

For ξ an orthogonal bundle (or spherical fibration), we write
A_ξ for the associated disc bundle,
S_ξ for the sphere bundle,

$T(\xi) = A_\xi / S_\xi$ for the Thom space,
B_ξ for the base.

For ξ the universal bundle over $B(C_n)$, these become $A(C_n)$, $S(C_n)$, $T(C_n)$.

$\Omega_m(X, \nu)$ is the set of normal cobordism classes of maps of degree 1 to X.
$P_m := \Omega_m(D^m, \varepsilon)$.
$\mathrm{Kerv}(\phi, \nu, T)$ is the Kervaire invariant of a normal cobordism class.
L_m is $\mathbb{Z}$, 0, $\mathbb{Z}_2$ or 0 according as $m \equiv 0, 1, 2$ or 3 (mod 4).
Ω_m^G is the cobordism group of m-manifolds with (weak) G-structure.
Ω_m^{fr} is the framed cobordism group.
Θ_m^k is the group of homotopy spheres $\Sigma^m \subset S^{m+k}$,
$F\Theta_m^k$ is the group of framed homotopy spheres $\Sigma^m \subset S^{m+k}$,
Σ_m^k is the group of embeddings $S^m \subset S^{m+k}$.
B_k is the kth Bernoulli number.

Homology theory

$H_r(X, Y; A)$ is the rth homology group of (X, Y) with coefficients in A.

If A is omitted, it is taken as $\mathbb{Z}$.

$\tilde{H}^k(X, Y)$ is the reduced cohomology group.
$K_k(M) := \mathrm{Ker}(\phi_* : H_k(M) \to H_k(X))$ for $\phi : M \to X$ a normal map.
$[M]$ is the fundamental homology class of the manifold M.
$\beta_p : H^k(X; \mathbb{Z}_p) \to H^{k+1}(X; \mathbb{Z}_p)$ is the Bockstein homomorphism.
$\mathcal{S}_p$ is the mod p Steenrod algebra,
χ its canonical anti-automorphism,
$\overline{\mathcal{S}}_p := \mathcal{S}_p / \langle \beta_p \rangle$.
$K(\pi, n)$ is the Eilenberg–MacLane space.
$J_k : \pi_k(SO) \to \pi_k^S$ is the stable J homomorphism.
$w_k(\xi), v_k(\xi) \in H^k(X : \mathbb{Z}_2)$ are the Stiefel–Whitney, Wu classes of a bundle ξ.
If $\xi \oplus \eta$ is trivial, $\overline{w}_k(\xi) = w_k(\eta)$.
$c_k \in H^{2k}(X; \mathbb{Z})$ is the kth Chern class,
$p_{4k} \in H^{4k}(X; \mathbb{Z})$ are the Pontrjagin classes.

Homotopy theory

$*$ is the base point.
X^+ is the disjoint union of X and $*$.
$X \wedge Y$ is the smash product of X and Y.
$X * Y$ is the joint of X and Y.

$[X : Y]$ is the set of (based) homotopy classes of maps $X \to Y$.
$\pi_r(X, Y)$ is the rth homotopy group of (X, Y).
$K^{(n)}$ is the n-skeleton of K.
$X^{\langle k \rangle}$ is a $(k-1)$-connected cover of X.
$SX := S^1 \wedge X$ is the suspension of X.
ΩX is the loop space of X.
$\{X : Y\} = lim_{n \to \infty}[S^n X : S^n Y]$.
$\pi_r^S(X) := \{S^r : X\}$.
$\mathbb{S}$ is the sphere spectrum.
$\mathbb{K}(A, k)$ is the Eilenberg–MacLane spectrum.
$\mathbb{TG}$ is the classifying spectrum of the stable group G (in the sense of §8.2).
$\mathbb{BU}$ and $\mathbb{BO}$ are the Bott spectra, with connective versions $\mathbb{BU}\langle k \rangle$ and $\mathbb{BO}\langle k \rangle$.

Index